Hazards in the Chemical Laboratory
Fifth Edition

LABORATORY ....

SAFETY OFFICER ....

DEPUTY SAFETY OFFICER OR PERSON RESPONSIBLE
FOR MAINTAINING FIRST AID BOX ....

NEAREST HOSPITAL FOR CASUALTIES ....

.... Telephone ....

NEAREST AMBULANCE SERVICE ....

.... Telephone ....

NEAREST FIRE SERVICE ....

.... Telephone ....

OR DIAL 999 FOR AMBULANCE/FIRE SERVICE;
THEN GIVE EXACT LOCATION OF CASUALTY/FIRE

NEAREST DOCTORS ....

.... Telephone ....

....

.... Telephone ....

PERSONS ON PREMISES TRAINED IN FIRST AID ....

....

....

PERSON TO WHOM ACCIDENTS MUST BE REPORTED

....

OTHER INFORMATION ....

....

....

Hazards in the Chemical Laboratory

Fifth Edition

Edited by
S. G. Luxon
Health and Safety Consultant

The publisher makes no representation, express or implied, with regard to the accuracy of the information contained in this book and cannot accept any legal responsibility or liability for any errors or omissions that may be made.

A catalogue record for this book is available from the British Library.

ISBN 0–85186–229–2

First published 1971.
Second edition published 1977.
Third edition published 1981.
Fourth edition published 1986.
Fifth edition published 1992.

Fifth Edition © The Royal Society of Chemistry 1992

Published by The Royal Society of Chemistry.
Thomas Graham House, The Science Park, Cambridge CB4 4WF

Typeset by Computape (Pickering) Ltd, North Yorkshire
Printed and bound in Great Britain by The Bath Press, Lower Bristol Road, Bath

Foreword

I welcome the fifth edition of Hazards in the Chemical Laboratory. It provides comprehensive information and practical guidance for all who work in chemical laboratories and for those who have responsibility for their safe management.

Legal requirements for health and safety at work impose duties on everyone working in laboratories. These duties are clearly explained in this book, but perhaps more significantly, the practical ways in which these duties can be sensibly met through safety management, planning, and training are fully explored.

Additionally, this book provides a readily accessible reference listing the hazardous properties of approximately 1400 flammable, explosive, corrosive, and/or toxic substances or groups of substances commonly used in laboratories.

This book can make an important contribution to safe working conditions. I urge those who work in chemical laboratories to use the information, and apply the principles contained in it.

A. J. LINEHAN
HM Chief Inspector of Factories, Health and Safety Executive

Preface to the Fifth Edition

The rapid development in many technical and legislative matters which have taken place since the publication of the fourth edition, has proceeded apace and calls for a major revision of many of the matters therein. The good reception accorded to the previous edition, coupled with the nature of these developments, lead the team of authors and myself, after discussions with the RSC management, to the view that its format and general contents are still sound and capable of major expansion and updating which has now been done.

The Introduction in Chapter 1 and Chapter 2 on Legislation have been revised to take account of recent developments and, in particular, the Regulations for the Control of Substances Hazardous to Health. This reflects the steady developments in the effects of key legislation upon laboratory life. Chapter 3 deals with Safety Planning and has been completely revised by Mr Everett. It discusses the general implications of the regulatory framework and, in particular, good laboratory design. Chapter 4, Fire Protection, has been updated by Mr Warwicker and gives detailed guidance on the important subject of fire prevention. The former Editor, Leslie Bretherick, has revised Chapter 5, Reactive Chemical Hazards, to take account of developments in the methods of predicting such hazards. Dr Magos has re-written Chapter 6, Chemical Hazards and Toxicology, to emphasize the importance of this subject. Chapter 7 in the fourth edition has been divided into two parts to give a clearer picture of both Control of Health Hazards and First Aid. Dr Murray has kindly written the section on First Aid and Dr Smith that on Control of Health Hazards. Chapter 9, Precautions against Radiation, has been completely re-written by Mr Beaver to include the detailed requirements of the new Ionising Radiation Regulations and the Approved Code of Practice. A new Chapter 10 has been added on Electrical Hazards to deal with the long awaited revision of the Electricity Regulations, and I am grateful to Mr Stephenson for his contribution on this subject. Mr Fawcett has completely revised Chapter 11, An American View, to reflect developments in safety and health in chemical laboratories in the USA.

The Hazardous Substances section which was Chapter 8 in the fourth edition, Hazardous Chemicals, the so called yellow pages, is of course the most important part of this work. It has now been placed at the end of the book for easy reference and has been made self-contained by recasting the explanatory foreword and including a list of the standard phrases used in the text. This

vii

Section has been extensively revised and extended to include what we believe to be the majority of commonly used chemical substances. I wish to acknowledge the valuable contribution of Mr Henning in making this major revision possible.

It is again a pleasure to record the great encouragement and assistance extended by the editorial and information services staff of the Royal Society of Chemistry, in particular, by Dr Philip Gardam and Dr Bob Andrews who were jointly responsible for converting our combined offerings into this new volume and for the technical assistance provided by Dr Mike Hannant in the compilation of the Hazardous Substances Section.

S. G. LUXON
May 1992

Preface to the First Edition

The present volume is a successor to the Royal Institute of Chemistry's 'Laboratory Handbook of Toxic Agents', first published in 1960, and issued in a revised edition in 1966. Before the second edition went out of print, the future of the publication was considered by both the Institute's Publications Committee, and by the previous editor, Professor C. H. Gray, and myself. It was generally felt that, rather than merely revise the existing material, it would be preferable to alter the underlying philosophy of the book by changing its scope from toxic hazards to a consideration of all hazards likely to be encountered in the chemical laboratory.

The general format remains the same, with the major part of the book (printed on tinted paper) being an alphabetical guide to hazardous chemicals and measures to be taken in the event of accidents in their use. However, whereas previous editions have included details of measures to be taken against the toxic hazards of such chemicals, the present edition also includes methods for spillage disposal and extinguishing of fires where appropriate. Once more, an alphabetical listing is adopted to enable the use of the book as a speedy reference in the case of emergency.

Much assistance was required in preparing this chapter, now extended to over 430 hazardous chemicals, and the preceding one on first aid. I must record my special thanks to my colleague Mr W. G. Moss for his collaboration when we prepared these chapters for the first edition of the earlier book, and BDH Chemicals Ltd. for permission to use their extensive records on the hazards, handling, and disposal of chemicals; also my colleague Dr P. Mostyn Williams who has added considerably to the earlier medical advice of Dr W. B. Rhodes.

Dr D. P. Duffield and Dr K. P. Whitehead of Imperial Chemical Industries Ltd. have also provided important medical advice on up-to-date first aid practice, particularly on the treatment of cyanide and phenol poisoning. The chemical world must always be grateful for the pioneer efforts of ICI in encouraging chemical factory and laboratory safety and we would record again our thanks to Dr A. J. Amor and Dr A. Lloyd Potter for their interest in the first book and to pay special tribute to the work of Dr L. J. Burrage who has contributed so much to promote laboratory safety in this country.

Many other firms have given us the benefit of special knowledge of certain of their products and we are also grateful to James North and Sons Ltd. for permission to reproduce their chart advising on the types of glove to be used when handling different classes of chemicals.

Despite extensive practical experience of chemical hazards, the writers of a book such as this lean heavily on the authors of major works on industrial toxicology. Not many may have had the privilege of knowing the charm and intelligence of that great lady, Dr Ethel Browning, who guided the Institute and some of the authors when the first book was conceived and drafted, and wrote two renowned works upon which we draft extensively, 'Toxicity and Metabolism of Industrial Solvents' and 'Toxicity of Industrial Metals', as well as editing the important series of monographs on toxic agents in which they appeared. Her death last year ended a long life of devoted service to industrial safety.

The valuable publications of the Chemical Industries Association— 'Marking Containers of Hazardous Chemicals' and 'Exposure to Gases and Vapours'—have been referred to frequently and we would like to thank the Association for the privilege of perusing the text of the latter at the proofing stage. The 'Laboratory Waste Disposal Manual' published by the Association's counterpart in the US, the Manufacturing Chemists' Association, was also consulted extensively in preparing Chapter 6, as were the following works:
'Dangerous Properties of Industrial Materials' by N. Irving Sax
'Extra Pharmacopoeia' (Martindale) edited by R. G. Todd
'Industrial Hygiene and Toxicology' edited by Frank A. Patty
'Industrial Toxicology' by L. S. Fairhall
'Poisoning by Drugs and Chemicals' by P. Cooper
'Poisons' by Brookes and Jacobs
'Toxicology of Drugs and Chemicals' by W. B. Deichmann and H. W. Gerarde.
Other acknowledgements appear in Chapter 5.

Of the remainder of the book, new chapters have been provided by Mr Ackroyd, Dr Taylor, and Mr Sheldon on fire protection and by Mr Neill and Dr Russell Doggart on the particular hazards facing chemical workers in hospital biochemistry laboratories. In addition, Mr Luxon, of the Department of Employment, has contributed an entirely new introduction to replace the one by the late Sir Roy Cameron which appeared in the earlier editions. Mr Beard and Dr Osborn have thoroughly revised chapters 2 and 7 respectively. To all these authors, I am extremely grateful for the time they have spent on and interest shown in this project.

Dr Farago and his staff in the Editorial Office of the Institute have my sincerest thanks for their painstaking help and I am particularly indebted to Dr Martin Sherwood for his close collaboration in co-ordinating our efforts, sharing in our proof reading, and carrying out all the necessary negotiations with the printers. With my co-authors he has made the task of editing not only light, but both stimulating and enjoyable.

Finally, I should like to pay tribute to the immense amount of work which Professor Gray put into the planning, and production of both editions of the 'Laboratory Handbook of Toxic Agents'. Although pressure of work prevented Professor Gray from taking a full part in the editing of this volume, I have had the benefit of his wise advice throughout its preparation. Without this and the substantial contribution he made to the earlier book, it is unlikely that this volume would have been possible.

G. D. MUIR
April 1971

Contents

Chapter 9   Precautions Against Radiation   123
P. Beaver, UK Health and Safety Executive

CHAPTER 1

Introduction

S. G. LUXON

There have been a number of important developments in the field of Health and Safety since the publication of the fourth edition. Perhaps the most important has been the increasing concern on the part of the public in respect of the manufacture, transport, and use of chemicals and in particular their ultimate disposal. This problem of waste and its ultimate fate typifies the other grave public concern—the long-term cumulative effect of chemicals on the environment. These latter concerns have led to the enactment of legislation the principal instrument being the Control of Pollution Act 1990. The scope of the work, in particular Chapter 12 entitled Hazardous Chemicals, has been extensively revised so as to take account of such considerations in as far as they affect good laboratory practice.

As was foreshadowed in the fourth edition, more detailed Regulations have now been made under the Health and Safety at Work Act to control hazards at the work place. They replace many of the earlier Regulations relating to specific processes and hazards some dating back to the beginning of this century. Of these Regulations, 'The Control of Substances Hazardous to Health' (COSHH) together with the associated Codes of Practice are the most important and provide a framework for good industrial hygiene practice in the laboratory.

For a broad outline, the general legal requirements of the Health and Safety at Work Act and the COSHH Regulations are set out in Chapter 2. For more detailed and specific information, reference should be made to the Regulations and Codes of Practice themselves.

While this book gives general guidance on the control of hazards associated with chemical substances in the laboratory, it must never be forgotten that a very important element in the control of such hazards lies in the training and instruction of the staff. In particular, attention should be paid to systems of work and to the clear delegation of specific responsibilities to those organizing units of laboratory activity. A further important aspect of the legislation is the requirement in respect of the assessment of risks and consultation with the workforce. These requirements place a particular responsibility on chemists

1

who are best able to advise the layman on the hazardous nature and properties of chemical substances and the precautions necessary to ensure their safe use. It is therefore important that everyone concerned makes an assessment of all aspects of health and safety and puts in train any necessary steps to ensure that his house is in order.

Regrettably perhaps, in the past, chemists who habitually handle dangerous substances have been inclined to disregard the hazards associated with their use particularly if such hazards were of a long-term nature. Every human being, and a chemist is no exception, tends towards the view that although an accident may happen to others it will never happen to him personally because he is too wise and knowledgeable. Experience shows that nothing can be further from the truth. During work in the laboratory, many persons have suffered injury to their health which, because of the insidious symptoms, may never have been associated with their work activity. It is only when permanent injury has occurred that many persons come to realize that the observance of even elementary precautions could have prevented such injury or, in extreme cases, premature death.

It is against this background that one should look at this edition of the handbook. The contributors have attempted to indicate and discuss the dangers likely to arise in the laboratory and have offered practical advice on their avoidance. The work will, I believe, also prove useful in devising precautionary measures in respect of the many reagents and substances which, for reasons of space, have not been included in the work.

At the same time the manual has become a much more useful and complete work not only for chemists in the laboratory but also for all those who handle hazardous substances on a small scale, *e.g.* in industry. Additionally, the work will be useful in schools and higher education establishments where training in the correct use and handling of these substances should be considered an integral part of the curriculum of students in science subjects.

The control of hazards in the laboratory is well known; namely: the enforcement of safe systems of work; the need for mechanical safety involving the guarding of dangerous parts of machines, even if driven by only fractional horse power motors, so that injury from contact with moving or trapping parts is prevented; the need to provide safe means of access to every place where anyone has to work even if the work is only undertaken on rare occasions; the need for good housekeeping to minimize the possibility of accidents occurring through persons striking or being struck by objects; the need for care in the handling of glassware; the need to protect electrical conductors and to provide or use low voltage supplies or adequate earthing; and, of course, the matters with which this handbook are intimately concerned—the prevention of injury from fire, explosion, or from exposure to hazardous substances.

1 Identification

Perhaps the most important single step we can take to ensure hazards are identified is using a system of labelling which draws attention to each and every hazard and sets out the simple precautionary measures to be followed. Since the publication of the fourth edition, there have been a number of EEC directives laying down requirements for the labelling of chemical substances. These take effect in the United Kingdom as the Classification, Packaging, and Labelling Regulations and subsequent amendments. For listed substances a warning symbol together with phrases to indicate the nature of the hazard and safety advice are mandatory. In the Section, Hazardous Substances, such symbols, phrases, and safety advice are set out in respect of every substance so listed.

Failure to give some such simple warning is inexcusable, particularly in the laboratory where many chemicals may at some time be handled by inexperienced and unskilled persons who are not members of the laboratory staff, *e.g.* cleaners and maintenance workers. Accordingly, all chemicals should be labelled even if they are not covered by statutory requirements following as far as practicable the general format and parameters of the system referred to above.

2 Management's Task

Safety and health is the responsibility of management and must be set out in a policy statement. Not only must the manager and all members of the staff know the hazard involved, but they must all be clearly seen to be directly interested and involved in the promotion of a safe and healthy environment. Strict procedures should be written into analytical and other methods. Where such methods do not exist, the work should be immediately supervised by a responsible person who is aware of the dangers and precautions to be followed both during normal working and in any emergency that may arise. In larger laboratories a safety officer and hygienist should be appointed to provide advice and general supervision and, not least, to look critically at the procedures involved from outside the group undertaking the project. Experience shows that such a view dissociated from the actual scientific work is invaluable in bringing to light relatively simple hazards which may have been overlooked.

General management and supervision must be tight to ensure that work is conducted in a predetermined and orderly manner, that unauthorized actions are checked, and that proper care and attention is given to the minute-by-minute operation of processes or experiments. In particular, at each meal or tea break and at the end of the day a thorough check must be made to see that everything can be left safely. If there is any doubt, arrangements should be made for the continued supervision of the operations still in progress.

The overall aim should be to design out hazards so that the whole system can operate in such a way that any possible human error is eliminated as far as is practicable. Chapter 3 deals with this general aspect of the problem.

3 Fire and Explosion Hazards

The dangers of fire are well known, but again we must remember the maxim 'familiarity breeds contempt'. The very large number of fires in laboratories proves the seriousness of this problem. Chapter 4 gives detailed advice on such risks and the text under each substance in the Section, Hazardous Substances, indicates properties on which an assessment of the fire hazard can be based and makes suggestions as to the selection of fire fighting equipment.

It should be clearly understood that when a liquid is used having a flash point below the highest normal ambient temperature it can, in suitable circumstances, liberate a sufficient quantity of vapour to give rise to a flammable mixture with air. This can accumulate in the workroom to such an extent as to give rise to the possibility of a serious explosion by ignition of the vapour/air mixture from an ignition source already present some distance away, causing a flash-back to the original source. There is then the consequent possibility of disastrous fire.

A flammable gas or vapour must be present in a concentration of the order of 1 per cent or more by volume if its mixture with air is to be flammable, so it is a relatively simple matter to check whether or not a dangerous concentration is likely to be present in closed plants such as ovens, *etc*. During normal working it is desirable to ensure that one-quarter of the lower flammable limit is never exceeded. The amount of flammable vapour or gas in the workroom air should, of course, never approach this concentration during normal working procedures. Account must, however, be taken of possible leakages and spillages so that although it is, perhaps, unnecessary to provide special precautions such as flameproof electrical equipment in normal circumstances, these may be very desirable to provide against contingencies arising from the unexpected loss of flammable solvents having a flash point of less than 32 °C and particularly those having a flash point of less than 21 °C. Our aim, therefore, should be to restrict the use of such liquids to situations in which they are absolutely necessary and even then to reduce the quantity involved as far as possible. This is particularly important with solvents used for routine operations where in every case we should carefully consider whether or not it is possible to substitute an alternative having a flashpoint above the highest possible ambient temperature (32 °C).

The quantity of all flammable materials and of solvents in particular should be kept to the absolute minimum. There is often a tendency to disregard this and allow large quantities of solvents that are used only occasionally to accumulate in the laboratory. When flammable substances are not in use there must be adequate supervision to ensure that they are kept in a properly constructed fireproof store.

Suitable fire fighting equipment should be readily available and adequate means of escape provided (see Chapter 4). All personnel should be trained and familiar with the use of the equipment so that a small fire can be quickly localized and prevented from spreading, while in the event of the fire getting out of hand everyone must know how to escape safely.

4 Reactive Chemical Hazards

Particular care must be exercised when using highly reactive or unstable substances that may be liable to cause an explosion. The quantities used should be kept to a minimum and, if necessary, several reactions carried out on a smaller scale. Consideration should be given to limiting the effect of the explosion—should one occur—by the provision of suitable reliefs venting to a safe place. In all cases protective screens should be provided or the experiment operated by remote control so as to ensure that the operator will not be injured. It should be remembered that such substances may be produced as a result of side reactions or in residues standing over a period of time.

Exothermic reactions should be carefully controlled and monitored to ensure that there is no failure of the cooling or stirring systems. Again quantities should be kept to a minimum and suitable screening provided. No operation of this kind should be entrusted to anyone other than a highly skilled and competent chemist knowledgeable in the dangers involved and the precautions to be taken. All these matters are discussed in detail in Chapter 5.

5 Toxic Hazards

Toxic substance can act in three ways causing poisoning by ingestion, percutaneous absorption, and inhalation. Our first thought should always be: can a harmless or less hazardous substance be used instead of the substance under consideration? Such a step removes or reduces the danger in an infallible manner and this possibility should therefore never be overlooked, or dismissed without very careful consideration.

The dangers of ingestion by contamination of the hands and food can be virtually eliminated if there is proper attention to personal cleanliness. Washing accommodation of a high standard should be provided together with a means of drying that is always available. This precaution is, of course, equally applicable to other health risks and promotion of personal cleanliness should never be neglected. Another common, but inexcusable, danger in this category is the use of mouth pipettes. Such methods should never be used for pipetting hazardous liquids and we should train school pupils in the use of rubber bulbs.

The contact of corrosive substances with the skin is generally obvious and so this is somewhat less dangerous than contact with percutaneous poisons. Nevertheless, gloves and, where necessary, protective clothing should always be worn where this hazard is present. If contact with the skin occurs, the affected parts should be washed immediately with soap and water. Special attention is necessary to protect the eyes, and where corrosive substances are regularly used, eye protection should be worn as a routine precaution. It must always be remembered that it is often a bystander and not the person actually carrying out the process who suffers injury.

Substances that can be imperceptibly absorbed through the skin present a more insidious hazard, particularly if the contact is repeated or prolonged. Great vigilance is necessary to ensure that the dangers are appreciated by everyone concerned. The aim should be to prevent contact. The use of protective gloves has been found in many cases to be of doubtful value as, unless strict working procedures are observed, contamination can occur on the inner surfaces through pinholes or by careless removal and replacement. Any splashes on the unprotected skin should be washed off immediately with soap and water. Where the more hazardous substances in this group are used regularly it is advisable to make arrangements for the periodic medical examination of the persons involved.

Inhalation of harmful vapours, gases, dusts, and liquid aerosols is another insidious and widespread danger in the laboratory. Many persons have at one time or another been exposed to excessive quantities of vapours such as mercury, benzene, and carbon tetrachloride or to dusts such as lead and beryllium. This is all the more serious because it is deceptive, there often being no sensible perception of danger. Additionally, since it is all-pervading, once it has entered the air of the workroom, the contaminant must be breathed. Everyone is inclined to judge the danger by the short-term effects whereas it is the long-term effects that are more serious and may give rise to permanent and irreversible injury. Unfortunately, such effects may not be directly attributable to exposure to toxic chemicals because the affected person may have changed his employment or be no longer working with the hazardous materials. Therefore, there is little statistical evidence as to the incidence of ill health brought about by such exposures in the laboratory and in consequence the very real and serious dangers tend to be disregarded.

It is the duty of every responsible person to see that substances having irreversible effects are only used when it is absolutely necessary. If such a hazardous substance must be used, then adequate instructions must be given, proper supervision assured, and the whole process carried out in such a way as to ensure that the material is contained and does not enter the air of the workroom where it may be breathed. A properly-designed fume cupboard should always be used and all steps in the process or experiment carried out therein. Where the hazardous substance giving rise to long-term effects is used regularly, the possibility of monitoring the atmosphere at regular intervals should be given serious consideration. If the sample can be taken in the breathing zone of the worker by means of a suitable personal sampler then confirmation can be obtained as to the efficiency of the precautionary measures and the worker reassured as to the absence of any long-term health risk. The assessment of toxicity is a complex matter, but general parameters are set out in Chapter 6 which, as indicated earlier, lock into the labelling system recommended in the Section, Hazardous Substances.

While the hazards from a toxic chemical may be difficult enough to assess it must always be remembered that two relatively harmless liquids or substances may, when in contact, liberate an unexpected poisonous gas. Perhaps the commonest example is bleaching powder and acid lavatory cleaners which liberate chlorine gas. Another common example is alkaline cyanides which, in contact with acid, liberate hydrocyanic acid gas. The phenomenon often occurs in sink traps and other parts of the drainage system or during the use of containers that have not been properly washed out from a previous operation. It is particularly important when any such substances are used by unqualified persons, *e.g.* cleaners, to consider in advance whether or not chance contact between such types of chemicals could produce a more serious hazard. If this is possible steps should be taken to avoid any possible contact. Working instructions should include the safe disposal of waste and require the routine cleaning of containers for chemicals.

The subject of carcinogenicity is a difficult one. Many substances if implanted repeatedly in animals will produce a carcinogenic reaction in the fullness of time, but they may not be hazardous when used in the normal way. On the other hand there are substances such as 2-naphthylamine that will almost certainly produce carcinogenic effects if ingested into the human body. Between these two extremes lie a large number of potentially hazardous materials. Where a known carcinogenic risk exists it is indicated in the text. Substitution of another reagent should always be considered, but, if any such substance must be used, all practical precautions should be taken to reduce all exposures to as low a level as is practicable.

6 General Environment

As with all chemical hazards good housekeeping, *i.e.* the cleaning up and removal of spillages and lost material, is a cornerstone to safe working. In this context all spilt material which may become a hazard must be completely removed or made chemically inert. Liquids may be removed with water if they are soluble. If insoluble, detergents or solvents may be used. In the last resort it may be necessary to react the material chemically to an inert form as a method of decontamination.

Toxic dusts are particularly hazardous in this respect since if lost material is not removed it will become repeatedly airborne whenever it is disturbed either by movement of persons or materials or by strong air currents. Such dust, if present, must be removed at frequent intervals by an industrial vacuum cleaner fitted with a high efficiency filter to prevent recirculation of the fine particles in a breathing zone.

Past experience clearly indicates that many cases of poisoning in laboratories are due to background contamination brought about by a neglect of these general principles.

Coupled with good housekeeping is the quality of the general working environment—in particular, lighting, ventilation, and heating. Good lighting is important because hazards become immediately apparent; there is less need when carrying out intricate manipulations for the operative to approach close to the danger area to obtain a better view of the operation and malfunction of the equipment or instrumentation is immediately apparent, permitting remedial measures to be taken before danger occurs.

Heating and ventilation go hand in hand. Good general ventilation is essential in laboratories where toxic materials are handled. A general standard of at least five air changes per hour should be the aim. Adequate heating must be provided during the winter months or such ventilation will not be used.

We may sum up all these general matters by saying that where hazardous materials are handled we should provide a pleasant, clean working environment properly planned with good access and means of escape, and a high standard of lighting, ventilation, and heating with regular cleaning to ensure removal of any spilt or lost material.

After a hazardous substance has been used there remains the problem of disposal. The first method to be considered is its return to stock via reprocessing. Highly hazardous chemicals should be reacted in a fume cupboard to break them down into less hazardous compounds which are more readily disposable. It must be clearly understood that the responsibility for safe disposal rests with the user and under no circumstance should substances be disposed of in such a way as to constitute a hazard either inside or outside laboratories. In the last resort specialist companies are available who will undertake the task of disposal for a fee.

Where laboratories are situated in a built-up area or where the effluent from exhaust systems must be discharged at a low level, steps must be taken to ensure that means of trapping the dangerous substances are provided so that the exhausted air does not contain a dangerous quantity of harmful substance. Care should be taken to discharge the effluent at a safe height where it will not recirculate into the laboratory or cause a nuisance or danger to other persons in the vicinity.

In all cases the disposal of hazardous substances should be the subject of prior discussion with the relevant local and water authorities so that authorizations can be agreed.

7 Radiation Hazards

7.1 Ionizing Radiation

There is increasing interest in the subject of radiological protection and the effect of low doses in particular. While this is in part generated by the nuclear power programme there is no doubt that any incident in which radioactive substances are involved will evoke a demand for a critical investigation.

As a result of initiatives by the European Community new Regulations together with Approved Codes of Practice have now come into force. These Regulations and Codes will apply to all work activities (see Chapter 2) and the documents give detailed practical guidance on the steps to be taken when using radioactive substances or machines generating ionizing radiation. Their general requirements are reviewed in Chapter 9.

The essential points to be noted are that notification to the Regulatory Authority is required unless only trivial amounts (as quantified in the appropriate regulatory documents) are used.

7.2 Non-ionizing Radiation

Since the publication of the previous edition much progress has been made in evaluating the hazards of non-ionizing radiation. Standards for exposure to microwaves, ultraviolet, and lasers have been published by responsible bodies.

8   Electrical Hazards

The danger of electric shock has been a particular problem in many laboratories where much temporary wiring is often in evidence. In the past such hazards have been the subject of Regulations (1908). Since the publication of the fourth edition, however, new Regulations have come into force and Chapter 10 sets out their general requirements and considers how they affect work in the laboratory.

9   Conclusions

With the passage of time knowledge of the toxicological effects of exposure to chemical substances is rapidly advancing and each year there is an improved understanding of the long-term hazards associated with their use. Massive documentation now exists and for the great majority of substances some general assessment of the hazard can be made. Nevertheless, the general policy should be to reduce exposure as far as is practicable in every case. This has become particularly important as the safe levels of substances are constantly being reduced and, in addition, new and unexpected toxic effects become apparent.

It must be the aim of everyone involved to ensure that all chemicals are handled safely without either immediate or long-term dangers. A clean, healthy general working environment must be provided and individuals encouraged to become safely conscious. The subsequent parts of this book provide detailed information on particular types of hazard.

In the Section, Hazardous Chemicals, details are given of the properties and dangers of each substance, together with the recommended limit and flashpoint where applicable. This will enable the reader to decide for himself what should be done. At the same time it must be emphasized that, given a wise, common-sense approach to the problem, the general elementary precautions indicated above will enable any normal operation to be carried out safely. One last piece of advice cannot be overemphasized. It is essential before carrying out any new procedure to stop, stand back, and consider what hazards may arise, what precautions should be taken, and what emergency procedures may be necessary.

CHAPTER 2

Legislation

S. G. LUXON

1 General

The principal legal instrument concerned with Health and Safety at the work-place is the Health and Safety at Work Act 1974. This Act was founded very largely on the principal recommendations of the Robens Committee on Safety and Health at Work whose report was published in 1972.

The main purpose of the Act is to provide one comprehensive and integrated system of law dealing with the health, safety, and welfare of workers, and the health and safety of the public as affected by work activities. The Act can be described as the most significant statutory advance in the field of health and safety at work since the Shaftesbury Factory Acts of 1833. It has changed radically not only the scope of the provisions, but also the way in which those provisions are enforced and administered. It thus extends obligations and protection to five million or more people who have never before come within the scope of this kind of legislation including workers in health, education, and research establishments. It also covers the self-employed.

The Act, of course, does not seek to cover every eventuality nor does it try to spell out rules for each and every work situation. It is an enabling instrument, whose foundation is the concept of a general duty of care in respect of people engaged in or associated with work activities. It adopts a flexible general approach and thus provides a legislative frame capable of being expanded and adapted to deal with the risks and problems associated with the current technological changes in industry at any particular time.

Thus the general umbrella of the Act provides for an interaction of responsibility between individuals and organizations associated with work or touched by its immediate consequences. The employer now has a duty to his employees with regard to their safety and health, and those employees in turn have a duty to one another. So, too, the self-employed person has a duty to other people around him. There are two significant advances which go much further even than the spirit of previous legislation. The general public is now entitled to a duty of care (in terms of safety and health) from people carrying out work activities, so that an employer, for example, not only has to ensure that his workers are safe, but also that members of the public who might be affected by any hazard from his work activities are not at risk. The legislation also includes an innovation requiring the incorporation of safeguards at an early stage— earlier than had been possible before—placing obligations on suppliers, importers, *etc.* of machinery, plants, substances, *etc.* to make sure that they will be safe when properly used.

2 Scope

The Act applies to all persons at work: employers, the self-employed, and employees, the only exception being domestic servants in private employment. Many of these people were, however, covered by earlier safety and health legislation and in particular the Factories Acts, which applied to process laboratories. The 1974 Act extends this cover to all types of laboratory. New Regulations and Approved Codes of Practice will largely replace most of the existing legislation on safety and health.

3 General Obligations

The obligations set out in the new Act, important to each and everyone working in the laboratory, are in addition to, and not in diminution of, the existing obligations under health and safety legislation, *i.e.* the Factory Acts will remain and will be enforced where they are applicable. In addition much earlier legislation may be taken as providing guidelines as to the interpretation of the general requirements of the 1974 Act in respect of other premises where similar hazards arise. Indeed the general obligations set out in Section 2 of the Act may be taken as giving effect to such requirements.

It will be seen that to obtain flexibility many of the duties imposed by the Act and related legislation are qualified by the words 'so far as is reasonably practicable'. If someone were to be prosecuted for failing to comply with a duty 'so far as is reasonably practicable', it would be the responsibility of the accused to show the court that it was *not* reasonably practicable for him to do *more* than he had in fact done to comply with the duty.

Although the expression is not defined in the Act, it has acquired quite clear meaning through long established interpretations by the courts. Someone who is required to do something 'so far as is reasonably practicable' must assess, on the one hand, the risks of a particular work activity or environment and, on the other hand, the physical difficulties, time, trouble, and expense which would be involved in taking steps to avoid the risks. If, for example, the risks to health and safety of a particular work process are very low, and the cost of technical difficulties in taking certain steps to avoid those risks are very high, it might *not* be reasonably practicable on balance to take those steps. However, if the risks are very high, then less weight can be given to the cost of measures needed to avoid those risks. The comparison does not take into account the current financial standing of the employer. A precaution which is 'reasonably practicable' for a prosperous employer is equally 'reasonably practicable' for the less

well off. The expression 'best practicable means' involves similar considerations: the person on whom the duty is imposed must use the most effective
means to comply with the duty, taking into account local conditions and
circumstances, the current state of technical knowledge, and the financial
implications.

4   Duties of Employers to Their Employees

The general duty imposed on every employer is to ensure, so far as is reasonably
practicable, the health, safety, and welfare at work of all his employees. It
applies to all employers, both those who already have certain duties under
earlier legislation, such as the Factories Act, and those who have never before
been covered by health and safety legislation, *e.g.* in research laboratories.

5   Employers' General Duties

The employer must, so far as is reasonably practicable, provide machinery,
equipment, and other plant that is safe and without risks to health, and must
maintain them in that condition. He must also ensure that, so far as is
reasonably practicable, the systems of work are safe and without risks to health.
'Systems of work' means the way in which the work is organized and includes,
for example, the layout of the workplace, the order in which work is carried
out, and any special precautions that may have to be taken before carrying out
hazardous tasks. This duty, therefore, means that, for example, a machine itself
and the way it is operated must both be safe.

6   Articles and Substances for Use at Work

The use of particular articles and substance at work, machinery and chemicals
for example, may give rise to risks to employees' health and safety. There is a
chain of responsibility involving various people with duties to ensure that those
risks are reduced as far as possible. In brief, the manufacturer must ensure that,
so far as is reasonably practicable, the materials are safe and without risks to
health when properly used, and users must be given sufficient information
about proper use and about any hazards. Once the materials have reached the
place of work there is also a duty on the employer to ensure that, so far as is
reasonably practicable, employees' health and safety are not put at risk by
contact with the materials. In particular, he is required to ensure that the ways
in which the materials are used, handled, stored, and transported are safe and
without risks to health. He should ensure that attention is paid to any information given by the manufacturer or supplier about safe handling and storage.
His duty extends to the end products as well as to the materials used during the
work process. He must ensure that the ways in which end products are
transported and stored before leaving the workplace are safe and without risks
to the health of his employees. He may of course also have duties as a supplier
in turn.

7 Training

The employer should provide for all his employees the information, training, and supervision necessary to ensure, so far as is reasonably practicable, their health and safety at work. The information to be supplied must include information about hazards at the workplace and methods of avoiding them. In particular, the employer should make sure that employees are given the information made available by manufacturers and suppliers of materials used at work about risks attached to the materials and about safe ways of handling them. Health and safety training might include such things as instruction in safety and emergency procedures such as routine checking of equipment, fire drills, and first-aid; special training for work involving a high degree of risk; and retraining when the work changes or new safety methods are introduced. It is the employer's duty to ensure that all his employees are competent to carry out their jobs in a safe manner, *i.e.* with the minimum of risk to themselves or others. Employers must ensure that managers understand their responsibility and have the necessary knowledge and skills to carry them out. Training is necessary not only for operatives, but also for supervisors and managers at all levels. Good supervision is vital in health and safety terms for spotting potential hazards and ensuring that safety rules are complied with.

The local emergency services should also be informed of any potential hazards which might affect their members so that they can provide suitable training and instruction.

8 Employers' Safety Policies

The Act requires every employer to prepare a written statement of the safety policy in his undertaking, except where there are less than five employees. However, it should be noted that 'undertaking' does not have the same meaning as 'establishment'. An employer may operate a number of small establishments, each employing less than five employees. If all the establishments form part of one undertaking and if the total number of employees is five or more the employer must prepare a policy statement.

The purpose of the safety policy requirements is to ensure that the employer carefully considers the nature of the hazards at the workplace and what should be done to reduce those hazards and to make the workplace safe and healthy for his employees. The statement should set out the employer's aims and objectives for improving health and safety at work. It should also set out the organization and arrangements currently in force for achieving those objectives. 'Organization' can be taken to mean people and their responsibilities and 'arrangements' systems and procedures.

Another purpose of the statement is to increase employees' awareness of the employer's policy and arrangements for safety. For this reason, the employer is required to bring the statement to the notice of all his employees. In some cases the best and easiest way of effecting this may be to give a copy to every employee and ensure that new employees are given copies during induction training.

Alternatively, copies may be posted on notice boards where they can be easily seen and read.

In most organizations, working conditions are continually changing. New hazards arise, control measures alter. The safety policy should therefore be kept up-to-date. All revisions in the statement must also be brought to the notice of employees.

9 Safety Representatives and Safety Committees

The 1978 Regulations give recognized trade unions the right to appoint safety representatives to represent the employees in consultations with the employer about health and safety matters. The Regulations also provide for the possibility of employers being required by safety representatives to set up safety committees that would have the job of keeping under review measures to ensure health and safety at the workplace. Full details are contained in the booklet Safety Representatives and Safety Committees published by HMSO which contains the Regulations, Approved Codes of Practice, and guidance notes. Approved Codes of Practice give practical guidance on the functions of safety representatives, the information to be provided to them by employers and the time off with pay to be allowed for training approved by the TUC or by individual unions.

10 Duties to People Who Are Not Employees

One of the major innovations of the Act is that an employer has duties not only to his own workers but also to outside contractors, workers employed by them, and to members of the public whether within or outside the workplace who may be affected by work activities. Both employers and the self-employed are required to carry out their undertakings in such a way as to ensure, so far as is reasonably practicable, that they do not expose people who are not their employees to risks to their health and safety. The duty extends to, for example, risks to the public outside the workplace from fire or explosion, or from the release of harmful substances into the atmosphere. The duty of employers and the self-employed also applies to people who may be inside the workplace, such as visitors, outside contractors, and their employees working on the premises on a permanent basis, for example, maintenance men or another employer's workers temporarily visiting the premises. It should also be noted that outside contractors, whether employed or self-employed, will themselves have responsibilities under this duty for the health and safety of workers and others on the premises they enter to carry out the contract work.

In general, the standard of protection required for visitors and others within a workplace will be similar to those an employer should give his employees. For example, if machinery and substances are used in such a way as to ensure the safety of employees, that will usually also be sufficient to ensure the safety of non-employees. There may be, however, a need to apply different criteria to achieve these standards in view of the fact that certain people, such as the very young or disabled, may be more vulnerable than others and that people visiting a workplace may have less knowledge of the potential hazards and how to avoid them.

Every employee must take reasonable care for the health and safety of himself and of other persons who may be affected by what he does or fails to do at work. This duty implies not only avoiding obviously silly or reckless behaviour, but also taking positive steps to understand the hazards in the workplace, to comply with safety rules and procedures, and to ensure that nothing he does or fails to do puts himself or others at risk.

11 Regulations and Approved Codes of Practice

The relationship between the Health and Safety at Work Act and earlier Acts dealing with health and safety, has been outlined above. Most of the provisions still in force are those laying down specific standards of health, safety, and welfare for particular circumstances. One of the most important aspects of the Act is that it contains powers to modify, replace, and repeal the earlier legislation. The eventual aim is to bring all health and safety requirements into a single system of regulations and approved codes of practice. The regulations specify requirements to supplement the general duties imposed by the Act, while approved codes of practice give practical guidance about how compliance with the general duties or regulations might be achieved.

Approved codes of practice do not themselves lay down legal requirements. No-one can be prosecuted for failing to follow the guidance contained therein. However, every approved code has a special legal status similar to the status of the Highway Code under road traffic laws. If someone is being prosecuted for breach of any requirement of the Act or related legislation, any approved code which appears to the court to be relevant to the case is admissible in evidence. If the guidance of the approved code has not been followed, it is up to the defendant to show that he has satisfactorily complied with the requirements in some other way.

Anyone who chooses not to follow the guidance in a particular approved code of practice must, therefore, realize that if legal proceedings are taken against him for breach of the requirements, that are the subject of the approved code, he must be able to prove to the court that he has nevertheless satisfactorily fulfilled those requirements by some other means. Such proof will often be very difficult. For example, the Safety Representatives Regulations require [Reg 7(2)] the employer to make available to safety representatives the information within his knowledge that is necessary to enable them to fulfil their functions. The Approved Code of Practice says that such information should include, for example, information about the plans and performance of the undertaking and any changes proposed insofar as they affect the health and safety at work of the employees. If an employer is prosecuted for failing to provide information in accordance with the Regulations, and if it were shown that he had information about plans and performance of the undertaking and had not disclosed it, he would probably find it difficult to show that he had complied with the Regulations in another way.

Since the fourth edition was published a number of important Regulatory Instruments have in fact made under the procedure referred to above. The most important of these are the Control of Substances Hazardous to Health Regulations, the Electricity Regulations, the Pressure Vessel Regulations, and the Ionising Radiation Regulations. The practical effect of these Regulations and their associated Codes of Practice as far as they affect laboratories have been reviewed in the text. Of these Regulations, the Control of Substances Hazardous to Health Regulations (COSHH) are so important that further mention of their general requirements is made here.

12   Control of Substances Hazardous to Health Regulations

As has been noted in the Introduction these Regulations and their associated Codes of Practice are of fundamental importance to laboratory work in that they set out general principles of good occupational hygiene practice which are of direct application to all hazardous chemical substances. The Regulations contain specific provisions regarding the handling of carcinogenic substances and should be consulted if such substances are used.

While it is obvious that carcinogens are hazardous to health it is not always clear what other substances might attract the requirements of the Regulations and Codes of Practice. Unfortunately no specific definition appears in the Regulations as to what substances are hazardous to health. Some guidance is provided in the Classification, Packaging, and Labelling Regulations, where parameters are set out for labelling substances which are toxic, harmful, irritant, or corrosive. The Regulations list many of the commoner substances which fall into these categories. For substances not so listed the parameters set out therein give a useful guide as to whether or not they might be included in one of these categories of hazard. Other useful and more specific information is set out in the Health and Safety Executive Guidance Note EH40, which lists occupational exposure limits for some commonly used substances. Substances appearing in any of these lists and those having similar properties must be considered as coming within the scope of the Regulations.

The COSHH Regulations contain an absolute requirement that an assessment of the risks to health must be made before any work activity is undertaken using a hazardous substance. It follows therefore that every chemist who works in or is in charge of a laboratory must be conversant with their requirements. Any such assessment must identify every hazardous substance, evaluate the degree of risk rising from any associated work activity, and make recommendations for the control of every such risk.

The assessment must be sufficient and suitable and the expertise and detail with which it is carried out must reflect the nature and degree of risk as well as the complexity and variability of the process. It therefore provides for a graded response commensurate with the hazard. Unless the risk is negligible the assessment should be in writing and kept as a factual record.

The Approved Code of Practice requires that the person making the assessment is competent to do so. More than one person may be involved in the task, particularly where the process is a complex one. Where an unresolved residual problem is found to exist outside expert help should be sought.

COSHH assessments should be regarded as establishing good working practices as set out in this handbook. In essence the Regulations in many laboratories require little more than is already expected of a competent and prudent chemist; a sound scientific approach, a modicum of common sense, a knowledge of the Regulations, and an ability to keep orderly records.

A simple method of evaluating the risks to health and the selection of appropriate control measures, by means of a matrix, is given in the booklet entitled 'COSHH in Laboratories' published by the RSC.

CHAPTER 3

Safety Planning and Laboratory Design

K. EVERETT

1 Introduction

In a world in which the only constant is change there will always be, to a greater or lesser extent, shortcomings in equipment and premises. It is the duty of management to minimize these shortcomings as part of the task of optimizing the operations of the unit concerned.

The operation of an efficient, safe laboratory depends on a number of factors:

  (i)   a management system which is suitable for the work of the laboratory;
  (ii)   clearly thought out and enunciated policies and targets;
  (iii) good internal communications to ensure that all grades of staff are integrated into a coherent unit;
  (iv)   effective systems of information and training;
  (v)   selection of staff who possess skills appropriate to their work;
  (vi)   equipment suitable to the intended purpose;
  (vii) well designed premises;
  (viii) efficient maintenance and support services.

Within the task of management a major component is the preparation and effective implementation of carefully thought out safety policies. These safety policies are not only strict legal requirements of the Health and Safety at Work Act 1974 and the Regulations made under that Act, but also the basis of work programmes uninterrupted by accidents.

A well administered safety policy engenders confidence in external agencies, *e.g.* insurance companies, the Health and Safety Executive, as well as contributing to good staff relations. An efficient information and training system ensures that the experience of senior staff is continually passed down to juniors. It is all too easy for middle-aged managers to forget that junior staff may be only a third to a half of their age and hence do not possess the unwritten knowledge their experience brings. Verbal communication is invaluable: a few minutes discussion and a well chosen anecdote can pay rich dividends. It is a subconscious memory of past events which is the basis of 'common-sense', and which leads to a better understanding of what is proper behaviour. Timely words of approval also make the occasional reproof more bearable.

2   Safety Policy and Implementation

To comply with the requirements of the Health and Safety at Work Act 1974, employers must provide their employees with clearly written and unambiguous statements of their current health and safety policy, and on how it is intended to be implemented in practice by both management and employees. This requirement has been amplified by subsidiary legislation, *e.g.* the Control of Substances Hazardous to Health Regulations 1988. The development of the legislation is a continuing process, strongly influenced by Directives of the European Commission.

Because the safety policy must apply to all employees, some or many of whom may not work in the laboratories but in the support services, the policy document must necessarily be concerned with general health and safety objectives, such as overall organization, training, areas of responsibility, and matters of consultation. In places of work where several distinct and specialized types of work are carried on it may well be necessary to provide more detailed implementation documents specific to each type of work, and a laboratory is an obvious case in point. The subsidiary regulations require detailed assessments to be made of many aspects of the work carried on, and continuous review is essential.

3   The Design of Laboratory Premises and Services

When modification of existing premises or the construction of a new complex is contemplated considerable planning effort is necessary. It is vital that the customer draws up a clear specification of the requirements and that not only immediate needs are catered for but also possible developments within the lifetime of the new construction. The need for flexibility of use is paramount, especially in large projects. The development of the organization and the discipline itself must not be compromised by short term personal fads and whims. Every project will be uniquely affected by local conditions and it is essential that from the outset there is a properly organized project team to ensure the correct questions are asked and an appropriate specification prepared.

It is vital that a project officer is appointed from within the customer department. This person should have immediate access to and have the full support of, but not be, the director of the customer unit. At a later stage he should be the liaison officer with the design team and architect. As well as having the confidence and respect of the staff who will initially work in the new premises, the project officer needs to be diplomatic and firm, but receptive to new ideas. Continuity is essential and the project officer should be expected to be in post from inception to completion. For a major building the period from

financial approval to final handover will be several years. This must be taken into account in choosing the appropriate person. The project officer must have good presentational skills and be prepared for the fact that many members of the design staff and team of architects may never have worked on a laboratory project before.

The preparation of the laboratory specification is basically an exercise in gathering information and identifying needs and future trends. The task of the project officer will include differentiating between the essential and the desirable, he must be able to recognize the difference between innovative ideas and personal whims and fashions and real technological developments.

The drawing up of a specification is an interactive process resulting from the identification of problems and constraints—physical, financial, and temporal. The last named constraint will determine when the specification is adopted. Selecting when the design philosophy and specification is to be finalized is an important and essential management decision, which must be taken before detailed work commences. Changes in design philosophy mid-project can produce dangerous or expensive illogicalities.*

The project officer must be prepared to ask and answer many questions. Some of the answers will be contained in the safety assessments prepared under the Control of Substances Hazardous to Health Regulations 1988, but some questions may also reveal when more assessments may need to be made.

The following list is only a guide. Many necessary questions may become apparent only as the answers to earlier questions are analysed. The review should not be limited to the laboratory area, but must start at the perimeter of the site, tracing the activities of both the laboratory and its support services, identifying any impact the work of the laboratory may have on the environment beyond the perimeter.

At an early stage of planning the preparation of a flow-sheet of the activities of the unit can be extremely helpful.

4   Some Questions To Be Addressed in Laboratory Planning

1   What sort of work is envisaged?
    (a)   in the immediate future;
    (b)   in the longer term.
2   What sort of material and equipment will be required in the laboratory? The answers to this question must specify both kind and quantity (see question 6). The list should include not only the inventory of chemicals but also *all* expendable stores used in the experimental areas *and* in the support areas, *e.g.* offices, workshops, cleaning, maintenance.
3   How, and where, will materials and equipment be delivered? Will there be more than one delivery point and who will be in charge of receipts and deliveries?

---

* The author recalls a laboratory which, as a result of a change in specification during building, was, at hand-over, equipped with flame-proof light-fittings, a mixture of flame-proof and standard electric-socket outlets, and gas taps for bunsen burners!

The answers to this group of questions will determine the access required for transport, both externally and on site. Where space is limited this is important. Suppliers delivery vehicles must be taken into account.

4   What sort of stores accommodation will be required and how extensive must this be? What legal requirements must be satisfied and what system of stores control is envisaged?
These questions should deal with the provision of gas cylinder stores, flammable liquid stores, and waste chemicals stores as well as more conventional materials. The siting, licensing, and control of these stores will need to be agreed with the appropriate enforcing authorities, Fire Brigade, Health and Safety Executive, *etc*.

5   What sort of materials handling system, if any, is needed?
Thought must be given not only to day-to-day problems but also to rare but important events, *e.g.* the delivery of new heavy research equipment. If fork lift trucks or electrically powered vehicles are envisaged, safe, weatherproof parking, and battery charging arrangements may be needed.

6   How large an inventory is required?
The scale of the work, as well as its nature, will affect this factor, which will reflect in the scale of the workshop and other support facilities.

7   Where will the experimental area(s) be located in relation to the workshops, stores, offices, *etc*.?
If the experimental areas are remote from the main support services it is possible that satellite workshops and stores may be needed.

8   Where should libraries and offices be in relation to the experimental area(s)?

9   What conference/colloquium/training facilities will be needed?

10 What staff welfare facilities will be required?

11 What occupational health, safety, and first aid arrangements are needed?

12 Where are cleaners and maintenance staff to be based?
At this stage of the planning process it is possible to review the constraints imposed by safety requirements.

13 What are the identified risks associated with the work of the laboratories?
The assessments prepared under the Control of Substances Hazardous to Health Regulations 1988 will be one source of information. The hazards arising during the use of radio-active sources, ionizing radiations, lasers, genetically modified products, flammable liquids, *etc*. will also need to be evaluated. Recent and forthcoming legislation on the use of electrical equipment, pressure vessels, noise, and mechanical handling must also be taken into account.

14 Can potentially dangerous activities be limited to separate areas, *e.g.* biological hazards, cryogenic hazards, explosion risks, fire, high pressures, radio-active hazards, toxic materials?

15 Is it possible to identify high risk areas, and if so, what and where are they?
These areas may include receiving bays, stores, experimental areas, in-house laundry facilities, *etc*.

16 Will it be necessary to provide separate stores for hazardous materials?

These materials include chemicals, cryogenic materials, explosives, gas cylinders, solvents, *etc.*

17 Do gas cylinders pose any special problems?

18 Do any activities require a licence from a public authority which might influence the building design?

What is basically a chemistry laboratory may well, from time-to-time, include activities covered by regulations relating to genetic modification, dangerous pathogens, radio-active substances, highly flammable liquids, petroleum spirit, pesticides, environmental protection, drugs, poisonous substances, *etc.*

19 What waste disposal facilities are required?

20 What design features will be required by
    (a) in-house safety rules;
    (b) statutory requirements?

21 Are there any special noise problems?

A nearby residential area can result in onerous restrictions on night-time noise. Low-level noise from ventilation plant, compressors, *etc.*; which is of no consequence during daylight hours may be the cause of considerable local ill-feeling if present at night-time.

It will be clear that in answering many of the above questions the project officer will have been liaising closely with in-house safety staff, security, and fire officers *et al.* Early consultation with the local authority and other interested parties, particularly enforcing agencies, will ensure that drains, exhaust ventilation, electricity, gas, and water can be supplied in a satisfactory manner. Of particular importance is a discussion with the local fire brigade and other emergency services, who should be kept informed through all stages of planning. Fire engine access and water supply problems should be resolved at the earliest possible stage (see below).

At this stage of the planning process sufficient information should have been gathered to prepare a description of the activities envisaged, perhaps with activity flow-sheets and an outline specification. Potential problems will be emerging and further questions may have to be resolved.

A second series of questions should now be addressed.

22 What services will be required?

This list should be divided into *essential* services and *desirable* services. It is very important to get as accurate an estimate as possible of the scale of these requirements. Over design can cause problems just as serious and expensive to correct as under design, particularly in relation to water supply. The ability to modify services at some unknown future date should be kept in mind.

23 What, if any, health and safety considerations affect the provision of services? Do any require integrity of supply?

Water supplies in general need to take into account risks of legionella during their use.

When the above considerations have been evaluated there should be a meeting between the senior maintenance engineer, and fire, safety, and security officers to review the data. Some of the important matters to be resolved are as follows.

24   Should services be placed overhead, underfloor, in service corridors, or in a mixture of these?

The answers may not be clear at this stage and final decisions may have to be made at a later stage when the design team have produced their first drawings.

25   Is it appropriate to design the building with inter-related safety zones?

The zones may be divided into three groups:

(a)   low hazard areas, containing offices, welfare facilities, inert materials stores, workshops, *etc.*;

(b)   medium hazard areas, containing ordinary laboratories, control panels, *etc.*;

(c)   high hazard areas, containing special experimental work and associated stores.

This design philosophy will probably involve building a plenum ventilation system feeding clean air into the lowest hazard areas and extracting from the highest hazard areas.

26   What staff facilities will be required?

It will be necessary to estimate staff numbers of all grades likely to be employed in the foreseeable future.

Staff will require offices, washing, changing, and toilet facilities. Whether to provide 'in-house' laundry facilities must be decided at this stage. Rest rooms where food may be eaten are essential and the details of any NO SMOKING policy must be settled.

27   What computer services are required?

28   What arrangements are needed for maintenance?

Zoning of building services can limit the disruption caused by maintenance or modification of services, as well as minimizing the effects of breakdowns, fires, *etc.* Routine cleaning should be seen as a part of maintenance and the policy in respect of the disposal of waste should be clarified at this stage in consultation with the relevant local authorities. It will be necessary to determine how many skills and trades will be involved and how many different workshops will be needed. Each trade will have its own supply and storage problems.

The question of 'permit-to-work' certificates will need to be reviewed to ensure that safe working is not jeopardized by loose management.

During this stage of planning it would be wise of the project officer to meet maintenance staff, tour plant rooms and similar areas, and hear from the staff of any difficulties they encounter. Well designed plant-rooms and service areas minimize down time.

A broad design philosophy can now be prepared. Having collected this information it may be worthwhile for senior management and the project officer to take a retrospective look over the past ten or twenty years to see whether a pattern of development can be recognized which will suggest future trends.

In a large scheme thought must be given to site access for pedestrians, cyclists, delivery vehicles, emergency services (ambulances, fire-engines, *etc.*), staff and visitors cars, and public service vehicles. The provision of a secure area within which stores and receiving bays are located should be considered.

5   Fire Precautions

Statutory and local authority regulations must be complied with and the views of the insurers should be canvassed. Large companies may have special rules which must be observed.

Consultation with the local fire service cannot begin too early. This may require considerable forebearance on the part of the project officer. When negotiating with fire officers it may appear that over severe standards are being set. This may well be so, and it is in order to question and examine each requirement. It must be remembered however that there are good reasons for high standards. If there is a major fire the lives most likely to be at risk will be those of the fire services officers colleagues in the operations branch. Also, they suspect, not without cause, that they are very vulnerable to public criticism based on ill-informed hindsight following a perceived major disaster. They will also be aware that corrosive and highly toxic gases and vapours, compressed gas cylinders, and concentrations of highly flammable liquids may be present, as well as possible explosion hazards. (An exploding Winchester of flammable solvent can produce a fire-ball up to two metres in diameter with a core temperature in the range of 700–900 °C, reinforced with flying glass, *e.g.* the equivalent of a large petrol bomb.) The reasoning underlying such restraints therefore lies in the belief that the easier it is for a fireman to fight the fire the less likely it is that it will get out of control or that people will get injured.

6   Fire Alarms and Detector Systems

A clear directive is necessary in respect of fire alarms and detector systems. Automatic detectors may be obtained for many purposes, *e.g.* fire, ionizing radiations, flammable vapours, toxic gases, equipment malfunction, services failure, security, and intruder detection.

The specification must indicate:
(a)   what, if any, automatic detectors are required;
(b)   whether the detectors should automatically operate the alarm system or merely give a local warning, *e.g.* in a security office;
(c)   whether a built-in fire extinguisher system, *e.g.* sprinklers, is required and if so whether it should be automatic or manually operated.

Where there is a central computer based data collection and control system it will probably be expedient to link the fire, safety, and security alarms into it.

There are four main functions of a detection and alarm system to be considered:

(a) the detection of malfunction of equipment due to fire or other unwanted event;
(b) the alerting of staff to the unwanted event or hazard;
(c) the control or suppression of the hazard;
(d) the evacuation of staff and visitors to safety.

A clear, preferably written, design philosophy is essential. The designer must have clear and unambiguous instructions.

6.1 Alarms

(a) When is an alarm to operate?
(b) What is an alarm to indicate? Is it to be an alert or is it to be the signal for evacuation of a specified area?
In the former case two separate signals are needed; an alert signal and an evacuate signal.
(c) Who is to be alerted by the alarm signal?
(d) Is the alarm signal to be aural, visual, or both?
Remember that there may be deaf people present.
(e) Is an automatic or manual alarm needed?
(f) What actions should follow an alarm?
The answers to this question may require a fairly detailed policy statement, setting out staff duties, assembly points, emergency services liaison, and support procedures, *etc.*
(g) How many types of alarm are needed?
(h) Is an all-clear required?
A policy statement setting out who has authority to declare an emergency over should be prepared.

If the signal is to be aural, there are four clearly distinguishable sounds to choose from: bells, klaxons, sirens, and warblers. These may sound continuously or intermittently. Special sounds may be required in animal houses, operating at frequencies which are inaudible to animals. If a visual alarm is being specified it may be a steady or flashing light. Because of the prevalance of colour blindness simple colour coded lamps should be avoided. Symbols, words, or shapes which are easily recognized and readily convey meaning should be used.

7 Disabled Persons

It is necessary to consider the problems of disabled persons. Again, a clear policy statement is helpful.

The definition of 'disabled person' should cover two distinct groups of people:

(a) permanently disabled;
(b) temporarily disabled.

The former group may include staff or visitors affected by colour-blindness, deafness, locomotory impairment, respiratory problems, *e.g.* emphysema, asthma, bronchitis, cardiac problems, haemophilia, *etc.* Epilepsy is less a disability than a condition of which colleagues should be aware. Amongst older staff the possible on-set of progressive disease may require special consideration.

The second group will consist mainly of victims of accidents and sports injuries, possibly with a period of convalescence following illness.

A building designed with the problems of the disabled in mind can reduce absence due to sickness and contribute both to the efficiency and peace of mind of the staff.

8   Miscellaneous Matters

The design engineers will need to know on what floor loading is likely to be, how much waste heat will be generated from freezers, refrigerators, ovens, furnaces, hot-plates, drying cabinets, autoclaves, bunsen burners, electric motors, *etc.*

Such features as fume cupboards, total containment glove-boxes, clean work areas, biological safety cabinets, animal houses, *etc.* all require the advice of specialist designers. They will have major implications for the design and scale of services, particularly ventilation. Adequate provision must be made taking all these matters into account and bearing in mind that the fume dispersal system may affect the aesthetics of the overall building design.

Apart from safety, health, and welfare, consideration at some convenient point in the preparation of the specification of a new building must be the general shape of the completed building. This will be, to a greater or lesser extent, dictated by the site available. The choice will be between a single building or a range of buildings, single or multi-storey buildings, and what provision must be made for future expansion.

So far no mention has been made of the appointment of architects, design consultants, or building contractors. The timing of these appointments will depend on the size of the project and the availability of 'in-house' expertise and services. The client, in particular the project officer, must be prepared to provide a considerable amount of expertise and guidance on both detailed and general matters to ensure that the specification is interpreted accurately. Nothing should be taken for granted.

It must be recognized that many compromises will have to be made, not least in response to financial imperatives, to the physical constraints of the site, engineering needs, and the requirements of planning authorities and other outside bodies.

9   Biology Laboratories

Where biology laboratories are closely associated with chemistry laboratories then special considerations arise. Special attention should be paid to the Health and Safety Executive (HSE) Guidance prepared by the Advisory Committee on Dangerous Pathogens and the Advisory Committee on Genetic Manipulation (the term genetic manipulation is being replaced by genetic modification to achieve uniformity with EC practice). Animal accommodation is a specialist subject outside the scope of this book.

10   Ventilation, Fume Cupboards, and Other Safety Cabinets

Detailed information on fume cupboards is contained in the Royal Society of Chemistry booklet, 'Guidance on Laboratory Fume Cupboards' and British Standard BS 7258, Laboratory Fume Cupboards, Parts 1, 2, and 3. Where appropriate British Standard BS 5726, Microbiological Safety Cabinets, may be consulted, but at the time of writing this is under revision.

11   Waste Disposal

There are now onerous legal controls on the disposal of waste and discharge to the environment and a clear policy statement is essential. It must be reviewed regularly in view of the rapidly increasing pressure of public opinion and the consequent development of legal requirements in this area. Materials may be segregated under such headings as:
flammable solvents;
halogenated solvents;
chemicals which can be incinerated safely;
heavy metal containing entities;
aqueous acids;
aqueous alkalis;
biological waste;
soft waste, *e.g.* paper tissue contaminated with chemicals;
sharps, *e.g.* needles, broken glass, metal items;
radio-active waste;
domestic/office waste.

The possibility of recovery and recycling of solvents must always be considered as a first step.

The fate of every type of waste must be studied from the point at which it is created to its ultimate disposal and impact on the environment. Of particular importance is the disposal of 'sharps'. Carelessly or thoughtlessly discarded sharps can be the cause of many injuries and needless anxiety amongst cleaning staff and others 'down-stream' of the laboratory in which they are produced. Responsibility for waste disposal must be clearly defined and the policy and rules strictly enforced.

It is important that the enforcing and inspection authorities are consulted and are made aware of the waste disposal policy. At the time of writing the principal piece of legislation is the Environmental Protection Act 1990 which sets up Waste Disposal Authorities.

The disposal of clinical waste should take account of the guidance issued by the Health Services Advisory Committee (HSAC) of the Health and Safety Commission in a note entitled, 'The Safe Disposal of Clinical Waste', HMSO, London 1982.

12   Electrical Safety

The Electricity at Work Regulations 1989 impose duties on managers, supervisors, technicians, and *users* of electrical equipment. The regulations require that persons engaged on electrical work are 'competent to prevent danger and injury'. Further more detailed advice is given in the chapter on Electrical Hazards. All newly acquired or locally made electrical equipment should be tested by a *suitably trained* member of the staff and should comply with the Low Voltage Electrical Equipment (Safety) Regulations 1989: 728 issued under the Consumer Protection Act 1987.

Although Great Britain and eighteen other countries adopted the following colour code for electrical wiring in 1970, *not* all major supplying countries adhere to it. The 1970 code:

Earth   —green and yellow (stripes)
Live   —brown (dark)
Neutral—blue (light)
is distinguishable by colour blind persons.

13   Compressed and Liquefied Gases

The Pressure Systems and Transportable Gas Containers Regulations 1989 come into force in stages up to 1994 and are backed up by several Approved Codes of Practice and HSE Guidance Notes.

14   Noise

The Noise at Work Regulations, 1989, are supported by an HSE booklet, 'Noise at Work: Guidance on the Regulations' (HMSO, London, 1989). The Regulations define three action levels:
  (i)   the First Action Level—a daily personal noise exposure of 85 dB(A);
 (ii)   the Second Action Level—a daily personal noise exposure of 90 dB(A);
(iii)   the Peak Action Level—a sound pressure of 200 Pa.

Assessments of exposure must be made by a competent person when it is likely that the first action level as peak action level may be exceeded. Regulation 7 requires that, so far as is reasonably practicable, the exposure shall be reduced by means other than the provision of ear protectors.

Continuous exposure to noise, particularly above 85 dB, can induce deafness. Because of the non-occupational exposure to noise, *e.g.* disco's, traffic noise, motor-cycle crash helmets, sports injury, normal illness, it is essential where such a level may be exceeded that management keep reliable records of the noise exposure of staff. The effects can be insidious and long-term and claims in civil law could appear many years after the alleged exposure. The measuring and recording of noise should be done by recognized professionally competent persons, whether 'in-house' or outside consultants.

15   Permit-to-Work Certificates

Whenever building work of maintenance is to be carried out it is essential that before the work commences formal safety clearance is obtained from all sections likely to be affected by the work.

The person issuing the 'Permit-to-Work' Certificate MUST be in a position of authority and responsibility, and preferably on a limited list of persons named by a Senior Manager or Director. The management structure of the organization will determine the necessary details of the scheme, but clear, written instructions should be given to *all* staff. The following points should be covered as a minimum:

  (i)   the Clerk of Works, foreman, or equivalent must ascertain, usually from the 'customer' department, whether a 'permit-to-work' is required, and ensure that, if it is necessary, it is completed, BEFORE work is scheduled to start;

  (ii) the 'customer' department is responsible for ensuring that the information on the 'permit-to-work' is correct;

 (iii) the customer must ensure that, once an area has been declared safe to work, no activities are allowed that would affect the terms of the certificate;

  (iv) the supervisor of the work being carried out MUST inform the customer department, and any other persons, of any and all services which will be affected. Particular attention should be given to electrical services and any safety and security systems;

   (v) matters of doubt should be referred UP to the appropriate manager and if appropriate to the professional safety advisors;

  (vi) any extensions and variations to working instructions may only be made by the person signing the certificate and a new certificate MUST be obtained if the original is no longer adequate;

 (vii) liaison must be with *authorized personnel* only;

(viii) workmen MUST be given clear instructions to ensure that they comply with the terms of the certificate. Particular attention must be paid to workmen employed by sub-contractors, to ensure adequate supervision is given.

CHAPTER 4

Fire Protection

L. A. WARWICKER AND M. SHELDON

1 Introduction

Chemical laboratories vary greatly in size and operation, but most are production, research, and development (including pilot plant), analytical, or teaching laboratories, and each of these present a variety of fire hazards. Therefore, each laboratory is unique and to implement fire protection an intimate knowledge of the working of the laboratory and the elements of fire prevention are required. Many research workers and laboratory staff, although highly trained in their particular fields are often unacquainted with the basis of fire prevention.

Managers must accept responsibility for providing a safe working environment, but those working in laboratories should appraise the fire hazards of their own particular work. If there is a safety committee, part of its task should be to establish what the fire hazards are. In a multi-occupancy building, a liaison committee should be set up for all the occupants to co-ordinate their fire precautions, because the spread of fire and smoke from one part of a building to another is the concern of all the occupants and if fire is to be contained or its spread delayed, concerted action is needed.

Action which may be required can vary from simple precautions, such as ensuring smoke stop doors are not jammed open, to complex precautions involving structural alterations. More elaborate precautions may require a detailed knowledge of the construction of the building and service systems in the building so that, for example, concealed spaces and ductwork do not permit fire or smoke to spread, perhaps unseen, vertically or horizontally through a building.

If fire should occur it may spread beyond the control of any one individual. Laboratory staff should, therefore, know how to summon the fire brigade, raise the alarm, and summon help.

2 The Nature and Source of the Fire Risk

The potential risk arises from the presence of combustible solids, liquids, or gases in conjunction with ignition sources. One or more of these classes is generally found in most laboratories.

2.1 Solids

Most combustible solids will not present a great fire risk unless they are ground into powder. Powders of combustible solids can be explosive when dispersed in air and if large quantities are ground to a fine state then precautions may have to be taken against dust explosion.[1] The main hazard will be those solids which are unstable and are likely to decompose explosively if they are heated or subjected to friction. Other solid materials may be hazardous due to their oxidizing properties so that a hazardous situation can result if they become contaminated with combustible material.

Another class of hazardous solids is those which will react spontaneously and exothermally with water or air. Obvious examples are alkali metals, metal hydrides, and certain organometallic compounds. Even aluminium powder can react with water and although the reaction is not very vigorous hydrogen may accumulate—which can be exploded by a small spark.

Special methods of storage of these materials are called for and periodic inspections are necessary to ensure that safe conditions are being maintained. For example, some unstable materials are kept damped down with water or a high flash point liquid. During a long period of storage there may be evaporation of the liquid and drying out so that the unprotected solid is exposed. When attempting to scoop out some of the unstable material, friction with the scoop may then ignite it, leading to fire or explosion.

2.2 Liquids

The most common of the hazardous materials to be found in laboratories are flammable liquids and it is essential to know their fire properties, such as flash point and ignition temperature. If the flash point is below room temperature then these liquids will always constitute a fire hazard and careful control should be maintained over them. Liquids with flash points well above room temperature will support a fire if heated to a temperature exceeding their flash point.

Before ignition can occur, a flammable vapour has to be heated, at least locally, to a temperature exceeding its ignition temperature. Almost invariably the ignition temperatures are well below those of common igniting sources such as flames, sparks, and incandescent surfaces. For some materials ignition temperatures can be extremely low. Carbon disulphide, for example, will be ignited by a source whose temperature just exceeds that of boiling water, while ethers and aldehydes usually have low ignition temperatures. Other properties of liquids have to be taken into account such as their propensity to form more hazardous substance after long periods. Ethers are one of such class of materials, and form highly explosive peroxides after long standing. The possibility of mutual interaction between flammable liquids should be considered when keeping them in laboratories or stores. Only the minimal amount of flammable liquids should be kept in laboratories. One days supply is often recommended but this may be inconvenient or impracticable. However, the requirements of the Highly Flammable Liquids and Liquefied Petroleum Gases Regulations 1972 may limit the total quantity of highly flammable liquids allowed to be stored in the laboratory to 50 litres with suitable containment. Nevertheless, the quantity of flammable liquid should be kept to a minimum.

Bottles containing flammable liquids should be positively identified by labels which can withstand the deleterious effects of any atmosphere likely to be present in the laboratory. Sand-blasted labels make loss of labelling impossible. Large containers of flammable liquids should never be carried in the hand and in particular Winchester bottles should not be carried by the neck. Suitable Winchester carriers are available and should be used for the transfer of liquids from one place to another. Only trained staff should refill bottles with flammable liquids in a special room or in a laboratory where all ignition sources have been removed.

Damaged glassware should not be employed in experiments as it may crack and spill the contents during the experiment. Poorly assembled or unsuitable apparatus may introduce serious fire risks. Ground glass joints are usually preferred to the classical use of corks and bungs, which if badly fitting will introduce a fire risk. The breakage of equipment by localized overheating using direct gas flames is a hazard which can be avoided easily by using water baths, hot plates, heating mantles, or sand baths.

2.3 Compressed and Liquefied Gases

Compressed or liquefied gases present hazards in the event of fire since heating will cause the pressure to rise and may rupture the container. Leakage or escape of flammable gases can produce an explosive atmosphere within the laboratory which can be ignited and result in a devastating explosion. Cylinders of gases should be provided with a suitable pressure regulating valve. The pressure at which the gas is to be used should be determined by this valve and not be operating the needle valve, as this leads to erratic control and the application of full pressure on any tubing should the exit of the tubing become blocked. Gas cylinders should preferably be placed outside and the gas piped into the laboratory. Properly designed compartments and reinforced walls should be used to protect personnel against possible explosion.

3 Structural Protection and Segregation

The concept of structural protection and segregation is to ensure that if a fire occurs, the performance of the elements of the building structure will not be sufficiently impaired to reduce their ability to act efficiently. Segregation is used to divide up or separate hazardous operations from one another so that any incident does not spread the conflagration but contains it in a known compartment. This makes fire fighting operations easier by not having to tackle the blaze on many fronts. The elements of building structure include walls, floors, columns, beams, ceilings, and roofs. Consideration should also be given to doors and the protection of lobbies and stairways.

The floor of the laboratory should be impervious to chemicals. It may be considered necessary to place a sill at the door which should be provided with a ramp which is not too steep. There should be at least two means of escape from the laboratory and benches should be laid so that people can escape easily if a fire occurs on the bench. This means that blind aisles should be avoided and the bench should more or less run parallel with the line of the exits. A space of at least 1.2 m should be provided between benches for passage of personnel and equipment. Reagent shelves should be situated on the bench so that it is unnecessary to lean across experimental apparatus in order to reach reagent bottles. Similar consideration should apply to services such as gas, electricity, and water supplies and drainage. If laboratories are situated in single storey buildings, then it may be sufficient to construct the buildings of non-combustible materials. If the laboratories are situated in multi-storey buildings then walls, floors, and ceilings should be of fire-resisting construction. Fire resistance should be of at least 120 min, but will vary according to the risk in adjoining compartments. Columns and beams should also be protected to the same standard of fire resistance. Openings made in the walls for the passage of ducts, pipes, and cables should be fire-stopped. Stairways should be enclosed by fire-resisting walls. Exit doors should be hung to swing outwards.

The quantities of flammable materials in the laboratory should be kept as small as possible. Where it is possible a separate building should be used for the storage of flammable liquids[2] and those laboratories subject to the Highly Flammable Liquids (HFL) and Liquefied Petroleum Gases (LPG) Regulations 1972 will have certain legal obligations regarding quantities stored. The building should have a single storey and be constructed from non-combustible materials. The roof should be of light construction and easily shattered or readily blown off in the event of an explosion. If a separate building cannot be provided the store should be on the ground floor and should be of fire-resisting construction. It should be well ventilated and unheated and have doors which open outwards. There should be a sill on the doorway in such a storage compartment so as to contain any spillages which may occur accidentally. As far as possible separate stores should be used for materials of differing hazards, and materials which react vigorously with one another should not be stored together (see p. 62). Pilot plant experiments are carried out principally because operations on this scale may indicate the possibility of unforeseen results. Laboratories which are used for pilot plant work are particularly hazardous as larger quantities of materials will be used and special buildings or areas should be set aside for these purposes. It is inevitable that flammable vapour or gases will be produced in the laboratory atmosphere from time to time. The best general precaution is to ensure that the laboratory is well ventilated. Windows and doors cannot be relied upon to provide adequate ventilation as they may well be kept closed much of the time and mechanical ventilation should be used. Any chemical operation which involves the possible production of flammable vapours should be carried out in a fume cupboard. The apparatus should be set up over a large metal tray to catch any flammable liquids which may escape due to accidents, such as breakage of equipment. The recommended air-flow velocity through the face of a fume cupboard is 30 m min⁻¹. The air-flow should be maintained by means of a fan and the fan motor should be placed outside the duct serving the fume cupboard, driving the fan by means of a shaft. The extraction duct should be made of non-combustible material and it is desirable that it should pass directly to the outside of the building without passing through ceilings. If, however, it must pass through one or more floors, or there is intercommunication with other fume cupboards, then careful consideration must be given to the control of spread of fire and toxic gases within the building. Several alternatives are available, *viz.*:

(1) The extraction duct-work should be of a fire-resisting construction.
(2) The extraction duct-work should be enclosed in a fire-resisting structure.
(3) Fire dampers may be fitted at positions where the duct passes through ceilings or partition-walls. However, the complexity of the problems associated with ducts intercommunicating and/or passing through buildings requires expert advice.

More detailed advice can be found in a British Standard Code of Practice.[3]

4   Controlling Ignition Sources

For a fire or explosion to occur it is necessary for a flammable or explosive atmosphere to exist and to have sufficient energy added to it. In the laboratory, as in other places, electric equipment, open flames, static electricity, burning tobacco, lighted matches, and hot surfaces can all cause ignition of flammable material.

Gas supplies to laboratory outlets should be by means of rigid permanent piping. A laboratory may have many outlets for gas supply and in this case a control valve should be placed just outside the laboratory so as to enable supplies to be cut off in the event of an emergency.

Electric wiring should, as far as possible, be of a permanent nature and installed in accordance with the latest edition of the IEE regulations.[4] Switches, sockets, and terminals should be placed where they are easily accessible and are safe from accidental wetting by water or other liquids. Temporary wiring should be installed by a competent electrician or at least inspected by one before use. High-sensitivity, current-operated, earth-leakage circuit breakers can provide an excellent means of protection against fires originating from earth-leakage faults, as well as protection against line-to-earth shocks.

Where it is likely that large amounts of flammable gases or vapours may be released into the atmosphere then it is necessary to install electrical equipment designed for use in explosible atmospheres. By various methods of design and construction electrical equipment suitable for employment in potentially flammable atmospheres can be manufactured. Probably the most well known of these are flameproof types of high power consumption equipment and intrinsically safe types where equipment requires only very low power for operation.

Flameproof equipment is not manufactured for all types of flammable vapours and gases and it may be necessary to use pressurized (purged) equipment. Care should be taken when using flameproof equipment that additions are not made which are below flameproof standard. A common mistake is to use domestic refrigerators for storage of materials and explosions have occurred when low flash point liquids have been placed in such refrigerators. Drying ovens are also a likely source of ignition and it should be remembered that, although the controls may be flameproof, the surface temperature may exceed the ignition temperature of some vapours. The many types of electrical equipment for use in potentially flammable atmospheres and their applicability to particular locations are summarized in the FPA Data Sheet[5] and given in detail in British Standard Specifications and Codes of Practice.[6,7] Thermostats on oil baths have often failed with consequent overheating and fire. Independent excess-temperature manual-reset cutouts should be incorporated to disconnect the supply if overheating should occur.

The use of organosilicon fluids in oil baths has much to commend it as they are highly stable and can have boiling points in excess of 400 °C. There are a number of general precautions which can be taken. Hotplates, furnaces, and ovens should stand on heat resisting surfaces. Where the heated unit is on a wooden bench top there should be an air space between it and the bench to prevent charring. Gas jets not in use should be turned off or adjusted to give a small luminous flame as the pale blue flame cannot be seen easily in bright sunlight.

The discharge of accumulated static electricity can provide a spark which will ignite flammable vapours. Static electricity is created by the relative movement of two materials. Non-polar materials such as hydrocarbon solvents accumulate static charges readily as they have high insulation values and do not allow the charge to leak away. It should be noted that a dispersion of an immiscible polar liquid in a non-polar liquid can generate static charges even more rapidly than a non-polar liquid alone. Some improvement can be obtained by the use of a small proportion of a conductive additive where this would not affect the chemical properties of the fluid. Even crystallization can produce static electricity, and the discharge of carbon dioxide has been known to produce static sparking. Laboratory coats and clothing made of certain synthetic fibres are prone to generate static electricity and should be avoided. Where possible all metalwork should be bonded and earthed when there is a static hazard. When no solution can be found to the static problem then all processes must be carried out as slowly as possible to give the accumulated charge time to dissipate.

Space heating should be safe and adequate. Hot water or low pressure steam radiators are desirable and installation and maintenance should be first class so as to reduce the likelihood of the introduction of uncontrolled and dangerous forms of space heating. Radiators should not be used for drying materials and should be provided with sloping wire mesh screen over them to prevent such abuse.

Good housekeeping will help to reduce the number of fires in laboratories in addition to reducing other types of accidents. Cleaning up as soon as possible after an experiment has been completed will help and the provision of adequate storage space beneath the bench is an advantage. A clear bench will enable the experimenter to see any dangerous procedure without being distracted by unnecessary equipment. The provision of suitable waste bins, preferably of non-combustible material such as sheet iron, will help to encourage a positive attitude to good housekeeping.

5   Emergency Action

In the event of fire or explosion occurring there should be a pre-arranged plan of the necessary action to be taken. All personnel must be made aware of this and fire drills should be carried out at least twice a year in order to familiarize staff with these procedures. The essential elements which should be covered by instruction are:

(1)   Raising the alarm.
(2)   Summoning the local fire brigade.
(3)   First aid fire fighting (including practice in use of extinguishers).
(4)   Evacuation.

5.1   Raising the Alarm

Personnel should be trained to recognize the severity of an outbreak of fire and the immediate danger presented by it. They should be able to report accurately on the fire situation, being able to discern whether a normal procedure or emergency procedure should be brought into operation. In the normal procedure the person discovering the fire may decide that it is relatively minor. He should report the fire immediately to the switchboard or instruct some other person to do so. A person in a position of responsibility should decide if it is necessary to evacuate the building. Meanwhile the switchboard operator should call the fire brigade and notify all persons who have been allocated special responsibility.

Emergency procedure is necessary when the fire has been found to have spread over a wide area or hazardous materials and processes are threatened. In such circumstances, the premises must be evacuated immediately and the manual fire alarm system should be operated by the person discovering the fire. Then, if possible, the switchboard operator should be notified. The switchboard operator should have standing instructions to call the fire brigade on hearing the fire alarm unless the alarm is connected to the fire brigade via a central alarm monitoring station.

5.2   Summoning the Fire Brigade

When the fire brigade is called by the switchboard operator it should be informed at the time which entrance is nearest to the fire. In all cases responsible persons should be sent to the entrance to give the brigade information it may need on the location and extent of the fire, water supplies, the nature of special risks, and details of casualties and trapped persons. If any particularly hazardous materials or processes have been brought into the laboratory the fire brigade should be forewarned. Ideally the fire brigade should have prior knowledge of the hazards through a liaison or safety officer who should be responsible for informing the local fire authority on hazards at the time they are introduced into the laboratory (see p. 19).

5.3   First Aid Fire Fighting

Provided that no danger is involved, the fire should be attacked with first aid fire fighting equipment as soon as it is discovered. It must be decided on relative merit whether attacking the fire takes precedence over reporting it. It can be dangerous to waste time trying to tackle a fire which cannot be controlled instead of immediately reporting the incident; on the other hand, it is undesirable to allow a small fire to obtain a hold through spending time reporting it. In most circumstances there is no conflict, for there is normally more than one person near the scene of the outbreak, and one person can report the fire leaving the others to try to extinguish or contain it. On returning that person can aid the other personnel.

Fire fighters should withdraw from the scene if the heat and smoke threatens to overcome them or if the fire endangers their escape route. They should also retreat if the fire spreads towards explosive materials or gas cylinders. On withdrawing, windows and doors should be closed in an effort to contain the fire.

5.4   Evacuation of Personnel

The essential features of evacuation are that it should proceed by a pre-arranged plan and that all personnel should be familiar with the escape routes to be used. It should be impressed on everybody that they should leave, without panic, immediately they have received instructions to evacuate the building. The assembly point should be fixed and known to all personnel, and the head of the department (or his deputy) should be responsible for ascertaining that all persons in his charge, including visitors, have been accounted for. To ensure that this is thoroughly carried out, one person must be delegated the responsibility of searching the department, including the lavatories and cloakrooms to ensure that nobody has been left behind. No person should attempt to re-enter the building without the expressed permission of the person taking the roll call. Escape routes should be clearly marked as such.

All responsibilities must have been allocated in advance in order to avoid delay and doubt. Moreover, each of the persons bearing responsibilities should designate deputies who could take over in the event of absence, illness, or accident. This delegation of responsibilities could easily be made to coincide with that of the delegation of normal administrative duties.

6   Classes of Fire and Suitable Extinguishing Agents

It is standard international practice to classify fires according to their nature.

Class A fires are those involving solid materials, usually of a carbonaceous nature, but excluding materials that readily liquefy on burning.

Class B fires are those involving liquids or liquefiable solids.

Class C fires are those involving gases (electrical equipment in the USA).

Class D fires are those involving burning metals, *e.g.* sodium, potassium, aluminium, titanium, magnesium, and calcium.

For Class A and B fires, portable hand extinguishers have been available for many years based on their content of suitable extinguishing agent, *e.g.* 9 l water and 0.9 kg dry powder. The current practice is to rate a particular extinguisher according to its ability to extinguish a standard fire under standard test conditions. Thus an extinguisher with a rating of 5A is capable of extinguishing a wood crib fire (Class A) having a cross-section of 0.5 m. A rating of 13B indicates that a Class B test fire of 13 l of the test fuel can be extinguished.

When the British Standard for rating extinguishers[8] was introduced in 1980 a time was allowed for unrated extinguishers to be given an assessed rating, but this time has now elapsed.

For a fire involving solid combustible material (Class A) such as wood, paper, or textiles, water is the most effective extinguishing agent when used as a jet or spray. The cooling power of water is unsurpassed by other media and is especially useful on materials which are likely to re-ignite and for penetration into deep-seated fires. Portable water extinguishers are usually of 9 l capacity but smaller 4–6 l extinguishers which are particularly suitable for use by women are available. Small hose reels which should be at least 19 mm (3/4 in) inside diameter and capable of delivering at least 30 l min^{-1} give a practically inexhaustible supply.

Flammable liquid fires (Class B) cannot normally be extinguished with water. Dry powder, foam, carbon dioxide, and vaporizing liquids can be effective. Dry powder rapidly extinguishes the flames over burning liquid and gives a quick 'knock down'. It is particularly effective in the case of spill fires, but gives little protection against re-ignition. It can be safely used on electrical equipment as dry powder is non-conductive. The capacity of portable dry powder extinguishers ranges from about 1–12 kg.

Foam extinguishes a fire by forming a blanket which floats on the surface of the liquid preventing the access of air or the escape of vapour. The foam blanket remains in position for sufficient time to allow the liquid and surroundings to cool, so preventing re-ignition. Foam is particularly effective in dealing with fires in containers which have become overheated and is more effective than dry powder. It is impossible to form a foam blanket over liquids flowing down a vertical surface and difficult when liquids are flowing freely over a horizontal surface. Polar liquids break down the normal protein based foam. An alcohol resistant foam has to be employed in fighting fires involving alcohol, acetone, *etc*. Foam is electrically conductive and should not be used when electrical equipment is involved in the fire. Foam extinguishers are usually

9 l capacity which will produce approximately eight times this volume of foam.

Carbon dioxide smothers flames mainly by excluding oxygen. It is effective in situations where it is not easily dispersed, and can be used on fires in refrigerators and ovens. It can also be used safely where there is danger of electric shock and where delicate equipment is involved since it leaves no residue. The extinguishers come in various sizes ranging from 1–6 kg in capacity.

Halons are vaporizing liquids containing a halogen which acts by inhibiting the flame reactions. A halon is defined as a 'halogenized hydrocarbon used as an extinguishing medium' and is given a number where the first digit is the number of carbon atoms in the molecule, the second digit is the number of fluorine atoms in the molecule, the third digit is the number of chlorine atoms in the molecule, and the fourth is the number of bromine atoms in the molecule. Halons work very effectively and are safe to use on electrical equipment, but some are very toxic and their use is not permitted. They are of course not 'environmentally friendly' being CFC's and it is possible that they will be phased out in the near future when suitable replacements are developed.

The principal ones in current use are:

Halon 1211 Bromochlorodifluoromethane (BCF)

Halon 1301 Bromotrifluoromethane (BTM) in fixed systems

Halon 1011 Chlorobromomethane (CBM).

All the halons produce toxic products on pyrolysis and they should not be used, or kept in confined spaces or any place where there is a risk that the vapours or their products could be inhaled. The capacity for hand halon extinguishers is 0.5–1 l.

Fires involving gases (Class C) may prove to be the most difficult to deal with, since the extinction of the burning gas, other than by cutting off the supply at source, will allow gas to build-up. If the gas is flammable a dangerous gas/air mixture may be produced which on reaching an ignition source will result in an explosion or re-ignition. A complete appraisal of the necessary actions to be taken in the event of a gas fire must be produced by the safety committee and the laboratory staff must be fully briefed on these actions.

Gas can be piped into a laboratory from a supply which is usually a town-main, bulk store, or cylinders. Often individual cylinders are used within the laboratory. When a piped supply is used a shut-off valve, or valves, should be incorporated into the system so that isolation of the supply can be achieved safely. Shut-off valves in easily accessible positions on both sides of the wall through which the supply is piped are the simplest solution.

Where individual cylinders are involved no attempt to extinguish the burning gas should be made unless it is certain that the cylinders can be turned off safely. If not it is better to extinguish the surrounding fires caused by the burning gas and keep the cylinder cool by spraying water over it and the surrounding area. Cylinders of inert gases, as well as flammable ones, when heated strongly present a potential explosion hazard.

Class D fires deal with special risks. Some of the substances used in laboratories such as sodium, potassium, and metal hydrides will react with extinguishing agents and it would be ineffectual or even dangerous to attempt to use these agents. Special dry powders are available to deal with such hazards, and manufacturers and suppliers of special risk chemicals will be able to assist in the selection of the appropriate dry powder for these substances.

6.1   Selection and Distribution of Extinguishers

A multiplicity of extinguishers is undesirable and probably all risks can be covered by two or three types. It is valuable to obtain extinguishers which are uniform in their method of operation; all personnel should be instructed in the use of these appliances and which type to use.

Extinguishers should conform with British Standard Specifications[9] and be approved by the Loss Prevention Council.[10] Further advice on choice of extinguishers may be obtained from the local fire authority or insurer. The number and size of the extinguishers required will vary according to the risk but there should be at least one of each required type in each laboratory. In large laboratories there should be a minimum of one water-type extinguisher for every 200 m^2 of area and the total should be not less than 18 l capacity on every floor. Further information regarding the use of water and other types of extinguishers can be found in the FPA Data Sheets.[11,12]

6.2   Fixed Fire Extinguishing Systems

Large laboratories and pilot plant areas can be protected by means of fixed installations. The system used can be automatic or manually operated and any type of agent can be employed in the system.

Hydrants, hose reels, or automatic sprinklers can provide general protection. Hydrants are advisable for premises covering a large area, remotely situated areas, and tall buildings. Hose reels have already been mentioned. They are permanently connected to the water supply, simple to use, and very effective for tackling fire at an early stage. Further details are given in the FPA Data Sheets.[13,14] Although water is the extinguishing agent, automatic sprinklers are acceptable in many laboratories. It is advisable to link the automatic sprinkler alarm to the local fire brigade. As the sprinkler heads open automatically in response to elevated temperature, the detection and operation is completely automatic. Automatic sprinklers are the most effective form of automatic protection (see Loss Prevention Council's Rules for automatic sprinkler installations).[15]

Protection for special risks can be given by water spray, foam, inert gas, and dry powder. Water spray systems consist of pipes with high or medium pressure outlets projecting sprays of a predetermined droplet size. This system is suitable for protection of large scale flammable liquid and liquefied gas risks. Such a system is not only used to extinguish a fire but to protect the plant, by cooling, from a nearby fire. Foam can be applied to a given risk by fixed pipework connected either to a self-contained foam generator or an inlet to which the fire brigade can connect their foam-producing equipment. Carbon dioxide or other inerting gases can be directed into the plant or rooms by pipework. Most of the gases used are primarily suitable for flammable liquids, electrical equipment, valuable water-sensitive equipment, and water reactive chemicals. The gas is delivered through pipes and the system involves automatically closing doors and ventilation ducts. Warning bells have to be sounded so as to warn occupants to make their escape as inert gases will asphyxiate them. If a total flooding system is contemplated, consultation at the design stage with the Health and Safety Executive and the insurers is strongly recommended.

A dry powder system consists of a dry powder container which is coupled to a gas cylinder, pipework, and outlets. The pressurized gas drives the powder to the outlets. The system is suitable for flammable liquids, electrical equipment, and materials where it is essential to avoid water contamination.

7 Working Practice

Considerable amounts of heat are released during some reactions. This is obvious in acid–base neutralization or oxidation reactions and can often be predicted when the mechanisms of the expected reaction are worked out beforehand. Such considerations may be second nature to experienced researchers but may not be so apparent to students and inexperienced laboratory assistants who should only undertake such work under the supervision of an experienced person.

When new work is being attempted, full consideration should be given to the possible dangers it presents. In these cases, before proceeding on the scale desired a very small scale experiment should be carried out to determine whether a large amount of heat is liberated, if gases are produced, or an unusually vigorous reaction occurs. These factors, although not serious on a small scale, may become extremely important in a large scale experiment. On scaling up a reaction, heat loss will increase according to a cube law. Additional means of cooling, such as cooling coils, may be required to take care of the extra heat produced.

The rate at which one reagent is added to another affects the rate at which heat is produced. In the case of a known exothermic reaction the reactants or reagents should be added as slowly as possible with adequate stirring to make sure that the heat is liberated slowly. One way in which too much reactant may be added at one time is when it is added to an overcooled mixture and accumulates in the reaction mix where the rate of reaction is artificially low. It then only requires a small increase in temperature, either due to the mixture warming up normally or to the gradual accumulation of heat of reaction, for a runaway condition to occur, leading to disastrous results. A better course is to allow the reaction to take place at a temperature where it can proceed at a suitable speed and add the reactant slowly with adequate stirring to make sure that it is consumed at the same rate at which it is added.

Flammable vapours can often escape into the atmosphere because they are not condensed as rapidly as they are produced. A balance must be struck between the rate at which the liquid is heated and the cooling provided by a water-cooled condenser.

Experiments should not be left unattended and if the person in charge of an experiment is called away he should ensure that whoever takes over is fully aware of the dangers involved and the precautions to be taken. No experiment should be left running at night unattended unless it can be ensured that somebody will inspect it at regular intervals.

The disposal of waste materials calls for special attention. There have been fatal accidents during the disposal of waste materials and the disposer should be fully aware of the risks involved. The overriding consideration should be not to try to dispose of too much material at any one time. This is particularly important in the case of heavy metals such as silver and mercury which can form explosive compounds of which fulminates are a well-known example. Such by-products can be produced accidentally without the intentions of experimenters (see p. 62) and full knowledge of the conditions under which explosive compounds can be produced should be available to such persons.

Untreated flammable liquids should not be disposed of down the drain, apart from *very* small quantities of water-soluble solvents. Limited spillage of flammable water insoluble solvents may be disposed of by the dispersion method described on p. 187. Larger quantities of flammable liquids should preferably be collected in sealed metal containers and recovered or disposed of in a safe fashion, for example by burning in a shallow metal tray in the open air. Drains should be properly trapped and vented and they should preferably discharge into an industrial waste sewer rather than a sanitary sewer. There are specially designed incinerators for burning flammable liquids and if one attempts to employ this method a full investigation should be made of the types of material it can handle.

Materials which are known to decompose spontaneously and explosively should be kept in a safe manner until they can be disposed of safely. Waste material should be removed daily and destroyed or disposed of in a safe manner. Filter papers, residues, and wiping cloths which have been in contact with unsaturated oils should be kept in covered receptacles of limited size to ensure frequent disposal of the contents. Bins used for disposal should be labelled appropriately and clearly so that no mistake is made in introducing the wrong type of material into the bin. Benches and glass apparatus should not be cleaned with flammable solvents after experiments.

8   Legislation

All statutory requirements cannot be covered in a survey of this nature. The original legislation should be consulted by interested parties since no attempt is made here to authoritatively interpret the requirements. An extensive guide to relevant legislation is contained in BS 5908.[3]

The legislation concerned with fire and explosion in laboratories is mainly contained in the Health and Safety at Work Act 1974 and the Factories Act 1961 when laboratories are concerned with process control, and the Petroleum

(Consolidation) Act 1928 and Regulations and Byelaws made under these Acts. The Highly Flammable Liquids and Liquefied Petroleum Gases Regulations 1972 (SI 1972: No. 917) made under the Factories Act regulate the use and storage of such materials.

The Health and Safety at Work Act 1974 has far reaching implications in that it enables regulations to be made to cover premises previously exempt from other legislation. It has transferred the implementation of the fire precautions requirements in the Factories Act 1961 to the Fire Precautions Act 1971 except for special premises, which are covered by the Fire Certificate (Special Premises) Regulations 1976 (SI 1976: 2003). In particular the following sections of the Factories Act have been repealed and incorporated in the Fire Precautions Act:

Sections 40–47—Provision and maintenance of means of escape.

Section 48—Safety in case of fire.

Section 49—Escape instruction.

Section 50—Enabling special regulations for fire protection to be made.

Section 51–52—Provision, maintenance, and testing of fire fighting equipment.

In addition to statutory requirements, consideration has to be given to Common Law duties of occupier and employer.

There is a strict duty in Common Law in respect of injury to one's neighbour's property by fire and employees. Also, in respect of the need to provide safe means of access, a safe place of work, a safe system of work and competent supervision; an obligation which almost certainly extends to fire protection measures.

8.1   The Factories Act 1961

The Factories Act deals with many matters having a direct bearing on fire protection. These are fire drills, fire warning, and means of escape as well as fire extinguishing equipment and fire prevention matters. The following sections of the Act are of special interest:

Section 31 Provisions for plant employing flammable gases, vapours, and dusts.

Section 54 Dangerous conditions and practices.

Sections 80 and 81 Notification of accidents and dangerous occurrences.

Section 148 Persons empowered to inspect premises.

Sections 120, 155, 160, and Second Schedule Persons held responsible for compliance with the Act.

8.2   Regulations Made or Deemed to Have Been Made Under the Factories Act 1961

Certain processes and materials are considered to be extra hazardous and extra legislation has been introduced to deal with these. The requirements of such legislation can be considered supplementary to the requirements of the Factories Act.

8.2.1   The Control of Substances Hazardous to Health Regulations (SI 1988: No. 1657). These Regulations have been made under the Health and Safety at Work Act 1974. The enactment of these Regulations included the revoking of the Chemical Works Regulations (SR and O 1922: No. 731) which had included requirements for electrical installations and the prohibition of naked lights, matches, *etc.* and special precautions for heating installations in any place where there is danger of an explosion from or ignition of flammable gas, vapour, or dust. COSHH deals with limiting exposure of employees to any substance hazardous to health. In relation to fire this would include the products of combustion (CO, CO_2, smoke, *etc.*), extinguishing agents (Halons), or other substances that fire fighters might encounter in the event of a fire. The regulations contain lists of hazardous substances and those which are prohibited for certain purposes.

8.2.2   The Highly Flammable Liquids and Liquefied Petroleum Gas Regulations 1972 (SI 1972: No. 917). These Regulations impose requirements for the protection of persons employed in factories and other places to which the Factories Act 1961 applies, in which any highly flammable liquid or liquefied petroleum gas is present for the purpose of, or in connection with, any undertaking, trade, or business. Regarding highly flammable liquids, the Regulations contain requirements as to the manner of their storage, the marking of storage accommodation and vessels, the precautions to be taken for the prevention of fire and explosion, the provision in certain cases of fire fighting apparatus, and the securing in certain cases of means of escape in case of fire. Regarding liquefied petroleum gases, the Regulations contain requirements as to the manner of their storage and marking of storage accommodation and vessels.

8.3   The Petroleum (Consolidation) Act 1928

The Act requires premises used for the bulk storage of petroleum spirit, defined as petroleum which evolves a flammable vapour at less than 73 °F tested as prescribed, to be licensed and empowers licensing authorities to attach to licences such conditions as are necessary to ensure the safe keeping of the petroleum spirit.

The Secretary of State is empowered to make Regulations governing the conveyance of petroleum spirit by road, and the keeping, use, and supply of petroleum for vehicles, motor boats, and aircraft. In fact for the latter use certain exemptions are made by the Petroleum Spirit (Motor Vehicles) Regulations (SR and O 1929: No. 952). Provision is made to enable Regulations to be made under the Act which can extend the Act or portions of the Act to substances other than petroleum spirit.

8.4 Petroleum (Transfer of Licenses) Act 1936

This Act empowers local authorities to transfer petroleum spirit licences granted under the Act of 1928.

8.5 Orders and Regulations Made Under the Petroleum (Consolidation) Act 1928

8.5.1 The Petroleum (Mixtures) Order 1929 (SR and O 1929: No. 993). This applies the whole of the Petroleum (Consolidation) Act to all mixtures of petroleum with any other substances, which possess a flash point below 73 °F.

8.5.2 The Petroleum (Carbon Disulfide) Order 1958 (SI 1958: No. 257). This applies the requirements of the 1928 Act in respect to labelling and conveyance by road to carbon disulphide. This Order is modified by The Petroleum (Carbon Disulfide) Order 1968 (SI 1968: No. 571).

8.5.3 The Petroleum (Compressed Gases) Order 1930 (SR 1930: No. 34). Air, argon, carbon monoxide, coal gas, hydrogen, methane, neon, nitrogen, and oxygen when compressed into metal cylinders are brought under certain sections of the 1928 Act.

Other relevant regulations are:

The Gas Cylinder (Conveyance) Regulations 1931 (SR and O: No. 679)

The Gas Cylinder (Conveyance) Regulations 1947 (SR and O: No. 1594)

The Gas Cylinder (Conveyance) Regulations 1959 (SI 1959: No. 1919)

The Compressed Gas Cylinders (Fuel for Motor Vehicles) Regulations 1940 (SR and O 1940: No. 2009).

8.5.4 The Petroleum (Organic Peroxides) Order 1973 (SI 1973: No. 1897). Specified organic peroxides and certain mixtures or solutions of them are brought under some sections of the 1928 Act.

8.5.5 The Petroleum (Inflammable Liquids) Order 1971 (SI 1971: No. 1040). This applies certain sections of the 1928 Act to a large number of flammable liquids, other than petroleum derivatives with a flash point below 70 °F, and also to certain other substances which are considered hazardous. The liquids and other substances are set out in Parts I and II of the Schedule attached to the Order.

8.6 Explosive Acts 1875 and 1923 and Explosive Substances Act 1883

These Acts, and Orders and Regulations made under them, govern the manufacture, sale, importation, and conveyance by road of all explosives.

8.6.1 Acetylene. Order of the Secretary of State No. 5 dated 28.3.1898
Order of the Secretary of State No. 5A dated 29.9.1905

These provide that under certain conditions compressed acetylene in admixture with oil–gas is not deemed to be an explosive within the meaning of the Act.

Order of the Secretary of State No. 9 dated 23.6.1919

This provides that when acetylene is contained in a homogeneous porous substance with or without acetone or some other solvent and provided certain prescribed conditions are fulfilled shall not be deemed to be an explosive within the meaning of the Act.

Order in Council No. 30 dated 2.2.1937 as amended by The Compressed Acetylene Order 1947 (SR and O 1947: No. 805)

This prohibits the keeping, importation, conveyance, and sale of acetylene compressed to over 9 psig. Certain exceptions are made among which is that acetylene at pressure not over 22 psig may be manufactured and kept under conditions approved by the Secretary of State. Subject to certain conditions, acetylene at pressures not exceeding 300 psig and not mixed with air or oxygen can be used in the production of organic compounds.

The Compressed Acetylene (Importation) Regulations 1978 (SI 1978: No. 1723).

These Regulations made under the Health and Safety at Work Act, indicate that acetylene at a pressure between 0.62 bar and 18.0 bar cannot be imported into the UK unless the Health and Safety Executive (HSE) allows it under licence.

8.7 Other Acts and Regulations Which May Be Applicable to Laboratories in Chemical Works

8.7.1 Fire Precautions Act 1971. Premises to which this Act can apply include those used for purposes of teaching, training, or research.

8.7.2 The Public Health Act 1961. Under this Act there have been made the very important Building Regulations 1991. These Regulations have far reaching provisions determining height, floor area, cubic capacity, and siting, and fire resistance of structural elements of new buildings. Scotland have their own legal requirements Building Standards (Scotland) 1990. The previous legal requirements for Inner London have in general been incorporated in the current Building Regulations since the repeal of the London Building Act.

8.7.3 Radioactive Substances Act 1948, Radioactive Substances Act 1960, The Ionising Radiations (Sealed Sources) Regulations 1969 (SI 1969: No. 808), and The Ionising Radiations (Unsealed Radioactive Substances) Regulations 1968 (SI 1968: No. 780). The Ionising Radiations Regulations (1985) and its accompanying Approved Code of Practice (1985) include the substance of the above four documents and supersede them.

8.7.4 The Classification, Packaging, and Labelling of Dangerous Substances Regulations 1984 (SI 1984: No. 1244). The Regulations were introduced as a result of EEC directives. They are to be applied in conjunction with the Authorized and Approved List: *Information approved for the classification, packaging, and labelling of dangerous substances for supply and conveyance by road*, Health and Safety Commission, HMSO, London, 1984. Approximately 2000 chemicals and mixtures appear in the list. The Regulations recognize substances as being toxic, harmful, explosive, corrosive, irritant, oxidizing, or highly flammable.

8.7.5 Notification of New Substances Regulations 1982 (SI 1982: No. 1496). These Regulations make provision for the notification of dangerous substances and their classification according to the type and degree of danger they present.

9 Insurers' Approach to Fire Protection

As previously advocated, consultation should take place with the fire insurers at the design stage of the laboratory. Whilst appreciating that safety of life is the prime objective of fire protection, fire insurers are principally concerned with material and consequential loss which may arise from damage to buildings and contents. The fire authority approach will be to ensure that legal requirements in relation to the safety of life are complied with, *e.g.* that there are sufficient means of escape provided and that the construction is capable of containing the fire long enough to allow the personnel to escape.

In some instances, insurers and brigades may put forward requirements in which the emphasis differs. In general, however, there is close liaison and difficulties are usually resolved and a solution found which is acceptable to both. This removes the dilemma of management having to make a decision between two differing requirements. In most cases the fire insurer's requirements will be more stringent, since they start from the basis that there is a statistical probability that a fire will occur at a particular risk some time. Their philosophy is to apply compartmentation by having the structure suitably divided by a fire resisting construction to contain a fire within a compartment to allow additional fire fighting equipment to be brought into use if necessary. By this means the premises at risk are sub-divided and they accept the possible loss of a portion of the premises, but seek to prevent any extension of the damage. In recognition of the fact that the extent of fire damage may be limited by the presence of suitably approved automatic sprinkler systems,[15] fire alarm systems,[16] or portable fire extinguishing equipment, the insurer may grant an allowance in the form of a reduced premium. The value fire insurers attach to automatic sprinkler systems is illustrated by the generous allowances which are given for this type of protection.

10 References

1. *Electrical apparatus with protection by enclosure for use in the presence of combustible dusts. Part 2: 1988 Guide to selection, installation, and maintenance.* BS 6467.
2. *Recommendations for the storage and use of flammable liquids.* London: Loss Prevention Council: 1990.
3. *Code of practice for fire precautions in the chemical and allied industries.* BS 5908: 1990.
4. *Regulations for electrical installations*, 16th edn. London: The Institution of Electrical Engineers, 1991.

5. *Flammable liquids and gases: electrical equipment* (revised 1983). Fire Safety Data Sheet No. 6014. London: Fire Protection Association.
6. *Electrical apparatus for potentially explosive atmospheres.* BS 5501.
7. *Code of practice for the selection, installation, and maintenance of electrical apparatus for use in potentially explosive atmospheres (other than mining applications or explosive processing and manufacturing).* BS 5345.
   Part 1: 1989: General recommendations.
   Part 2: 1983: Classification of hazardous areas.
8. *Code of practice for fire extinguishing installations and equipment on premises. Part 3: Portable fire extinguishers.* BS 5306: 1985.
9. *Specification for portable fire extinguishers.* BS 5423: 1987.
10. *List of approved products and services.* Loss Prevention Council: (Annual).
11. *Portable fire extinguishers.* Data Sheet PE4. London: Fire Protection Association.
12. *First aid fire fighting: training.* Data Sheet PE5. London: Fire Protection Association.
13. *Fixed fire extinguishing equipment: hose-reels.* Data Sheet PE7. London: Fire Protection Association.
14. *Hydrant systems.* Data Sheet PE8. London: Fire Protection Association.
15. *Rules for automatic sprinkler installations* (incorporating BS 5306: Part 2). Loss Prevention Council, London: 1990.
16. *Rules for Automatic Fire Detection and Alarm Installations for the Protection of Property RLSI.* Loss Prevention Council, London, 1991.

11   Bibliography

R. R. Young and P. J. Harrington, 'Design and Construction of Laboratories', (Lecture series 1962, No. 3), Royal Institute of Chemistry, London, 1962.

D. D. Libman, 'Safety in the Chemical Laboratory', *Laboratory Equipment Digest*, October, 1967.

Mathew M. Braidech, 'Fire and Explosion Problems in Laboratories and Pilot Plants', *J. Chem. Educ.*, 1967, **44**, A319.

'Safety in the Chemical Laboratory', Vol. 1 1967, Vol. 2 1971, Vol. 3 1970, ed. N. V. Steere, Chemical Education Publishing, Easton, Penn., 1967.

'The Care, Handling, and Disposal of Dangerous Chemicals', Northern Publishers, Aberdeen, 1970.

'Guide to Fire Prevention in the Chemical Industry', Chemical Industries Association, London, 1983.

'What You Should Know About Lab. Safety', Scriptographic Publications, London, 1983.

'Design, Construction, and Refurbishment of Laboratories', ed. R. Lees and A. F. Smith, Ellis Horwood, Chichester, 1984.

'Safe Practices in Chemical Laboratories', Royal Society of Chemistry, London, 1989.

'Safety in Laboratories', Ciba-Geigy, London, 1986.

N. I. Sax and R. J. Lewis, 'Dangerous Properties of Industrial Materials', 7th Edn., Van Nostrand and Reinhold, New York, 1989.

'The Storage of Flammable Liquids in Fixed Tanks (Up to 1000 m^3 Total Capacity)', Health and Safety Executive HS(G) 50, HMSO, London, 1990.

'The Storage of Flammable Liquids in Containers', Health and Safety Executive HS(G) 51, HMSO, London, 1990.

'Recommendations for the Storage and Use of Flammable Liquids', Loss Prevention Council, RC 20, London, 1990.

'Recommendations for the Storage, Use, and Handling of Industrial Gases in Cylinders (excluding LPG)', Loss Prevention Council, RC 8, London, 1992.

'Recommendations for the Storage, Use, and Filling of Liquefied Petroleum Gas in Containers', Loss Prevention Council, London, 1989.

'Recommendations for the Fire Protection of Laboratories', Loss Prevention Council, RC 5, London, 1992.

'Recommendations for the Prevention and Control of Dust Explosions', Loss Prevention Council, RC 12, London, 1992.

CHAPTER 5

Reactive Chemical Hazards

L. BRETHERICK

1 Fundamental Causes

All chemical reactions necessarily involve changes in energy, usually evident as heat. This is normally released during exothermic reactions, but occasionally may be absorbed into the products in endothermic reactions, which are relatively few in number.

Reactive chemical hazards invariably originate from the release of energy in a quantity or at a rate too great to be dissipated by the immediate environment of the reacting system, so that destructive effects appear. To try to eliminate such hazards from chemical laboratory operations, attempt to assess the likely degree of risk involved in a particular operation and then plan and execute the operation in a way which will minimize the foreseen risks. It may, of course, be necessary to accept that a certain degree of risk is likely to be attached to a particular course of action, but in this case, personal protection appropriate to the risk will be necessary.

This chapter is concerned with various aspects of the recognition and assessment of reactive hazards, with practical techniques for reaction control, and with personal protection where this is deemed necessary. Several references to sources of more detailed information are given by superscript numbers in the text, and where appropriate, page references in parentheses are also given to a recently updated monograph on reactive hazards.[1]

2 Physicochemical Factors

Many of the underlying causes of incidents and accidents in laboratories which have involved unexpected violent chemical reactions are related to a lack of appreciation of the effects of simple physicochemical factors upon the kinetics of practical reaction systems. Probably the two most important of these factors are those governing the relationship of rate of reaction with concentration, and with the rise in temperature during the reaction. The latter is the major factor.

2.1 Concentration of Reagents

It follows from the law of mass action that the concentration of each reactant will directly influence the velocity of reaction and the rate of heat release.

It is, therefore, important not to use too-concentrated solutions of reagents, particularly when attempting previously untried reactions. In many preparations 10 per cent is a commonly used level of concentration where solubility and other considerations will allow, but when using reagents known to be vigorous in their action, 5 per cent or 2 per cent may be more appropriate. Catalysts are commonly employed at these latter or even lower concentrations.

Of the many cases where increasing the concentration of a reagent, either accidentally or deliberately, has transformed a safe procedure into a hazardous event, three examples will suffice. When concentrated ammonia solution was used instead of the diluted solution to destroy dimethyl sulfate, explosive reaction occurred (305). Omission of most of the methanol during preparation of a warm mixture of nitrobenzene and sodium methoxide led to rupture of the containing vessel (604). Catalytic hydrogenation of *o*-nitroanisole at 34 bar (34×10^5 Pa) under excessively vigorous conditions (250 °C, 12 per cent catalyst, no solvent) ruptured the hydrogenation autoclave (710).

2.2 Reaction Temperature

According to the Arrhenius equation, the rate of a reaction will increase exponentially with increase in temperature, and in practical terms an increase of 10 °C roughly doubles the reaction rate in many cases. This has often been the main contributory factor in accidents where inadequate temperature control had caused exothermic reactions (normal, polymerization, or decomposition) to run out of control.

An example relevant to the first two reaction types is the explosive decomposition which occurred when sulfuric acid was added to 2-cyano-2-propanol with inadequate cooling. Here, exothermic dehydration of the alcohol produced methacrylonitrile, acid-catalysed polymerization of which accelerated to explosion, rupturing the vessel (1217). An example of an exothermic decomposition reaction is the violent explosion which occurred during storage of *m*-nitrobenzenesulfonic acid at 150 °C under virtually adiabatic conditions. It was subsequently found that exothermic decomposition of the solution set in at 145 °C, and the pure acid decomposed vigorously at 200 °C (607).

Many other examples of various types of hazardous and unexpected reactions have been collected and classified.[1,2]

3 Operational Considerations

Effective control is essential to minimize possible hazards associated with a particular reaction system and, to allow you to achieve such control, relevant knowledge is necessary for you to assess potential hazards in the system.

Try to find what is already known about the particular procedure or reaction system (or a related one) from colleagues, or from existing literature.[1-4]

If no relevant information can be found, or the work proposed is known to be original, it will be necessary to conduct cautiously a very small scale preliminary experiment, to assess the exothermic character and physical properties of the reaction system and its products.

When subsequently planning and setting up larger-scale reactions or preparations, attention to many practical details may be required to ensure safe working as far as possible. Relevant factors include:

- adequate control of temperature, with sufficient capacity for heating, and particularly cooling for both liquid and vapour phases;
- proportions of reactants and concentrations of reaction components or mixtures;
- purity of materials, absence of catalytic impurities;
- presence of solvents or diluents, viscosity of reaction medium;
- control of rates of addition (allowing for any induction period);
- degree of agitation;
- control of reaction atmosphere;
- control of reaction or distillation pressure;
- shielding from actinic radiation;
- avoiding mechanical friction or shock upon unstable or sensitive solids, and adequate personal protection if such materials will be isolated or dried (without heating).

Further details of specialized equipment, techniques, and safety aspects are to be found in the publications devoted to preparative methods.[5-7] Information on recently published hazards is now readily accessible.[8]

3.1   Pressure System

Some of the above considerations assume greater significance when conducting reactions in closed systems at relatively high pressures and/or temperatures. In high pressure autoclaves, for example, the thick vessel walls and generally heavy construction necessary to withstand the internal pressures implies high thermal capacity of the equipment, and really rapid cooling of such vessels to attempt to check an accelerating reaction is impracticable. This is why bursting discs or other devices must be fitted as pressure reliefs to high pressure equipment. A brief account of autoclave techniques in high pressure hydrogenation is readily available.[9]

A further important point specific to closed systems which will be heated is to make adequate allowance for expansion of liquid contents. Several cases are recorded in which cylinders or pressure vessels partially filled with liquids at ambient temperature have burst under the hydraulic pressure generated when heating caused expansion of the liquid to fill completely the closed vessel. This is also likely to happen if cylinders of liquefied gases are exposed to fire conditions.

3.2   Adiabatic Systems

If exothermic reactions proceed under conditions where heat cannot be lost (*i.e.* in adiabatic systems) they readily may accelerate out of control.

Although, in general, few laboratory situations will approach adiabatic conditions, occasionally the combination of a uniform heating system which is unusually well insulated (such as a thick heating mantle with top jacket to surround completely a flask), or which is of unusually high thermal capacity and inertia (such as a deep, well lagged oil-bath), coupled with a strongly exothermic reaction may approximate to an adiabatic system.

If it is really necessary for technical reasons to use such systems, provision must also be made for application of rapid cooling in the event of an untoward rise in internal temperature. This may be effected by lowering the jack-supported heating mantle or bath, and application of an external air blast, or of cooling to an internal coil.

4   Types of Decomposition Reactions

Several types of decomposition reactions may be distinguished, with a wide range of subsequent effects possible. The effects of slow reactions, such as the progressive hydrolysis of water-sensitive compounds by atmospheric moisture in storage or auto-oxidation of air-sensitive materials, are not immediately evident, but may become so after weeks or months have elapsed. During this period kinetic energy (in gas pressure) or endothermic energy (in peroxidized materials) may accumulate to a significant level. Practical aspects of such energy accumulation are detailed later.

Fast reactions, usually those accelerated by increase in temperature, may be subdivided into those in which no combustion occurs during or after the decomposition, and those in which combustion is involved, either from an external ignition source, or from the reaction temperature exceeding the auto-ignition temperature of the reaction components or products. Of those combustion reactions, the rates of energy release and the violence of the effects increase in the order deflagration, explosion, and detonation. The two former types are most likely to be experienced in the laboratory, and usually arise from combustion of a flammable vapour or gas mixed with enough air to give a composition within the flammable limits.

If the fuel–air mixture is relatively small in volume and virtually unconfirmed when ignition occurs, a deflagration or 'soft' explosion will occur. Persons in the close vicinity may suffer flash burns from such an incident, but material damage other than scorching by the moving flame-front will be minimal. This is because the unconfined rapid combustion will give no significant pressure effects. Such a situation might arise from a small release of gas or vapour and its ignition in a relatively large room, or from a spill of flammable liquid under an open-fronted hood or lean-to building. There is, however, a volume effect, and large-scale incidents of this type can produce significant destructive effects from over-pressure.

The effects of explosions under conditions of confinement are invariably more serious, and if the confinement is relatively close, and the fuel–air mixture is nearly stoichiometric, instantaneous pressures several times that of the normal atmosphere may readily be produced. Such pressures are sufficient to demolish a laboratory building of normal construction. It is for this reason that operations of high potential hazard (involving highly energetic substances, and/or high temperature and pressure) are conducted in isolation cells of reinforced construction designed to withstand possible pressure effects.[10]

Explosion situations may also arise if a reaction accelerates out of control and to the point where the containing vessel fails and/or the vaporized contents reach their auto-ignition temperature. Fire will then definitely occur, but explosion may not.

Detonation is the name applied to a particularly severe form of explosion where the velocity of flame propagation and the associated decomposition temperature and pressure are much higher (by up to two or three orders of magnitude) than in deflagrations. Under some circumstances, a gaseous deflagration can accelerate into detonations, but the necessary conditions (physically long vessels or pipelines) are seldom present in laboratories. However, explosive decomposition of unstable solids or liquids may occasionally involve detonation (during a few μs and with propagation velocities up to 8 km s^{-1}) with the associated violent shattering effects. A more detailed analysis of explosive phenomena is available.[11]

Note that the term 'deflagration' used above for combustion of gaseous fuel–air mixtures has a rather different meaning from deflagration as applied to the sparking decomposition of a heated solid, such as ammonium dichromate (1074).

5   Chemical Composition, Structure, and Reactivity

In the last few years instrumental methods have been developed to determine the stability or otherwise of a compound or reaction mixture under a wide variety of possible processing conditions. These methods range from simple (and cheap) equipment which will usually give a general qualitative indication of likely instability, to complex and sophisticated sensitive calorimetric instrumentation of very high cost, but which will determine the hazard potential quantitatively and with great precision. Such equipment is, however, unlikely to be found in a small or educational laboratory.

It may eventually be possible to calculate the hazards related to the stability and reactivity of chemical compounds and reaction mixtures in advance of experience. However, at the moment and in most cases, quantitative or qualitative assessments based on known examples represent the best practicable means of assessment of such hazards for small laboratories. Assessment may be based either on overall composition or on detailed structure of the materials involved.

5.1 Overall Composition and Oxygen Balance

One of the fundamental factors which may determine the course of a reaction system is that of the overall elemental composition of the system. It is a fact that the majority of reactive chemical accidents or incidents have involved oxidation systems and, especially in organic systems, the oxygen balance is an important criterion.

Oxygen balance is the difference between the oxygen content of a system (a compound or a mixture) and that required to oxidize fully the carbon, hydrogen, and other oxidizable elements present to carbon dioxide, water, *etc.* If there is a deficiency of oxygen, the balance is negative, and if a surplus, positive. Oxygen balance is often expressed as a weight percentage with appropriate sign.

5.1.1   Mixtures. In laboratory oxidation reaction systems, one should plan the operations to keep the oxygen balance as negative as possible to minimize the potential energy release. This consideration will dictate, therefore, that wherever possible an oxidant will be added slowly (and with appropriate control of cooling, mixing, *etc.*) to the other reaction components, to maintain the minimal effective concentration of oxidant throughout the reaction. It is important to establish as early as possible, from the physical appearance or thermal behaviour of the system, that the desired exothermic reaction has become established. If this does not happen, relatively high concentrations of oxidant may accumulate before onset of the reaction, which may then become uncontrollable.

Two relevant examples may be quoted. Mixtures of several water-soluble organic compounds (ethanol, acetaldehyde, acetic acid, acetone, *etc.*) with aqueous hydrogen peroxide show clearly defined limits within which the mixtures are detonable (1212). Oxidation of 2,4,6-trimethyltrioxane ('paraldehyde') with nitric acid to glyoxal is subject to an induction period, and the reaction may become violent if addition is too fast. Presence of nitrous acid eliminates the induction period (1177).

In other cases it may be necessary for practical reasons to add one or more of the reaction components to the whole (or preferably part) of the oxidant, but the other considerations will still apply.

5.1.2   Compounds. The concept of oxygen balance has more usually been applied to individual compounds rather than to reaction mixtures as mentioned above. A fairly rapid appreciation of any potential tendency towards explosive decomposition may be gained by inspection of the empirical formula of a particular compound.

If the oxygen content of a compound approaches that necessary to oxidize the other elements present (with the exceptions noted below) to their lowest state of valency, then the stability of that compound is doubtful. The exceptions are that nitrogen is excluded (it is usually liberated as the gaseous element), and halogen will go to halide if a metal or hydrogen is present. Sulfur, if present, counts as two atoms of oxygen.

This generalization is related to the fact that most industrial high-explosives are well below zero oxygen balance, and some examples follow.

- Compounds of negative balance include:

  trinitrotoluene (-67.4 per cent, 693)

  $$C_7H_5N_3O_6 + 10.5O \rightarrow 7CO_2 + 2.5H_2O + 1.5N_2 \qquad (1)$$

  peracetic acid (-50 per cent, 279)

  $$C_2H_4O_3 + 3O \rightarrow 2CO_2 + 2H_2O \qquad (2)$$

  Presence of an oxidant will decrease the negative balance

- Zero balance:

  performic acid (151)

  $$CH_2O_3 \rightarrow CO_2 + H_2O \qquad (3)$$

  ammonium dichromate (1074)

  $$Cr_2H_8N_2O_7 \rightarrow Cr_2O_3 + 4H_2O + N_2 \qquad (4)$$

  Energy release is maximal at zero balance

- Compounds of positive balance include:

  ammonium nitrate ($+50$ per cent, 1247)

  $$H_4N_2O_3 \rightarrow 2H_2O + N_2 + O \qquad (5)$$

  glyceryl nitrate ($+5.9$ per cent, 371, 1500)

  $$C_3H_5N_3O_9 \rightarrow 3CO_2 + 2.5H_2O + 1.5N_2 + 0.5O \qquad (6)$$

  dimanganese heptoxide ($+133$ per cent, 1331)

  $$Mn_2O_7 \rightarrow Mn_2O_3 + 4O \qquad (7)$$

  Presence of a fuel or reductant in such materials will increase the potential energy release.

Compounds with unusually high proportions of nitrogen and N—N bonds are also suspect (1678). Hydrazine (87.4 per cent nitrogen) and hydrogen azide (97.6 per cent) are both explosively unstable, but not ammonia (82.2 per cent).

In practical terms, the margin between potential and actual hazard of explosive decomposition may be very narrow or quite wide, depending on the energy of activation necessary to initiate the decomposition. Performic acid is treacherously unstable (low energy of activation), and a sample at $-10\,°C$ exploded when moved (152). TNT, on the other hand, is relatively stable and will not detonate when burned, or under impact from incendiary bullets, but requires a powerful initiating explosive ('detonator') to trigger explosive decomposition.

5.2 Molecular Structure

Instability and/or unusual reactivity in single compounds is often associated with a number of molecular structural features, which may include the specific bond systems given below.

C≡C	Acetylenes (1479), haloacetylenes (1610), metal acetylides (1653).
CN_2	Diazo compounds (1560).
C—NO	Nitroso compounds (1703).
C—NO$_2$	Nitro compounds (1700).
C—(NO$_2$)$_n$	*gem*-Polynitroalkyl compounds (1763).

C—O—NO	Alkyl or acyl nitrites (1501, 1486).
C—O—NO$_2$	Alkyl or acyl nitrates (1500, 1486).
C=N—O	Oximes (1733).
C=N—O—C	Isoxazoles (1644).
C≡N→O	Metal fulminates (1665).
N—NO	*N*–Nitroso compounds (1703).
N—NO$_2$	*N*–Nitro compounds (1701).
C—N=N—C	Azo compounds (1524).
C—N=N—O	Arenediazoates (1516), bis(arenediazo) oxides (1526).
C—N=N—S	Arenediazo sulfides (1516), xanthates (1823), bis(arenediazo) sulfides (1527).
C—N=N—N—C	Triazenes (1817).
N$_3$	Azides (1523).
N=N—NH—N	Tetrazoles (1808).
C—N$_2$$^+$	Diazonium salts (1652).
N—C(=N$^+$H$_2$)—N	Guanidinium oxosalts (178, 180).
N$^+$—OH	Hydroxylaminium salts (1634).
N—Metal	*N*–Metal derivatives (heavy metals) (1661).
N—X	*N*–Halogen compounds (1619), difluoroamino compounds (1572).
O—X	Hypohalites (1634).
O—X—O	Chlorites (1538), halogen oxides (1622).
O—X—O$_2$	Halates (1659, 1669, 1678).
O—X—O$_3$	Perhalates (1503, 1509, 1562), halogen oxides (1622).
C—Cl—O$_3$	Perchloryl compounds (1745).
N—Cl—O$_3$	Perchlorylamide salts (1744).
Xe—O$_n$	Xenon–oxygen compounds (1823).
O—O	Peroxides (1746).
O$_3$	Ozone (1418).

A further large group of compounds which cannot readily be represented by line formulae, and which contains a large number of unstable members is the amminemetal oxosalts (1511). These are compounds containing ammonia or an organic base coordinated to a metal, with coordinated or ionic chlorate, nitrate, nitrite, nitro, perchlorate, permanganate, or other oxidizing groups also present. Such compounds as dipyridinesilver perchlorate (802), tetra-amminecadmium permanganate (932), or bis-1,2-diaminoethanedinitrocobalt(III) iodate (506) will decompose violently under various forms of initiation, such as heating, friction, or impact.

An interesting application of personal computers to stability considerations is a revised program[12] which calculates the maximum possible energy release for a compound or mixture of compounds containing up to 23 elements, and then gives an overall hazard assessment. No information other than the chemical structure is necessary, and the result, which is semi-quantitative, is used as a screening guide to decide which reaction systems need more detailed and/or experimental investigation. A further program will calculate the maximum reaction heat possible from mixtures of 2 or 3 compounds, the composition for maximum heat release, and then assess the probability of ignition of that mixture.[13]

5.3   Redox Compounds

When the coordinated base in an amminemetal oxosalt is a reductant (hydrazine, hydroxylamine), decomposition is extremely violent. Examples are bis-(hydrazine)nickel(II)   perchlorate   (1014)   and   hexahydroxylaminecobalt(II) nitrate (1060), both of which have exploded while wet during preparation.

Other examples of highly energetic and potentially unstable redox compounds, in which reductant and oxidant functions are in close proximity in the same molecule, are salts of reductant bases with oxidant acids, such as hydroxylaminium nitrate (1253), hydrazinium chlorite or chlorate (968), or double salts such   as   potassium   cyanide–potassium   nitrite   (reductant   and   oxidant respectively, 184).

5.4   Pyrophoric Compounds

Materials which are so reactive that contact with air (and its moisture) causes oxidation and/or hydrolysis at a sufficiently high rate to cause ignition are termed pyrophoric compounds. These are found in many different classes of compounds, but a few types of structure are notable for this behaviour.

- Finely divided metals: calcium (916), titanium (1462).
- Metal hydrides: potassium hydride (1142), germane (1139).
- Partially- or fully-alkylated metal hydrides: diethylaluminium hydride (494), triethylbismuth (665).
- Alkylmetal derivatives: diethylethoxyaluminium (663), dimethylbismuth chloride (292).
- Analogous derivatives of non-metals: diborane (70), dimethylphosphine (311), triethylarsine (664).
- Carbonylmetals: pentacarbonyliron (511), octacarbonyldicobalt (729).

Many hydrogenation catalysts containing adsorbed hydrogen (before and after use) will also ignite on exposure to air.

Where such materials are to be used, an inert atmosphere and appropriate handling techniques and equipment are essential to avoid the distinct probability of fire or explosion.[14]

5.5   Peroxidizable Compounds

A group of materials which react with air at ambient temperature much more slowly and less spectacularly than pyrophoric compounds, but which give longer term insidious hazards, may now conveniently be described.

Auto-oxidation usually takes place slowly when the liquid materials are stored with limited access to air and exposure to light, and the hydroperoxides initially formed may subsequently react to form polymeric peroxides, many of which are dangerously unstable when concentrated and heated by distillation procedures.

The common structural feature in organic peroxidizable compounds is the presence of one or more hydrogen atoms susceptible to auto-oxidative conversion to the hydroperoxy group —OOH. Some of the typical structures susceptible to peroxidation are:

O—C—H   in ethers, cyclic ethers, acetals

H_2C
  \
    C—H   in isopropyl compounds, sec-alkyl compounds, decahydronaph-
  /   thalenes
H_2C

C≡C—C—H in allyl compounds

C≡C—H   in vinyl compounds, dienes (*i.e.* monomers)

C—C—Ar   in cumene, tetrahydronaphthalenes, styrenes
  |
  H

Several commonly used organic solvents including diethyl ether, tetrahydrofuran, dioxan, 1,2-dimethoxyethane ('glyme'), and bis-2-methoxyethyl ether ('diglyme') are often stored without inhibitors being present, and are therefore susceptible to peroxidation, and many accidents involving distillation in use of the peroxide-containing solvents have been reported. It is essential to test these solvents for peroxide (with acidified potassium iodide solution) *before* use and, if present, peroxides must be eliminated by suitable means[15] before proceeding.

Less susceptible solvents, such as 2-propanol and 2-butanol, may become significantly peroxidized in prolonged storage under poor conditions, and some explosions on distillation of old samples have been reported (391, 485).

Di-isopropyl ether must be mentioned as particularly dangerous for two reasons. Its structure with 2 susceptible H atoms is ideal for rapid peroxidation, and the peroxide separates from solution in the ether as a readily detonative crystalline solid. Several fatal accidents have occurred, and it should not be used.

If allyl and, particularly, vinyl monomers become peroxidized, they are potentially dangerous for two related reasons. The peroxides of some vinyl monomers, such as 1,1-dichloroethylene (236) or 1,3-butadiene (434) separate from solution and are extremely explosive. Even when this does not happen, the peroxide present may initiate the exothermic and sometimes violent polymerization of any vinyl monomer during storage. Reactive monomers must therefore be inhibited against oxidation and stored cool, with regular checks for presence of peroxide.[15] However, air in the container should not be displaced by nitrogen, as some oxygen is essential for effective inhibition.

A few inorganic compounds, such as potassium and the higher alkali metals and sodium amide, are subject to auto-oxidation and production of hazardous peroxidic or similar products. Many organometallic compounds are also subject to auto-oxidation and require handling in the same way as pyrophoric compounds.[14]

5.6   Water-reactive Compounds

Second to air (oxygen), water is the most common reagent likely to come into contact, deliberately or accidentally, with reactive chemical compounds. Some of the classes of compounds which may react violently, particularly with a limited amount of water, are:

- alkali and alkaline earth metals (potassium, 1292; calcium, 916);
- anhydrous metal halides (aluminium tribromide, 35; germanium tetrachloride, 1049);
- anhydrous metal oxides (calcium oxide, 928);
- non-metal halides (boron tribromide, 58; phosphorus pentachloride 1054);
- non-metal halide oxides (*i.e.* inorganic acid halides, phosphoryl chloride, 951; sulfinyl chloride, 1023; chlorosulfuric acid, 951);
- non-metal oxides (acid anhydrides, sulfur trioxide, 1425).

Some of the reactive halogen-containing compounds above may pose long term problems if stored in closed containers which are not resistant to ingress of water vapour. Under these conditions, slow hydrolysis may cause a progressive accumulation of significant pressures of hydrogen halides (sometimes with sulfur dioxide), which may burst the container. Several organic halomethyl derivatives (benzyl chloride, 2-bromomethylfuran) which do not react rapidly with water, may similarly decompose with accumulation of gas pressure in storage unless special precautions are taken (702, 1623). Traces of catalytic impurities may hasten the decomposition of such materials.

Concentrated solutions of some acids and bases also give an exotherm when diluted with water but this is a physical effect.

5.7   Endothermic Compounds

Most chemical reactions are exothermic, but in the relatively few endothermic reactions heat is absorbed into the reaction product(s), which are thus endothermic (and energy-rich) compounds. These are thermodynamically unstable, because no energy would be required to decompose them into their elements, and heat would, in fact, be released.

There are a few endothermic compounds with moderately positive values of standard heat of formation (benzene, toluene, 1-octene) which are not usually considered to be unstable, but the majority of endothermic compounds do possess a tendency towards instability and possibly explosive decomposition under various circumstances.

Often the structure of endothermic compounds involves multiple bonding, as for example in acetylene, vinylacetylene, hydrogen cyanide, mercury(II) cyanide, dicyanogen, silver fulminate, cadmium azide, or chlorine dioxide, and all these compounds have been involved in violent decompositions or explosions. Examples of explosively unstable endothermic compounds without multiple bonding are hydrazine and dichlorine monoxide.

In general terms, endothermic compounds may be considered suspect with regard to stability considerations (1582).

6 Hazardous Reaction Mixtures

Although the number of combinations of chemical compounds which may interact is virtually unlimited, the combinations which have been involved in hazardous incidents are limited to those which led to an exotherm too large or too fast for effective dissipation under the particular experimental conditions.

The exotherm may have arisen directly from the primary reaction, but a two-stage exotherm is also possible. This will arise when the undissipated primary exotherm leads to instability and subsequent further exothermic decomposition of a reaction intermediate or product.

The great majority of incidents of these types have involved a recognizable oxidant admixed with one or more oxidizable components (or fuels). Examples are the vigorous combustion of glycerol in contact with solid potassium permanganate (1297, the high viscosity of glycerol prevents effective heat transfer), and the violent or explosive oxidation of ethanol by excess concentrated nitric acid (involving formation of unstable fulminic acid, 1150). Some further common examples are given in Table 1, and many others are available in classified form.[1,2]

A smaller group of incidents has involved recognizable reductants admixed with materials capable of oxidation. Examples here are the explosive decomposition of a heated mixture of aluminium powder and a metal sulfate (29), the shock-sensitivity of sodium in contact with chlorinated solvents (1372), or the violent interaction of powdered magnesium with moist silica when heated (1325).

There is an unusual potential hazard associated with use of the versatile reductant lithium tetrahydroaluminate, which has limited thermal stability. Unless care is taken to prevent local overheating, by using gentle stirring and oil-bath heating, decomposition may occur to give very finely divided aluminium metal. This is capable of undergoing violent thermite-like reactions with the oxygenated solvents (ether, THF, dioxan, glyme) commonly used for these reduction reactions. This is thought to have been the cause of several unexplained accidents during such reductions (42). Close attention to experimental procedure may thus prevent one reducing agent degrading to give a more powerful one.

As might be expected, interaction of obvious oxidants and reductants is always very energetic and potentially hazardous, and must be conducted under closely controlled conditions with ample cooling capacity available. Attempted reduction of dibenzoyl peroxide by lithium tetrahydroaluminate led to a fairly violent explosion (870), and hydrazine is decomposed explosively in contact with chromium trioxide (1245). Rocket technology furnishes further examples of the extreme energy release possible in undiluted redox systems (1788).

Table 1 *Partial List of Incompatible Chemicals (Reactive Hazards)*

Substances in the left hand column should be stored and handled so they cannot possibly accidentally contact corresponding substances in the right hand column under uncontrolled conditions, when violent reactions may occur.

Acetic acid	Chromic acid, nitric acid, peroxides, and permanganates.
Acetic anhydride	Hydroxyl-containing compounds, ethylene glycol, perchloric acid.
Acetone	Concentrated nitric and sulfuric acid mixtures, hydrogen peroxide.
Acetylene	Chlorine, bromine, copper, silver, fluorine, and mercury.
Alkali and alkaline earth metals, such as sodium, potassium, lithium, magnesium, calcium	Carbon dioxide, carbon tetrachloride, and other chlorinated hydrocarbons. (Also prohibit water, foam, and dry chemical on fires involving these metals—dry sand should be available.)
Aluminium powder	Halogenated or oxygenated solvents.
Ammonia, anhydrous	Mercury, chlorine, calcium hypochlorite, iodine, bromine, and hydrogen fluoride.
Ammonium nitrate	Acids, metal powders, flammable liquids, chlorates, nitrites, sulfur, finely divided organics, or combustibles.
Aniline	Nitric acid, hydrogen peroxide.
Bromine	Ammonia, acetylene, butadiene, butane, and other petroleum gases, sodium carbide, turpentine, benzene, and finely divided metals.
Calcium oxide	Water.
Carbon, activated	Calcium hypochlorite, other oxidants.
Chlorates	Ammonium salts, acids, metal powders, phosphorus, sulfur, finely divided organics, or combustibles.
Chromic acid and chromium trioxide	Acetic acid, naphthalene, camphor, glycerol, turpentine, alcohol, and other flammable liquids.
Chlorine	Ammonia, acetylene, butadiene, butane, other petroleum gases, hydrogen, sodium carbide, turpentine, benzene, and finely divided metals.
Chlorine dioxide	Ammonia, methane, phosphine, and hydrogen sulfide.
Copper	Acetylene, hydrogen peroxide.
Fluorine	Isolate from everything.
Hydrazine	Hydrogen peroxide, nitric acid, any other oxidant, heavy metal salts.
Hydrocarbons (benzene, butane, propane, gasoline, turpentine, *etc.*)	Fluorine, chlorine, bromine, chromic acid, conc. nitric acid, peroxides.
Hydrogen cyanide	Nitric acid, alkalis.
Hydrogen fluoride	Ammonia, aqueous or anhydrous.
Hydrogen peroxide	Copper, chromium, iron, most metals or their salts, any flammable liquid, combustible materials, aniline, nitromethane.
Hydrogen sulfide	Fuming nitric acid, oxidizing gases.
Iodine	Acetylene, ammonia (anhydrous or aqueous).
Mercury	Acetylene, fulminic acid*, ammonia.
Nitric acid (conc.)	Acetic acid, acetone, alcohol, aniline, chromic acid, hydrogen cyanide, hydrogen sulfide, flammable liquids, flammable gases, and nitratable substances, fats, grease.

Table 1 *continued*

Nitromethane, lower nitroalkanes	Inorganic bases, amines, halogens, 13X molecular sieve.
Oxalic acid	Silver, mercury, urea.
Oxygen	Oils, grease, hydrogen, flammable liquids, solids, or gases.
Perchloric acid	Acetic anhydride, bismuth and its alloys, alcohol, paper, wood, grease, oils, dehydrating agents.
Peroxides, organic	Acids (organic or mineral), avoid friction, store cold.
Phosphinates	Any oxidant.
Phosphorus (white)	Air, oxygen.
Potassium chlorate	Acids (see also chlorates).
Potassium perchlorate	Acids (see also perchloric acid).
Potassium permanganate	Glycerol, ethylene glycol, benzaldehyde, sulfuric acid.
Silver	Acetylene, oxalic acid, tartaric acid, fulminic acid*, ammonium compounds.
Sodium	See alkali metals (above).
Sodium nitrite	Ammonium nitrate and other ammonium salts.
Sodium peroxide	Any oxidizable substance, such as ethanol, methanol, glacial acetic acid, acetic anhydride, benzaldehyde, carbon disulfide, glycerol, ethylene glycol, ethyl acetate, methyl acetate, and furfural.
Sulfuric acid	Chlorates, perchlorates, permanganates.
Thiocyanates	Metal nitrates, nitrites, oxidants.
Trifluoromethane sulfonic acid	Perchlorate salts.

* Produced in nitric acid ethanol mixtures.

7 Potential Storage Hazards

Most of this chapter has been devoted to the possible outcome of the deliberate interaction of chemicals, but it must not be forgotten that hazardous reactions may occasionally arise from accidental contact of chemicals due to breakage, spillage, or more seriously, from fire in a chemical store. This possibility will dictate that a good deal of thought must be given to segregation of different materials in storage to minimize the effects of accidental contact.[16]

Tables 1 and 2 give a selection of materials which need to be segregated on the grounds of potential reactive or toxic hazards, respectively.

8 Protection from Reactive Hazards

Where it has been decided after assessment of the various factors discussed in this chapter that a potential reactive hazard may exist in work being planned, consideration must also be given to the level of personal protection which may be required to allow the work to be executed safely or with minimum risk.

There is a considerable range of possibilities which may be involved, depending on the scale of operations and the type of assessed hazard, but in all cases eye protection will be mandatory, and that from approved safety spectacles may be supplemented with a visor or full face mask.

Table 2   *Partial List of Incompatible Chemicals (Toxic Hazards)*

Substances in the left hand column should be stored and handled so they cannot possibly accidentally contact corresponding substances in the centre column, because toxic materials (right hand column) would be produced.

Arsenical materials	Any reducing agent*	Arsine
Azides	Acids	Hydrogen azide
Cyanides	Acids	Hydrogen cyanide
Hypochlorites	Acids	Chlorine or hypochlorous acid
Nitrates	Sulfuric acid	Nitrogen dioxide
Nitric acid	Copper, brass, any heavy metals	Nitrogen dioxide (nitrous fumes)
Nitrites	Acids	Nitrous fumes
Phosphorus	Caustic alkalis or reducing agents	Phosphine
Selenides	Reducing agents	Hydrogen selenide
Sulfides	Acids	Hydrogen sulfide
Tellurides	Reducing agents	Hydrogen telluride

* Arsine has been produced by putting an arsenical alloy into a wet galvanized bucket.

A small-scale reaction not involving toxic hazards could be run on a wall-facing laboratory bench behind a portable safety screen, but a larger scale reaction, or one also involving toxic materials would best be run with the greater degree of protection afforded by a good fume cupboard with lowered sash. Where exceptionally reactive materials are involved, provision of a specific fire extinguisher additional to the general purpose laboratory extinguishers may be necessary. Analogously, where materials of high toxicity (especially cylinder gases) are involved, a specific respirator, or air breathing set may be required to be available.

Where the possibility of explosive decomposition has been assessed, and in any case where high-pressure reactions are being used, an isolation cell of suitable design is appropriate,[10] and the Safety Code[17] should be applied.

In the absence of laboratory facilities appropriate to the degree of hazard assessed, operations must be deferred until such facilities can be made available.

One topic related to several of the hazardous possibilities in Table 1 is the use of powerful oxidants or mixtures (nitric acid, or mixtures of dichromates or chromium trioxide with sulfuric acid) to clean glassware from traces of reaction residues, especially tars of organic derivation. In most cases this is not only potentially hazardous, particularly with large containers or residues, but unnecessary, as the specially formulated laboratory detergents which are available are just as effective and free of hazard.

9   References

1. L. Bretherick, 'Bretherick's Handbook of Reactive Chemical Hazards', Butterworths, London, 4th edn., 1990. There is also a fast electronic version, 'Reactive Chemical Hazards Database', Butterworth-Heinemann, London, 1991, which is searchable on a personal computer.

2. 'Manual of Hazardous Chemical Reactions', 491M, National Fire Protection Association, Boston PA, 6th edn., 1985.
3. A. I. Vogel, 'A Text-book of Quantitative Inorganic Analysis', Longmans, London, 6th edn., 1987.
   A. I. Vogel, 'A Text-book of Macro and Semi-micro Qualitative Analysis', Longmans, London, 4th edn., 1976.
4. A. I. Vogel, 'A Text-book of Practical Organic Chemistry', Longmans, London, 5th edn., 1989.
5. 'Organic Syntheses', various eds., Wiley, New York, Coll. Vols. 1–7, 1944–1990, annual volumes thereafter.
6. 'Inorganic Syntheses', various eds., McGraw-Hill, London, Vols. 1–26, 1939–1989, annual volumes thereafter.
7. G. Brauer, 'Handbook of Preparative Inorganic Chemistry', Academic Press, London, 2nd edn., Vols. 1 and 2, 1963, 1965.
8. M. Hannant, *Laboratory Hazards Bulletin*, Royal Society of Chemistry, Cambridge, issued monthly since July 1981.
9. Ref. 4, pp. 87, 97.
10. W. G. High, 'The Design of a Cubicle for Oxidation or High-pressure Equipment', *Chem. Ind.*, 1967, 899.
    A. L. Glazebrook, 'Safety in the Study of Chemical Reactions at High Pressure', Tech. Bull. No. 100, Autoclave Engineers, Erie, 1974.
11. V. J. Clancey, 'Explosion Hazards', *Protection*, 1971, **8**(9), 6.
12. CHETAH, 'ASTM Chemical Thermodynamics and Energy Release Evaluation Program', American Society for Testing and Materials, Philadelphia, release 4.4, 1989.
13. T. Yoshida, *Chem. Abstr.*, 1983, **95**, 7293.
14. D. F. Shriver, 'Manipulation of Air-sensitive Compounds', McGraw-Hill, London, 2nd edn., 1986.
15. H. L. Jackson, W. B. McCormack, C. S Rondevsted, K. C. Smeltz, and I. E. Vicle, 'Control of Peroxidizable Compounds', *J. Chem. Educ.*, 1970, **47**, A175. An addendum (G. S. Mirafzal and H. E. Baumgarten, *ibid.*, 1988, **65**, A226), claiming that primary alcohols are susceptible to peroxidation, has been refuted (L. Bretherick, *ibid.*, 1990, **67**, A230).
16. L. Bretherick, 'Safe Storage of Laboratory Chemicals', (ed.) D. A. Pipitone, Wiley, Chichester, 2nd edn., 1991, Chapter 4, p. 89.
17. 'High Pressure Safety Code', High Pressure Association, London, 2nd edn., 1989.

CHAPTER 6

Chemical Hazards and Toxicology

L. MAGOS

1 The Basic Rules of Toxicology

1.1 How Toxic is 'Toxic'?

In everyday language terms like poison, toxic, or toxicity imply a self explanatory and absolute quality which separates one group of substances (the toxic) from another (the non-toxic). This is a misconception. Every chemical has the capacity to cause injury and that is why the rating of toxicity does not start with zero, but with the *practically* non-toxic category. Though there is no chemical which does not have the capacity to cause harmful effects, even the most toxic chemicals can be used harmlessly. Exposure to a chemical is harmless when it does not disturb the state of equilibrium of biological functions. To say that water is harmless is true only in the sense that water consumption to a certain limit has no adverse effect, but forced drinking may be lethal and in a hot environment uncontrolled consumption of non-salted water can cause severe muscle cramps. The replenishment of essential elements or molecules depleted either by defective nutrition (*e.g.* low in iron, zinc, vitamin C) or by impaired synthesis (*e.g.* hormones, neurotransmitters) is beneficial, but overdose can provoke adverse reactions. Another group of drugs are given for their selective toxicity on invading organisms (antibiotics), on the central nervous system (general anaesthetics), or on neurotransmitters (β-blockers). However selectivity is relative and does not exclude the occurrence of adverse side effects even within the therapeutic dose range and severe intoxication by overdose. Side effects may be mild and reversible, but they may be serious enough to require the suppression of therapy. The key issue is the risk to benefit ratio which must be low to permit the prolonged administration of any drug. Pesticides are another group of substances which are used for their toxicity. Alkylmercury fungicides added to grain protect sown seeds against fungal infection, but the use of the same grain for making bread resulted in mass epidemics. An even larger group of toxic substances is manufactured and used for non-biological purposes. They become health hazards through regular or accidental release into the working or general atmosphere and through food contamination. However toxicity is not the prerogative of man-made chemicals. Several of the most potent toxins are of plant and animal origin. Thus the worldwide yearly occurrence of poisonings caused by the consumption of fish or shellfish fed on toxic dinoflagellates is around 100 000.

The toxicity of one chemical relative to another may be different for single or repeated exposure, for oral or inhalation exposure, for the young or the old, for the mother or her foetus, and naturally it can change from one species to another. Consequently, to say that a compound is toxic or non-toxic conveys no more information than to say that a chemical is soluble without defining its physical state, the solvent, temperature, and pressure. It sounds even more absurd to make a general statement about solubility of a group of compounds, but it is not without precedent that no distinction is made between the toxicities of an element and its compounds. Statements made by a Professor of Chemistry in a scientific journal will help to pinpoint the most frequent mistakes and will illustrate the most basic rules of toxicology. He criticized the need for preventive measures aimed at decreasing exposure to methylmercury, a neurotoxic agent, which had already caused mass epidemics in Japan and Iraq.

Statements by the Professor:
1. He had suffered no adverse effects, with the exception of loose teeth, after a three hour exposure to butylmercury.

Comment:
Butylmercury is not neurotoxic and less toxic than methylmercury. Most people would find unacceptable an exposure which resulted in loose teeth.
2. He distilled dimethylmercury for two days and his technician worked with the same compound for 200 days without adverse effects. Rats exposed to dimethylmercury for 20 days in a static inhalation chamber of 4 m^3 where 25 g dimethylmercury was evaporated showed no visible signs of intoxication.

Comment:
Experiment was meaningless because exposure was not measured and rats might have had co-ordination disorders only recognizable by trained observers. Rats may not mimic the human sensitivity to dimethylmercury. In 1863 two laboratory technicians who worked with dimethylmercury died of methylmercury poisoning and in 1971 a chemist died of methylmercury poisoning in Czechoslovakia after synthesizing 600 g of dimethylmercury.
3. During a mercury scare, which occurred 49 years ago, a test carried out on himself showed that he eliminated per day as much mercury as he absorbed, and therefore mercury does not accumulate.

Comment:
Supposing the balance study was carried out correctly, it indicated that his exposure to mercury was long enough to result in a steady-state body burden and it did not prove that mercury does not accumulate.

4. The dominant use of organomercurial diuretics, before 1952, to combat salt and fluid retention proves that the fuss about methylmercury as an environmental hazard is groundless.

Comment:

Organomercurial diuretics, unlike methylmercury, do not damage the sensory part of the nervous system, but frequently caused diarrhoea, vomiting, irritation, fever, and heart injury. These side effects prompted a search for safer diuretics.

The errors in these statements are manifold. Firstly no distinction is made between the different forms of mercurial compounds, though their toxicity differs both quantitatively and qualitatively. Secondly a change in the molecular structure changes not only toxicity, but also physical and chemical properties which influence evaporation and absorption. Thirdly, the experience of occupational exposure cannot be extrapolated directly to the non-occupational exposure of a less homogeneous population partly because the route of intake may be different and partly because the general population includes children, pregnant women, *etc*. Fourthly, dose or exposure was never quantified.

1.2   Molecular Structure and Physicochemical Properties

Without proof one should never state that different compounds of an element have the same potential to be harmful or hazardous and that their toxicities are identical. In every statement concerning toxicity the exact molecular structure of a chemical must be identified. Change in molecular structure caused by oxidation, reduction, cleavage of a bond, introduction of a new ligand, or replacement of one ligand with another can have profound effect on the potential of a compound to be hazardous. Some of these changes might occur during storage or by the use of a chemical and in these cases consideration must be given to the properties of the possible derivatives.

A change in the molecular structure can alter the physical properties. Thus dimethylmercury has a higher vapour pressure than methylmercury chloride. The volatility of methylmercury depends on the anion; methylmercury chloride is 350 times more volatile than is the corresponding dicyanodiamide salt. The second important physical property is solubility. The solubility of mercury metal in water and biological fluids is very low, which explains why 2 lb of mercury taken orally in four divided doses by a young man caused no ill effects. The solubility of mercurous chloride is low compared with mercuric chloride, and in the case of oral administration the more soluble mercuric chloride is 60 times more toxic than the mercurous chloride. Other examples are: lipid solubility favours absorption through the outer layer of the skin and water solubility helps the transport through the lower layer. Consequently, of two lipid soluble compounds, like aniline and nitrobenzene, the absorption of aniline is somewhat faster due to its higher water solubility. The third physical property that can influence the potential hazard of a chemical is viscosity. Increasing viscosity would diminish the danger of swallowing the chemical when pipetting, although pipetting by mouth is undesirable.

Change in the molecular structure can affect the reactivity of a compound. Thus dimethylmercury having no charge, behaves like a neutral gas and becomes toxic only after decomposition to methylmercury. In mice within 2.5 hours of the intravenous administration of dimethylmercury 85% of the dose was exhaled and only 10% was converted to methylmercury, *i.e.* to the neurotoxic agent.

Dibutylmercury behaves in a similar way to dimethylmercury, but the metabolic product, butylmercury, is less toxic than methylmercury and seems to lack neurotoxic properties. Thus one cannot even make generalizations for the toxicity of alkylmercury compounds, let alone other organomercurials. Organomercurial diuretics are rapidly metabolized in the body and the role of the organic part of the molecule is to carry mercury and release it as mercuric ions at renal receptor sites involved in diuresis. This effect is qualitatively different from the neurotoxic effects of methylmercury or mercury vapour. The former mainly affects the sensory nerves, while mercury vapour intoxication is characterized by tremor and psychotic disturbances. An extreme form of mercury vapour intoxication had been epidemic in the hat industry (mad as a hatter) before the abandonment of mercury nitrate for processing felt.

1.3   Biological Variation in Response to Chemicals

Not every species reacts with the same type or same degree of response to equal doses of a chemical. This presents problems when animal experimental data is extrapolated to man. The cause of species difference can be anatomical. The absence of the iridescent membrane which makes the eyes shine in the dark excludes the extrapolation of the toxic damage observed in dogs or cats to man. More important are those physiological and biochemical species differences which result in quantitative changes in toxicity. An example for physiological differences is the role of nasal breathing in the induction of nasal cancer by formaldehyde. Formaldehyde is carcinogenic in obligatory nasal breathers, like rats and mice, but there is no evidence for similar effect in man. There is also a difference between the carcinogenic responses of the two rodents because mice respond to the irritant effect of formaldehyde with reduced respiration at a lower concentration.

Biochemical differences are the most frequent source of variation in toxic response. Metabolic rates, like the minute volumes of alveolar ventilation and cardiac output, are related to a fractional power of body weight that is close to 3/4. Consequently the weight specific metabolism is higher in small than in large animals. Thus it is to be expected that relative to a small animal the larger animal is at a greater risk when the parent compound is toxic and at a smaller risk when stable metabolites are toxic. An additional problem is that none of the experimental animals mimic all human metabolism patterns. Owing to differences in the metabolic handling of acetaminophen (paracetamol), men and cats react to an overdose differently. Men develop hepatic damage with hardly any methaemoglobin formation (inactivation of haemoglobin by the oxidation of its iron) and cats react with severe methaemoglobinaemia without hardly any liver injury.

Species differences in carcinogenicity are frequent but the cause is rarely identified. Carcinogenicity studies on male rats and mice showed that out of 311 chemicals 54 chemicals caused cancer only in rats and 41 only in mice.

Another difficulty in the interpretation of toxicological data is the biological variation within a single species. Strain differences are the expression of genetic variation. Fischer rats are more sensitive to the nephrotoxicity of acetamino-phen than Sprague–Dawley rats. Fischer rats are also more sensitive to carbon tetrachloride induced liver damage than Wistar rats. Examples for the sig-nificance of genetic differences in humans are glucose-6-phosphatase deficiency resulting in hypersusceptibility to the haemolytic effect of several chemical agents (mainly aromatic amino and nitro compounds, but also acetylsalicylic acid) and serum antitrypsin deficiency resulting in increased sensitivity to respiratory irritants. Age, sex, and nutrition can also influence toxicity.

Clinical studies indicated that the foetus is more sensitive to the neurotoxicity of methylmercury than the mother. The sensitivity of rat pups to the nephro-toxicity of mercuric chloride increases with maturation from nil after birth to adult sensitivity. At birth 55–85% of the haemoglobin is in a form which is more easily oxidized to an inactive form (methaemoglobin) than adult haemo-globin. Epidemiological studies seem to indicate that some effects of lead on the central nervous system are age dependent and children are at a greater risk than adults.

Male rats are more sensitive to the renotoxicity of mercuric chloride or the carcinogenic effect of aflatoxin than female rats. Females are more sensitive to the neurotoxicity of methylmercury or to the acute toxicity of organophosphate insecticides than male rats. Men are more sensitive to cigarette smoking induced lung cancer or aflatoxin induced liver cancer than women. Carcinoge-nicity tests on 311 chemicals showed that 37 chemicals induced cancer only in male and 16 only in female rats. The corresponding numbers for mice were 13 and 29.

Susceptibility to lead depends on a wide variety of dietary factors like calcium, iron, protein, vitamin D, ascorbic acid, nicotinic acid, and alcohol consumption. Restricting the diet of mice to 75% of the freely available food decreased the incidence of liver tumours by about 80% and all malignant tumours by 65%.

The elimination of known influences on toxicity by using an inbred group, which is also homogeneous in age, sex, and nutrition, will decrease but does not eliminate variation in response. Thus individual observations, though they may be clinically interesting, do not allow the quantification of risk. This needs greater numbers for suitable statistical analysis and also detailed data on kinetics and metabolism when extrapolation is required.

2 Kinetics and Metabolism

2.1 Absorption

The toxic response in any cell, tissue, or organ is a function of the concentration of the causative agent at the site of action. Chemical burns and irritation of the mucous membrane or skin are a result of external penetration into the layers of skin, but most chemicals act internally. This explains why knowledge of the absorption of toxic compounds from the site of administration, its subsequent distribution, metabolism, and excretion are so important for the interpretation of data obtained from toxicity tests. Some conclusions can be drawn from single dose experiments: nearly identical LD_{50}s (dose which kills 50% of the group) for oral and parenteral administration indicate that the compound is almost completely absorbed from the gastrointestinal tract while incomplete gastrointestinal absorption requires substantially higher oral doses than the parenteral ones. Dermal application of the test material also gives some information on the ability of the tested chemical to pass through the skin. With increasing lipid solubility and decreasing molecular weight the absorption of non-polar substances increases through the outer layer of skin, though further transport into the blood vessels is promoted by water solubility. Abrasion of the skin increases absorption for both polar and non-polar compounds.

Lung offers a very much larger surface for absorption than either the gastrointestinal tract or skin as the thin walls of tiny air chambers, completely filled with a network of capillary blood vessels, cover an area of 90 m^2. Between the alveoli and the blood capillaries there is only a 0.36 to 2.5 μm thick membrane which allows the free equilibration of gases and vapours between the alveolar space and blood. Equilibration between alveolar air and alveolar blood is also responsible for the exhalation of volatile substances formed *in vivo*. Examples are the exhalation of mercury vapour after the injection of $HgCl_2$ under the skin, the exhalation of CS_2 after treatment with antabuse, or the garlicky smell of breath in intoxications caused by arsenic, selenium, tellurium, or organophosphates.

The alveolar region is the main deposition site for particles of 0.1 to 5 μm diameter or relatively water insoluble gases and vapours (ozone, nitrogen dioxide, chloroform). Large particles ($> 5\ \mu$m) or water soluble gases and vapours (SO_2, formaldehyde) are mainly deposited in the nasal passages and other parts of the upper respiratory tract. The main deposition of middle sized particles or moderately water soluble substances shift in the direction of the tracheobronchial part of the respiratory tract. Change from nasal to oral breathing removes nasal deposition and increases the exposure of other sites.

2.2 Distribution

Once a chemical is absorbed, blood plays an important part in its distribution. Methylmercury and atomic (elemental) mercury easily cross membranes, including the membrane of red blood cells and the blood–brain and placental barriers. After exposure to methylmercury or mercury vapour the red blood cell to plasma mercury ratio, brain uptake, and foetal concentration are very much higher than after the administration of inorganic mercury salts. The easy transfer of methylmercury and mercury vapour through the blood–brain barrier explains why the central nervous system is their main target. Passage of toxic compounds through the placenta is the prerequisite of direct toxic actions on the foetus, though placental deposition can also interfere with the supply of essential nutrients to the foetus.

The passage of chemicals from the blood into organs is a complex process. An uncharged molecule can diffuse easily through membranes, but the transport of other molecules usually requires reaction with physiological carriers. Thus the transport of organic anions from plasma into the liver is facilitated by a transport protein called ligandin. Diffusion and metabolically-dependent transport are the two forms of transport. An example for diffusion is the distribution of hydrocarbons. There is some linear relationship between the anaesthetic potency of chlorinated hydrocarbons and their serum/air or oil/air partition coefficient. High lipid solubility increases uptake and decreases elimination from organs like the brain which has a high lipid content. Metabolically-dependent transport contributes to the accumulation of mercury in the kidneys. For example, dinitrophenol given before the administration of mercuric chloride decreases the kidney uptake of mercury.

2.3 Metabolism

The absorbed chemicals are in contact with highly reactive biological compounds and catalytic systems. The oxidation of metallic mercury by catalase in red blood cells reduces the possibility of diffusion from blood to brain. However, the main site of oxidation is in the liver where other toxic substances are also transformed by oxidation, reduction, hydrolysis, and conjugation. The result of these metabolic processes is usually a compound which can be easily excreted in bile or urine. As all the toxic substances which are absorbed from the gastrointestinal tract must pass through the liver before they enter the systemic circulation, the intervention of liver decreases the exposure of other organs and tissues to the parent compound and increases exposure to metabolites. Chemicals absorbed from the oral cavity, the lower rectum, and anal canal, avoid the first-pass elimination by the liver. The influence of first-pass elimination by liver is so high that when the aim of drinking whisky is to become intoxicated, the best method is to increase oral absorption by drinking it by the teaspoonful.

The biotransformation may increase or decrease toxicity. Alcohols are converted to the corresponding aldehyde and acid but with widely different consequences. The conversion of ethanol to acetaldehyde is a detoxification step, but the conversion of methanol to formaldehyde increases toxicity 33-fold. Aniline is converted to *p*-aminophenol which is a potent methaemoglobin forming agent. From carbon tetrachloride highly reactive free radicals are formed, which are at least partly responsible for its hepatotoxicity. Parathion is a weak acetylcholine esterase inhibitor but it is oxidized to a powerful inhibitor, paraoxon.

Phenobarbitone potentiates the metabolism of many foreign chemicals. After phenobarbitone treatment carbon tetrachloride, carbon disulfide, chloroform, or nitrosamine become more, but aflatoxin less, toxic. Chemicals that inhibit the enzymes responsible for the formation of a more toxic compound have the opposite effect and they decrease the risk. Thus carbon disulfide or diethyl-dithiocarbamate decreases the toxicity of carbon tetrachloride.

There are other forms of interaction which do not fit into the normal biotransformation processes, such as the protective effect of methaemoglobin forming agents against cyanide. Interaction can change even the toxicity of metals, thus cadmium pretreatment has a protective effect against the renotoxic doses of mercuric chloride, and selenium protects against cadmium and mercury and *vice versa*.

2.4 Body Burden and Biological Half-time

Whether exposure is single or repetitive, whether it continues or is terminated, each day a constant proportion of the body burden is excreted. Thus the elimination constant is used to calculate the time needed to excrete half of the total body burden after the termination of exposure. The higher the elimination constant the shorter is the half-time. When exposure is constant and continuous the body burden increases until daily uptake equals elimination. That condition is called the steady state. For example, methylmercury has an approximately 70 day half-time so if a person ingests 2 mg methyl-mercury per week of which 95% is absorbed from the gastro-intestinal tract, one can calculate that after six months the body burden of mercury will be 24 mg, after one year 28 mg, and 28.9 mg at a steady state when exactly 2 mg of mercury will be excreted weekly.

As far as half-time is concerned, methylmercury belongs to the group of compounds with a single dominant half-time, and thus daily body burden plotted on a semilog paper against time gives a straight line. There are other compounds, such as lead or mercury vapour, that have more than one half-time, a short half-time followed by one or more longer half-times. The biological half-time of lead in bone is about 10–20 years, but in blood and soft tissues it is only 20–30 days.

This indicates that the half-time for different organs or tissues is also different and as time passes, tissues with a longer half-time dominate the elimination curve. Another complicating factor is the metabolism of the chemical yielding derivatives which might have a longer or shorter half-time than the parent compound. Thus the half-time of dimethylmercury is very short and the half-time of its derivative, methylmercury, very long. Phenylmercury, alkoxy-alkyl mercury salts, or organomercurial diuretics are rapidly metabolized to inorganic mercury and afterwards the half-time of the mercury component is identical with the half-time of mercuric mercury.

2.5   Excretion

The three main routes of excretion are exhalation, urinary, and faecal excretion.

As a general rule every chemical that can be inhaled as vapour or gas can also be exhaled, and the concentration in the exhaled air depends on exposure and metabolism. However, not only inhaled chemicals can be exhaled. Exhalation is noticeable even when a solvent is given orally or by injection. The exhalation of carbon disulfide after the injection of dithiocarbamates or their disulfide (like disulfiram = antabuse) shows that some non-volatile compounds can be metabolized to volatile ones.

Excretion in the faeces or urine is the most important route for the majority of toxic chemicals and non-volatile compounds are excreted only by these routes, though small quantities might appear in tears, sweat, or milk.

A proportion of an ingested chemical can be excreted with faeces without being absorbed. Less than 10% of toxic inorganic metallic compounds are absorbed, though with an organic radical their absorption (like that of methylmercury) might be nearly complete. The chemical which is absorbed into the blood stream might occur in the gastro-intestinal lumen through the normal shedding of the inner surface of the gut, but the most important source of faecal excretion is the bile. After oral exposure biliary secretion is helped by the fact that intestinal blood first passes through the liver before it reaches the general circulation. Biliary secretion is influenced by the molecular weight of the compound and by its polarity. The optimum molecular weight is 500–1000; between 300 and 500 mainly polar compounds are excreted. There are mechanisms in the liver which facilitate the biliary secretion of compounds by increasing their polarity and their molecular weight. A secreted compound might be excreted with the faeces or reabsorbed from the gut. The reabsorption of biliary methylmercury contributes to its long biological half-time. When biliary secretion is the major route of elimination, diseases of the liver which affect the function of liver cells or obstruct the bile flow, will increase the half-time of the chemical and increase its toxicity.

Extraction of chemicals by the kidneys is facilitated by the fact that one-quarter of the blood pumped into the circulation by the heart passes through the kidneys and about half of this is submitted to a filtration mechanism which is the first step in urine formation. Not all the chemicals that appear in the filtrate (tubular urine) are excreted, some are reabsorbed into the tubular cells. Urinary excretion depends on the ability of the kidney to extract the chemicals from the blood, and on the ability of tubular cells to accumulate or release the toxic compound. Chemicals with a high lipid/water partition coefficient are passively reabsorbed from the kidney tubules, while polar compounds and ions, if they are to remain in these forms in the tubular urine, are excreted. The availability of small molecules that form complexes with toxic chemicals can increase the urinary excretion because they compete for the chemicals with protein binding sites in the cells. Thus the excretion of heavy metals can be increased very significantly by the administration of complexing agents, such as the excretion of inorganic mercury by BAL or D-penicillamine, or the excretion of lead by EDTA.

3 Dose and Damage

3.1 Critical Organ

The rate of blood flow through organs and the affinity of tissues to the chemical and its metabolite(s) are responsible for differences in organ (or tissue) concentrations. Whether an organ becomes the target of a chemical depends on the tissue concentration of the active form and the sensitivity of the tissue to this form. The organ which first attains a concentration that affects its function is called the critical organ or, in other words, if the injury of the critical organ is prevented, intoxication is prevented.

For a toxic chemical there might be more than one critical organ. Acute exposure to cadmium fumes damages the lungs, whereas long-term exposure to the same metal results in kidney damage. The critical organ is not necessarily the organ which contains the highest amount or concentration of the chemical. The main storage organ for lead is in bone, but the critical organ is the haemopoetic system which forms the blood and, in infants, the critical organ can be the central nervous system. Irrespective of the chemical form of mercury the kidneys are always the organs with the highest concentration, but the kidneys are the critical organs only for mercuric mercury. For metallic mercury or methylmercury the critical organ is the nervous system.

The identification of the critical organ for a toxic chemical is based on the dose–effect relationship which gives the sequence of toxic effects in relation to dose.

3.2 Dose–effect Relationship

Evaluation of a chemical hazard is based on two relationships, one is the dose–effect relationship, the other is the dose–response relationship. The methylmercury epidemic in Iraq, which involved many thousands of people, proved that the daily consumption of certain amounts of methylmercury-contaminated bread caused paraesthesia (sensory disturbances), higher levels of exposure caused ataxy (unco-ordinated movements), and even higher daily uptake resulted in the loss of vision, deafness, or even death. Thus methylmercury, depending on the exposure, is able to cause effects of varying severity. Another example is the narcotic and hepatotoxic effect of carbon tetrachloride. In the case of lead, porphyrin in the urine, anaemia, abdominal pain, palsy, and, mainly in children, encephalopathy are the main constituents of the dose–effect relationship. Rats given $1–2$ mg kg^{-1} amphetamine run around the walls of the cage, but when treated with 6 mg kg^{-1} amphetamine they soon become stationary and move their heads repetitively. Doses higher than this produce cardiovascular disorders. Listing effects against the corresponding dose gives the dose–effect relationship. If one effect, like anaemia or loss of nerve function, can be measured on a graded scale of severity, the gradation of this effect in relation to dose is used also as a dose–effect relationship.

3.3   Dose–response Relationship

If one effect is selected as a response, the percentage of animals giving this response in every dose group is used to produce a dose–response curve. From the dose–response curve one can calculate the dose that is able to produce a response in 50% of the animals and this dose is called ED_{50} (ED = effective dose). If the response is death the dose that killed 50% of the animals is called LD_{50} (LD = lethal dose). Both ED_{50} and LD_{50} can be calculated from single or multiple administration experiments, but in every case the route of administration, dose, treatment, and observation period must be stated. In the case of inhalation exposure, the atmospheric concentration of the test compound and the length of the exposure are the essential data. Both ED_{50} and LD_{50} are valid only for the species tested with the qualification of age, sex, and nutrition. Even when men or animals can be killed by a single dose, this does not necessarily mean that by lowering the dose, a dose–effect relationship which includes all the possible effects can be established. In the case of methylmercury, the typical neurological symptoms in rats can be produced only after repeated administration. Many of the chemical carcinogens must be given for an extended period of time and the observation period might cover the whole lifetime of the animal.

The dose–response relationship of a chemical administered by a certain route to a species may not be valid for another route or for another species because the relationship between administered and delivered dose at the target site may be different. Extrapolation is helped when in the dose–response curve the integrals of tissue exposure to the chemical or its metabolite(s) are substituted for administered dose.

4   Toxicity Testing

4.1   Acute (Single Dose) Toxicity Tests

The most frequently used procedure in the assessment of acute toxicity is the LD_{50} test which is the estimation of the single dose lethal to 50% of the test animals. The chemical is most frequently administered orally, but occasionally it is given parenterally (intraperitoneally, intravenously, or subcutaneously). For volatile compounds a more appropriate, and for gases the only, way of administration is by inhalation. The aim of the inhalation toxicity test is to calculate LC_{50}, *i.e.* the atmospheric concentration which results in 50% death. The aim of the acute dermal tests is usually restricted to the detection of local effects (*e.g.* irritation) or whether by dermal application the compound can cause systemic toxic reactions. However, when the absorbed dose is measured, a dermal test can be used for the quantitation of acute toxicity. In the dermal toxicity test the compound is spread evenly on a skin area shaved 24 hours earlier. After treatment the area is bandaged for 24 hours with an aluminium foil or polythene lining. The dressing is removed 24 hours later and the treated area is washed and cleaned with soap and water. From the quantitative analysis of the compound in the washing liquid and bandage it is possible to calculate how much of the applied chemical penetrated the skin.

In acute toxicity tests the compound is given to two species at four dose levels. Each dose group should contain an equal number of males and females. The maximum dose usually does not exceed 2.0 g kg^{-1} orally or 2.8 g dermally. The maximum concentration for a 6 hour inhalation exposure is approximately 8 mg l^{-1} when hazardous chemicals are screened for the possible consequences of occupational exposure. Animals are observed daily for 14 days. Some dying and some other less affected animals at day 14 are autopsied for gross pathological abnormalities and depending on the result of this examination organs are taken for histological examination. LD$_{50}$ or LC$_{50}$ are calculated for 24 hours, 4 days, and 14 days. Some additional numbers of animals dosed at proven toxic levels are autopsied 2–3 days after dosing. The gross pathology and histology might reveal early or temporary changes not seen at 14 days.

Daily observations made on treated animals and the result of pathological and histological tests are of great importance for the design of further toxicity studies. They help to select more sensitive responses than death for ED$_{50}$ studies (effective doses which produce a certain response in 50% of the animals) and for sub-acute tests. They also help to narrow the dose range in these studies. Thus a well organized LD$_{50}$ test can result in a significant saving in both the number of animals and work. When no substantial information is available on the toxicity of a chemical, in order to avoid death from causes other than cancer, the first step in the selection of realistic doses for long-term carcinogenic tests is also an acute LD$_{50}$ test. LD$_{50}$ values are therefore the first estimate of toxicity which help to classify toxic compounds into broad categories. The categorization shown in Table 1 is widely used and indicates some relationship between a single oral dose which may cause death in man and the single oral LD$_{50}$ for rats. Naturally if the toxic compound accumulates in the body, has a cumulative effect, is carcinogenic, or can produce sensitization, then the potential hazard bears no relationship to the above toxicity rating. Thus, based

Table 1 *Categories in Relation to Relative Acute Toxicities*

Toxicity rating	Term of toxicity	Probable human lethal dose for a 70 kg man	Compound belonging to the group (oral LD$_{50}$ for rats in mg kg^{-1})
1	practically non toxic	> 15 g kg^{-1}	propylene glycol (26000)
2	slightly toxic	5–15 g kg^{-1}	sorbic acid (7400)
3	moderately toxic	0.5–5 g kg^{-1}	isopropanol (5800)
4	very toxic	50–500 mg kg^{-1}	hydroquinone (320)
5	extremely toxic	5–50 mg kg^{-1}	lead arsenate (100)
6	supertoxic	< 5 mg kg^{-1}	nicotine (50)

on a single dose, methylmercury is less toxic than mercuric chloride, but in the case of repeated administration animals develop a resistance to the renotoxic effect of mercuric chloride, but not to the neurotoxic effect of methylmercury. Di-isocyanates are only slightly toxic according to the toxicity rating, but it is not uncommon that following acute symptoms, an asthma-like syndrome is precipitated even by exposure to a minute amount of di-isocyanate.

4.2   Sub-acute Toxicity Test

The acute toxicity test gives the relevant information for the organization of sub-acute tests in which exposure lasts from a few days up to three months. In sub-acute or short-term tests the highest daily dose must be sufficiently below the single dose LD_{50} to avoid the early occurrence of severe toxicity, but must be high enough to cause intoxication. Usually three dose levels are used on male and female animals belonging to two species, and the route of administration should be relevant to the expected form of exposure. During the whole test time animals are regularly observed for their clinical conditions. Body weight and food intake are recorded and laboratory investigations are carried out for blood and urine chemistry and haematology. At the end of the test period animals are killed and subjected to full post-mortem and histological examination. Brain, heart, lungs, liver, kidneys, spleen, gonads, adrenals, thyroid, pituitary, and pancreas are removed and weighed. Autopsy carried out at a later date gives information on the reversibility of damage. One of the objectives of the sub-acute toxicity test is to demonstrate some toxic response and the second objective is to provide an indication on the maximum dose which given for a certain time does not produce any adverse effects. The comparison of single dose LD_{50} with the repeated dose LD_{50} also helps to establish whether the compound has a cumulative effect. For example, if the LD_{50} of 28 daily doses is 28 times less than the single dose LD_{50}, this indicates that the compound is completely cumulative, and the closeness of the two LD_{50}s indicates that one daily dose and its effect is eliminated before the next treatment is given.

For many chemical agents, daily oral treatment or inhalation exposure lasting one-tenth of the lifespan of a species (90 days for rats) allows the setting of the non-adverse effect level for long term exposure. Of 122 toxic agents studied in rats only 3 (2.45%) and of 566 dosage levels only 15 (2.65%) induced effects after, but not before, 90 days exposure. Thus with the exception of carcinogenic effects, it is possible to predict from the negative outcome of a 90 days test on rats that even the life-time administration of the compound will not produce adverse effects.

4.3   Chronic Toxicity Test for Carcinogenicity

The determination of carcinogenic potential requires chronic toxicity testing. The most widely used species are the rat and the mouse, but Syrian golden hamsters proved to be an excellent test animal to reveal bladder and lung cancer.

The route of administration is designed to follow the route of human exposure and can be oral (in food or drink), dermal, or inhalation, though it can be administered also by gavage or by parenteral injection. Oral administration is carried out 7 days per week and the others 5 days per week. Treatment starts a few weeks after weaning and lasts for 2 years in rats and 18 months in mice.

The test is carried out on two species at two, but preferably more, dose levels. Doses must be selected on the basis that even the highest dose should not shorten the lifespan by non-tumourogenic lesions. Each group must consist of at least 30 male and 30 female animals, preferably more. For every group equal numbers of vehicle controls are included, that is animals which receive the vehicle without the test material. All animals in carcinogenicity tests are subjected to complete gross pathological and histological examination.

The sensitivity of the test depends on the frequency of tumours in the controls and on the number of animals per group. Thus when no tumour occurs in controls at a site, significant tumourogenic effect requires a more than 50% tumour frequency in 10 animals per group and slightly more than 10% in 75 animals per group. When in controls the probability of tumours is 20%, the corresponding threshold frequencies for detection are 87% and 43%. Dose related increase in malignant neoplasms or benign neoplasms which have the ability to progress to malignancy is the clear evidence of carcinogenic activity. When 27 proved or probable human carcinogens were bioassayed 25 produced cancer at least in one species. However the *predictive value* of rodent bioassays is less satisfactory because low carcinogenic potency may remain undetected and high dose carcinogenic potency may remain undetected and high dose carcinogenicity which acts through prolonged disturbances of the physiological status is irrelevant for the condition of human exposure.

4.4 In Vitro *Tests for Carcinogenicity*

Chronic animal tests for cancer are extremely expensive (about £100 000 per chemical). That is one of the reasons for the interest in predictive *in vitro* tests. The most widely used predictive test is the *Salmonella*/mammalian–microsome mutagenicity test named after its developer, Dr. Bruce Ames. This test identifies not only the mutagenicity of a chemical but also the mutagenicity of its metabolite(s). Other less frequently used tests aim to detect gene mutation, chromosome damage, or cell transformation in mammalian cells and a wide variety of genetic damage in *Drosophilia melanogaster*.

Early tests indicated that 85% of known rodent/human carcinogens had been mutagenic in the Ames test. However, when a larger group of rodent carcinogens, including those without DNA reactivity, were tested, positive response in the Ames test was given by 30% of the non-carcinogens and about 50% of the carcinogens. These results indicate that the assumption regarding the mechanistic relationship between carcinogenesis and mutagenesis may be flawed. Thus benzene and acrylamide are negative in the Ames test but the induction of chromosomal breaks indicate that they are genotoxic in mammalian cells. Moreover the absence of any proof for the genotoxic activity of several rodent carcinogens (*e.g.* aldrin, dieldrin, thiourea, di-2-ethylhexyl phthalate) indicates epigenetic mechanism. Epigenetic mechanism implies that a non-genotoxic carcinogen can produce genetic damage indirectly through chronic tissue injury and gross disturbances in the physiological status.

4.5   Special Tests

Special tests are those which measure skin irritation, eye irritation, sensitization, and reproductive disorders.

Skin and eye react to contact with an irritant with redness, swelling, and blisters. These signs of primary irritation occur rapidly and restricted to the site of contact. One or repeated contact to concentrations which may not be irritant can lead to sensitization reactions precipitated years later by the same or a related agent. The reaction most frequently occurs after 12 to 48 h delay at the site of original exposure but it may spread to other parts and becomes general.

Reproduction tests aim to estimate effects on male and female fertility, gestation, and the offspring. Toxic effects on the embryo or foetus can be direct or through damage to the placenta and maternal toxicity. The outcome can be developmental disorder or death with partial or complete resorption. Teratological effects are congenital malformations caused by direct toxic action on the developing organs. The index of teratogenic potential is the ratio of lowest adult toxic (lethal) dose to lowest developmental toxic dose. Low ratio indicates that the protection of the mother will protect the foetus, high ratio indicates high teratogenic potential.

Genetic damage caused by toxic compounds in male germ cells is demonstrated with the dominant lethal test in which treated male rats are mated with untreated virgin females. The failure of the fertilized egg to implant in the uterine wall or to survive to mid-pregnancy is detected by the necropsy of females. In fertility tests male rats are treated for 60 days before mating and female rats from 15 days before mating to the end of gestation. In teratology tests treatment is restricted to the embryonic stage when cells differentiate and organs develop. This period in rats corresponds to days 5–15 after fertilization. The time of administration is crucial. For example, 25 mg kg^{-1} lead nitrate given to pregnant rats on day 9 results in malformations in the foetuses, while on day 10 the same dose is lethal to them. Treatment given later can cause only functional defects, most importantly in the central nervous system.

5   The Predictive Value of Toxicity Testing

5.1   Extrapolation from Animal Experiments to Man

Extrapolation from animal experimental data to man is difficult for many reasons. Man has different body surface:body weight and respiratory minute volume:body weight ratios than the test animals. Also the metabolism of the compounds might be quite different. The extrapolation is relatively easy if a reference point is established both for the experimental animal and man. For example, if the dose required to elicit a well defined response is established for man and the experimental species, there is a scientific base for the extrapolation from the experimental dose response curve to safe exposure level in man. Data on the half-time and the metabolism of the toxic chemical in man and in experimental animals also helps extrapolation. Without parallel data, extrapolation from animal experiments is only guesswork based on analogies and probabilities with bias on the side of safety. Thus the threshold limit for uranium established on the basis of the urinary catalase activity of rabbits exposed to this metal was reduced four-fold when the same response had been established in hospitalized volunteers. In physiologically based pharmacokinetic models extrapolation from one species to another species requires data on: (i) body and organ weights; (ii) flows (alveolar ventilation, cardiac output, and organ blood flows); (iii) partition coefficients between blood and air and between blood and tissues; (iv) and metabolic constants.

A simple extrapolation procedure that starts with a no observed adverse effect level (NOAEL) established for non-carcinogens in 90 days tests considers the difference between the respiratory volumes of the test animal and man. The respiratory volume in m^3 for an 8 h period is proportional to 0.176 times the 3/4 power of the body weight in kg. Thus the respiratory volume of a 0.2 kg rat is 0.053 m^3 and that of a resting man of 70 kg body weight is 4.3 m^3. However in the extrapolation procedure the respiratory volume is increased to 12 m^3 which is near to the respiration of a man performing physical activity comparable to walking at a speed of 6.5 km h^{-1}. Thus the respiration on the per kg basis is 1.5-fold (0.263 : 0.171) higher in rats than in the working man. As it is assumed (if there is no evidence for the contrary) that the fraction absorbed is the same for all species, the rat NOEAL (exposure concentration) multiplied by 1.5 gives the human NOEAL. A further division by a safety (or uncertainty) factor gives the tentative permitted concentration. An additional safety factor of 10 should be considered when the toxicant bio-accumulates, causes cumulative damage, or decreases resistance.

It has been assumed that correction of dose to surface area improves the conversion of animal NOAEL to human NOAEL at least when the parent chemical is responsible for the toxic action. The surface area being proportional to the 2/3 power of body weight (W) in kg, the mg kg^{-1} dose of a 0.2 kg rat multiplied by 0.14 ($[0.2 : 70]^{1/3}$) gives the surface corrected equivalent human dose in mg kg^{-1}. Based on surface correction, the human NOAEL concentration is not 1.5-fold higher but 4.6-fold (7/1.5) lower than the rat NOAEL concentration.

The extrapolation has more uncertainties when only oral NOAEL or LC_{50} (the acute median lethal concentration) is known. From an established relationship between inhalation NOEAL and oral NOAEL (or LC_{50}) for other compounds, the unknown inhalation NOEAL (surrogate NOAEL) of a substance can be calculated from its oral NOAEL (or LC_{50}).

This extrapolation method is not suitable for chemicals that are able to cause cancer. The toxicological evaluation of toxic compounds which might cause cancer requires longer tests than the 90 days test. These tests not only cover nearly the whole lifespan of the species, but their extrapolation to man requires a more cautious approach and depends on whether there is a threshold limit for a carcinogen at all.

5.2   Threshold and Risk

The threshold concept implies that a certain effect of a compound can not be produced below a certain dose (threshold dose). This concept is generally accepted for most toxic effects but there is a controversy concerning the acceptance of threshold for carcinogenesis. The non-threshold approach is justified by the need to be conservative when the outcome is a dreaded irreversible reaction. Moreover even if threshold exists the estimation of the threshold which can protect the most sensitive members of a population (universal threshold) poses logistic problems. Variation in individual sensitivity and therefore threshold is great and the combined effects of interacting carcinogens can be multiplicative (*e.g.* smoking and asbestos). These problems explain why in spite of experimental evidence of threshold for non-genotoxic carcinogens, the US regulatory agencies use non-threshold models for the low dose extrapolation of both genotoxic and non-genotoxic carcinogens.

5.3   Low Dose Extrapolation

The frequencies of response recorded in acute, sub-acute, or chronic experiments are usually plotted on a linear scale against dose. The scale of dose is governed by convenience and by the view on threshold. Thus when the dose is increased by a constant factor, the linear dose scale may be inconvenient owing to large gaps between neighbouring higher doses and overcrowding between neighbouring lower doses. The log dose scale avoids this inconvenience and it gives an approximately straight line with a hockey stick end at the lower end. The straight line helps to define LD_{50} or ED_{50} and it intercepts abscissa or the baseline frequency. One characteristic of this plot is that it always gives a threshold even when the linear dose plot intersects the ordinate (see Figure 1).

The opposite is true to the Probit plot. In a Probit plot log dose is plotted against standard deviates (Probits) of response frequencies. On the Probit scale 50% frequency gets the arbitrary value of zero (or 5) and one standard deviation above and below 50%, that is $50 \pm 34\%$, is converted to ± 1.0 (or 6 and 4), $50 \pm 47.7\%$ to ± 2.0 (or 7 and 3), and so on. The Probit plot gives a straight line. Because normal distribution values never reach zero (or 100%), Probit does not indicate threshold even when the experiment shows zero response at a wide range of doses. To make this model more conservative experimental points are replaced by their upper (95 or 99%) confidence limits (see Figure 2).

Several other models give response frequencies on the linear scale against linear dose but avoid threshold by arbitrarily converting the lower part of the dose response curve to a straight line which ends in the origo. The slope depends on the chosen mathematical model. Owing to the use of upper con-

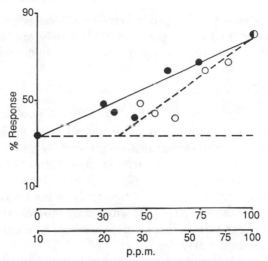

Figure 1 *The prevalence of liver neoplasms in mice at the end of 33 months dietary exposure to 2-acetylaminofluorene. Response is either plotted against linear dose (solid circles, upper abscissa) or against log dose (empty circles, lower abscissa).*
(Data of N. A. Littlefield *et al.*, *J. Environ. Pathol. Toxicol.*, 1979, **3**, 17–34)

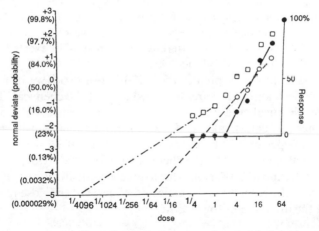

Figure 2 *The tumourogenic effect of a single injection of methylcholanthrene in mice. One dose unit equals 0.078 mg/mouse. Dose (log scale) is either plotted against linear frequency (solid symbols in insert where dose is lined up with the main abscissa) or against normal deviate (Probit) units without correction (empty circles) or corrected to the upper 99% confidence limit (empty squares). Note that without correction both zero and 100% responses are ignored and correction shifts zero responses upward.*
(Data of N. Mantel and W. R. Bryan, *J. Natl. Cancer Inst.*, 1961, **27**, 455–470)

fidence limits the real life-time risk can be lower (may be zero) but unlikely higher than the upper bound. A popular model is the linearized multistage model. The US Environmental Protection Agency has been using this model for a considerable time and regularly lists upper-bound slopes for a great number of chemicals. In a model-free approach response frequencies in the low dose range are converted to their upper confidence limits and each of these is connected with the origo. The straight line with the minimum slope is used for low dose extrapolation.

5.4 *Specific Problems in Risk Estimation*

In risk assessments interaction between chemicals is rarely included, though there are examples which show that such an interaction can cause a substantial shift in risk. Thus it has been documented that cigarette smoking increased lung cancer 10.8 times, exposure to asbestos by 5.15 times, but in the population of cigarette smoking asbestos workers the frequency of lung cancer increased by 53.24 times. Thus for those carcinogens which may promote the effects of each other, the setting of exposure limits, be they threshold or acceptable risk limits, always leaves a question mark.

A similar problem is presented by compounds which accelerate the ageing process. Ageing is a normal physiological process, but due to exposure to certain chemicals it becomes evident earlier than usual. For a long time it was believed that carbon monoxide has only acute effects which are present if at least 20% of the haemoglobin is made unsuitable for oxygen transport. Recently it has become apparent that exposure resulting in less than this level is able to accelerate the ageing process of the cardiovascular system. A similar effect—though probably by a different mechanism—is exerted by another simple compound, carbon disulfide.

No model is available to predict the safe level of exposure if, as a result of antigen formation, antibodies are produced against the chemical. The reaction between the chemical and the antibodies in the sensitized individual can produce rashes (urticaria), allergic dermatitis, or asthma. Sensitization is quite different from hypersensitivity. In the case of sensitization the response differs from the normal toxic effect of the chemical and its severity depends much less on the level of exposure than on the degree of sensitization. In the case of hypersensitivity the reaction is the expected effect, but the dose response curve is shifted in the direction of lower doses.

6 Exposure as a Measure of Relative Hazard

6.1 *Toxicity and Exposure*

The main aim of toxicology testing is to establish how toxic the tested compound is in relation to other well known occupational chemical hazards. However toxicity is only one of the factors that determine the degree of hazard for occupational exposure. The probability that the toxic chemical will produce injury depends not only on its toxicity but also on the conditions of exposure. The results of the toxicity test already give the information as to whether the compound is irritative to skin or eye, whether it is a sensitizer and consequently if there is a need to protect the eyes with goggles and the hands with gloves.

Toxic effects after the dermal application of the compound will reveal the hazard of absorption from the contaminated skin. The difference in oral and parenteral LD_{50} values will indicate how hazardous it is to risk ingestion by smoking or having a meal in a contaminated environment. It is necessary to know the physical form of the toxic substance, whether it is solid, liquid, gas, or vapour. Liquids can be splashed or sprayed, solids might form dust, and gas, vapour, spray, and dust are the source of inhalation exposure.

6.2 The Problem of Complex Exposure

Exposure to a single chemical is not the rule but an exception. Exposure to more than one compound might not change the risk presented by each of them, but there are examples of additive, antagonistic, and synergistic effects. In the case of an additive effect, the critical organ or, in a broader sense, the critical mechanism is the same for the two compounds. Thus the effect of two narcotic solvents may add to each other. Contrary to additive effects, an antagonist decreases the toxicity of another compound. Thus methaemoglobin-forming agents act as antagonists against cyanide, azide, or hydrogen sulfide. Another interaction is synergism. A synergist potentiates the effect of another compound either by interfering with its metabolism or by interfering with a basic biochemical process on which the toxic action depends. An example for the first mechanism is the potentiation of the toxicity of malathion by another pesticide, EPN, and for the second mechanism is the increase in the behavioural effect of amphetamine by exposure to CS_2.

6.3 Physical Conditions and Other Factors in Relation to Exposure

The physical state, vapour pressure, and solubility of a chemical are the factors which determine the type of exposure. The volume of the chemical, dilution, temperature, and ventilation are the external factors which influence the degree of hazard. The knowledge of each of these factors is important in the assessment of possible danger to health.

When the toxic chemical is a gas or a vapour, inhalation is the most important route of uptake. However vapours can irritate the skin and gaseous hydrogen cyanide can penetrate the skin to such an extent that in spite of respiratory protection it can cause severe or even lethal intoxication. Fumes and solid particles floating in the air are also taken up by inhalation. Evaporation depends on the vapour pressure of the chemical and on its temperature. The higher the temperature the higher the rate of evaporation. Surface area is a very important factor—that is why spilled solvents evaporate so rapidly.

From the physical conditions of the chemical one cannot predict the volatility. Thus methylmercury chloride, which is a solid, is 57 times more volatile than the liquid metallic mercury. Compared with an undiluted chemical, evaporation is less from its solution, though in an enclosed space the final air concentration depends only on the vapour pressure at a given temperature.

Fumes in the laboratory mainly derive from acids and their formation is also temperature dependent (external heat or reaction heat). The grinding of solid chemicals, and all solid chemicals in powder form, can form solid aerosols. Ventilation of the premises decreases inhalation exposure by the replacement of contaminated air by fresh air, though too strong a draught can increase evaporation or the formation of solid aerosols (dusting of the chemical).

6.4   Working Processes as the Source of Exposure

A high proportion of exposure to toxic chemicals is of accidental origin. Fire promotes the evaporation of volatile chemicals and assists formation of toxic fumes (*e.g.* oxides of nitrogen) and toxic gases (carbon monoxide, phosgene). Breakage of glass containers and spillage of volatile liquids also results in increased evaporation.

Pipetting by mouth can result in swallowing of the toxic chemical, but if the chemical is volatile, vapours dissolve in the saliva or can be inhaled. Grinding of solids or measuring out light powders in a draught increases the danger of the formation of solid aerosols. Inhalation exposure is a likely consequence when chemicals are heated outside a fume cupboard or in a fume cupboard with inadequate ventilation.

In analytical work chemicals are used in small quantities and thus the potential hazard is reduced, though for many purposes solvents, acids, and alkali are used in large quantities without dilution. If one considers only the health hazards, the preparation of a stock solution is always more dangerous than the final analytical reagent.

Distillation or other purification procedures and synthesis are the working processes where larger quantities of toxic chemicals might be handled. All the known laboratory methylmercury poisonings affected chemists who synthesized the compound.

6.5   Exposure and the Route of Uptake

In the case of occupational exposure to toxic substances the exact dose is unknown and sometimes it needs a thorough examination even to identify the route of exposure. Intake is most frequently by inhalation followed by dermal absorption and occasionally by ingestion or the combination of these routes. The respiratory tract is by far the most important route by which toxic chemicals enter the body. If atmospheric contamination is responsible for the exposure, the difference in the concentration of the chemical in the inhaled and exhaled air multiplied with the respiratory volume and time gives the dose. However, without any attempt to calculate the exact dose, the atmospheric concentration can be related to one or more responses. This is the basis for RL (recommended limit) or CL (control limit), TLV (threshold limit value), or MAC (maximum allowable concentration) figures.

It is more difficult to obtain the measure of exposure when oral or percutaneous absorption is the route of uptake or at least contributes to the total exposure.

In occupational circumstances oral or percutaneous exposure are usually interlinked. Contamination on the hand is the usual source of oral uptake through contamination of food and cigarettes. Absorption through the skin depends on the physical and chemical qualities of the toxic substance, but the surface area contaminated and time allowed between contamination and decontamination are equally important factors. Inside contamination of rubber gloves presents the most favourable conditions for absorption: increased temperature, vapour pressure, sweating, and softening of the epidermis. Skin and the mucous membranes can be directly affected by the chemical. Chemical burns are usually caused by liquids.

In laboratories where the working procedures are alternating, it is unlikely that the occasional measurement of the atmospheric concentration of a chemical will give a truthful picture about the level of exposure. That is why it is important to judge whether appreciable exposure can exist at all. The awareness of the presence of toxic chemicals and of the conditions which might transform a potential hazard into an actual hazard is the best safeguard against harmful exposure. Nevertheless, in experimental plants or when a process is carried out for extended periods, *e.g.* synthesis, distillation, *etc.* it might be necessary to analyse the working environment for possible contamination.

There is no similar method of measuring absorption through the skin or through the gastrointestinal tract. If the chemical has a colour, discoloration of the skin can indicate the contact. Similarly the analysis of the skin wash may show how carefully the toxicants are handled. A very useful indicator of exposure is the concentration of a chemical or its metabolite in blood or urine samples. These biological tests give an overall picture of exposure independent of the route of entry.

6.6 The Permissible Limits of Exposure

The correlation between the atmospheric concentration of a toxicant and its health effects is the basis of exposure limits. The first comprehensive list of exposure limits, called Threshold Limit Values (TLV), was the result of the pioneering work of a non-governmental organization, the American Conference of Governmental Industrial Hygienists (ACGIH). TLVs set the pattern for other exposure limits, like Permissible Exposure Limits, Maximum Allowable Concentrations, and in the UK Maximum Exposure Limits (MEL) and Occupational Exposure Standards (OES). There is a substantial overlap between the different standards as they are all 8 h Time Weighted Average (TWA) concentrations and expected to prevent adverse effects in nearly all individuals exposed daily (8 h per day, 40 h per week) during the whole working life. Nevertheless it is important to read the exact definition of these and similar other terms because for example the difference between the short list of MEL and the long list of OES values published by the Health and Safety Executive

(HSE) is that the first is mandatory while compliance with the second depends on the urgency and cost of the necessary action.

Though for most toxic substances the 8 h TWA is adequate, the more acute is the effect of a substance the more dangerous is an upward excursion from the average. Thus for substances which exert their effects (irritation, sensitization, or acute intoxication) immediately or shortly after the onset of exposure, the TWA is replaced with a ceiling value. Examples in the ACGIH listed standards are hydrogen bromide, chloride, cyanide, or fluoride. For a second group of compounds upward excursion from the daily average concentration is limited by a 15 min (ACGIH) or a 10 min TWA concentration (HSE). Thus for acetone the TLV is 750 p.p.m., while Short-Term Exposure Limits (STEL) are 1000 p.p.m. and 1500 p.p.m. respectively. When the induction of a toxic effect requires repeated long-term exposure, the reference period of averaging can be increased. Thus the technical reference concentrations (TRC) for carcinogenic substances in Germany and the European Community's directive for vinyl chloride refer to an averaging time of one year. For proven carcinogens with high potency no exposure is the only satisfying solution. Thus in the UK the manufacture and use of β-naphthylamine, benzidine, 4-aminodiphenyl, and 4-nitrodiphenyl is prohibited.

6.7   Biological Monitoring

The estimation of exposure frequently requires prolonged and complex environmental assessment. When intake of the toxic substance is not restricted to inhalation, biological monitoring, that is the estimation of the concentration of the toxic agent in biological samples, has many advantages over environmental tests because it uses the exposed person as a sampling device. Biological monitoring uses not only direct tests, that is the estimation of the toxic substance or its metabolite in blood, urine, hair, or exhaled air, but also indirect tests which detect biochemical effects sensitive to exposure below or around the accepted TLV or MAC value. The demarcation line between direct and indirect tests is not always clear. For example, the estimation of carbon monoxide (CO) in blood is clearly a direct exposure test. However blood CO is associated with haemoglobin and the estimation of this abnormal blood pigment (COHb) fits into the category of indirect tests. In addition to being both a direct and indirect test, the estimation of COHb is also a diagnostic test because the impairment of oxygen transport through the formation of COHb is the essential step responsible for the toxic effects of CO. Though periodic medical examination does not fit into the category of biological monitoring, it may reveal the transgression of exposure above the acceptable limit, especially when employees are screened for the possible health effects of their particular exposure. Thus urinary porphyrin, an abnormal by-product of haemoglobin synthesis, and anaemia are indicators of toxic exposure to lead.

Procedures for the identification of individuals exposed to carcinogens exist but their practical application is not yet at hand. Mutation, aberrant chromosomes, or DNA synthesis in lymphocytes (one form of white blood cells) are non-specific biological markers of exposure to possible carcinogens. There are also specific methods for the detection of macromolecular adducts with electrophilic carcinogens (*e.g.* alkylating agents). The DNA is the obvious test material but the removal of the reaction products by normal repair processes gives some advantage to haemoglobin over DNA. Both the non-specific and specific methods are exposure markers without being the indicator of cancer induction.

7   The Consequences of Toxic Exposure

The symptoms and signs of intoxication are diverse and are neither always specific for, nor characteristic of, a toxic chemical. The following discussion might help to illustrate how the body responds to toxic chemicals. It also helps the understanding of the meaning of toxic effects in relation to the listed chemicals in the Hazardous Substances section.

7.1   The Respiratory Tract and the Lungs

An inhaled toxic substance may have its critical effect on the respiratory tract, or it may exert systemic effect, or both. The localization of local effect depends on deposition. Particles of carcinogenic metals of 0.2 to 0.5 μm diameter most likely cause tumours in the deep lung, while exposure to 2–3 μm particles which are predominantly deposited in the conducting airways leads to the formation of tumours in the middle part (bronchus) of the respiratory tract. The rapid dissolution of water soluble SO_2 or formaldehyde in the watery layer of the upper respiratory tract results in an immediate irritative effect on the nasal passages. The less water soluble nitrogen dioxide usually causes only mild discomfort but after a latent period, which may be 48 h, lethal pulmonary oedema (retention of water in lungs) may develop. Lung oedema and respiratory irritation have a pronounced detrimental effect on gas exchange. If the respiratory irritation is too severe, the lumen of the bronchial system constricts and increases the resistance to the respiratory airflow. Irritative chemicals such as sulfur dioxide, sulfuric acid mist, ozone, formaldehyde, phosgene, bromine, methyl bromide, and nitrogen dioxide are all able to increase the respiratory resistance and many are able to cause oedematous changes in the lungs.

The bronchial system has a mechanism which removes particles from the lungs and any damage to this escalator promotes sensitivity to infections or dusts. These secondary effects with the primary irritant effect can lead to irreversible structural changes in the lungs manifested by a decrease in the surface area available for gaseous exchange.

7.2   Skin and Eyes

In the laboratory environment contact of the skin with chemicals is a frequent though not an unavoidable risk. Compounds that irritate the skin or mucous membranes in the vapour form can cause more severe damage if splashed on the skin. Irritation is a general term which covers a multiplicity of processes. For example the removal of the lipid layer by solvents or alkaline detergents; dehydration of the skin by acids, anhydrides, and alkalis; precipitation of proteins by heavy metals; oxidation by acids, peroxide, or chlorine; the dissolution of keratin by soap, alkalis, or sulfides. If the contact lasts longer or is repeated frequently the injury called contact dermatitis becomes more and more severe.

There are so-called sensitizers which have the same effect as the primary irritants but the first signs appear only after 5–7 days exposure. Chromate, formalin, phenylenediamine, and turpentine belong to this group.

Halogenated compounds act on the skin where there is discontinuity: chloracne is composed of plugged sebaceous glands and suppurative inflammation of the hair follicles. There are substances such as coal tar, creosote, Rhodamine N, and bergamot oil which sensitize the skin to sunlight.

Skin can be affected by a chemical taken orally or by other routes. Thus thallium results in the loss of hair, and arsenic the proliferation of the keratin layer.

Vapours that irritate the skin are more likely to irritate the conjunctiva or other mucous membranes. The effect on conjunctiva or cornea is more pronounced when the irritant is rubbed into the eye with a dirty hand, or when acids, alkalis, lime, solvents, or detergents are splashed into the eye. Acids coagulate and alkalis dissolve the corneal epithelium. Visual function can be permanently impaired by scarring, vascularization, and opacification. Solvents can also cause loss of corneal epithelium, but the injury regenerates completely in a few days.

Chemicals can affect not only the surface of the eye; chronic exposure to dinitrophenol is able to cause opacity in the lens and thallium or methanol can damage the optic nerve.

7.3   *Gastrointestinal Tract and Liver*

The gastrointestinal tract can be the route of uptake of a toxic chemical, but it can also be the site of action if corrosive chemicals are swallowed. Every chemical that can cause chemical burns on the skin, can injure the oesophagus and the stomach. Bleeding, perforation, and deformation are the outcome of such a contact effect.

Other chemicals like arsenic, barium, chloromethane, fluorides, or mercuric chloride lead to gastroenteritis and increased motility of the intestines.

The liver is the first organ exposed to a chemical absorbed from the gastrointestinal tract. However, its vulnerability is based more on the extraordinary metabolic activity of the liver cells which can structurally modify many foreign chemicals. If the chemical is detoxified, the liver is the last organ to be exposed to the more toxic parent compounds; if the change is in the opposite direction, the liver is the first organ to be exposed to the more toxic forms.

The uptake of non-polar chlorinated hydrocarbons by the liver is facilitated by their lipid solubility, but their toxicity mainly depends on the metabolic transformation with the formation of highly reactive metabolic products.

Chemicals can injure the liver in many ways; they can increase its water and fat content and they can destroy the cells. The liver has the ability to regenerate by replacing dead cells, but repeated toxic injury results in disorganization of liver architecture by scars of fibrous sheath. Owing to the loss of metabolically active liver cells and inadequate blood flow normal physiological functions are impaired. These functions contribute to the general metabolic processes of the whole organism (lipid, protein, and sugar metabolism), the formation of the bile and the extraction of chemicals from the plasma for bile secretion. Hepato toxic agents, besides causing morphologic changes, can depress the metabolic activity of the liver, the bile flow, and the excretion of bilirubin and other compounds. Bilirubin is a metabolic product of haem, the active part of haemoglobin and different haem enzymes. Approximately 300 mg bilirubin is formed per day (more when haemolysis occurs) and a failure of its biliary excretion due to 1,1,2,2-tetrachloroethane, carbon tetrachloride or phosphorus can cause severe liver damage resulting in jaundice. Nearly all the solvents have the potential for causing some degree of liver injury which might be noticed only if enzymes, leaked from the liver cells into the blood, are estimated or some functional tests are carried out. If the liver once suffers chemical damage or an infective disease, its threshold to the effect of hepatotoxic agents is decreased.

7.4   Blood

The effect of a chemical on blood can be direct or indirect. In the former case the chemical acts on the blood itself and in the latter case on the formation of the formed elements or the plasma. The formed elements, red blood cells, white blood cells, and platelets, are suspended in a volume ratio of 1:1 in the plasma, which is an aqueous solution of proteins, amino acids, salts, and other small molecules. The main function of 4.5–5 million red blood cells per μl is the transport of oxygen from the lungs to the tissues. This function is carried out by a special protein called haemoglobin which makes up 97% of the dry weight of these cells. The number of white cells is very much smaller, only 7000 per μl, their function is the defence against infection. The smaller platelets, which number 300 000 per μl, contribute to another defensive mechanism: blood coagulation.

One of the most common effects of toxic chemicals on blood manifests itself in a change of the number of formed elements. There are so-called haemolytic poisons, like arsine, phenylhydrazine, or a large group of aromatic amino and nitro compounds which are able to haemolyse the blood. In the case of arsine the effect can be so dramatic that the large quantity of haemoglobin released from the cell aggravates the effect of arsine on the liver and kidneys. Anaemia caused by lead is the consequence of its interference with the synthesis of haem, *i.e.* the non-protein part of haemoglobin. The first block in the synthesis of haem results in the increased urinary excretion of δ-aminolaevulinic acid, the substrate of one of the inhibited enzymes. Other faults in haem synthesis result in the urinary excretion of coproporphyrin and the accumulation of zinc protoporphyrin in red blood cells. The estimation of these three compounds is widely used in both clinical diagnosis and the biological monitoring of lead exposure.

The aromatic amino and nitro compounds are able to oxidize the ferrous iron of the haem part of haemoglobin. This causes, through the formation of an inactive haemoglobin called methaemoglobin, a deterioration in the supply of oxygen to the tissues. Sometimes methaemoglobin formation is useful, as methaemoglobin competes for cyanide with essential ferric enzymes engaged in the utilization of oxygen.

One of the most common forms of inactive haemoglobin is carboxyhaemoglobin, the reaction product of carbon monoxide and haemoglobin. The inactivation is based on the affinity of carbon monoxide for haemoglobin which is approximately 240 times higher than that of oxygen. Carboxyhaemoglobin is fully dissociable and when the carbon monoxide concentration decreases in the inhaled air, carbon monoxide is replaced on the haemoglobin by oxygen. The reactivation of methaemoglobin needs the participation of reductive enzymes present in the red blood cells.

The best known toxic chemical affecting the white blood cells and platelets is benzene. Benzene can shift their numbers in two directions. At first the numbers of white blood cells and platelets decrease, as after ionizing radiation. The most common manifestation at this stage is leakage of blood from the capillaries even after minor trauma. If the exposure is high and long lasting, the effect of benzene can move in the opposite direction resulting in the proliferation of the tissues which produce the white blood cells.

Cell death caused in any part of the body, but mostly in the liver, can leak enzyme proteins into the plasma, and thus their activity in blood can be used as a measure of tissue damage. Chronic liver damage caused by alcohol or other solvents might result in a disorder in the synthesis of plasma proteins with, as the first sign, a decrease in the albumin concentration. A general decrease in the concentration of plasma proteins might be the consequence of loss of proteins through the kidneys.

7.5   Nervous System

The nervous system is divided into two parts: the central and the peripheral. The central part is the brain and the spinal cord; the peripheral part consists of the motor and sensory nerves. The uptake of chemicals by the central nervous system depends on the vascularization of the brain and the characteristics of the chemical. Non-polar, lipid soluble compounds, such as halogenated solvents easily penetrate the barrier between blood and brain tissue. Their effect, like that of chloroform, is anaesthetic-narcotic, that is they depress the function of the central nervous system. This, in severe cases, leads to respiratory and/or cardiac failure. Though the biochemical mechanism is quite different, the general effect of anoxaemic poisons is very similar to the narcotic ones. Chemical anoxaemia is caused either by the failure of the oxygen transport in the blood by the inactivation of haemoglobin (carbon monoxide, aromatic amino, or nitro compounds) or by the inhibition of the utilization of oxygen in the tissues (cyanide, azide).

Though it is a solvent, carbon disulfide has a more selective effect on the central nervous system than other solvents. It interferes with the metabolism of catecholamines, which belong to those compounds which participate in the transmission of an impulse from one nerve cell to the next. This effect is more pronounced in a certain part of the brain, which is affected also by manganese. Cerebral oedema caused by triethyltin, encephalopathy caused by lead, ataxy, blindness, deafness, *etc.* caused by methylmercury, and behavioural disorders and tremor caused by mercury vapour are based on more selective interactions between the toxic chemical and biochemical processes in the brain.

Many of the chemicals like carbon disulfide which affect the central nervous system also affect the peripheral nerves. Central nervous symptoms like drowsiness and headaches are the consequence of mild intoxication with anticholinesterase organophosphorus compounds, but some of them are able to cause delayed peripheral neuropathy in the form of paralysis. Tri-*o*-cresyl phosphate though it is not a potent anticholinesterase inhibitor caused a mass epidemic of paralysis in Morocco when engine-oil treated with tri-*o*-cresyl phosphate was sold and used as cooking oil. Acrylamide, a different type of compound, injures not only the motor but also the sensory nerves. Contamination of the skin with phenol results in a chemical burn which may go unnoticed because of the concurrent anaesthetic effect.

7.6 Kidneys

One of the main functions of the kidneys is the excretion of metabolic products (mainly from protein metabolism) and the regulation of the salt and water balance. Both processes are linked to the formation of urine which is also an important route for the clearance of toxic compounds and their metabolites. During the process of urinary excretion the chemicals filter, diffuse, or are transported through cells in the kidneys, but some remain temporarily bound to cell components.

Heavy metals such as mercury, or chlorinated hydrocarbons like carbon tetrachloride, are the most prominent groups of compounds which can damage the kidney cells. The consequence is either a change in the composition or in the volume of urine or both. For example, in severe cases of mercuric chloride intoxication there might be no urine at all, in milder cases the occurrence of cell debris and protein in urine indicates renal damage.

Heavy metals like cadmium, mercury, uranium, and lead are also able to increase the excretion of amino acids, glucose, or phosphate as a result of an impairment in the function of tubular cells to reabsorb these small molecules from the preformed urine. As the half time of cadmium is 10–40 years in the kidneys, even a short-term exposure to cadmium will increase the kidney concentration for life. Without occupational exposure cadmium is taken up only from food and the cadmium content of the kidneys increases up to 50–60 years of age. Then the ageing process probably results in loss of kidney cells with their cadmium.

Excretion of toxic compounds with urine is a powerful protective mechanism. In some cases the concentration of toxic chemicals in the urine exposes the lower part of the urinary tract to higher concentrations than from the blood. 2-Naphthylamine and benzidine proved to be so potent in causing bladder tumours by this way that their use in the UK is now prohibited.

7.7   Cardiovascular System

Heart or respiratory failure is the cause of death in many acute intoxications. Chemicals such as carbon monoxide, phosgene, nitrogen dioxide, hydrogen sulfide, chlorinated compounds, or solvents, which can injure cells in other organs can also injure the heart. In the not too severe cases of anoxaemia caused by carbon monoxide, cyanide, or methaemoglobin-forming agents, the toxic effect is shown only by a decrease in blood pressure or an increase in the frequency of heart beat.

Ageing of the cardiovascular system is one of the cardinal processes of ageing in general. Lead, carbon monoxide, or carbon disulfide are able to speed up atherosclerosis and increase the possibility and severity of coronary disease. Fluorocarbons, trichloroethylene, and some other halogenated alkanes can produce, in addition to nausea and dizziness, irregular heartbeats. Heart failure may be precipitated after a symptom-free period by physical stress. Exposure to organic nitrates, like nitroglycol and nitroglycerine, results in skin flushes and throbbing headache. These vasodilatory effects subside and disappear when exposure is repeated regularly, but after a few years the interruption of the exposure, that is the daily supply of vasodilators, precipitates chest pain and even sudden death.

7.8   Bone and Muscles

The whole body is supported by the skeletal system. Loss of calcium by exposure to cadmium, when aggravated by nutritional deficiency, can result in the softening and deformation of the bones. The compression of nerves by deformed bones can be painful. In a district in Japan where river water contaminated with cadmium was used for irrigating paddy fields this disease (called Ouch Ouch) reached epidemic proportions. In contrast to cadmium, fluoride and phosphorus can increase the fragility of bones. Interestingly, lead, which is mainly stored in the bones, has no similar effects.

In severe carbon monoxide intoxication the loss of muscle power may trap the victim at the site of exposure, as he will be unable to move.

8   Bibliography

M. O. Amdur, J. Doul, and C. D. Klaassen (eds.), 'Casarett and Doull's Toxicology. The Basic Science of Poisons', 4th Edn., Pergamon Press, New York, 1991.

American Conference of Governmental Industrial Hygienists, Threshold limit values for chemical substances and physical agents, 1990–1991.

D. B. Clayson, I. C. Munro, P. Shubik, and J. A. Swenberg (eds.), 'Progress in Predictive Toxicology', Elsevier, Amsterdam, 1990.

R. B. Conolly and M. E. Andersen, *'Biologically Based Pharmacodynamic Models. Tools for Toxicologic Research and Risk Assessment'*, Ann. Rev. Pharmacol. Toxicol., 1991, **31**, 503–523.

P. Grasso, M. Sharratt, and A. J. Cohen, *'Role of Persistent, Non-genotoxic Tissue Damage in Rodent Cancer and Relevance to Humans'*, Ann. Rev. Pharmacol. Toxicol., 1991, **31**, 253–287.

Health and Safety Executive, *'Occupational Exposure Limits 1991'*, WH40/91.

J. Huff, J. Haseman, and D. Rall, *'Scientific Concepts, Values, and Significance of Chemical Carcinogenesis Studies'*, Ann. Rev. Pharmacol. Toxicol., 1991, **31**, 621–652.

D. Krewski, D. Gaylor, and M. Szyszkowicz, *'A Model-free Approach to Low-dose Extrapolation'*, Environ. Health Perspect., 1991, **90**, 279–285.

F. R. Lu, 'Basic Toxicology. Fundamentals, Target Organs, and Risk Assessment', 2nd Edn., Hemisphere Publ. Corp., New York, 1991.

R. O. McClellan and R. F. Henderson (eds.), 'Concepts in Inhalation Toxicology', Hemisphere Publ. Corp., New York, 1989.

National Research Council, 'Pharmacokinetics in Risk Assessment. Drinking Water and Health', Vol. 8, National Academy Press, Washington, 1987.

M. V. Roloff (ed.), 'Human Risk Assessment. The Role of Animal Selection and Extrapolation', Taylor and Francis, London, 1987.

J. F. Stara and L. S. Erdreich (eds.), 'Advances in Health Risk Assessment for Systemic Toxicants and Chemical Mixtures', Princeton Sci. Publ., Princeton, New York, 1985.

R. G. Tardiff and B. D. Goldstein (eds.), 'Methods for Assessing Exposure of Humans and Non-human Biota', SCOPE Vol. 46, John Wiley and Sons, Chichester, 1991.

J. A. Timbrell, 'Introduction to Toxicology', Taylor and Francis, London, 1989.

V. B. Vouk, G. C. Butler, D. G. Hoel, and D. B. Peakall (eds.), 'Methods for Estimating risk of chemical injury: human and non-human biota and ecosystems', SCOPE Vol. 26, John Wiley and Sons, Chichester, 1985.

CHAPTER 7

Control of Health Hazards

D. M. SMITH

Physical hazards at work are relatively easy to understand and precautions often obvious, for example, guarding machinery or keeping flames away from flammable liquids. Hazards to health are less familiar, but can be eliminated or controlled by a similar simple hierarchy of controls and checks.

Laboratory managers should be aware of what chemicals are used and stored and their toxic properties and routes of absorption. This may appear a daunting task, but in most cases information is readily available from suppliers, in published literature, and in Government publications. The Hazardous Substances section of this book gives a valuable summary of information on many laboratory chemicals, and gives some other references.

In the UK, the Health and Safety Executive (HSE) publishes an annual list of occupational exposure limits, Guidance Note EH40 1992.[1] It contains lists of maximum exposure limits (MELs). These limits must not be exceeded and exposure should be kept as low as reasonably practical. There is a longer list of occupational exposure standards, to which a less stringent standard applies; it is sufficient to keep exposure below the limit or to show why it has been exceeded and remedy the situation as soon as reasonably practicable.

The American Conference on Government Industrial Hygienists (ACGIH) gives a list of threshold limit values (TLVs)—values which may be safely tolerated for up to 8 hours daily (40 hours per week) by most people, which is widely used by enforcement agencies.[2]

The USSR, Germany, and other EC Countries, use maximum allowable concentration, again for 8 hour exposures.

All these relate to environmental measurements which can only give guidance about the risk of inhalation. Many chemicals are absorbed through the skin or by ingestion and these measurements take no account of these routes. Laboratories use many chemicals which are toxic by ingestion. Standard precautions such as good personal hygiene, prohibition of mouth pipetting, and avoidance of contamination of surfaces and eating, drinking, and smoking at work will prevent ingestion.

In the UK the Health and Safety at Work Act[3] places obligations on employers to ensure safe working conditions. Controls of chemical hazards are spelt out in more detail in the Control of Substances Hazardous to Health Regulations 1988 (COSHH).[4] The mandatory steps to be taken are as follows.

1 Assess the Hazards

- List the chemicals used and their toxic properties;
- Consider the method of use—how much, how often, by whom, in what degree of containment;
- Is the chemical
  —volatile?
  —respirable (gas, vapour, dust)?
  —absorbed via intact skin?
  —irritant or corrosive?
  —known to cause sensitization—either skin or respiratory?
  —carcinogenic?

(See Figure 1.)

2 Eliminate or Control the Hazards

Always consider eliminating toxic or carcinogenic substances by substituting a less toxic one or by changing the process or procedure using it. For example, toluene or xylene may be used in place of benzene, which is known to cause aplastic anaemia and leukaemia. Where this cannot be done, controls will vary with the type of substance, route of absorption, and method of use. They will include the use of fume cupboards and remote handling methods. Only if these are insufficient should personal protective equipment be considered. There will, however, be cases where skin absorption or irritation is important; gloves and protective overalls will then be needed.

3 Monitoring

Environmental monitoring may be needed as part of the assessment to establish risk and degree of risk, or as part of the control—in order to make sure that controls are working.

This does not mean that environmental hygiene tests are always needed. It is often enough to test extraction systems, *e.g.* by a smoke emitter, to see that they are working. One of the silliest laboratory accidents I have seen resulted in quantities of chlorine being evolved in a classroom, from failure to try the switch on the fan before starting the experiment in the fume cupboard. Such precautions should become a routine part of the procedure.

In assessing the laboratory environment, it may be prudent to make simple environmental tests for the presence of common solvents and any other toxic substances habitually used. Their presence at levels above prescribed levels (OES, MEL, TLV, or MAC) would indicate the need for a more detailed survey, which should be carried out by a person qualified in occupational

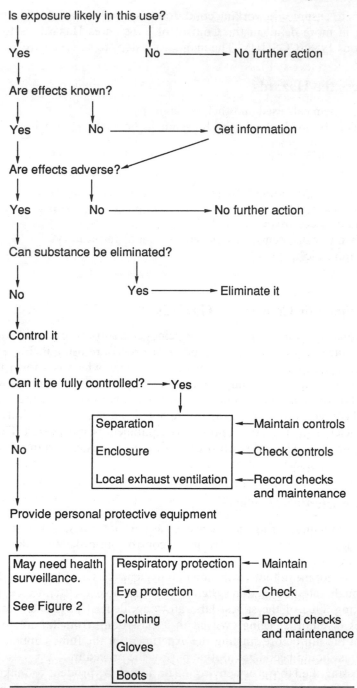

Figure 1 *Assessment of a substance hazardous to health*

hygiene. This will deal only with inhalation exposure, and biological monitoring may be needed in order to assess absorption by other routes.

Biological monitoring may also be used in assessment of exposure, *e.g.* to lead; blood lead levels of less than 40 μg dl^{-1} after exposure would indicate no significant absorption and no need for continued surveillance, whereas levels of 40 μg dl^{-1} or more would point to a need for greater control and a need for continued health surveillance.

4 Health Surveillance—Exposure, Effects, and Techniques of Detection

Health surveillance does not always mean medical examination. It may or may not include biological monitoring. The aim is always to benefit employees and self-employed workers by ensuring that the work place is safe for each of them.

To medically examine everyone in a general way is wasteful of money and medical time and does not lead to increased protection. Any health surveillance should be specific, and targetted to the exposure (see Figure 2). At its simplest, it can consist of keeping records of exposure of each person exposed, so that these can be referred to in any case of illness which might be casually related to exposure. This may be the only health surveillance possible with some carcinogens. The COSHH Regulations require such records to be kept whenever health surveillance is used and its Approved Code of Practice (ACOP)[4] specifies what it should contain.

It should contain as a minimum:
- identification details;
- date of start of present employment;
- historical record of jobs involving exposure in this employment.

If other health surveillance is carried out, it should also note the dates:
- by whom carried out;
- conclusions, *i.e.* fitness or otherwise for his/her job.

It should not contain confidential, clinical data. In other words it is a record of:
- (a) identification;
- (b) exposure;
- (c) fitness for the task.

Fitness means medical suitability for the particular exposure, *e.g.* a person suffering from asthma should not be exposed to irritant or sensitizing chemicals such as isocyanates, even in very small amounts, but a paraplegic in good respiratory health could work with these substances providing exposure was kept below the statutory limit.

The next stage in health surveillance is to ask questions about general health. Simple questions as to the presence or absence of symptoms or known disease of the lungs or skin will indicate those people who should be referred for a professional opinion. Respiratory questionnaires and lung function testing, when needed in the case of sensitizers, should be carried out in standardized form by health professionals. Skin inspection is used in respect of those who

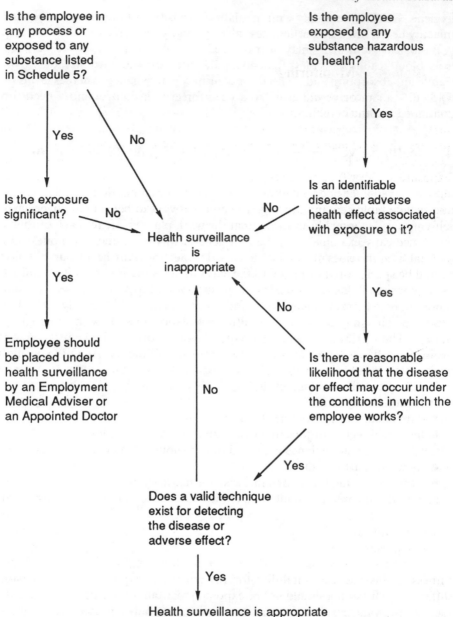

Figure 2 *Where health surveillance is appropriate*
(Reproduced by permission HMSO.)

work with chrome salts, to detect ulceration; it may also be used to monitor the occurrence of dermatitis from a variety of exposures.

The OELs are set at levels thought to prevent sensitization, but those already sensitized may react to much lower levels and may need to be removed from even minimal contact. For example, people sensitized to di-isocyanates have reacted to the small quantities emitted by a neighbouring factory.

5 Biological Monitoring

Biological monitoring will give the only true measurement of absorption, which may result from a combination of inhalation, ingestion, and skin absorption. It may be a direct measurement of a substance or metabolite, *e.g.* blood lead, urinary mandelic acid after styrene exposure, or a measurement of biological effect such as the lowering of acetylcholine esterase through organophosphorus exposure, or liver function tests for solvents. Blood, breath, and urine samples may be tested. The ideal biological monitoring test would be non-invasive, convenient, specific, and sensitive. Unfortunately such tests are available for only a few substances and no biological monitoring tests are available for a great many.

It is not sufficient merely to measure the amount of the substance in blood or urine—this has to be compared with levels known to occur in non-exposed people, and with levels known to cause toxicity. If these data are absent or not known, measurements may simply cause needless anxiety in workers and management. In other words, results are worse than useless when their significance is not known. This may sound obvious, but I have known such measurements to be taken more than once. In the case of new tests, full co-operation with a good clinical laboratory will establish baselines before any reliance is placed upon test results.

The desirability of biological monitoring will depend not only on the properties of the substance, but on the circumstances of exposure, whether daily, regular, infrequent, or occasional. Any biological monitoring undertaken should be under the supervision of an occupational health doctor. He/she need not carry out all the tests personally, but should be consulted when decisions are needed as to whether an individual should start or continue to be exposed to a particular hazard.

If, for example, only records are kept, they should be scrutinized from time to time to determine whether there is increased sickness absence, even deaths, from a particular cause. When questionnaires disclose symptoms, or biological monitoring or lung function tests show signs which might be related to exposure, a full medical evaluation by a qualified medical adviser is invaluable. Many symptoms and signs are common to illnesses caused by a variety of agents, for example, headaches may well result from infection, carbon monoxide, or solvent intoxication. This is where expert medical help is valuable.

6 Occupational Health Services

Most undertakings in the UK are still without occupational health services. However, many large firms and most National Health Service staff have in-house services, and group services and commercial agencies provide help to smaller firms. The Employment Medical Advisory Service (EMAS) has doctors and nurses in each of 20 area offices in the Health and Safety Executive. Although they cannot undertake day-to-day services, their statutory duties include investigation of poisoning incidents and other ill-health effects, and supervision of statutory medical surveillance. They can provide advice on how to set up or organize an occupational health service, on first aid, and on specific problems. EMAS can provide a list of local doctors suitably qualified to undertake occupational health surveillance.

It is helpful for managers to know what they may expect from an occupational health service:

(a)   It should provide advice on health risks and should be consulted before introducing new materials or procedures so that the extent of precautions needed can be evaluted and the best way of proceeding decided upon.

(b)   It should institute and maintain systems of health surveillance when necessary, advising on the type and extent of these and consulting with managers about who is to be included. Needless surveillance of those only marginally exposed can lead to complacency and carelessness, as well as adding to expense.

(c)   It should carry out pre-employment, pre-placement, or interval checks when needed.

(d)   It should advise on health problems arising at work, whether caused by work or not. For example, managers often need advice about the employment or return to work of people with disabilities, or common illnesses such as diabetes and heart attacks. The service should maintain medical confidentiality, while advising on the capacity of the individual for the particular task.

(e)   When there is an occupational health service, its staff should be responsible for arranging first aid cover and purchase of stocks. They should ensure that first aid is of a standard to meet any national, state, or other legal requirement.

(f)   It should liaise with local hospitals to ensure adequate emergency treatment at work and hospital.

(g)   If statutory health surveillance is required, the doctor may apply to be appointed under the lead, asbestos, ionizing radiation, or COSHH Regulations, to carry out examinations under the general supervision of EMAS.

Some laboratories may wish to carry out additional procedures. For example:

(h)   Arranging confidential counselling for health problems.

(i)   Arranging promotional activities dealing with factors such as smoking, alcohol, and stress and providing life-style counselling and exercise programmes. As occupational health risks diminish, these become more important features because of the effect of these factors on the health of people at work. For example, alcohol will potentiate the effects of solvents.

7   First Aid

7.1   Training

Appropriate training must be given by a person or organization approved for the purpose by the Health and Safety Executive and the Approved Code of Practice (ACOP)[5] describes the syllabus for general and specific first aid training. This ACOP should be available to all laboratory managers. In the UK, training for first aid at work by the St John Ambulance Association, St Andrews Ambulance Association, and the British Red Cross Society is available in all areas. Other organizations also offer training; these include group occupational health services and some large firms, as well as some technical colleges.

The new ACOP lays more emphasis on risks at work than on numbers of employees and it is certainly advantageous to have trained people available. Not only may this be life-saving in emergencies, but far more commonly, will prevent minor injuries from becoming serious and save time spent in unnecessary medical consultation.

In most countries similar advice and books are available, for example, 'Standard First Aid and Personal Safety'—American Red Cross,[6] 'First Aid Manual'—Industrial Non-Pharmaceutical Chemical Substances.[7]

First aid is designed to save life in emergency, to take charge, and render aid until medical aid can be obtained. In addition, at work, first aiders undertake treatment of minor injuries which would be treated at home without reference to a doctor or nurse. Their normal training does not, however, include any instruction on the use of medications, whether tablets, medicines, or external applications. They should not, therefore, be asked to prescribe or administer medications, which have no place in emergency treatment. However, such training may be advisable in the few cases where an immediate remedy for poisoning or skin burns is available and may save life or prevent damage to tissue. In the latest UK ACOP,[5] hydrofluoric acid—danger of burns—and cyanides—danger of poisoning—are singled out as examples of this class. In addition, where there is need for oxygen as an adjunct to resuscitation, appropriate training is necessary.

7.2   Selection of First Aiders

Choose people who are likely to remain calm and calm others; people who can leave their post in an emergency (not teachers or instructors who could not safely leave a class in a laboratory). They should be able to convey information to emergency services, hospital, or occupational health department and be quickly available on site.

7.3   Equipment

Boxes must be marked according to national regulations (in the UK green cross on a white background). They should:

- contain only first aid items;
- be kept at hand;
- be kept clean;
- contain contents as listed.

Supplements:
- cyanide kit should be included if necessary (Kelocyanor or equivalent);
- calcium gluconate gel—for hydrofluoric acid burns;
- oxygen.

7.4   Eye Washing

All chemical laboratories should have adequate eye-wash facilities on the spot. It is worth noting that medical advice on this has recently changed and it is no longer acceptable to store unsterile water for this use. Use clean, running tap water. If this is not readily available, use sterile disposable bottles of water or normal saline, containing at least 300 ml. Do not use eye baths or eye cups; they may retain the chemical in contact with the eye and are in any case too small and slow for efficient dilution and elimination of the chemical. For larger sites, special mains-fed eye-wash stations are available commercially. Systems which store non-sterile water should be avoided.

7.5   Eye Protection

Regulations on eye protection vary in different countries. Eyes should be fully protected from chemical splashing by shields, the use of remote handling, or if necessary, protective eye-wear. This should be to the relevant British, DIN, or other standard for chemicals. Protective glasses can be made up to individual prescriptions for regular users. Side shields should be used.[8] Face protection may be needed as well.[9]

7.6   Washing Facilities

Douche showers are available for emergency decontamination, and should be installed where there is a hazard of gross contamination with corrosives.

8   Skin Hygiene

It is important to have easily accessible hand-washing facilities separate from laboratory sinks and with hot and cold water, soap, and towels. One of the hazards of laboratory work is irritant eczema/dermatitis from frequent hand washing without adequate drying. A simple emollient after-work cream may be provided. Barrier creams may facilitate the removal of some contaminants such as resins but should not be relied on to give full protection.

9 Dermatitis

'Dermatitis' is the commonest occupational disease in many countries, causing discomfort, disability, and sometimes permanent loss of employment. It can result from contact with substances in 2 main ways—irritation and allergy. Proper diagnosis is essential because the advice given about employment will differ.

9.1 Irritant Contact Eczema

Many substances will cause drying and de-fatting of the skin which reduces its defences and allows penetration of contaminants. This will happen to most people, although some are more vulnerable than others, those with fair or red hair often being more sensitive than those with darker skins. Constant washing without adequate drying will also leave the skin vulnerable to irritants. Care should, therefore, be taken to minimize skin contact with any irritant or solvent, by no touch techniques or by protective gloves and clothing. Any spillage should be removed by copious washing with clean tap water, continuously for at least 10 minutes.

9.2 Allergic (Hypersensitivity) Eczema

Some substances are prone to set up an allergenic response, *i.e.* one contact causing the body to manufacture antigens which then react with the provoking substance on further contact. Not everyone will react in the same way, some individuals being more vulnerable than others. It is, however, not possible to predict who will become sensitized so it is necessary to protect everyone. Laboratory chemicals which commonly produce sensitivity include:

p-phenylenediamine
nickel
chrome
amines—especially rubber accelerators
epoxy resins
phenol–formaldehyde resins
pharmaceuticals, *e.g.* antihistamines
1,2-diaminoethane.

As with respiratory sensitizers, it is important that those sensitized should not be exposed again, as very small exposure can cause severe dermatitis. For this reason, people with these allergies should not be employed in tasks which will expose them.

In addition, some substances can cause sensitization to ultraviolet light (sunshine). These include paraphenylenediamine dyes as well as many plants and extracts. Of these, oil of bergamot is well known, causing 'Berloque Dermatitis'—typically at sites where perfume is applied. They contain psoralens which sensitize the skin to ultra-violet light, causing in effect small burns.

This will happen to everyone exposed, unlike allergic eczema. Laboratory workers monitoring pesticide residues may be exposed to such plants.

It is not easy even for doctors to sort out the causes of eczema and dermatitis. All cases should be seen by a doctor, and if necessary referred to a dermatologist or occupational physician. Specialized testing may be needed; patch testing should be undertaken only under supervision or by someone experienced in its use, because inappropriate dilution or timing will affect the result and irritant reactions may cause confusion.

It is important to distinguish because one can continue working with irritants provided adequate precautions against overexposure are introduced and maintained, but it is unwise to continue once sensitized, as small exposures can set up serious reactions.

10   Training and Information

Everyone using the laboratory should have sufficient information and training about health hazards to safeguard his or her health. The degree of training will, of course, vary with the status and tasks performed. It is important to remember maintenance and cleaning staff, who should be safeguarded from substances intentionally or accidentally present in pipelines and other systems, as well as spillages. I have known a maintenance engineer suffer a severe asthma attack from the unexpected presence of di-isocyanate in a pipeline; he knew he was sensitized and wore breathing apparatus when he knew he was dealing with this chemical.

In the UK, guidelines from the Department of Education and Science[10] are applicable to students and pupils. Teachers and instructors have a valuable opportunity to make safety a part of all procedures taught, with the reasons for it. It is nearly half a century since my chemistry teacher said 'I have a great respect for hot, strong sulfuric acid', but his Scottish tones still ring in my ears. Familiarity too often leads to cutting corners—safety considerations should be a natural and integral part of any operation.

11   Summary

Control of health hazards entails consideration of:
   (a)   Substances used;
   (b)   Methods of work;
   (c)   Health surveillance, including biological monitoring;
   (d)   Provision of first aid;
   (e)   Washing facilities, including eye-wash facilities;
   (f)   Training and supervision.
It does not entail untargetted medical examinations nor a slavish adherence to a rigid system to the exclusion of common sense. It should foster in everyone using laboratories respect for the hazards and respect for their own and other people's health. If this is backed up by adequate information, safety will become a natural part of the work and not a burden separately tacked onto it.

12 References

1. 'Occupational Exposure Limits', Health and Safety Executive Guidance Note EH40 (1992) (updated annually), London, HMSO.
2. 'Threshold Limit Values for Chemical Substances and Physical Agents in the Workroom Environment', The American Conference of Government Hygienists (updated annually).
3. 'The Health and Safety at Work Act 1974', London, HMSO.
4. 'Control of Substances Hazardous to Health Regulations 1988. Approved Code of Practice', Health and Safety Commission.
5. 'Health and Safety (First Aid) Regulations 1981 and Guidance 1990', London, HMSO.
6. 'Standard First Aid and Personal Safety', American Red Cross, Washington DC.
7. M. J. Lefevre, 'First Aid Manual—Industrial Non-Pharmaceutical Chemical Substances', J. Duculot, Gembloux, Belgium, 1974.
8. 'Protection of Eyes Regulations 1974', London, HMSO.
9. N. V. Steere, *J. Chem. Educ.*, 1964, **41**, pp. A936, A939, A942, A944, and A948.
10. Department of Education and Science, 'Safety in Science Laboratories', DES Safety Series No. 2 revised 1978, London, HMSO.

CHAPTER 8

First Aid

V. S. G. MURRAY AND H. M. WISEMAN

The object of this Chapter is to give general guidance on First Aid. It is not intended to replace professional handbooks such as the 'First Aid Manual' of the Voluntary Aid Societies.[1] The reader should consult such books for detailed instructions on methods of resuscitation and treatment of unconsciousness and should attend a practical course. A list of agencies is provided at the end of the chapter.

1 Introduction

In a laboratory the accidents most likely to occur are:
- splashes of chemical on the skin;
- splashes in the eyes;
- inhalation of gases, fumes, or dusts;
- ingestion of chemicals;
- heat burns.

Remember that chemicals may enter the body by more than one route at the same time or consecutively. For example,

* liquid spilt on the skin may not only be absorbed through the skin but may be inhaled if the liquid evaporates easily;

* gases and vapours may not only be absorbed into the lungs by inhalation but may be absorbed through the skin, especially if the victim has been sweating or is wet;

* a dust or powder may get into the casualty's mouth and nose, and may be swallowed as well as being inhaled into the lungs. If the casualty doesn't spit out material coughed up from the lungs, it will end up in the stomach.

If there is a risk of absorption by more than one route, the priority should be to terminate the exposure which will most quickly cause harm. Absorption through the lungs is faster than absorption through the skin, so the exposure to vapours or gases should be stopped before the skin is washed. If chlorinated solvent is splashed on the trouser leg there is a risk of inhalation as well as skin contact: remove contaminated clothing and put outside in the open air, or in a sealed container, then flush the skin with water. Warn people in the laboratory that there has been a spill of chemical and that there may be a risk of inhaling poisonous vapours.

Chemicals damage the eye more rapidly than the skin, so in a case of a chemical splash affecting both the eye and the skin, the eyes must be washed immediately and take priority over the decontamination of the skin.

2 Rescue

If you are first on the scene your responsibility is:

- to protect yourself from danger;
- to move the victim away from contact with hazardous chemicals or other dangers if it is safe for you to do so;
- to give emergency aid until the casualty is seen by a doctor or other health-care provider;
- to warn others of the danger and call for help.

If there is someone else within hearing, call for help immediately, otherwise attend to the casualty first. Act as quickly as you can, but stay calm.

When you help a victim of an accident, fire, or explosion the first thing to do is to assess the situation and the dangers. Then before you do anything else, make the situation safe so that you are safe and the casualty is safe from additional injury. Never go to help a casualty without being sure it is safe for you to do so. If you become another victim there may not be anyone to help you.

If you rescue or help someone who may have been poisoned, it is particularly important to wear appropriate and adequate protection. There may be solid or liquid chemical on the floor or work-surfaces, or vapours or gases in the atmosphere.

In the rare event of a laboratory accident resulting in someone being overcome by toxic chemicals and trapped in an enclosed space no-one should attempt to rescue the casualty unless they have breathing equipment and lifelines and are trained in their use, as are firemen, for example.

2.1 Secondary Contamination

The casualty may be contaminated with toxic chemicals on his skin, clothes, hair, or personal belongings. The amounts present on clothing, skin, hair, and personal belongings may be sufficient, in the case of some highly toxic chemicals (Table 1), to present a danger to first aiders and other people coming into contact with the casualty or the environment around them. Such contact must be kept to a minimum and proper decontamination may have to take priority over giving artificial respiration and heart massage. Artificial respiration should be carried out using a method which lessens the risk of the first aider being exposed, if possible.

Other chemicals (Table 2), although highly toxic in high concentrations, are not likely to be a hazard to first aiders in the amounts remaining on the

Table 1 *Substances Which Present a Risk of Secondary Contamination*

acids, alkali, and corrosives (if concentrated)
cyanide salts, hydrogen cyanide gas, and related compounds
hydrofluoric acid solutions
nitrogen containing compounds (aniline, aryl amines, aromatic nitro-compounds)
chlorates
pesticides
phenol and phenolic compounds
many oily or adherent toxic dusts and liquids

Table 2 *Substances With a Low Risk of Secondary Contamination*

most gases and vapours unless they condense in significant amounts on the clothing,
skin, or hair
weak acids, weak alkalis, and weak corrosives in low concentrations (except
hydrofluoric acid)
arsine gas
carbon monoxide gas
petrol, petroleum distillates, and relates hydrocarbons
phosphine gas
smoke/combustion products (except chemical fires)
small quantities of common hydrocarbon solvents (*e.g.* toluene, xylene, paint thinner,
ketones, chlorinated degreasers)

casualty. The first priority is then resuscitation and after that the chemicals can
be washed off the skin and contaminated clothing removed.

If the nature of the chemical is unknown, and the casualty was exposed to
large amounts of chemical, assume there is a risk of secondary contamination
and consider decontaminating the casualty before resuscitation. If the casualty
was only exposed to a small amount of gas or vapour the risk is probably quite
small.

3 General First Aid

1 Check if the casualty is conscious.
2 Open the airway and make sure the tongue is not blocking the throat.
3 Check if the casualty is breathing.
4 Clean out the mouth and clear the throat.
5 Give artificial respiration using mouth-to-mouth respiration or other
   methods if appropriate.
6 Check if the heart is beating.
7 If the heart is not beating give heart massage and artificial respiration, or
   if the heart is beating but the patient is not breathing, carry on with
   artificial respiration.
8 An unconscious casualty who is breathing, should be turned on one side,
   in the recovery position.

Each of these steps is explained below. Decontamination of skin and eyes is
explained in detail in the standard protocols.

General warnings

Do not give anything by mouth if the casualty is unconscious, drowsy, having fits, or unable to swallow.

Do not give alcohol, tea, or coffee to drink or give any drugs, medicines, or stimulants.

3.1   Check if the Casualty is Conscious

Try to make the casualty wake up. Shout 'Are you all right?' and gently shake his shoulders, but take care not to make any injuries worse.

3.2   Open the Airway

The casualty will die within 4 minutes if he cannot breathe. In an unconscious casualty the tongue may block the throat and block the airway. Make sure the airway is open and air can get down the throat:

- Place the casualty on his back;
- Tilt the head back and lift the chin up with the finger and thumb of one hand on the bony part of the chin, while pressing the forehead back with the other hand.

This will stop the tongue blocking the throat.

3.3   Check Breathing

After opening the airway quickly check if the casualty is breathing.

Look: for the abdomen or the chest moving up or down; *Feel:* the chest moving up and down and *feel* the casualty's breath on your cheek; *Listen:* for breath sounds. Put your cheek close to the casualty's mouth.

Use all three checks. Remember that the chest may move up and down even though the throat is completely blocked and air cannot get to the lungs.

3.4   Clean Out the Mouth and Clear the Throat

If the casualty is not breathing after you have tilted the head back it is possible that something is blocking the throat.

Wipe or rinse all traces of chemical from in and around the casualty's mouth. Turn the head to one side and with one finger scoop deeply round the mouth and throat to clear any blockage such as vomit. Take out false teeth if they are broken or have moved out of place.

If the casualty starts breathing turn him on one side into the recovery position (see below 3.8). Check breathing and pulse frequently.

If the casualty does not start breathing you must act immediately and help the casualty to breathe.

3.5   Give Artificial Respiration

Put the patient flat on his back. Wipe or rinse all traces of chemical from in and around the casualty's mouth.

Use mouth-to-mouth respiration if you know the nature of the chemical that the casualty was exposed to, and know that there is little risk of secondary contamination.

If corrosive chemicals have burnt the lips and chin, give mouth-to-nose respiration instead of mouth-to-mouth respiration. Close the mouth and cover it with a cloth to protect yourself from getting poison on your lips or hands.

If you do not know the nature of the chemical or if you believe there is a risk to you from secondary contamination, use, if possible, a method which will lessen the risk such as a bag-valve mask or a pocket one-way valve mouth-to-mouth resuscitation device, or give oxygen by mask.

If the casualty has inhaled an irritant gas, the mouth and throat may be full of froth. You cannot remove this froth by wiping, so do not waste time trying to remove it. They are air bubbles and all you have to do to move air in and out of the lungs is to blow the froth into the lungs.

3.6   Check that the Heart is Beating

Feel the pulse in the neck, in the hollow between the voice box and the muscle. Place two fingers on the Adam's apple and slide your fingers into the groove under the jaw. Keep your fingers there for at least 5 seconds to feel if there is a pulse.

If you cannot feel a pulse then the heart has stopped. The casualty will be unconscious and will probably have a blue-grey colour to the skin and large pupils. (When the casualty has a black or brown skin look for blue colour to the nails, lips, and the inside of the lower eyelids).

3.7   If the Heart is Not Beating Give Heart Massage

If the heart stops the breathing will stop also and the casualty will need both heart massage and artificial respiration. Always start artificial respiration before heart massage.

Do not give heart massage if the heart is beating, even faintly. Stop as soon as you feel a pulse in the neck, but carry on artificial respiration if the casualty is still not breathing.

If the heart is beating, but the casualty is still not breathing, carry on with artificial respiration.

3.8   When the Casualty is Breathing Again Turn Him on One Side in The Recovery Position

An unconscious casualty should be turned to lie on one side to stop the tongue blocking the throat and to allow fluid to come out of the mouth. The casualty may vomit when breathing starts again but the vomit will not block the throat if the casualty is lying on his side. Let the vomit come out and clear it out of the mouth with your finger.

The casualty should be turned on one side, so that:

- the head, neck, and body are in a straight line;
- the head is placed so that the tongue will not block the throat and vomit or saliva will come out of mouth;
- the arms and legs are placed so that the casualty stays in the same position

This is called the recovery position.

Do not leave the casualty alone. Check breathing and heartbeat every few minutes and be prepared to give heart massage and artificial respiration.

4 First Aid for Convulsions

Convulsions may occur as a complication of serious poisoning. First aid aims to protect the casualty from injury and to keep the airway open.

1   Let the casualty lie down in a safe place; make sure there are no hard or sharp objects nearby and protect the casualty from injury.
2   Turn the casualty to lie on one side so that the tongue comes to the front of the mouth and froth can come out of the mouth easily.
3   Fold a cloth and put it under the casualty's head, or hold the head so that it does not bang on hard things.
4   **Do not** try to stop the shaking movements.
5   Loosen any tight clothing.
6   **Do not** put anything in the casualty's mouth or try to open it.
7   After the fit keep the casualty in the recovery position to rest.

5 Standard First Aid for Chemical Exposure and Burns

5.1 *Chemical Contamination of the Skin and Skin Burns*

Many chemicals burn the skin. Some chemicals are absorbed through the skin and may cause poisoning. Absorption is likely to be quicker and more significant if the skin and tissues have been damaged. Chemicals which vaporize quickly may cause poisoning by inhalation.

Heat burns may be dry burns from flames or hot surfaces, or wet burns (scalds) from hot liquids or vapours such as steam.

Mild burns often cause pain, swelling, and redness at the site of contact. More serious burns cause severe pain in and around the injured area, blistering or loss of skin and/or underlying tissue. There may be symptoms and signs of shock which may be delayed.

The severity of chemical burns, and the amount absorbed into the body, depends on the concentration of the chemical and the length of time it is in contact with the skin. Rapid removal of the chemical may prevent serious damage to skin and absorption of a toxic dose.

5.1.1 Objectives. For chemical contamination:
  • to wash chemical off the skin and remove contaminated clothing as quickly as possible;
  • to watch for signs of poisoning and resuscitate the casualty if necessary;
  • to reduce the effect of heat from heat burns and scalds;
  • to prevent infection, relieve pain, and minimize shock.

5.1.2 Immediate Action.
Chemical on the Skin
If the contaminated area is small:
  1   Take the casualty immediately to the nearest source of clean water.
  2   Immediately wash the affected part of the body under cold or lukewarm running water, using soap if you have some. Do it quickly and use a lot of water.
  Do not try to neutralize acid burns with alkali or alkali burns with acid. There is no evidence that this helps and it may cause more damage.
If large areas of the body are contaminated with chemical:
  1   Wash the casualty under a cold or lukewarm shower if there is one nearby. There is usually a special shower in a workplace where corrosive chemicals are used. If there is no running water use jugs or buckets of water. **Do not** rub or scrub the skin. Remember to clean the hair and under the fingernails, in the groin, and behind the ears, if necessary.
  2   Protect yourself from splashes of chemical with gloves and an apron if needed. Some chemicals give off vapour, be careful not to breathe it in. If you think there is a significant risk to yourself of secondary contamination more adequate protective clothing may be appropriate while decontaminating the casualty.
  3   At the same time quickly remove any of the casualty's clothes contaminated with chemical or vomit, as well as his shoes and wrist watch if necessary. Speed is important—cut the clothes off if the chemicals are very poisonous or corrosive.
  4   Continue to pour water over the casualty for 10 minutes, or longer if you can see chemicals still on the skin. If the skin feels sticky or soapy, wash it until the feeling disappears. This may take an hour or more.
  5   Make sure the water drains away freely and safely as it will have chemical in it.
  6   Dry the skin gently with a clean, soft towel. If clothing is stuck to the skin after pouring water over it, do not remove it.
  7   Put the dirty clothes outside the room, in a sealed bag if possible. Dispose of contaminated clothing with care if chemical is known to be flammable, explosive, or vaporizes easily.
  8   Many chemicals can pass through the skin very quickly. Look for signs of poisoning, check breathing, and heartbeat every few minutes and be prepared to give heart massage and artificial respiration.
Dry Burns from Fire or Heat
  1   If the burnt area is small immediately cool the burn by holding under cold

running water from a tap for at least ten minutes to reduce pain and damage.

2   **If the casualty's clothes are burning:** flames and heat rise, so pull the casualty to the floor, make him lie down and cover with a heavy blanket or coat to extinguish any flames. Use water to put out the flames if a bucket of water is handy, but do not waste time filling one.

   Do not use nylon or other inflammable materials to smother the flames. **Do not** roll the casualty along the ground as this can result in burns on previously unharmed parts of the body.

   If clothing is smouldering or hot apply large amounts of cold water.

3   If the casualty is badly burnt and the burn is large make him lie down and pour water from a jug or bucket over the burn for 10 minutes. **Do not** remove burnt clothing or anything that is sticking to burnt skin. Immobilize a badly burnt limb by placing something soft on either side.

4   Give frequent sips of cold water (up to about 100 mls) if the casualty wants it and is fully conscious.

Scalds

1   Rapidly remove clothing that is saturated with boiling liquid or steam.
2   Scalded flesh swells up so remove watches, belts, ties, and boots from scalded areas as quickly as possible.
3   If the scalded area is small, cool it by holding under cold running water from a tap for at least ten minutes.
4   If a large area is scalded pour water from a jug or bucket over the scalded part. Make the casualty lie down. Immobilize a badly scalded limb by placing something soft on either side.

What to do if the Casualty has Severe Burns

Severe burns cause both pain and distress. Full thickness burns, if present are pain free, but all casualties will be frightened and distressed and constant reassurance and communication are vital. Remember that even casualties with very severe burns are usually conscious and sentient.

1   If the casualty shows signs of shock (weak, rapid pulse, cold hands and feet, pale clammy skin) make him lie down keeping his head to one side and cover him with a blanket. Raise his feet 8–12 inches, so that his head is lower than his legs.

2   Dress the casualty in clean clothes or cover him with a sheet. Ensure that the casualty is kept warm and quiet.

   Do not apply wet sheets, towels, or ice packs or use them during transit as this will not provide any pain relief for casualties with full thickness burns and can cause profound hypothermia and may even result in death.

3   Keep burnt limbs raised above the heart.

4   *Replace fluid loss*: if a large area is burnt give half a cup of water every 10 minutes if the casualty is conscious.

5   **Do not** break open blisters or remove skin. Do not rub skin or put oily substances on skin. Do not use adhesive dressings. Where the skin is red and painful or raw, cover it and the skin round it with a sterile, dry dressing and bandage. Keep the bandage loose.

5.2   Chemical Contamination and Heat Burns of the Eye

Contamination of the eye with chemical may cause burns. A few chemicals are absorbed through the eye and cause poisoning. Scalds may be caused by hot liquids or vapours such as steam.

Effects may range from mild injury with irritation, watering and slight redness of the eye, to severe pain, with profuse watering and spasm of the eyelids, and loss of vision. Severe burns may be painless if the corneal nerves have been destroyed.

Organic solvent vapours characteristically cause blurred vision and irritation which may be delayed until several hours after exposure, and resolve without treatment.

5.2.1   Objectives.
- To wash chemical from the eyes and face without delay;
- To watch for signs of poisoning and resuscitate the casualty if necessary;
- To reduce the effect of heat from scalds;
- To prevent infection, relieve pain, and minimize shock.

5.2.2   Immediate Action.
1   Immediately wash the eye(s) with plenty of cool, clean water to remove any solid or liquid chemical from the eye, or to reduce the effect of heat from scalds. Even a delay of a few seconds increases the likelihood of severe damage.

   Do not try to neutralize acid burns with alkali or alkali burns with acid. There is no evidence that this helps and it may cause more damage.

2   Make the casualty sit or lie down with the head tilted back and turned towards the worst affected side. Gently open the eyelids of the affected eye(s) and run cold water over it from a tap or pour water from a jug. Make sure the water drains off the face without going into the unaffected eye. Do not use hot water.

   The casualty may be in great pain and may want to keep his eye(s) closed, but you must rinse out the chemical in order to prevent permanent damage. Gently pull the lids wide open, and keep the lids apart.

3   Gently brush or wipe any liquid or powdered chemical off the face.

4   Continue to wash out the eye(s) for 15–20 minutes, timed by a watch if possible. Carry on longer in the case of strong acids or alkalis.

5   While you rinse the eye(s) check that the inside of the eyelids have been well washed. It is most important to check that there are no solid pieces of chemical in the folds of skin round the eye(s), or on the eyelashes or eyebrows. If you are not sure whether all the chemical has been removed, wash out the eye(s) for ten more minutes.

6   **Do not** let the casualty rub the eye(s).

7   If the light hurts, cover the eye(s) with a sterile eye pad, a dry gauze pad, or a pad of clean cloth. Bandage in place securely, but not too tightly.

All casualties will be frightened and distressed and constant reassurance and communication are vital.

8 **Do not** put oils or creams in the eye.

9 Some chemicals are absorbed through the eye very quickly. Check breathing and heartbeat every few minutes and be prepared to give heart massage and artificial respiration.

10 The casualty should have his eyes examined by a doctor even if there is no pain, because damage may be delayed.

5.3   Inhalation of Chemicals

Some gases and vapours are irritant, causing irritation to eyes and nose, coughing and choking, and possibly difficulty in breathing; they may or may not cause poisoning. Some gases and vapours may cause poisoning but have no irritant effects. Poisonous effects may include collapse and unconsciousness, damage to the lungs, fits, and affects on the heart. Gases such as carbon dioxide, nitrogen, methane, and hydrogen are dangerous because they displace oxygen in the atmosphere and cause asphyxia (suffocation).

Chemicals are more rapidly absorbed through the lungs than through the skin or gastrointestinal tract, so the effects of inhaling poisonous or asphyxiating gases may be very rapid. High concentrations of toxic gas in an enclosed space may very rapidly cause collapse and unconsciousness and would-be rescuers who enter such areas without adequate protection may themselves be overcome. Many irritant gases cause almost immediate effects on eyes, nose, and throat thus providing a warning of their presence. However, with some chemicals effects may be delayed for many hours.

5.3.1   Objectives.
- To remove the casualty from exposure;
- To give artificial respiration and heart massage if necessary;
- To wash chemical from the skin and eyes and remove contaminated clothing, if necessary.

5.3.2   Immediate Action.
In most cases resuscitation of the casualty is the priority. However, if you think that there is a significant risk to yourself from secondary contamination you may have to decontaminate the casualty first, while wearing adequate protection yourself.

1 Remove casualty to fresh air, while protecting yourself with a respirator and protective clothing if necessary.

2 If the casualty is unconscious, open the airway, check whether the casualty is breathing, and give artificial respiration, if necessary.

3 Check if the heart is beating and give heart massage if necessary and continue artificial ventilation.

4 If the casualty is unconscious but breathing, place in the recovery position.

5 Remove contaminated clothing and equipment while wearing gloves

and other appropriate protection as necessary. Put clothing in a sealed
bag.

6   If the casualty is conscious but is coughing or short of breath, keep him
lying down, but with his head and chest propped up.

7   Wash chemicals out of the eyes and off the skin if necessary (see above
5.1 and 5.2).

8   Keep the casualty warm and quiet and do not leave him alone. Check
breathing and heartbeat every few minutes and be prepared to give heart
massage and artificial respiration.

5.4   *Ingestion of Chemicals*

Ingestion of a corrosive or irritant chemical may cause immediate pain in
mouth, throat, and stomach. The skin, lips, inside of the mouth, and throat may
be inflamed and in serious cases there may be ulceration, salivation, and pain
on swallowing, so that saliva dribbles out of the mouth. The throat may swell
and block the airway so that the patient has difficulty breathing. There may be
abdominal pain, vomiting, shock, and collapse.

Signs and symptoms of acute poisoning are often non-specific with nausea,
vomiting, dizziness, weakness, or drowsiness. Serious poisoning may result in
unconsciousness, convulsions, or effects on the heart. Absorption of chemical
from the gastrointestinal tract may be slow, especially if there is food in the
stomach, so effects of poisoning may be delayed several hours after ingestion.

5.4.1   *Objectives.*
- To give artificial respiration and heart massage if necessary;
- To wash chemical from the skin and eyes and remove contaminated
clothing, if necessary;
- To prevent vomit blocking the airway.

5.4.2   *Immediate Action.*
1   If the casualty is unconscious, open the airway, check breathing, and give
artificial respiration if necessary.

2   Check if the heart is beating and give heart massage if necessary and
continue artificial ventilation.

3   If the casualty is unconscious but breathing, place in the recovery position.

4   Remove contaminated clothing and equipment while wearing gloves and
other appropriate protection as necessary. Put clothing in a sealed bag.

5   Wash chemicals off the skin if necessary (see above 5.1).

6   **DO NOT under any circumstances make the casualty vomit.**

7   **Do not give anything by mouth if the casualty is unconscious, or having fits.
Someone who is unconscious or having fits may choke if they are given
anything by mouth.**

Do not give anything by mouth if the casualty finds it painful to swallow.

8   If the casualty is awake and not having fits, ask him to rinse out his mouth
several times with cold water and spit it out. Give small sips of water (up
to a total 100 mls) if the casualty wants to drink.

Do not make the casualty drink a lot of liquid at once: the casualty may vomit and this may be dangerous. Stop if the casualty feels sick.

Do not force the casualty to drink: if he has swallowed a corrosive substance and has burns inside the mouth he will not be able to swallow. In this case water will not help the burns and may make the damage worse.

Do not give alcohol, tea, coffee, fizzy drinks, drugs, or any stimulant.

Do not give acids to neutralize an alkali, or give alkalis to neutralize an acid.

9   Keep the casualty warm, and quiet. Check breathing and heartbeat every few minutes and be prepared to give heart massage and artificial respiration.

6   Special Treatments

There are very few poisons with specific antidotes and fewer still where it is appropriate for the antidote to be given outside hospital. This section describes specific first aid that is available for a few chemicals.

6.1   Hydrogen Cyanide, Cyanide Salts, and Other Dangerous Nitriles

Hydrogen cyanide gas is extremely toxic and may cause death within minutes of exposure, but after ingestion of cyanide salts, as solid, or as solution, effects may be delayed.

Solid or liquid cyanide salts on the casualty's skin or clothes may contaminate rescuers and first aiders unless they are adequately protected. If there are large amounts of liquid or solid cyanide on the casualty's clothes or skin, decontamination may have to take priority over giving artificial respiration and heart massage.

Inhalation or Skin Contamination
1   Remove the casualty from the contaminated area only after protecting yourself from exposure by wearing appropriate protective clothing such as a respirator and occlusive clothing.
2   If the casualty is unconscious quickly assess the risk of secondary contamination and whether the priority is to decontaminate the casualty or give resuscitation if necessary.
3   If there are large amounts of liquid or solid cyanide on the casualty's clothes or skin remove contaminated clothing and equipment and wash the casualty with large amounts of water, as described above (5.1), while yourself wearing appropriate protective equipment and clothing.
4   If the casualty is unconscious open the airway, check whether the casualty is breathing, wipe mouth and face, and give artificial respiration if necessary. If possible use a method which lessens the risk of the first aider being exposed, such as a bag-valve mask or a pocket one-way valve mouth-to-mouth resuscitation device. Give oxygen by mask if equipment and personnel trained to use it are available.

5   Check if the heart is beating and give heart massage if necessary, and continue artificial respiration.

6   The antidote is amyl nitrite. It should be given as soon as possible, to any unconscious casualty with suspected cyanide poisoning, but start resuscitation first if needed.

    Break a perle or ampoule of amyl nitrite into a handkerchief and hold it under the victim's nose for 30 seconds every minute, so that he inhales the vapour, or under the intake valve of the bag-valve mask. Repeat with a fresh ampoule at 3 minute intervals using up to 5 ampoules.

7   If the casualty is unconscious but breathing place in the recovery position.

8   Remove contaminated clothing and wash the casualty if you have not already done so.

9   Keep the casualty warm and quiet. Check breathing and heartbeat every few minutes and do not leave him alone.

Ingestion

1   If the casualty is unconscious proceed as for inhalation. Assess the risk of secondary contamination, and then give artificial ventilation, heart massage, and amyl nitrite as necessary. Remove contaminated clothing and wash the casualty while wearing appropriate protective equipment.

2   If the casualty is conscious, have him rinse out his mouth several times with cold water and spit out.

3   Give activated charcoal, 60–100 grams mixed with water to drink. This is probably more effective than making the casualty vomit.

4   Keep the casualty warm and quiet. Check breathing and heartbeat every few minutes and do not leave the victim alone.

6.2   Hydrofluoric Acid

On the skin this penetrates deep into the tissues where the fluoride ion is released and destroys the cells in soft tissues and bone. Weak solutions (5–15%) may cause very little pain on contact but may cause serious injury after a delay of 12–24 hours. Solutions from 20–40% cause little initial pain but produce injury in 1–8 hours, strong solutions (50–70%) cause immediate pain.

Skin Contamination

1   Take the casualty immediately to the nearest source of clean water.

2   Immediately wash the affected part of the body under cold or lukewarm running water, using soap if you have some. If large areas of the body are contaminated with chemical wash the casualty under a cold or lukewarm shower or use jugs or buckets of water. **Do not** rub or scrub the skin.

3   At the same time quickly remove any of the casualty's clothes contaminated with chemical. Speed is important—cut the clothes off if necessary.

4   Protect yourself from splashes of chemical with gloves and an apron if needed.

5   Continue to pour water over the casualty for 10 minutes, or longer if you can see chemicals still on the skin.

6   Apply calcium gluconate gel on and around the affected area and massage it continuously until pain is relieved. If the acid has penetrated below the nails, massage the gel over and around the nail. This will take at least 15 minutes and can be continued during transport to hospital. Cover the area with dressing soaked in gel. Bandage lightly.

7   Everyone who is exposed to hydrofluoric acid should be taken to an Accident and Emergency Unit without delay even if there is no apparent damage to the skin.

Eye Contamination
Irrigate the eye(s) with clean cool water for at least 20 minutes. This can be continued during transport to hospital. Send the casualty to the accident and emergency unit or local eye hospital, in all cases informing the hospital of the cause of the injury.

6.3   Phenol
Skin Contamination
Small amounts of water may increase absorption. Deluge with large amounts of water. If available, swab repeatedly with polyethylene glycol, then with soap and water.

Eye Contamination
Wash eyes as described above (5.2) with large amounts of water.

6.4   Chemicals Which Ignite in Air or Water
Chemicals that spontaneously and almost instantaneously burst into flames as soon as they come into contact with atmospheric oxygen or water should still be washed from the skin or eyes with water.

Protect yourself. **Do not** touch the casualty or handle contaminated clothing or other articles with bare hands. Wear appropriate protective clothing including gloves.

The casualty may panic so it is important to restrain him and prevent him running away.

Skin Contamination
Wash the casualty under a cold or lukewarm shower if there is one nearby. If there is no running water deluge the casualty with large amounts of cold water using jugs or buckets. The flames will increase in intensity when water is first applied but will quickly die out. Continue to pour water over the casualty for 10 minutes follow the standard first aid procedure.

Eye Contamination
Pour large amounts of cold water over the casualty's head, using a hose if possible. The flames will increase in intensity when the water is first applied but will quickly die out. Continue to wash out the eye for 15–20 minutes as described in the standard first aid procedure (5.2).

7   Agencies Providing First Aid Training
St. John Ambulance, 1 Grosvenor Crescent, London. SW1X 7EF

St. Andrew's Ambulance Association, St. Andrew's House, Milton Street, Glasgow. G4 0HR
The British Red Cross Society, 9 Grosvenor Crescent, London. SW1X 7EJ

8   References

1. 'First Aid Manual'. The authorized manual of St. John Ambulance, St. Andrew's Ambulance Association, and the British Red Cross Society, 5th Edition, Dorling Kindersley, London, 1987.

CHAPTER 9

Precautions Against Radiation

P. BEAVER

1 Introduction

For the purposes of discussing risks from radiation it is convenient and conventional to distinguish between ionizing radiation and non-ionizing radiation. Ionizing radiation comprises gamma rays, X-rays, or corpuscular radiations which are capable of producing ions either directly or indirectly. Corpuscular radiations include alpha particles, beta particles, electrons, neutrons, and positrons. Non-ionizing radiation is the rest of the electromagnetic fields. Thus, intense light emissions, ultra-violet, infra-red, microwaves, static electric fields, and static magnetic fields are included in this description.

Work in chemical laboratories may involve work with radioactive substances which may emit gamma rays and any of the corpuscular radiations and work with machines and apparatus embodying X-ray generators. The work of chemical laboratories could also involve exposure of workers within the laboratory to non-ionizing radiations.

This chapter will focus principally on work with ionizing radiations, an area of work activity where hazards have been well defined, researched, and extensively regulated. Nevertheless, exposure to hazardous levels of non-ionizing radiation should always be considered. In this field the volume of literature is extensive and considerations within the European Community may well lead to a Directive and hence domestic legislation.

2 Ionizing Radiations

2.1 Characteristics

The characteristics of the main types of ionizing radiation are as follows:
X-*rays*. These are electromagnetic radiation as is light but have a much shorter wavelength and hence higher energy. They are produced whenever a charged particle, usually an electron, is stopped as for instance when accelerated through an electrical potential difference in a vacuum tube into a solid metal

target. This is the principle of the X-ray tube used to produce X-rays for radiography in research and medicine;

Alpha-rays. Alpha particles are emitted by some radioactive materials, particularly those of higher mass. Examples are uranium, thorium, and plutonium. The alpha particle range in air is very small and it is readily stopped by paper or skin. It follows therefore that alpha emitting radioactive materials can only present a hazard, *i.e.* impart their energy to the human body, if the emitter is taken into the body by ingestion or inhalation;

Beta-rays. Many radioactive materials emit beta-rays. Beta-rays are essentially the same as electrons, have modest range in air, and can penetrate skin. Thus beta emitters can present a hazard when external to the body as well as when absorbed into the body. In the same way as stopping electrons produces X-rays, stopping particles in a heavy target will produce similar radiation, this time it is called bremsstrahlung;

Neutrons. Some radioactive materials produce neutrons spontaneously but more often neutrons are generated for work in laboratories by source containing a mix of an alpha emitter with beryllium. The most common alpha emitters for this purpose are polonium and americium. Neutrons can also be produced electrically by accelerating atoms of deuterium or tritium within a vacuum tube into a target of deuterium or tritium absorbed into a metal such as zirconium. Neutron penetrating power can be high in some materials but they can be absorbed by materials containing atoms of light elements such as hydrogen. Hence neutron shields are often constructed of water, wax, or concrete;

Gamma rays. Gamma rays are emitted to some extent by almost all radioactive substances. They are identical physically to X-rays but are often of even greater energy. Gamma rays are highly penetrating, hence their use for radiography of articles containing heavy elements such as castings. Gamma emitters may present significant hazards to humans when external to the body.

2.2 Quantities and Limits

Radiation quantities and units are defined by the International Commission on Radiation Units and Measurements. Their reports should be consulted if a full description is sought.[1,2] The International Commission of Radiological Protection (ICRP) (q.v.) also provided recommendations on the quantities which apply only to radiation protection.[3]

The radioactivity of a quantity of material is stated in bequerels (Bq), *i.e.* an amount of radioactivity in which one radioactive disintegration occurs per second having an activity of 1 Bq. This unit is extremely small, the amounts of radioactive material being dealt with in laboratories will often be in the MBq range (*i.e.* 10^6 Bq). Radioactive concentrations, sometimes called 'specific activities' will be quoted in bequerels per gram of material.

The rate of disintegration of radionuclides vary over a wide range and for any single nuclide is constant and cannot be affected by outside factors, such as temperature, or by chemical composition. It is expressed in terms of its half-life, *i.e.* the time taken for the activity of a sample to decrease to one-half of its initial value. Half-lives can be as short as 10^{-9} seconds and as long as 10^9 years. When radiation is absorbed in a material the *absorbed dose* is average energy imparted by the ionizing radiation to unit mass of the material. The unit of absorbed dose is joule per kilogram (JKg^{-1}) and its special name is the gray (Gy). Absorbed dose rates are measured in grays per hour (Gyh^{-1}).

The biological effect of a given dose of radiation depends on the rate at which the dose is delivered, the variation of loss of energy with distance that the ionizing radiation travels in tissue, and on the nature of the body tissue involved. The loss of energy with distance is described by a quantity called the linear energy transfer and for a given body tissue the quality factor which describes the probability of harmful effects occurring in the tissue can be related to the linear energy transfer.

Rather than considering the complexities of the relationship between quality factor and linear energy transfer, the current way of describing the detriment resulting from exposure to various radiators of different energies (hence different linear energy transfers) is by use of the radiation weighting factor, W_R. Values currently recommended by ICRP are given in Table 1.

Table 1 *Radiation Weighting Factors*

Type -v- Energy Range	Radiation Weighting Factor, W_R
Photons, all energies	1
Electrons and muons, all energies	1
Neutrons less than 10 Kev	5
10 KeV to 100 KeV	10
100 KeV to 2 KeV	20
2 MeV to 20 MeV	10
greater than 20 MeV	5
Protons, energy greater than 2 MeV	5
Alpha particles, fission fragments, heavy nuclides	20

The term 'equivalent dose' is used to indicate the biological implications of a radiation exposure to a given tissue or organ and is the product of the absorbed dose and the appropriate radiation weighting factor. Since radiation weighting factors are dimensionless, the unit of equivalent dose is joules per kilogram. Its special unit is the Sievert (Sv).

The relationship between probability of a long-term effect and equivalent dose is found also to vary with the organ or tissue irradiated. A further quantity is therefore derived from equivalent dose to indicate the combination of different doses to several different tissues in a way which is likely to correlate well with the overall long term effect in the human body as a whole. The factor by which the equivalent dose in a tissue or organ is weighted is called the tissue weighting factor, W_T. Equivalent doses weighted in this way are called *effective doses*. The unit is JKg^{-1} and the special name is the Sievert (Sv).

Finally, the other factor previously mentioned, that of variation of biological effect with dose rate, is not regarded as being of significance at the dose rates encountered in normal practice. It only becomes worthy of consideration when extrapolating for radiation effects from high dose rate to low dose rates, either for epidemiological purposes (*e.g.* studies of atomic bomb survivors), when considering the consequences of massive accidents, or when considering the effect of radiation administered during radiotherapy.

To complete mention of quantities and units, it is necessary to consider the special circumstances when radioactive material has become incorporated in human tissue—say following inhalation or ingestion. In these cases the dose will be delivered over a period of time dependent on the rate of elimination from that organ or tissue, whether that be biological elimination or physical (by decay of the radionuclide). The quantities *committed effective dose* and *committed equivalent dose* are the time integrals over the 50 years (70 years for children) following the intake. In legislation and common parlance, the term *dose* is used to mean effective dose, equivalent dose, committed effective dose, or committed equivalent dose as the context so requires.

2.3   Effects of Radiation

The effects of radiation may be separated into two classes:
(a)   *Stochastic effects*. This category covers those effects which are probabilistic in character, *i.e.* where the probability of the effect occurring is a function of the dose which causes the effect. The severity of the effect is not related to the magnitude of the initiating dose. Examples of this effect are radiation induced cancers and radiation induced genetic mutations;
(b)   *Deterministic effects*. The other category of effect is that whose severity is directly related to the magnitude of the initiating dose. Such effects are characterized by a threshold below which the effect is not observed. Examples are skin reddening or erythema and eye cataracts.

Deterministic effects usually require quite high exposures—for example, erythema may show itself after skin doses of 3 Gy delivered over a short period of time.

2.4   ICRP Standards

The International Commission on Radiological Protection (ICRP) is an independent scientific body set up under the auspices of the International Congress of Radiology. Although independent of national bodies and international agencies it is supported financially by various governments and international agencies. The Commission comprises twelve members (plus emeritus members) elected on the basis of the scientific contributions they are able to make. The headquarters of the Commission, which essentially consists only of a scientific secretary, is at Chilton in Oxfordshire, United Kingdom. The output of the Commission takes the form of Recommendations which appear under the title of 'Annals of the ICRP' published by Pergamon Press.

The importance of ICRP and its recommendations in the field of radiation protection is that the recommendations are widely accepted and form the basis of international and multinational agencies' guidance and in turn many countries base their national legislation upon these recommendations.

The recommendations of ICRP which address numerical values of dose limits depend heavily on evidence obtained from a range of scientific studies of the survivors of the Japanese atomic bombings. This evidence has recently been subject to revision, firstly the doses received by those exposed have been reassessed in the light of new knowledge about the yields of the weapons and the shielding afforded by the atmosphere between the exposed person and the weapon. More importantly, however, the longer follow-up of the remaining survivors has led to the adoption of a different biological risk model for the prediction of late effects such as cancer. ICRP now works on the basis of a greater effect of radiation on a population but that effect occurring at a later stage. The revised recommendations were published early in 1990 as ICRP 60.[3] These recommendations not only deal with dose limits but take a fresh look at the conceptual basis of radiation protection. At the time of writing the relevant intergovernmental organizations are developing but have not yet published their revisions of radiation standards. Work is proceeding within the European Community on revision of a Directive on Basic Safety Standards for radiation protection of workers and the public. That Directive, when adopted, will be binding on Community Member States as to the result to be achieved but will leave to national authorities the choice of form and methods.

The present Directive, which dates for the bulk of its content from 1980, was implemented in the UK in the main by the Ionizing Radiations Regulations 1985 (IRR 85) and it can confidently be expected that a revision of those regulations will follow on after adoption of a revised Directive.

In the international field, the International Atomic Energy Agency (IAEA), in co-operation with the World Health Organization (WHO), the International Labour Office (ILO), the Food and Agriculture Organization (FAO), the Nuclear Energy Agency of the OECD (NEA/OECD), and the Pan-American Health Organization (PAHO), is revising its Basic Safety Standards (BSS) for Radiation Protection. These are only binding to the extent that they are binding on the Agencies' own activities; principally, the BSS are commended to Member States and if desired could form the basis for their own radiation protection legislation.

2.5 General Principles of Radiation Protection

The conceptual framework of radiation protection falls into two different compartments. The first (called 'practices' by ICRP) is when the work adds further radiation exposure to that which occurs naturally. This is often called the 'normal' situation as, for example, when a laboratory worker uses a radionuclide as a tracer in examining the efficiency of a process. He will receive some additional exposure—possibly quite small over and above the approximately 2 mSv per year he will be receiving as a result of exposure to cosmic rays, gamma rays from natural radioactivity in rocks, soil, and building materials, and radon emanating from the same natural radioactivity. The recommen-dations of ICRP for 'practices' only apply to the additional exposure arising because of that practice.

The system for radiation protection in 'practices' is based on the following general principles:

(a) No practice involving exposure should be adopted unless it produces sufficient benefit to individuals or society to offset the detriment it causes. ('Justification');

(b) In relation to any particular source within a practice, the magnitude of individual doses, the number of people exposed, and the likelihood of incurring exposures where these are not certain to be received, should all be kept as low as reasonably achievable (ALARA); economic and social factors should also be taken into account. This procedure should be constrained by restrictions of doses to individuals (dose constraints) or risks to individuals in the case of potential exposures so as to limit the inequity likely to result from the inherent economic and social judge-ments. ('Optimization');

(c) The exposure of individuals resulting from the combination of all relevant practices should be subject to dose limits or some control of risk in the case of potential exposures. These should be aimed at ensuring that no individual is exposed to radiation risks that are judged to be unacceptable in normal circumstances.

The second compartment in the framework of radiation protection relates to situations (called 'intervention' situations by ICRP) where the action which could be taken would reduce existing radiation exposure of people. Examples of such situations are those existing after a serious radiological accident where evacuation of people living nearby would reduce the exposure they might otherwise receive. Another common situation is that of exposure to naturally occurring radon where installation of remedial measures, such as underfloor ventilation would reduce exposure that occupants of dwellings might otherwise receive. The general principles of radiation protection for 'intervention' situ-ations are:

 (i)   the proposed intervention should do more good than harm; and

(ii)   the form, scale, and duration of intervention should be optimized.

It is important to note that dose limits are not recommended for intervention situations.

The remainder of the sections in this Chapter will consider how the general principles are carried through in practice and how procedures are regulated to ensure that the general principles are met. There is little value however in addressing intervention situations because in the circumstances to which this chapter apply, the need to intervene should not arise; that does not mean that the possibility of mishaps and ways by which they are coped with should not be addressed, merely the likelihood of disrupting the living of the general public as a result of the work is extremely low. The obvious contradiction would be the case of a source being lost in the public domain; it is vitally important that effective security procedures are in place to prevent the loss of radioactive material.

It must also be emphasized that compliance with dose limits, *i.e.* general principle (c) above is not enough, the overriding principle is that of keeping doses ALARA. In fact, normal practices do not give rise to a significant probability of doses exceeding dose limits in any case. Statistics over recent years demonstrate that the number of workers from all industries exceeding the limit per year is in single figures.

2.6 As Low As Reasonably Practicable (ALARP)

The central message of ICRP, carried over into domestic legislation and reflected in day-to-day procedures, is that doses should be kept ALARA (or in UK legislative parlance 'ALARP'). What this means is that for any particular case the risks of the work activity should be identified, quantified, and the costs—whether in financial terms or social terms, of averting part or all of the risk—likewise evaluated. The criterion of reasonable practicability is regarded as satisfied if further expenditure to reduce risks is found to be grossly disproportionate to the risk thereby averted.

In really practical cases the difficulties of putting all the parameters—risk, financial cost, and social cost—into the same quantities and evaluating them numerically are immense and where the risks are small anyway, it is probably not justifiable to even make the attempt. What can and should be done is a critical appraisal of the proposed work activity by all the parties to it, *e.g.* the worker himself, his supervisor, management, and management advisers, to explore how best radiation risks might be reduced in an efficient and cost effective way. Comparisons can be drawn with similar tasks performed elsewhere, but this will require good assessment of exposures received or likely to be received. Although the involvement of professional advice (see later paragraphs) is often very effective, a lot can be done by involving workers in considerations of their own exposures or likely exposures and this is particularly rewarding where the workforce understand the basic practical principles underpinning radiation protection.

2.7 Practical Radiation Protection Principles

It is a cardinal principle that protection should first be considered as a source related procedure. Only if all that is reasonably practicable has been done in that respect should one turn to protecting the individual personally—say by protective clothing or breathing apparatus. For all radiation sources the three simple guides are:

(a) *Distance.* The radiation intensity will fall off with distance from a source of radiation. If the source is essentially a point source, the fall-off will be proportional to the square of the distance. Thus, increasing the distance from the source by any remote control device from robots, handing tongs, or tweezers will help considerably. The type of remote device used will have to be considered in relation to its cost, any loss of dexterity, and hence increase in time of exposure;

(b)   *Time*. The amount of exposure is directly proportional to the duration of the exposure. Anything that can be done to speed up the work will therefore benefit radiation protection as well as improving productivity. However, caution is called for lest speeding up of the work brings with it an increased probability of minor accidents, *e.g.* spillages, lost or dropped sources;

(c)   *Shielding*. Depending on the nature of the radiation emitted and its energy, shielding which might range from aluminium for beta particle sources to lead or concrete for *X*- and gamma sources, can be highly effective. Shielding for large sources, such as industrial chemical changes (*e.g.* polymerization), is a science in itself and choosing the optimum requires a good deal of skill. On the other hand shielding of small sources such as those used in analytical instruments is relatively easy. The suppliers should of course have addressed the matter in the first instance.

It is worthwhile considering if doses could be further reduced by local shielding when the equipment has been installed. This means that the radiation survey which should be carried out on installations will also sever the purpose of giving a starting point for assessing the need for further practical measures;

When the source is radioactive material capable of being dispersed (*i.e.* unsealed radioactive material) two other principles should be considered:

(d)   *Containment*. As far as possible radioactive substances should be contained in a way which minimizes the possibility of dispersal. The techniques include well-designed bottles or vessels for solutions of radioactive materials, fume cupboards, or glove boxes for larger quantities, particularly if they are likely to become airborne, trays for work done on the bench to capture spillages, *etc.*;

(e)   *Good housekeeping*. Radioactive material is best dealt with in laboratories where the standard of construction is high, where thought has been given to the ability to clean, and where there is a good approach to the general standards of tidiness, minimizing of surplus materials, accountancy for substances, storage, and use of protective clothing.

2.8   Legislation

Work with ionizing radiation is a comprehensively regulated matter. The principal legal instruments are the Ionising Radiations Regulations 1985 (IRR 85)[4] and the Radioactive Substances Act 1960 (RSA 60). Associated with IRR 85 are four parts of an Approved Code of Practice,[5] and a substantial amount of guidance material addressing specific topics. IRR 85 is enforced in the main in the areas covered by this Chapter by the Health and Safety Executive (HSE) from its regional offices of the Field Operations Division (FOD). The RSA 60 on the other hand is enforced by HM Inspectorate of Pollution (HMIP) or HM Industrial Pollution Inspectorate in Scotland (HMIPI). This Act deals with the storage, accumulation, and disposal of radioactive substances and requires registration of all users and authorizations of all disposals of radioactive substances; however, various exemptions and exclusions are current and hence any intending user of radioactive substances should make early contact with the nearest office of HMIP or HMIPI as the case may be for further guidance.

2.9 Essential Features of the Ionising Radiation Regulations 1985

There can be no substitute for a study of the Regulations, the Approved Codes, and other guidance. However, this section of the Chapter attempts to identify the essential features which are relevant to the subject matter under consideration.

Once an employer intends to use radioactive materials or apparatus embodying machines which emit ionizing radiation, he is required to notify the local office of the HSE. There are exemptions for trivial quantities and special uses—such as smoke detectors. Notification of intending use is not required to include a comprehensive safety analysis; a simple letter (or use of the form included at the back of Parts 1 and 2 of the Approved Code) should suffice.

The next step should be to determine whether the scale of intended use demands the appointment of a Radiation Protection Adviser (RPA). If the work is sufficient to require the creation of a controlled area which persons will enter or creates a dose rate in excess of 7.5 μSv h^{-1} then the employer of the worker concerned (or the worker himself if self-employed) is required to identify an RPA and to notify the particulars, qualifications, and experience of the proposed appointee to the HSE one month before the appointment is finalized.

The RPA may be drawn from the employer's own workforce, if there is someone suitably experienced and qualified, or a candidate can be sought from radiation safety consultancies operating commercially. Trade Associations or professional societies (such as the Society for Radiological Protection) may be able to assist in locating suitable consultants. The sort of experience and qualities required for an RPA and the tasks which he might be expected to undertake are listed in the Approved Code, Part 1. Clearly the intention is that the person appointed is a professional and the advice he is called upon to give should be authoritative.

Nevertheless, the employer cannot abrogate his responsibilities to comply with legal requirements and there are no legal duties on an RPA which are enforceable under the Health and Safety at Work Act 1974.

An RPA will advise on all aspects of safe working with ionizing radiation and on compliance with legal requirements and necessarily will make that advice specific to the work which the employer is conducting or proposes to conduct. Certain issues are common to all work with ionizing radiation, for example, the appointment of a Radiation Protection Supervisor (RPS) (or Supervisors if necessary). The role of the Supervisor, who is required to be drawn from the employer's workforce is to supervise the work at the operational level and to ensure that the work is done in accordance with legal requirements and good radiation practice. Both, in respect of the particular work activity, should be described in local rules and these rules therefore are the guidelines to which the supervisor will act. The appointment of supervisors, their training and development of local rules are all matters which fall within the remit of the RPA.

Advice from the RPA will be necessary for carrying out prior assessments and continuing assessments of the work and the likely impact of that work on the radiological conditions and exposure of workers. Assessments should consider both normal situations and possible mishaps and accidents. In the latter case action should be taken, where reasonably practicable, to reduce the probability of the mishap or accident, and to reduce its consequence. If there is still a residual risk of accidents then a contingency plan should be devised to minimize the effect on workers and any members of the public. When the assessment of radiological conditions for normal situations has been completed, it provides a base line from which to develop further ways of reducing exposure and as a base line for comparisons with future assessments.

All assessments should be periodically reviewed and brought up to date to account for changes in procedures and practices and for any new knowledge about radiation and its effects which has been gained.

In addition to assessments, special investigations should be carried out whenever a worker reaches certain specified doses. These are:
(a)   30 mSv in any calendar quarter;
(b)   15 mSv in any calendar year and;
(c)   75 mSv in any period of five calendar years.
Essentially these investigations are aimed at examining whether doses are being kept as low as reasonably practicable. The latter investigation (c) is however designed to address the circumstances of the individual worker on the basis that continued accumulation of dose at that rate would lead to an unacceptable level of risk for that worker. All the investigation levels may be regarded as 'generic constraints' and employers will in practice use them as upper bounds to individual dose and hence avoid the need to conduct any special investigations.

Nevertheless, employers should arrange, perhaps in conjunction with their RPA, to keep a continuing watch on personal doses, to look for trends of increasing exposure and, if appropriate, to institute measures to reduce exposures.

Practical radiation protection is very much about identifying workers at greatest risk and places where risk is greatest.

This is done by designating workers as classified workers if their individual dose is likely to exceed 15 mSv in any calendar year. There is no need to consider each worker and his work procedures individually when coming to decisions about classification. The rule is that, with certain exceptions, anyone who works in a controlled area should be regarded as classified. The UK legislation has a detailed schedule which lays down in very specific terms how a controlled area should be defined. The simplest rule is that a controlled area encompasses all areas where the dose rate within its boundary is likely to be greater than 7.5 μSv h^{-1}. However, the schedule contains clauses which make allowance for limited occupation of the area, variable dose rates within the area, although the RPA should be involved in applying these concessions.

There is also a very useful way of determining when areas need not be designated which depends on the inventory of radioactive material in the area. For example, when less than 50 MBq MeV of gamma emitter is present, the area need not be designated as a controlled one.

A supervised area does not carry with it the same requirements for classification of workers within it and can be entered freely by any worker although of course the requirements for local rules and general supervision still apply. Supervised areas are required at levels of dose three-tenths of the level for controlled areas.

Hence, much work with small sources can be done without the need to set up controlled areas but even if the criteria are achieved and controlled areas set up it is possible to avoid classification of workers if sufficient control can be exercised to keep individual doses below three-tenths of the dose limit.

The control is demonstrated by way of a 'written system of work' which should be designed and sufficiently detailed to ensure adequate control of the workers or groups of workers covered by it. Controlled areas should be delineated by barriers, markings on the floor, or for preference, by the natural boundaries to the workplace such as walls and doors. Warning notices should be placed at the boundaries and the local rules should draw attention to the meaning of the warning signs and the requirements to observe them.

Classification of workers carries with it a requirement to measure the dose received by the worker for it to be recorded and for the worker to be subject to medical surveillance. Dosimeters to measure the dose accumulated by workers should be obtained from a dosimetry service which holds a certificate of approval from the HSE. The same service will make arrangements to issue and maintain records and provide suitable summaries to the employees concerned. Medical surveillance should be arranged with a medical practitioner who holds a Certificate of Appointment from the HSE and will comprise a pre-employment medical examination and annual reviews of health.

Accountancy and security of radioactive substances are matters which are very important. Numbers of accidents and incidents are reported every year in which sources are lost or stolen and the results are often significant exposures of persons inadvertently coming into contact with such sources. Therefore, a strict accountancy system should be set up whereby the whereabouts of each sealed sources of radioactive substance is known and the fate and location of all unsealed radioactive substances can be demonstrated. Accountancy records should be kept and stock-takes conducted at regular intervals to verify their accuracy.

All sources when not in use should be kept in secure storage which should guard against radiation emissions from the source, leakage or dispersal of radioactive material, and possible attempts to steal or remove when not authorized. For the smaller user standard steel safes or cabinets are often suitable provided the need for ventilation of the store has been assessed. The store should be labelled to indicate that it contains radioactive materials and the store should be reserved for that purpose; in particular, the store should not contain flammables or explosives.

Where the radioactive material is a specimen which requires refrigeration, the refrigerator if capable of being locked would then constitute a store. However, the temptation to store non-radioactive materials (especially foodstuffs) should be resisted. Security shielding and containment should be maintained when sources of radioactive material are being moved or transported. When being transported outside the establishment then the requirements for packaging, labelling, consignment information, *etc.* follows the recommendations of the International Atomic Energy Agency (IAEA).[6] Within the establishment or institution, the package should be suitably labelled and the person receiving it should be informed in writing about the nature and quantity of the radioactive substance which he is receiving.

When radioactive materials are being used as unsealed radioactive materials it will be necessary to provide washing and changing facilities and some measure of personal protective equipment. The latter might simply be a laboratory coat but if there is any likelihood of floor contamination, overshoes should be provided, and a shoe barrier should indicate where they should be worn. In extreme cases of possible airborne radioactivity, respirators should be provided and worn. However, the standard of such respiratory protective equipment should be that approved by the HSE. Furthermore, the equipment should be regularly examined and tested and records kept of such examinations and tests.

Changing facilities should provide accommodation for clothing not being worn in the workplace separate from workplace clothing which might be contaminated by radioactive materials. The washing facilities should be of a good standard, taps operated by foot or elbow with adequate supplies of soap, nail brushes, and drying facilities such as disposable towels.

In addition to personal monitoring of workers using the services of an approved dosimetry service, the workplace should be regularly monitored to verify that the levels of radiation and contamination remain satisfactory, that nothing untoward has occurred, and that conditions are consistent with prior assessments. The persons carrying out the monitoring should be properly trained and equipped with monitoring instruments of a design that will detect and measure the radiation or contamination that is likely to be present. There is considerable merit in training operators to make their own measurements so that they do not always have to call on the services of someone in the organization who is specially trained.

Where contamination is possible there should be facilities for the monitoring of hands, shoes, and clothing. These latter facilities may be purpose-built hand and clothing monitors in large establishments, but in smaller units a multi-purpose monitor will serve to check hands as well as benches and other surfaces. All monitoring instruments whether for radiation or contamination should be scaled in such a way that the user can determine readily whether or not a non-acceptable level has been reached.

All instruments used for monitoring should be adequately tested and calibrated. The HSE has issued detailed guidance[7] on this topic which should be consulted by those who intend to carry out their own testing and calibration. However, the majority of users will verify that the instrument is working on each occasion it is used by exposing it to the radiation from a small check source, and arranging for an annual test by an organization which is equipped to carry out such tests and will issue certificates to that effect.

2.10 Practical Precautions with X-Ray Machines

X-Ray machines may be used for radiographic purposes, *e.g.* for examining an article to determine flaws in construction. The image may be received and recorded on film or examined by way of an image intensifier. The latter is particularly suited to routine tasks and cabinets embodying the X-ray machine and the examination equipment are available. The commonest examples are of course X-ray baggage examination facilities seen at all airports but the same general layout can be used for letters, postal packets, production items, such as castings or even in foodstuffs looking for foreign bodies. There is considerable pressure on suppliers and manufacturers to supply complete, properly shielded units so that all the users has to do is to follow the instructions for correct use and carry out a survey with a monitoring instrument from time to time to check that there has been no loss or displacement of shielding.

The equipment should be examined by the supplier upon installation and information should be supplied to the user to enable him to use the equipment without risk.

More operational precautions are necessary when X-ray machines are used free-standing with articles being brought to them for examination or, on occasions, when the articles are large, the X-ray machine being taken to them. For preference, X-ray machines used for radiography should be used in a . walled enclosure which provides adequate shielding—the door of which being interlocked to the operator of the X-ray set so that opening the door interrupts operation. Re-operation should only be possible after re-setting and going through the radiation initiation procedure from the beginning. It is necessary to ensure that warning is given of the intent to operate the machine and another warning to indicate that the machine is actually emitting X-rays. In this way persons are excluded from the enclosure, or if caught inside are given adequate warning to allow them to make their way out without being exposed.

X-Ray crystallographic and spectroscopic analytical equipment incorporate X-ray sources which produce highly intense and highly collimated beams. The principles of ensuring safety are to prevent access to the beam of X-rays and to warn operators and others when the X-ray apparatus is energized and the beam is being generated. Practical ways of doing this vary widely from one type of manufacturer's equipment to another and many ways are described.[8]

2.11 Practical Precautions with Sealed Sources of Radioactive Material

Small sealed sources find a variety of uses, such as incorporation into chromatographs and other inspection or analytical equipment, checking of instruments, for static elimination (for example, in weighing balances). The principles of practical protection are to enclose and shield any beam of radiation that might be emitted from the source, and to ensure that the equipment and source are suitably marked and that indications are given if any shutter or shield is removed. The sources used in such equipment are generally of relatively small size and making adequate provision therefore not onerous. Such equipment is most likely to have been made commercially and therefore users can expect (but should check) that proper radiation protection features have been incorporated.

2.12   Practical Precautions with Unsealed Radioactive Substances

Unsealed radioactive substances are most likely to be used in tracer experiments or process control where a small quantity of a suitable radionuclide is added to the process and its fate determined by measurements downstream of the addition. In this way efficiency of chemical reactions, loss of fluids from notionally enclosed systems, *etc.* can be very precisely discovered. Such procedures require scrupulous care to avoid cross-contamination otherwise the results will be meaningless. This is in favour of good radiation safety practice since the essential principles are to avoid inhalation, ingestion of the substance by persons, and to avoid personal contamination. The techniques require sensitive measuring equipment and skill and expertise on the part of practitioners.

Any organization contemplating taking advantage of these techniques would do well to consult the specialist suppliers of radionuclides which could be used for the purpose. Their advice would include the necessary radiation protection precautions for the application in mind.

Unsealed radioactive materials might also be encountered as a natural constituent of other materials—for example, rock and ores which might contain natural uranium or thorium. The levels of these radionuclides can be quite high in many minerals and therefore a simple check using a commercially available geiger counter would be well justified if any significant quantity of minerals are being handled. Concentrations of a few Bqg^{-1} are sufficient to give rise to the need to take precautions if the total quantity is of the order of kilograms.

Minerals containing uranium can exude the radioactive gas called radon. Any unventilated storage of such minerals is very likely to present high concentrations of radon and hence a radiation risk to workers. Indeed, ill-ventilated workplaces in areas of the country (such as Devon and Cornwall) where the natural subsoil contains uranium bearing minerals may well have elevated levels of radon. Very simple passive radon measuring devices are now available commercially and should be deployed, according to the suppliers' instructions if the workplace is in a part of the UK designated by the Department of the Environment as an 'affected area' (currently Devon and Cornwall) and is not well ventilated.

2.13　General Advice

The emphasis nowadays is much more on users of ionizing radiation control-ling their work with the advice of a Radiation Protection Adviser rather than the user or intending user consulting text books and/or advisory documents and attempting to work out what he should do from the written word. Having said that, there are a wide range of booklets, leaflets, guidance documents, *etc.* which serve the various purposes of training workers, advising safety repre-sentatives, instructing managements, and updating Radiation Protection Advisers, *etc.* and all these are regularly being added to as time passes. Rather than attempt to list these here, the user or intending user will do better to seek book/literature lists from the undermentioned bodies, according to the use he has in mind:

International Atomic Energy Agency—publications available from HMSO.
National Radiological Protection Board, Chilton, Didcot, Oxon OX11 0RQ.
Health and Safety Executive, Enquiry Point, Baynards House, 1 Chepstow Place, London W2 4TF.

3　Non-ionizing Radiations

3.1　Introduction

For the purposes of this short résumé of the hazards arising from non-ionizing radiation, the various sections within the electromagnetic spectrum will be discussed separately. This is justified because different properties and hence different protective measures are appropriate to each sector. Unlike the field of ionizing radiation where the unifying approach of ICRP has led to harmonized standards and a common approach, there has been no group which has achieved international recognition in quite the same way. However, the Inter-national Non-Ionizing Radiations Committee which was founded under the auspices of the International Radiation Protection Association, (the Inter-national Federation of Radiation Protection Professional Societies) is fast assuming a leading role. Within the UK the National Radiological Protection Board remit was extended to include non-ionizing radiation by an amendment to the Radiological Protection Act 1970. It now conducts extensive research, offers a variety of training courses, and publishes a wide selection of guidance material and reports.

3.2　Ultra-violet

Intense sources of ultra-violet find application in the curing or acceleration of chemical reactions (as well as applications in sun parlours which will not be further discussed here). Ultra-violet radiation may, in addition to causing skin effects, lead to eye damage or irritation.

Conventionally the ultra-violet spectrum is divided thus:

UVA	400–315 nm
UVB	315–280 nm
UVC	280–110 nm

This division is particularly useful since the biological hazard from UVA is relatively low. The maximum permissible exposure for UVA is set at 10 Wm^{-2} for periods greater than about 16 min; for shorter periods the exposure should not exceed, in total, 10^4 Jm^{-2}.

UV radiation hazard and hence setting of limits for the maximum permissible exposure in the UVB and UVC regions is heavily dependent on the wavelength of the radiation; the most hazardous occurring at about 270 nm (very near the principal emission of mercury vapour lamps at 254 nm). At this wavelength the total exposure over 8 hours should not be allowed to exceed 30 Jm^{-2}. (For a complete list of maximum permissible exposures see reference 8.)

The principal method of protection must be good engineering design of the equipment whatever it might be—whether a comparatively simple light source, an insect killing device, or more complex equipment for polymerization of adhesives. Equipment should be designed, constructed, and maintained to keep all worker exposures well below the appropriate Maximum Permissible Exposures. Manufacturers and suppliers should provide adequate information to users to allow them to maintain and use the equipment to the relevant standard. In particular, care must be taken to ensure proper maintenance, for example immediate replacement of broken glass envelopes which may be providing a filtering out of the more harmful short wavelength ultra-violet. Window glass and perspex are excellent materials for shielding and it may be worth investigating whether further shielding should be installed for non-standard applications. Where interlocks are fitted, as occurs with many commercial equipments, between the radiation source and shielding or housing, a quality assurance or maintenance schedule procedure should be instituted to ensure that the interlocks work effectively and have not in any way been defeated.

Where there is any substantial use of ultra-violet sources the advice of a competent safety adviser should be sought. The advice preferred would cover such topics as demarcation hazard areas, provision of warning signs, training and information to workers, wearing of protective clothing, and, if necessary, goggles or eye shields.

3.3 Lasers

The principle of classification and labelling of lasers into categories is now well established. The scheme used is as follows:

Class 1. Inherently safe lasers where the maximum permissible exposure cannot be exceeded; that is no matter how long the exposure there is no danger to the eye. Class 1 lasers also include those which are safe by virtue of engineering design.

Class 2. Low power visible light lasers where the output power is limited to the accessible emissions limits (AEL) of Class 1 lasers but for exposure times of up to 0.5 s. For continuous wave lasers the power is limited to 1 mW. These lasers are too bright for direct viewing but momentary exposure is not regarded as hazardous.

Class 3A. Lasers which exceed the AEL for Class 2 products and have power limited to 5 mW for continuous wave lasers or five times the limit for Class 2 for pulsed or scanning equipment. If the laser operates in the visible area the irradiance must not exceed $25\ \mathrm{Wm}^{-2}$. Protection in the visible region is considered to be provided by the natural 'blink' reaction. If the laser operates in the non-visible region the radiation must not exceed five times the AEL for Class 1 lasers.

Class 3B. Lasers of any wavelength with a maximum emission power of 0.5 W for continuous sources or $10^5\ \mathrm{Jm}^{-2}$ for pulsed sources. Direct viewing is hazardous.

Class 4. Lasers of any wavelength with emission powers greater than Class 3B. Here even diffuse reflections can be a health and fire hazard. Extreme caution is necessary for their use.

Manufacturers should identify the class of laser and should provide information on labelling safe use and maintenance, beam containment, interlocking, warning signals, and alarms.

As far as possible lasers should be operated in well designed enclosures appropriate to the class of the lasers. Nevertheless with higher class lasers it may be necessary to provide workers with protective eye wear. This must be chosen specifically for the particular wavelength of laser and the advice of a specialist is essential. Laser Safety Officers should be appointed where Class 3A, 3B, and 4 lasers are in use. They should be involved in matters such as drawing up of local rules, designation of hazard areas, and investigation of any inadvertent exposures.

3.4 Radio Frequency (including Microwave)

The most likely applications which may be encountered in laboratories relate to heating, either industrial type dielectric or microwave ovens. In all cases proper equipment design with a sufficiency of information on proper usage are the keys to dealing with potential hazards. Exposure limits are recommended by both the International Non-Ionizing Radiation Committee (INIRC) and the NRPB and differ only in relatively minor respects. The limits are expressed in terms of electric and magnetic field strengths and equivalent plane wave power densities. The references should be consulted for details.

Since there are many factors influencing instrument response and measurement uncertainty, any attempt at making measurements of field strengths for comparison with limits should be carried out by experts. Most important are checks of equipment to ensure for example that interlocks are working properly, and that door seals on ovens are not damaged.

3.5 General

The whole area of non-ionizing radiation falls within the scope of a European Directive which is currently being discussed between experts of the Community Member States. Necessarily some time will elapse before the Directive, known as the 'Physical Agents Directive' is adopted and even more time before its terms are transposed into national legislation. While it is unlikely that markedly different standards will feature in the Directive, the fact that they will appear in national legislation and be enforceable at law will inevitably cause greater pressure to keep exposures substantially below the specified limits.

4 References

1. Radiation Quantities and Units, ICRU Report 33, International Commission on Radiation Units and Measurements, Bethesda, 1980.
2. Determination of Dose Equivalents Resulting from External Radiation Sources, ICRU Report 39, International Commission on Radiation Units and Measurements, Bethesda, 1985.
3. 1990 Recommendations of the International Commission on Radiological Protection ICRP 60, Pergamon Press, Oxford, 1990.
4. The Ionising Radiations Regulations 1985, HMSO, London, 1985.
5. ACoP Parts 1 and 2, The Protection of Persons against Ionising Radiation Arising from any Work Activity, HMSO, London, 1985.
   ACoP Part 3, Exposure to Radon, HMSO, London, 1988.
   ACoP Part 4, Dose Limitation—Restriction of Exposure, HMSO, London, 1991.
6. Regulations for the Safe Transport of Radioactive Material, IAEA Safety Series No. 6, 1985.
7. The Examination and Testing of Portable Radiation Instruments for External Radiations, HMSO, Health and Safety Executive, London, 1990.
8. A Guide to Radiation Protection in the Use of X-Ray Optics Equipment, Bines and Hughes (eds.), Science Reviews Ltd., 1986.
9. Threshold Limit Values for Chemical Substances and Physical Agents in the Workroom Environment, American Conference of Governmental Industrial Hygienists, Cincinnati, Ohio.

CHAPTER 10

Electrical Hazards

M. R. STEPHENSON

1 What are the Electrical Hazards?

In order to appreciate the need for precautions and to understand the underlying logic behind statutory requirements it is necessary to have a basic
knowledge of the nature of electrical hazards. The dangers arising directly from
electricity are those of shock, burn, and the effects of ultra-violet radiation
generated by arcing. Indirect dangers include those resulting from fire and
explosion involving the electrical equipment itself or from other fires or explosions where the electricity is the source of ignition. Less frequent but equally
serious dangers can be caused by loss of electrical functions which contribute to
the control of non-electrical dangers.

Hazards originating from various forms of electromagnetic radiation and
static electricity are outside the scope of this chapter.

1.1 Electric Shock

The severity of an electric shock and its effects are largely dependent on the
magnitude and duration of the current passing through the body. Other factors
having an effect include the particular current path through the body and the
characteristics of the current itself such as wave form, and whether it is
alternating or direct, continuous or pulsed. The sensation of current passing
through the human body ranges from the imperceptible to extreme pain.
Possible lethal effects include ventricular fibrillation, asphyxia, and cardiac
arrest. Muscular contractions which result in the inability to let-go have the
effect of extending the shock duration until the current flow is interrupted by
some other means. This increase in shock duration is likely to increase the
probability of a fatal outcome.

When a person receives an electric shock the current will depend on the
voltage and the body resistance. The voltage will be related to the operating
voltage of the equipment and the particular circumstances in which the shock is
received. A significant proportion of the resistance of the body is within the skin
and therefore wet skin or skin contaminated with conductive solutions will

141

increase the severity of shock. Body path resistance will also depend on the individual, the contact area, the current path, and the touch voltage since body resistance tends to decrease with increasing voltage. A typical value of body resistance where the touch voltage is 240 volts (V) might be 1000 Ohms (Ω). These values would result in a current through the body of 240 milliamperes (mA). To put these figures into perspective, a current of 50 mA maintained for as short a period as two seconds could prove fatal. It is virtually impossible to specify a voltage that will be safe in all circumstances and the best advice must be to minimize the probability of electric shock at any voltage. One of the most widely recognized publications on the subject of electric shock and its effects on the human body is the International Electrotechnical Commission Publication 479-1.[1]

1.2   Burns

Burns can be caused by the heat generated by the passage of electric current through the body tissue or by thermal radiation from power arcing which results from short circuits between live conductors or between live conductor and earth.

The temperature of an electric arc is extremely high and consequently the amount of heat radiated considerable. In addition heat may be transmitted by convection and ejection of molten metal. Not all short circuits are equally dangerous since the severity of the arc will depend upon several factors:

- the nature of the short circuit itself;
- the voltage of the system;
- the fault current (the current that flows under short circuit conditions which could be several hundred times the normal full load current);
- the duration of the fault current.

On 240/415 V systems the danger from burns resulting from the heating effects of current passing through tissue is likely to be less than the accompanying electric shock. Where higher voltages are involved the internal heating of body tissue may be significant. This can damage muscles beneath viable tissue and results in renal failure or the need to amputate affected limbs.

1.3   Ignition

Electrical equipment is capable of becoming a source of ignition as a result of surface temperature, arcing, or sparking which may occur in normal operation or because of malfunction. Excessive surface temperature may arise as a result of overload, high resistance connections, or because environmental conditions have interfered with the normal arrangements for dissipating heat.

Arcing, unless it is an intended function of the equipment, usually results from insulation failure or short circuit inadvertently caused by some work activity. Sparking (in effect very small arcs) is often inherent in the normal functioning of equipment but can also be caused by bad connections.

1.4 Electrical Functions

Electrical equipment is often used to eliminate or reduce non-electrical hazards. Obvious examples are artificial lighting, monitoring and control of laboratory process variables, air conditioning, and fume extraction. Loss or aberrant behaviour of such function could cause significant hazard to personnel.

2 Statutory Legislation

The requirements of The Health and Safety at Work Act 1974 (HSW Act) apply to electrical hazards just as they do to other hazards associated with work activities. The Act makes no specific reference to electricity or to electrical hazards but it imposes broad duties on employers and others. The Electricity at Work Regulations 1989 (the EAW Regulations) are more specific about the hazards against which precautions must be taken and about the nature of the precautions. Subject to the issue of any Exemption Certificates, and none has been issued to date, the EAW Regulations apply to virtually all work premises and activities to which the HSW Act applies except off-shore oil rigs and pipelines.

The Regulations ensue together with a commentary on their objectives. Some regulations or parts of regulations have been excluded where these apply only to mines or where they are not directly relevant to electrical standards in chemical laboratories. Reproduction of the Regulations is by kind permission of Her Majesty's Stationery Office. Reference should be made to Statutory Instrument 1989 No. 635[2] for the full text of the Regulations.

2.1 Regulation 1: Citation and Commencement

These Regulations may be cited as the Electricity at Work Regulations 1989 and shall come into force on 1st April 1990.

Commentary. Although the Regulations did not come into force until 1st April 1990 they apply in full to any electrical equipment or installation even though the equipment or installation may pre-date the Regulations.

2.2 Regulation 2: Interpretation

In these Regulations, unless the context otherwise requires—
'circuit conductor' means any conductor in a system which is intended to carry electric current in normal conditions, or to be energized in normal conditions, and includes a combined neutral and earth conductor, but does not include a conductor provided solely to perform a protective function by connection to earth or other reference point;
'conductor' means a conductor of electrical energy;
'danger' means risk of injury;
'electrical equipment' includes anything used, intended to be used, or installed

for use, to generate, provide, transmit, transform, rectify, convert, conduct, distribute, control, store, measure, or use electrical energy;

'injury' means death or personal injury from electric shock, electric burn, electrical explosion, or arcing, or from fire or explosion initiated by electrical energy, where any such death or injury is associated with the generation, provision, transmission, transformation, rectification, conversion, conduction, distribution, control, storage, measurement, or use of electrical energy;

'system' means an electrical system in which all the electrical equipment is, or may be, electrically connected to a common source of electrical energy, and includes such source and such equipment.

Commentary. Subsequent regulations require measures to be taken to avoid danger but the definition of danger only includes those dangers which result in a risk of injury as defined. This has the effect of excluding, for example, danger arising from loss of control of an exothermic chemical reaction where that loss of control is due to the malfunction of an electrical control system but such danger is covered by the duties imposed by Sections 2 and 3 of the HSW Act. Note, however, that danger arising from the electrical ignition of a flammable atmosphere is included within the definition.

2.3   Regulation 3: Persons on Whom Duties are Imposed by these Regulations

(1)   Except where otherwise expressly provided in these Regulations, it shall be the duty of every—

(a)   employer and self-employed person to comply with the provisions of these Regulations in so far as they relate to matters which are within his control; and

(b)   manager of a mine or quarry (within in either case the meaning of section 180 of the Mines and Quarries Act 1954) to ensure that all requirements or prohibitions imposed by or under these Regulations are complied with in so far as they relate to the mine or quarry or part of a quarry of which he is the manager and to matters which are within his control.

(2)   It shall be the duty of every employee while at work—

(a)   to co-operate with his employer so far as is necessary to enable any duty placed on that employer by the provisions of these Regulations to be complied with; and

(b)   to comply with the provisions of these Regulations in so far as they relate to matters which are within his control.

Commentary. In some regulations or parts of regulations the duties are subject to the qualification 'so far as is reasonably practicable'. Where there is no such limit the duty is said to be absolute. Some of the regulations which impose absolute duties are subject to the provisions of Regulation 29.

Virtually everyone who is involved with the activities of the laboratory will have some duties imposed by the Regulations. In order to avoid a situation in which everyone assumes that particular duties are the responsibility of somebody else, duty holders should be appointed and advised as to their specific responsibilities.

2.4 Regulation 4: Systems, Work Activities, and Protective Equipment

(1) All systems shall at all times be of such construction as to prevent, so far as is reasonably practicable, danger.
(2) As may be necessary to prevent danger, all systems shall be maintained so as to prevent, so far as is reasonably practicable, such danger.
(3) Every work activity, including operation, use, and maintenance of a system and work near a system, shall be carried out in such a manner as not to give rise, so far as is reasonably practicable, to danger.
(4) Any equipment provided under these Regulations for the purpose of protecting persons at work on or near electrical equipment shall be suitable for the use for which it is provided, be maintained in a condition suitable for that use, and be properly used.

Commentary. This regulation has very broad application and covers most aspects of electrical safety. Any danger not covered by subsequent more specific regulations is likely to be covered by Regulation 4. Regulation 4(1) recognizes the need to give due consideration to the electrical system as a whole. The safety of a system depends on the compatibility of the component parts with each other and with system parameters such as voltage, load current, and the magnitude and duration of fault currents.

2.5 Regulation 5: Strength and Capability of Electrical Equipment

No electrical equipment shall be put into use where its strength and capability may be exceeded in such a way as may give rise to danger.

Commentary. This regulation requires that every item of electrical equipment shall be capable of withstanding, without giving rise to danger, the electrical effects imposed on it by the system of which it forms part. This requirement covers not only normal operation but also performance under fault conditions and the equipment to be considered includes protective conductors such as earthing conductors.

2.6 Regulation 6: Adverse or Hazardous Environments

Electrical equipment which may reasonably foreseeably be exposed to—
  (a) mechanical damage;
  (b) the effects of the weather, natural hazards, temperature, or pressure;
  (c) the effects of wet, dirty, dusty, or corrosive conditions;
  (d) any flammable or explosive substance, including dusts, vapours or gases;
shall be of such construction or as necessary protected as to prevent, so far as is reasonably practicable, danger arising from such exposure.

Commentary. Regulation 6 requires that electrical equipment shall be capable of withstanding, without giving rise to danger, virtually any effect of its environment. Electrical equipment that may be safe in one environment may be totally inadequate for another. In general those responsible for electrical installation have two options. To situate equipment in an arduous environment and ensure that it is designed and constructed or protected to withstand the environment or to install the equipment in an area outside the arduous environment. The latter may prove to be less expensive and maintenance conditions are likely to be better. The overall result may be more reliable performance. The first option will be unavoidable in some cases. Later sections on Environmental Conditions and Electrical Equipment in Potentially Explosive Atmospheres give further information.

2.7 Regulation 7: Insulation, Protection, and Placing of Conductors

All conductors in a system which may give rise to danger shall either—
   (a) be suitably covered with insulating material and as necessary protected so as to prevent, so far as is reasonably practicable, danger; or
   (b) have such precautions taken in respect of them (including, where appropriate, their being suitably placed) as will prevent, so far as is reasonably practicable, danger.

Commentary. The objective of the regulation is to prevent dangers arising as a result of electric shock, arcing, and sparking. It is important to note that the minimum basic insulation necessary for an electrical system to operate is unlikely to be adequate in any practical situation. The integrity of the insulation must be assured because of the seriousness of the potential hazards. In virtually all cases either double insulation (as in flexible leads and installation wiring cables) will be required unless some other form of protection such as secure enclosure is provided. Experimental installations are not exempt from this regulation and exposed dangerous conductors are not acceptable. The argument that the installation was only temporary and that the provision of effective insulation was not reasonably practical in the circumstances is unlikely to be convincing if there has been a fatality.

2.8 Regulation 8: Earthing or Other Suitable Precautions

Precautions shall be taken, either by earthing or by other suitable means, to prevent danger arising when any conductor (other than a circuit conductor) which may reasonably foreseeably become charged as a result of either the use of a system, or a fault in a system, becomes so charged; and, for the purposes of ensuring compliance with this regulation, a conductor shall be regarded as earthed when it is connected to the general mass of earth by conductors of sufficient strength and current-carrying capability to discharge electrical energy to earth.

Commentary. This regulation applies particularly to conductors such as metallic enclosures of electrical equipment and also to extraneous conductive parts such as metallic water, gas, and oil pipes, ventilation ducting, and structural steelwork. Such items could reasonably foreseeably become electrically charged if they are near or otherwise associated with electrical equipment. In the chemical laboratory the techniques most frequently employed to satisfy this regulation are likely to be earthing and bonding, double insulation, and reduced voltage.

2.9 Regulation 9: Integrity of Referenced Conductors

If a circuit conductor is connected to earth or to any other reference point, nothing which might reasonably be expected to give rise to danger by breaking the electrical continuity or introducing high impedance shall be placed in that conductor unless suitable precautions are taken to prevent that danger.

Commentary. This regulation is particularly important in the case of referenced circuit conductors which function as combined neutral and protective conductors. An open circuit in such circumstances would result in the neutral and protective conductors on the load side of the open circuit becoming live when equipment is energized. Combined neutral and protective conductors should not normally be used in laboratory distribution systems. The regulation is, however, relevant to distribution systems having separate neutral and protective conductors since an open circuited neutral could result in the neutral conductor on the load side of the open circuit becoming live and this may have serious consequences particularly if work is being carried out on the system. For this reason fuses should not be placed in the neutral of the fixed installation. Double pole fusing is becoming common in electronic measuring equipment, particularly imported equipment. This should not create danger if the equipment is self contained and easily isolated from all supplies by means of a plug and socket. Any person carrying out maintenance, calibration, or testing with the equipment live must be aware of the double pole fusing and its significance.

2.10 Regulation 10: Connections

Where necessary to prevent danger, every joint and connection in a system shall be mechanically and electrically suitable for use.

Commentary. This regulation applies to joints and connections in the whole of the system and therefore covers not only the permanent fixed installation but also portable equipment and temporary or experimental installations. Neither this regulation nor any other gives exemption to temporary electrical installations or equipment. The requirement that joints and connections should be mechanically suitable is particularly relevant to experimental rigs. Such installations often only involve small currents and there is a tendency to use wiring of small cross-sectional area and to neglect to prevent mechanical stress on wiring and cables being transmitted to connections which are not designed to take such stress. In such circumstances there is a risk of wiring becoming partially detached and live ends making contact with extraneous conductors which may not have been earthed or bonded so creating the risk of potentially fatal electric shock.

2.11 Regulation 11: Means for Protecting From Excess of Current

Efficient means, suitably located, shall be provided for protecting from excess of current every part of a system as may be necessary to prevent danger.

Commentary. The provision of electrical protection to ensure safety in the event of short circuits or overload conditions needs to be considered when the fixed installation is being designed and when any modifications or extensions are being planned. It is a specialized field and should be entrusted to persons competent in such matters. Those using electrical facilities in the laboratory should recognize the need to advise those with electrical responsibilities of any significant planned changes in circuit loading. Any operation of overload, short circuit, or earth fault protective devices (fuses or circuit breakers) should be brought to the attention of an electrically competent person. Operation of protective devices should be investigated and the underlying problem eliminated before fuses are replaced or circuit breakers reset.

The protection provided in respect of portable equipment will generally be in the form of a fused plug. If sensitive earth leakage protection is not provided by the fixed installation then a plug and earth leakage circuit breaker combined may be necessary. Again it is vitally important that the operation of such devices be investigated before equipment is re-energized.

2.12 Regulation 12: Means for Cutting Off the Supply and for Isolation

(1) Subject to paragraph (3), where necessary to prevent danger, suitable means (including, where appropriate, methods of identifying circuits) shall be available for—

(a) cutting off the supply of electrical energy to any electrical equipment; and

(b) the isolation of any electrical equipment.

(2) In paragraph (1), 'isolation' means the disconnection and separation of the electrical equipment from every source of electrical energy in such a way that this disconnection and separation is secure.

(3) Paragraph (1) shall not apply to electrical equipment which is itself a source of electrical energy but, in such a case as is necessary, precautions shall be taken to prevent, so far as is reasonably practicable, danger.

Commentary. This regulation distinguishes between the means of simply switching off and the means necessary to disconnect equipment so reliably that it is safe to touch circuit conductors which have thereby been made dead. A domestic type light switch is quite adequate to switch off the light but it is not a suitable means of isolating circuit conductors in order to work on them. This is so for two reasons, firstly, the integrity of the switch itself is unlikely to be adequate (to prevent possible death), and secondly, it is not provided with any means of preventing it being inadvertently switched on during the course of the work.

When designing an installation and deciding on the need for isolators the circumstances of normal operation and routine maintenance need to be considered. It may not be desirable or practical to shut down the whole of an air conditioning system when a single extraction fan needs to be replaced or to shut down all fume cupboards when one is being maintained. If inadequate numbers of isolators are provided this can result in pressure being placed on electricians to work with equipment live in order to keep other 'essential' parts of the plant running. This in turn can lead to a contravention of Regulation 14.

The 'suitable means' required by this regulation need to be suitable not only with regard to their electrical capabilities and location in the system but also with regard to their location in the building, accessibility, and the ease with which the circuits they control can be identified.

2.13 Regulation 13: Precautions for Work on Equipment Made Dead

Adequate precautions shall be taken to prevent electrical equipment, which has been made dead in order to prevent danger while work is carried out on or near that equipment, from becoming electrically charged during that work if danger may thereby rise.

Commentary. The object of this regulation is to require that the means of isolation provided in compliance with Regulation 12(1) is used in such a way that the equipment being worked on or near will remain dead for the duration of the work. Unless the means of isolation is completely and solely under the control of the competent person in control of the work, the act of switching the isolator off will not in itself be sufficient. Additional means such as locking will be required to ensure that the supply can not be re-connected so causing danger. If locking facilities are not available the removal of links or fuses or other effective means must be used. Whatever method is employed it is recommended that the means of isolation should bear a temporary label indicating that work is in progress. (See also the Section on Electrical Maintenance.)

2.14 Regulation 14: Work On or Near Live Conductors

No person shall be engaged in any work activity on or so near any live conductor (other than one suitably covered with insulating material so as to prevent danger) that danger may arise unless—

    (a)   it is unreasonable in all the circumstances for it to be dead;

    (b)   it is reasonable in all the circumstances for him to be at work on or near it while it is live; and

(c) suitable precautions (including where necessary the provision of suitable protective equipment) are taken to prevent injury.

Commentary. The effect of this regulation is to prohibit work on or near
dangerous live conductors unless it can be justified and only then, if suitable
precautions are taken to prevent injury. It is suggested that live work in the
chemical laboratory is rarely if ever justified. The only exception is likely to be
fault finding on permanently installed equipment and only then by competent
persons specifically trained to do the task in question and providing suitable
precautions are taken. Live working should not be undertaken simply because
the need has arisen. The method of working needs to be planned in advance and
the necessary training provided. It may be that more than one person is
required in order to assist in taking the necessary precautions. In addition the
environment in which the work is undertaken needs to be suitable. There
should be adequate working space (see Regulation 15) free from other (non-
electrical) hazards which need care and serve to distract attention. In the case of
voltages up to 240/415 V a minimum clear space of about one metre is
recommended. If this space includes areas used as walk-ways or corridors then
such use should be prohibited by barriers or other effective means for the
duration of the work. Those not involved should be excluded from the area.

Routine maintenance functions such as calibration of laboratory equipment
should, if the equipment is well designed, be possible without risk of injury. If
the work needs to be done with the equipment live then the necessary precautions should be incorporated during design and manufacture.

2.15   *Regulation 15: Working Space, Access, and Lighting*

For the purposes of enabling injury to be prevented, adequate working space,
adequate means of access, and adequate lighting shall be provided at all
electrical equipment on which or near which work is being done in circumstances which may give rise to danger.

Commentary. The objective of this regulation is self-evident. There is no
obligation for electrical equipment to be illuminated at all times but in the
majority of cases permanently installed lighting will be required unless access is
very infrequent. Electrical controls and switchgear need to be easily accessible
and lighting immediately available.

Sufficient working space must be available to enable injury to be avoided and
this is particularly vital if work needs to be carried out on or near exposed live
conductors. (See Commentary under Regulation 14.)

2.16   *Regulation 16: Persons to be Competent to Prevent Danger and Injury*

No person shall be engaged in any work activity where technical knowledge or
experience is necessary to prevent danger or, where appropriate, injury, unless
he possesses such knowledge or experience, or is under such degree of supervision as may be appropriate having regard to the nature of the work.

Commentary. Perhaps the first point to note is that the application of the regulation is not limited to electrical work. Competence is not defined in the Regulations but the wording of this regulation is such that competence must include a degree of knowledge of the technicalities of electricity, some practical experience, and adequate knowledge of the equipment and system which is the source of the danger. Since there are few, if any, individuals who are competent in respect of all aspects of everything electrical an important attribute of competent persons is that they should be able to recognize the limits of their own competence and know whether it is safe to proceed.

When an item of laboratory equipment fails to function there may be a temptation, particularly when it is required urgently, to remove the cover, switch on, and prod around with a screw driver in the hope of finding a loose connection. Such temptation should be resisted and indeed prohibited unless the person is competent and unless the work is carried out in a suitable environment and with effective precautions.

2.17 Regulations 17 to 28 inclusive

These apply only to mines.

2.18 Regulation 29: Defence

In any proceedings for an offence consisting of a contravention of regulations 4(4), 5, 8, 9, 10, 11, 12, 13, 14, 15, 16, or 25, it shall be a defence for any person to prove that he took all reasonable steps and exercised all due diligence to avoid the commission of that offence.

Commentary. This defence applies in criminal proceedings and relates to the duties imposed by the regulations listed.

3 Guidance on Electrical Systems and Equipment

3.1 Memorandum of Guidance on the Electricity at Work Regulations

Probably the most comprehensive guidance on the requirements of the EAW Regulations is contained in the Memorandum of Guidance[3] (the Memorandum) prepared by the Health and Safety Executive (HSE) to assist duty holders to meet the requirements of the Regulations. It is strongly recommended that all those with any responsibility for electrical safety in the working environment should read this publication. It is particularly directed towards electrical engineers, electricians, technicians, and managers. The Memorandum does not, however, give detailed design information.

3.2 The Institution of Electrical Engineers Wiring Regulations

The IEE Wiring Regulations—Regulations for Electrical Installations[4] provide comprehensive guidance on the design, selection, erection, inspection, and testing of electrical installations in and about buildings generally. Certain equipment and aspects of installation are excluded from their scope. In particular, the methods of dealing with the explosion hazard in potentially explosive atmospheres, are not covered.

3.3 British and International Standards and Codes of Practice

Specific standards are referred to elsewhere in the text and other standards on electrical equipment and practices are far too numerous to mention here. Many European and British Standards have been harmonized. A comprehensive catalogue and sectional lists of standards are available from the British Standards Institution.[5]

3.4 Health and Safety Executive Guidance

The HSE publishes a wide range of guidance notes and booklets covering many aspects of electrical safety. Those of particular relevance are mentioned elsewhere in the text. The HSE publishes a comprehensive catalogue of all its publications.[6]

3.5 The Status of Guidance

There can be no guarantee that compliance with the foregoing guidance will ensure compliance with the requirements of the Law. However, all the publications mentioned have been prepared by bodies with high national and international reputations for their expertise on electrical safety matters and for their professionalism in preparing standards with industrial and commercial application. During the preparation and drafting of such publications there has been consultation between equipment manufacturers and users, professional bodies, trade and employer's association, and, in many cases, the HSE.

Despite their title the IEE Wiring Regulations are not Statutory requirements but they are widely recognized in the United Kingdom and compliance with them is likely to achieve compliance with those requirements of the EAW Regulations to which they are relevant.

4 The Design of the Electrical Installation

It will generally be cheaper in the long run to install a system which initially has spare capacity than to install the bare minimum and then discover the need for expensive modifications and extensions after a few months. The designer should be supplied with as much information as possible about present and foreseeable electrical requirements.

4.1 Permanently Installed Equipment

Details of all the permanently installed equipment (as opposed to portable and transportable plug-in type equipment) will need to be made available to the designer. The information needed about each item of equipment will include, location, electrical load, and load characteristics (*e.g.* intermittent or continuous), and any special requirements relating to earthing. The latter may be particularly important in the case of computer and other electronic equipment. This information should be available from those who will be responsible for the day-to-day operation of the laboratory and manufacturers of the equipment. It is also important to advise the designer about any equipment which will have high utilization and be essential to the smooth running of the laboratory. The function and utilization of types and groups of equipment should be considered. For example, if all fume cupboards are supplied from one circuit with a single means of isolation, then any maintenance requiring the electrics associated with one fume cupboard to be made dead will result in all fume cupboards being out of use. The safety implication is that pressure may be put on maintenance staff to work with equipment live when live working is not justified (Regulation 14). Depending on the importance attached to having laboratory facilities available there may be justification for having the facility to supply the laboratory from more than one source (*e.g.* an alternative load centre, sub-station, or standby generator).

The main distribution boards and switchgear should be situated in a suitable environment, conveniently close to the laboratory, and easily accessible from floor level.

4.2 Socket Outlets for Portable or Transportable Equipment

Portable equipment which is rated at not more than 3 kW will often be supplied via socket outlets on a ring main but care should be taken to ensure that there will be sufficient diversity between loads. This will be particularly important if there are items of equipment having ratings of the order of kilowatts which are operating for extended periods. Where equipment, although portable, is virtually permanently plugged-in, the contacts in the plug and socket do not benefit from the cleaning action of regular operation and dirty contacts that are heavily loaded may overheat. In these circumstances permanent connection may be more appropriate. In large laboratories several ring mains will be needed to supply socket outlets but even in the smallest laboratory reliance on a single ring circuit is not recommended since maintenance of the ring or extensions will require it to be isolated and this could bring laboratory work to a standstill.

Adequate numbers of socket outlets should be installed to avoid the need for multi-way adapters, flexible cables longer than about 1.5 m, and cables running along benches, across apparatus, or the floor. Socket outlets should not be situated on bench tops or other places where they are likely to be subjected to wet conditions or other contamination. The most suitable place for socket outlets supplying equipment used on benches will usually be on a vertical surface at the back of the bench and at a height to avoid mechanical damage and the effects of spilt liquid. Areas around sinks and other potentially wet areas should be avoided. Wall mounted sockets should be provided for equipment such as floor polishers and transportable equipment which is used at floor level. Such socket outlets should be below the level of bench tops so that cables do not trail across benches but high enough above floor level to avoid mechanical damage or moisture during floor washing operations. It is strongly recommended that socket outlets should be supplied through a sensitive residual current circuit breaker especially where flexible cables are liable to mechanical damage or where there are damp or otherwise onerous conditions.

4.3   Earthed Equipotential Bonding

The need for earthing and bonding was discussed briefly in the Commentary under Regulation 8. The need to earth the exposed conductive parts of electrical equipment in chemical laboratories is no different from most other installations. There are, however, likely to be a variety of extraneous conductive parts such as water pipes, gas pipes, ventilation ducting, and possibly exposed metallic structural parts of the building. Should any of these become live, possibly because of a fault on an associated item of electrical equipment, then serious danger could result particularly if two different extraneous conductive parts can be touched simultaneously and if they are at different potentials (*e.g.* one live and the other earthed). This danger can be significantly reduced if the extraneous conductive parts are electrically bonded together and earthed. This has the effect of creating an equipotential zone. The IEE Regulations give comprehensive advice on the circumstances in which this technique is necessary and also on design criteria.

4.4   Environmental Conditions

Electrical equipment including cables should preferably be installed in a dry, dust-free unpolluted environment where the ambient temperature is constant at around 15–20 °C and where the equipment is not subjected to mechanical stress or physical damage. Where this is not practical equipment has to be selected accordingly and further protected as may be necessary to prevent danger.

Ambient air conditions within the body of the laboratory should not, if suitable for personnel, pose any special problems for electrical equipment. Where there are higher temperatures particular attention will need to be paid to the electrical rating of the equipment (*e.g.* cables may have to be down rated). Where temperatures are extreme, specially designed equipment may be necessary. Electrical equipment is available which is designed to withstand wet and dusty environments. The indication of the degree to which it is protected is signified by its IP Classification which is a two digit number. The first digit is an indication of the degree of protection against entry by solid objects (*e.g.* 0 represents no protection and 6 represents dust-tight). The second digit is an indication of the degree of protection against harmful ingress of water (*e.g.* 0 represents no protection and 8 represents protection against submersion). This system of classification is the subject of International Electrotechnical Commission Publication 529[7] and British Standard 5490.[8] These levels of protection will only be achieved if the equipment is properly installed and cable entries made through appropriate glands. Protection against the ingress of water should not be assumed to be valid for other liquids. Manufacturers should be consulted where protection against other liquids or chemicals is required.

5 Electrical Equipment in Potentially Explosive Atmospheres

The extent of potentially explosive atmospheres in most laboratories should be very limited and in a high proportion of the cases that do occur it should be possible to avoid the need for electrical equipment to be situated in the hazardous zone. It is clearly the responsibility of chemists or chemical engineers rather than electrical engineers to define the existence and extent of any such zones.

The following comments relate to explosive atmospheres arising as a result of a mixture of gas or vapour with air. Mixtures with oxygen pose more severe problems and specialist advice should be obtained. Electrical equipment can be a source of ignition in several ways. The surface temperature may exceed the ignition temperature of the gas or vapour; sparking may occur in the course of normal operation or arcing may occur as a result of a fault. Several techniques are employed to reduce the probability of equipment becoming a source of ignition. These techniques or concepts of protection include:

- intrinsically safe, (ia or ib);
- specially protected, (s);
- flameproof, (d);
- pressurized, (p);
- increased safety, (e);
- type N, (N);
- oil immersed, (o);
- powder filled, (q).

Application standards recognize that these protection concepts do not all provide the same degree of protection and attempt to correlate equipment integrity with the degree of hazard. The most common approach is to 'zone' the hazardous area and specify the higher integrity equipment for the more onerous zones:

- Zone 0—explosive gas/air mixture is present continuously or for long periods.
- Zone 1—explosive gas/air mixture is likely to occur in normal operation.
- Zone 2—explosive gas/air mixture is not likely to occur in normal operation but if it does it will only be for a short period.

United Kingdom and Continental practice does vary in detail but the objective should be to ensure the precautions match the potential hazard.

The selection of suitable equipment may be quite complex and depending on the protection concept adopted may involve:

- matching the protection concept to the hazard (zone);
- ensuring that the temperature classification of the equipment is suitable for the gas or vapour involved;

- ensuring that the equipment is suitable for the particular gas or vapour and for other environmental conditions to which it may be subjected.

The need to use explosion protected electrical equipment in the laboratory may arise in measurement and control applications. In such situations the intrinsic safety concept may prove to be the most flexible and economic. Intrinsic safety is achieved by restricting the electrical energy in that part of the electrical system that is in the hazardous zone so that it is insufficient to ignite the gas or vapour. This is often achieved by the use of 'safety barriers' which are designed to restrict the current and voltage available in the hazardous zone even under certain fault conditions. The barriers themselves must be situated in a 'safe zone' but equipment in the hazardous zone supplied only via the barrier does not have to be explosion protected. There are some constraints, however, as there is a need to limit the energy that can be stored in the intrinsically safe circuits.

Although the EAW Regulation do not specifically require certification, it is strongly recommended that only equipment certified as being designed and constructed to comply with an appropriate standard should be used in potentially explosive atmospheres. The need for suitably protected equipment extends to cables, their termination and glands, and if due attention is not paid to these aspects the high integrity of the equipment itself will be lost.

Comprehensive guidance on the selection, installation and maintenance of electrical apparatus for use in potentially explosive atmospheres is given in British Standard Code of Practice BS 5345 Parts 1 to 9.[9] Another useful source of information is the Health and Safety Executive publication HS(G)22.[10]

Potentially explosive atmospheres that arise as a result of flammable dust in air will also require appropriately protected electrical equipment. The problems are not quite so complex and the solution is based primarily on limiting the external surface temperature and ensuring that dust cannot enter enclosures. British Standard 6467[11] gives advice on this topic but specifically excludes dust from materials that are themselves regarded as 'explosives'.

6   Portable Equipment

The following advice applies equally to portable and transportable equipment and includes equipment that is generally used on the laboratory bench. There are several techniques used individually or in combination to help make portable equipment safe. These include the use of:

- safety extra low voltage (SELV) which is a voltage not exceeding 50 V which is isolated from earth and usually derived from the mains via a safety isolating transformer;
- reduced voltage which is a voltage not exceeding 110 V usually derived from an isolating transformer with the mid-point of its secondary winding earthed so reducing the shock to earth voltage to 55 V;
- double insulation which is high integrity insulation provided by two discrete insulating barriers or some form of reinforced insulation between live circuit conductors and unearthed exposed conductive parts;

- earth continuity monitoring achieved by using an additional core in the supply cable to pass an extra low voltage source monitoring current through the protective (earthing) conductor and its connection to the exposed conductive parts. If the earth connection is lost, the monitoring current is interrupted and the equipment is automatically disconnected from the mains supply. This system has the disadvantage that it requires the extra core and non-standard plugs and sockets;
- sensitive earth leakage protection achieved by using a residual current circuit breaker which will disconnect the mains supply if earth leakage current is detected. To be effective such circuit breakers should require no more than 30 mA to trip. They should be installed to protect the circuit supplying the socket outlet or, failing this, a plug-in device can be used at the socket outlet itself.

Experience suggests that, where practical, SELV and reduced voltages are to be preferred. SELV is particularly suitable for relatively low power applications such as soldering irons where there is a high risk of damage to the flexible cable by the heat of the iron. Reduced voltage has also proved its worth for supplying portable tools used in wet and arduous environments such as construction sites. Double insulation has proved effective in reducing accidents due to faults within the equipment itself but the flexible supply cable, although itself double insulated, is still vulnerable to damage in arduous conditions. Earthing is effective in reducing the probability of electric shock providing the earth continuity can be assured but here again damage to the supply cable is a potential source of danger.

For the foregoing reasons it is recommended that mains voltage equipment when used in damp or otherwise conducting or arduous conditions should be supplied via a sensitive earth leakage circuit breaker (residual current device). Alternatively, in the case of earthed (Class 1) equipment, circulating current earth monitoring can be employed but this does not provide protection against the effects of damage to the supply cable.

The Health and Safety Executive Guidance Note PM 32[12] on the safe use of portable electrical apparatus gives further information.

7 Fume Cupboards

Any electrical equipment within a fume cupboard is, over a period, likely to be subjected to wet or humid conditions; the corrosive effect of chemicals; solvents; electrolytes; dusts (which may be flammable, conducting, or non-conducting); flammable materials and atmospheres; and to temperature variations. To anticipate all the possibilities in advance is difficult and to obtain suitable equipment would be extremely difficult. The best advice must therefore be that no electrical equipment should be permanently installed within the fume cupboard.

Illumination can be provided by means of an externally mounted light shining through a suitably sealed toughened glass panel. Socket outlets for supplying portable equipment temporarily inside the fume cupboard should be situated outside the cupboard possibly on the front side facias above the level of the sash so that flexible leads run down from the plug and socket into the fume cupboard. This avoids the possibility of liquid splashes within the cupboard running down the flexible cables out of the fume cupboard and into the plugs.

Any portable equipment used within the fume cupboard should be suitably protected for the environment that exists during its use. When not in use such equipment should be removed so that it is not adversely affected by environments for which it is not suitable.

8   The Need for Secure Supplies and the Integrity of Controls

There will be many laboratory facilities that need electricity to function. Each of these facilities should be considered to determine whether its loss would be likely to create a hazard either directly or indirectly through loss of control of experimental or other reactions. Examples of possible safety related facilities include:

- lighting;
- water supply, if pumped;
- compressed air including instrument air;
- vacuum;
- ventilation and fume cupboard extraction systems;
- alarm systems and annunciators;
- instrumentation and control of processes.

Without specific knowledge of working conditions and processes it is impossible to give detailed advice, however, some general guidance is offered for consideration. If loss of electrical supplies would mean that staff had to take some remedial action before vacating the laboratory then emergency lighting should be provided. If loss of any electrically dependent function would result in the realization of a potential hazard then it will be necessary to have alternative high integrity supplies to the safety critical equipment or provide some other independent safeguard that will maintain safe conditions. Clearly where the hazards are more serious additional precautions may be necessary.

Hazards can be created not only by loss of electrical supplies but also as a result of loss of control of potentially dangerous processes. Failure of instrumentation, control valves, and control system logic are a few examples. The introduction of programmable logic controllers which are normally very reliable has introduced a new dimension as their failure modes are difficult to predict and there is a tendency for all control functions to be routed through one controller. This may mean there are no independent control loops. Unless the potential hazard is relatively minor some back up protection of adequate integrity should be provided. The Health and Safety Executive has published advice specifically on Programmable Electronic Systems[13] but this also provides useful guidance on the analysis of safety related control systems generally.

9   Electrical Maintenance

Two aspects of electrical maintenance will be considered, the need for maintenance of electrical systems and equipment and the need to do that maintenance in a safe manner.

9.1   The Need for Electrical Maintenance—the Fixed Installation

The laboratory electrical distribution system if well designed and installed should require minimal maintenance. There is a tendency to assume that because everything is working then everything is satisfactory and this is not necessarily the case. Over a period of time connections can become loose or otherwise deteriorate, sealing gaskets can deteriorate and allow ingress of dust. If discovered early enough such faults are comparatively easily and cheaply corrected. If left they could prove fatal. For this reason a scheme of preventive maintenance should be implemented. The frequency of routine inspection and testing is best determined by experience. Records of faults found and the results of tests provide a good guide. A long list of faults or significant changes in test results indicate the need for a shorter interval between inspections or perhaps the need for some fundamental changes. It is suggested that the fixed installation should be inspected and tested after it has been in operation about a year and that subsequent inspection could be extended to say five years if experience so dictates. It must be stressed, however, that a system of planned maintenance by electrically competent persons does not mean that the installation can be ignored between inspections. There are simple checks and observations that can and should be carried out by laboratory staff in the course of their regular activities. These include the following which should be investigated and corrected by an electrically competent person:

- any suggestion that someone may have received an electric shock however minor it may appear to have been;
- exposed cables showing evidence of physical damage;
- insecure or inadequately protected cable entries to equipment;
- evidence of equipment overheating particularly socket outlets;
- changes to the local environment that cast doubt on the adequacy of the protection provided by or for electrical equipment;
- intermittent faults such as a flickering light or temporary loss of supply which may be evidence of a bad connection.

9.2   The Need for Electrical Maintenance—Portable Equipment

Modern portable electrical equipment is generally manufactured to high safety standards yet it probably still accounts for more fatal accidents than any other single cause. Accidents usually occur because the equipment is used in conditions for which it was not intended or, most likely, because it has not been maintained over years of use or mis-use. Extension leads can be a particular problem because they are often used in arduous environments for which they are not suitable. They are inevitably subjected to regular mechanical stress and it is vital that all extension leads are included with other portable equipment in a routine preventative maintenance scheme.

It is strongly recommended that each item of portable equipment should be identified with a number and that number and a brief description of the equipment recorded in a register. This arrangement means that every item can be called in for routine inspection and provides a means of identifying any item which is not on the register. It is important to record defects found and the results of any tests since this can be a useful indication of slow deterioration or abuse or use in conditions for which the equipment is not suitable. Total reliance should not rest on the periodic inspection. Those using the equipment should report any suspect equipment and take it out of service pending investigation and any necessary repair. The Health and Safety Executive Guidance Note on the Safe Use of Portable Electrical Apparatus[12] provides a useful check-list of the routine inspections and tests that should be carried out by a competent person.

9.3   Safe Electrical Maintenance

No electrical work should be done in the laboratory without the authority of the person in charge of the laboratory who has responsibility for ensuring that those carrying out electrical work are not subjected to other hazards. Agreement is needed as to where, when, and how the work is to be carried out.

Electricians working in laboratories may be at risk from both chemical and microbiological materials and it is essential that the Control of Substances Hazardous to Health Regulations assessment (of both the host and employing organizations) take these risks into account.

Assistance from other trades or professions may be needed if hazards other than electrical hazards need to be isolated or protected against. In some cases even the electrically competent person may not have the experience or local knowledge to make the necessary electrical isolations. This situation may arise in the case of a manufacturer's service technician who may not be familiar with or have the authority to operate local switchgear. It is suggested that the following features will each add to the need to implement a permit to work system:

- hazards other than electrical involved requiring action by persons other than the one carrying out the work;
- the means of isolating the hazards (including the electrical hazards) are remote, not straightforward, or not under the control of the person doing the work;
- some hazards remain after isolation of others and therefore special precautions are needed;
- the work requires precise definition in order to avoid danger;
- the work involves several trades or professions;
- the work will extend over more than one work period or shift;
- the work may adversely affect the safety integrity of processes or systems.

If a permit to work system is used then the electrical aspects that it covers should include:

- the names of the persons issuing and receiving the permit and their signatures;
- a clear unambiguous statement defining the work to be done and its limitations;
- details of the means of isolation, discharge of any stored energy, and earthing;
- an indication of how the equipment to be worked on, or near, has been proved to be dead;
- details of any special precautions to be taken to minimize the risk from remaining hazards;
- the time the permit is issued;
- provision for subsequent clearance and cancellation including an indication of whether or not the equipment can be returned to service, the time, and the appropriate signatures.

In virtually all circumstances conductors which have been made dead should be proved dead before work commences. The voltage indicator should itself be proved before and after the test. An exception to this would be for work on a self-contained item of portable equipment which has been unplugged and where the plug and all the flexible cable are in the permanent view of the person working on the equipment and where the equipment is incapable of storing electrical energy. Suitable test equipment for use by electricians is described in an HSE Guidance Note[14] on the subject.

Maintenance involving the fixed installation will clearly have to be undertaken *in situ*, however, portable and transportable equipment should preferably be removed from the laboratory to a suitable electrical repair workshop. There are two reasons for this. The laboratory environment could well be a source of contamination for the internal components which are unlikely to be protected once the equipment has been dismantled and, if any work is to be done in which dangerous live conductors are exposed, (*e.g.* testing or calibration) then the likely conducting and earthy conditions to be found in many laboratories will increase the risks. Note the relevance of Regulation 14 to testing and calibration.

10   References

1. Effects of current passing through the human body, International Electrotechnical Commission Report, Publication 479–1, 1984.
2. Statutory Instrument 1989, No. 635, Health and Safety, The Electricity at Work Regulations 1989, HMSO.
3. Memorandum of Guidance on the Electricity at Work Regulations 1989, Health and Safety Executive, HMSO.
4. The IEE Wiring Regulations, Regulations for Electrical Installations, The Institution of Electrical Engineers.
5. BSI Catalogue, British Standards Institution.
6. Publications in Series: List of HSC/HSE publications, Health and Safety Executive.

 7. Classification of Degrees of Protection Provided by Enclosures, IEC Standard Publication 529, International Electrotechnical Commission.
 8. Specification for Degrees of Protection Provided by Enclosures, British Standard 5490.
 9. Code of Practice for the Selection, Installation, and Maintenance of Electrical Apparatus for Use in Potentially Explosive Atmospheres (other than mining applications or explosive processing and manufacture), British Standard 5345: Part 1, Basic Requirements for All Parts of the Code.
10. Health and Safety Series Booklet HS(G)22, Electrical Apparatus for Use in Potentially Explosive Atmospheres, Health and Safety Executive, HMSO.
11. Electrical Apparatus with Protection by Enclosure for Use in the Presence of Combustible Dusts, British Standard 6467: Part 1, Specification for apparatus.
12. The Safe Use of Portable Electrical Apparatus (electrical safety), Health and Safety Executive Guidance Note PM 32, HMSO.
13. Programmable Electronic Systems in Safety Related Applications: An Introductory Guide, 1987, and also: General Technical Guidelines, 1987, Health and Safety Executive, HMSO.
14. Electrical Test Equipment for Use by Electricians, Health and Safety Executive Guidance Note GS 38, HMSO.

CHAPTER 11

Chemical Laboratories—An American View

H. H. FAWCETT

1 Introduction

Chemistry is a basic physical science, which is both dynamic and subject to change as new frontiers are explored or discovered. Laboratories, which a decade or two ago were relatively conservative in their approaches, have evolved in unexpected directions. The US Occupational Safety and Health Administration has recently noted 34 214 laboratories of widely diverse size and missions, with 934 000 scientists engaged in these laboratories in the United States.

Chemical health and safety, once considered a hindrance or deterrent to laboratory work, is now slowly being recognized and observed as an essential part of the operation. Table 1 shows that firms alone accounted for some $32 000 000 of damage over a four year period. This recognition is especially important since chemistry no longer has the high priority as in previous years in attracting students into scholastic and academic circles. A variety of plans and techniques has been suggested for presenting chemistry as a more attractive career to the oncoming generation of students. Professional organizations and large corporations sponsor appeals through the electronic and print media. The American Chemical Society, for example, has instituted an attractive program for Prehigh School Chemistry in the elementary grades, in which experienced chemists and engineers are invited as guest speakers into the classroom, to acquaint students with the fundamentals of chemistry, and they usually supervise simple 'hands-on' experiments, supplementing and reinforcing the regular instructor's viewpoint. Properly presented, programs from more mature scientists are well received by most students. Other technical societies which have been conducting active programs to interest and confirm young chemists and engineers include the American Institute of Chemists, American Association for the Advancement of Science, and the National Science Foundation. Posters on laboratory hazards and many chemical safety articles have been published since 1940 in the *HEXAGON* magazine by the Safety Committee of Alpha Chi Sigma.

Table 1 *US Laboratory Fires Originating in Laboratory Areas Reported to US Fire Departments 1984–1988, Annual Average*

Fixed property use	Fires	Civilian injuries	Direct Property Damage
Laboratories; insufficient information to classify further	25	0	$497 000
Chemical or medical laboratory	67	9	$2 029 000
Personnel or psychological laboratory	1	0	$9 000
Radioactive materials laboratory	3	0	$40 000
Electrical or electronic laboratory	14	0	$69 000
Agricultural laboratory	7	0	$29 000
General research laboratory	53	4	$290 000
Laboratories not classified above	9	0	$123 000
Total	209	15	$32 242 000

Source: 1984–1988 NFIRS, NFPA Survey

As chemistry and laboratories have changed, so have the personnel. The percentage of female chemists has increased, as has the number of chemists and engineers from other countries where English is not the native tongue.

Since World War II, many highly competent and trained chemists and engineers who have emigrated to the United States are considered deficient in communication skills. One major research laboratory conducts an on-the-job course for technical employees with language skills considered inadequate to communicate clearly both orally and in writing. Proper presentation and reception of ideas is essential to operations both for efficiency, cost/benefit ratio, and health and safety.

The discovery and registry of new chemical compounds continues. Chemical Abstracts Service reports nearly 600 000 new listings per year; the total as of late 1990 is over 10 000 000. Unfortunately, registration of a compound alone does not consider the 'real-world' needs of the chemists and others who will ultimately make and use the compound. However, the Chemical Abstracts Service has recently made available much updated information on a subscription basis, including properties of the compound such as toxicity, flammability (or inflammability), reactivity, effects on humans as well as animals and the environment; all such aspects are often neglected or ignored, even today, until mandated by law, when the producer attempts to market the compound, or to have the substance accepted as not hazardous by the ultimate user and dispenser. The first researchers and users may thus still be at risk until more is known and published about the compound, its reactivity, and human/environmental effects. Such testing and eventual certification are essential to insure that proper warning labels and precautionary statements on the product are both accurate and complete, and should be printed in letters large enough to be read and clearly understood.

The appearance in the laboratory of plastic bottles and other containers as a supplement or replacement for glass should be noted with interest. While it is true that most plastic bottles offer protection from simple fracture, care must be exercised to prevent contact with sharp objects or exposure to temperatures which may soften, melt, or ignite the plastic. Certain plastic and also a considerable number of glass bottles for reagent are now fabricated with a special chemically-inert liner to prevent contamination of the product. This lining procedure could be used in a practical real-world way; lead crystal decanters and glasses, widely used in many social circles for dining may leach relatively high levels of lead into wine and spirits, as was recently noted and reported. The highest lead levels occurred in wine that had been stored in crystal made of 32% lead oxide. Only laboratory examination would protect the user from the possibility of lead poisoning. Lead is a matter of much concern in the US. Tetraethyl lead is no longer permitted in motor fuel (petrol or gasoline) in the US. However, an estimated 12 million children under 7 years old are believed exposed to high lead levels from paint in their homes.

Care in the packaging and handling of chemicals continues to be important. Recently, a 500 ml bottle of peracetic acid was packaged in what appeared to be paper absorbent in a metal shipping container. The container had been placed on its side, allowing liquid to leak through the bottle's vented cap (peracetic acid bottles are vented to allow release of the pressure which builds up from the normal slow decomposition of the material). The spilled liquid reacted exothermically with the absorbent, heating the metal can until it became hot to the touch. Luckily, the heat generated did not trigger any further reaction. Heating peracetic acid above 65 °C may cause self-accelerating decomposition, and temperatures above 110 °C will cause violent explosion. This incident emphasizes the importance of ensuring compatibility between packaging materials and reactive chemicals such as peroxides. Tanks, barrels, bottles, and bags made of materials that resist chemical attack are often lined with thermoplastic polymers.

Many laboratories are engaged in activities for purposes other than traditional experimentation and instruction. Regulations involving chemicals in America are becoming more complex and numerous, as the Federal government, the states, cities, and even counties attempt to control chemicals in various ways. We note that similar questions are arising in the United Kingdom as the Control of Substances Hazardous to Health (COSHH) develops controls for specific compounds.

As will be discussed later in this chapter, the OSHA Occupational Exposures to Hazardous Chemicals in Laboratories Act was promulgated to be in effect in most chemical laboratories by 31 January 1991. The broadened base and interest in chemically-related science, including the environment as well as safety and health, have increased demands for more 'testing for compliance'. These interests have brought into the laboratory highly sophisticated and expensive equipment with a wide range of use, from monitoring odours to identifying and measuring trace amounts of chemicals into the parts per trillion. These precisely small analytical limits of detection are made possible by high performance liquid chromatography (HPLC) and other analytical equipment which are important workday tools. To cite recent examples where such equipment found important uses, benzene (C_6H_6), in very low concentrations, was detected in a widely-consumed, imported, bottled drinking water. The product was recalled and replaced with water exhibiting, on analysis, no contamination, taste, or odour, but at a high cost and much confusion to all concerned. However, on 10 April 1991, the Food and Drug Administration seriously questioned accuracy of claims of purity in bottled water sold widely in the US. Not long before, the question of the pesticide Alar (succinic acid 2,2'-dimethylhydrazide) in apples eaten by children, and of cyanide suspected in two grapes from a very large shipment, gave the public much concern until analyses showed the risks were insignificant. Of much greater significance was the detection of aflatoxin-tainted corn, since aflatoxin is a proven carcinogen which is produced by natural processes. Chemical laboratories have a unique service in protecting the public by analyses of food and beverages, and their reaction-breakdown products.

2　Orientation and Reviews

The increasing number and complexity of regulations at Federal, state, city, and local levels involving chemicals make it mandatory that even chemists and engineers who perform 'routine' assignments be aware of the implications of their actions and work.

Many laws and regulations regarding chemicals in various aspects, have been enacted by Congress and the regulatory agencies in their attempts to control some specific incidents or undesired phases of chemical manufacture, transport, use, or disposal. While most of these are not aimed toward the laboratory, the implications of these laws are important to even the 'bench chemist' who may feel no direct day-by-day involvement but in fact may well have a significant role.

Until recent times, OSHA (Occuptional Safety and Health Administration) rules were largely directed to the industries and commercial operations while the EPA (Environmental Protection Agency) approached chemical control in terms of effects on the environment and ecosystem (air, water, land, and the upper atmosphere). However, recently, especially since the Chemical Laboratory Standard was introduced by OSHA, all chemists and engineers should be alert to the laws and how they relate to the profession and practice of chemistry. For example, during the past decade, the Material Safety Data Sheet (MSDS) has become an important instrument for informing userts of chemicals of the potential as well as real hazards which a chemical can present if improperly shipped, poorly labelled, badly handled or reacted, spilled, or improperly disposed. The MSDS is now considered a 'must' for consideration of all who may have contact with the chemical in question. The supplier of the chemical is required to develop and transmit such a document with each shipment. It is intended to highlight (preferably with references) those physical and chemical properties which should be considered, especially as to the human/chemical interface. Unfortunately, not all MSDSs now in laboratories and industry are without significant flaws. Most are written for users of large amounts, or written and reviewed by lawyers to minimize liability. Some are incomplete, inaccurate, and misleading; all should be revised frequently as new data become available, especially in respect of toxicity, reactivity, first-aid, or medical aspects. The Chemical Manufacturers' Association (US) has proposed an improved format and data form for the MSDS which should assist in presenting important information in a more understandable way. The MSDS sheets received by every laboratory with each shipment of a chemical should be carefully reviewed for accuracy and completeness, and must be plainly displayed for easy and quick reference, not just to laboratory workers, but also to emergency services including fire and paramedical personnel who may arrive during non-working hours when supervisory and laboratory researchers are not available. Dr Wayne Wood, Safety Manager of McGill University, Montreal, Canada has recently commented on the relatively poor quality of the information in many MSDS sheets, as well as the necessity of buying the MSDSs from some suppliers—clearly not the intent of the laws either in Canada or the US.

Sgt Mark E. Davis of the Montgomery Country, Maryland (MD) fire service has presented an analysis of chemical laboratories from the responders' view. 'Basically there are two type of chemistry labs. that can be found in educational occupancies, the learning laboratory and the research laboratory. The major difference between these two types of labs. centers around the personnel involved in the operations of such facilities. In the simplest of learning labs., the high school chemistry lab., a significant number of inexperienced students are guided through experiments by an experienced leader. In the simplest of research labs., a university research lab., a single or group of experienced, knowledgeable researchers work under varying guidelines toward a hypothetical solution. As one can see, a key difference between the two types of labs. is the type of personnel involved in the experiments.

When we talk about a school chemistry lab. we are talking about an important learning environment that is also full of hazards. We have students operating in these labs. who have not yet mastered lab. safety skills. We have a wide variety of common chemicals in relatively small quantities. It is also quite possible that we may have some unique chemical surprises tucked away in the corner of the storage room. And most likely of all, we will have a fairly low instructor-to-student ratio so that not all students are supervised equally.

Accidents can be a common occurrence in school chem. labs. Broken glassware, spilled liquids, incorrect mixing of chemical substances, overheated instruments—all can mean embarrassing moments for the novice chemist. They also all can mean injury or harm to the student. A good instructor who follows the rules of lab. safety will have a significant positive impact on the safety of everyone in the lab.

Two common scenarios involving the fire department and the school chem. lab. are the chemical spill and the out-dated, unstable chemicals often found in the storage room. The first scenario usually involves a strange, but irritating, odour somewhere in the building, which turns out to be a broken 5 litre bottle of sulfuric acid down in the science wing. The second scenario usually involves the trusting custodian who suddenly locates an old container of highly unstable or reactive chemical behind some dusty boxes in the storage room. Both of these scenarios can turn into messy embarrassments for emergency personnel if handled incorrectly. Picric acid is another chemical which has caused much fear and concern if it has not been properly wet during its storage life, since it is well documented as an explosive when dry. A drawback of picric acid is its tendency to form impact-sensitive metal salts (picrates) when in direct contact in explosives with shell walls and metal caps.

While lab. accidents may happen sooner or later, a lab. can be made safer through the use of good safety practices and protective equipment. Chemistry labs. should have adequate fixed protection systems to handle the flammable hazards present. Fire extinguishers suitable for the particular hazards of the lab. should be installed, and lab. personnel trained in their use. A lab. should be well ventilated allowing fresh air to circulate about. Fume cupboards should be used to remove dusts, vapours, mist, and fumes produced from laboratory operations. A lab. should also have a safety shower and eye bath for the protection of personnel from acids, caustics, corrosives, and other emergencies in which large volumes of water are needed for removal of the chemical from eyes, skin, and clothing, followed, if necessary, by immediate medical attention. Lab. incidents involving the mixing of chemicals, often incompatible substances, will require special handling. The synergistic effects of these mixtures can sometimes be much worse than the effect of the individual substances themselves. A student who accidentally (or deliberately) creates a wrong mixture may not know or care exactly what was combined in that mixture. Care must be exercised when handling all chemical substances from the time of arrival to the time of proper disposal. Records of each container should be available showing date of arrival and date and method of disposal. Even materials not usually of concern should be considered; a new labelling standard which became effective November 18, 1990 will encourage safer use of art and craft materials by students and artists. These products include solvents, spray paint, silk-screen inks, adhesives, and any other substance marketed or represented as suitable for use in any phase of the creation of visual or graphic art.

A school chemistry lab. is full of novices who can easily make an error which could become extremely interesting very fast. The safety of a lab. depends upon everyone's attitude, the safety procedures of the lab., and the protective equipment located and immediately available in the facility. Spills and the locating of unknown substances tend to rank high in the frequency of chem. lab. incidents. It often seems as though the fire service is called upon simply because the public believes firefighters can do anything. Unfortunately, too many firefighters believe that they can and end up in a perilous situation.'

Programmes for formal on-the-job training sessions, often revised and reviewed on a semi-annual basis (and especially important for all new personnel, both technical and non-technical) are also recommended to bring all employees up-to-date on the newer rules and facilities. This is especially important for persons joining the laboratory, since knowledge of the facility, exits, and emergency equipment, supplemented by appropriate personal action, is essential for safe conduct in unusual incidents. All personnel must know the standard operational procedures designed to prevent unwanted incidents and when undesirable situations occur, to minimize confusion. Among the elements in such plans and procedures must be clearly marked exits, the location of alarms, emergency telephone numbers, and equipment, the availability, proper selection, and use of fire extinguishers for the incident (more than one fire control agent may be on hand), the location and prompt use of the proper respiratory protective devices and specific information, and appropriate alerts to the trained emergency responders who may arrive. The outside responders must have visited the facility on a regular basis to become familiar with the lay-out, evacuation routes, chemicals, and ventilation. Even in a small laboratory, it is highly important to have key personnel (preferably organized into a formal brigade) to assist other personnel in prompt action. An emergency medical service (EMS) should be an integral part of the hazardous materials response teams (HAZMAT). In this connection, several chemical companies in the US have recently found that peer pressure from fellow-employees, supplemented by technical 'safety' supervision and management support, has had a very positive effect on reducing or preventing accidents. Sprinklers or other water applicators should be carefully examined, especially if the facility has water-reactive chemicals or radioactive isotopes. Halon extinguishing systems have been installed in many large computer or electrical systems rooms with a built-in warning system. Responders should know which of the 22 different agents available in the US are on site for fire suppression and their effect on personnel. The National Fire Protection Association symbol 704-M, which has provision for indicating the fire potential, the health hazard, chemical reactivity, and water-incompatibility or radioactivity is a very valuable aid to responders as well as for normal operations. CHEMTREC, a service of the Chemical Manufacturers' Association, offers emergency information regarding chemicals by phone on a 24 hour basis by calling 1-800-424-9300. Information from the American Chemical Society Library Service is reached on 1-800-227-5558, option 6. The Division of Chemical Health and Safety of the American Chemical Society can be reached on (314) 334-3827 or Fax Number (314) 324-9972.

In a university laboratory recently, an unidentified person discarded an opened metal can of metallic sodium scrap into a waste paper container. The paper ignited; the security and fire responders, not suspecting chemicals in the burning paper, poured water onto the container, with near-blinding effects on the four responders when the burning mixture, which evolved hydrogen from the reaction of water on the sodium, exploded. In another laboratory, five university students were injured recently when a bottle of sulfuric acid apparently exploded during a chemical laboratory experiment. It is well known that the introduction of even relatively small amounts of volatile organic solvents into concentrated acid can produce exothermic reactions. Systematic separation of chemical wastes followed by proper disposal according to the law is essential if such incidents are to be avoided.

Many unusual incidents occur during 'off-hours' or at night when regular experienced personnel are not on duty. Careful instructions, as well as posted emergency telephone numbers, and, if available, radio communication techniques, must be given in advance in writing to both the security guards and to the responders. A notice posted on the laboratory door or in the adjacent hallway specifying who is responsible for the experiment and how to contact him or her is highly important to the responders and security personnel.

Many laboratories attempt to prohibit scientists and others from working alone or during off-hours, but occasionally such actions cannot be avoided. Operations and equipment which have been cited as especially hazardous in such situations include high-energy materials, flammable liquids, toxic gases, toxic liquids, or hazardous solids, high-pressure systems, cryogenic materials, rotating parts of equipment and machinery, electrical systems, absorption towers, cold rooms, and any confined space, vessel, or chamber from which escape could be difficult or delayed. In such cases, the scientist should check in with the security personnel, and re-check on a regular schedule, in order to insure that all operations are normal. Power failures or other loss of essential services may create serious incidents from routine emergencies. Time is very important in any incident, and even more so when help is delayed or not immediately at hand.

Experimental efforts in the laboratory or elsewhere can never be completely 'safe', since by the very nature of the operation, unknowns are present. Many misunderstandings can arise when a dedicated safety or health professional approaches a scientist with specific suggestions for safer procedures or arrangements. The scientist may not fully understand or appreciate the importance of safety to his or her personal health, to his or her career, or to the success of the experiment at hand. One suggestion often accepted is to reduce the size of the experiment, especially if documentation can be produced to suggest possible danger. Another idea is to contain the experiment within a strong, well-ventilated fume cupboard or area. A complete and update literature search of the safety aspects should be made before the experiment is set up. In this context, it should be noted that not all 'safety data bases' are fully reflective of the literature; many of the older but significant references may not have been entered onto computerized data bases. It is wise to inquire just what time period and references the compilation includes, since chemical safety literature existed before computers. Even a hundred-year-old reference may be of specific value today. The author shares the view of Dr P. Lewis who writes 'the fundamental requirement for success in raising standards of health (and safety) is to improve attitudes to—and awareness of—good practice by individuals at all levels in companies (and academic circles). This cannot be achieved solely by imposing detailed regulations. European legislation therefore consists of simple enabling instruments covering principles of good practice. Attention should be paid to cost/benefit considerations so that health resources can be targeted sensibly to areas of highest priority. The Chemical Industry Association (UK) believes that the Control of Substances Hazardous to Health (COSHH) philosophy should be applied beyond the European Community, in other parts of the industrial-

ized world—particularly the US and Japan—as well as in downstream industries which use hazardous substances. Improved standards will be developed
more effectively by applying peer pressure, resulting from a sharing of industry
experiences, rather than reliance on a detailed regime of government regulations that cannot be properly enforced. The regulatory system should thus be
seen as enabling rather than prescriptive.' (Words in parentheses added by this
author.) Quote is by permission of Dr Lewis.

In an earlier study, we reviewed basic essential information regarding
reagents, conditions, and reactions needed for reasonable assurance of safety
and health. These are listed below in a condensed form:

1   If a gas or liquid, what is flammability (inflammability) in the presence of
    air, or with oxygen, chlorine, fluorine, and other gases? Over what
    range? If a liquid, what are the flash- and fire-points? By what method
    was the data obtained?

2   If flammable (inflammable), what are the auto-ignition or self-ignition
    temperatures? What are the products of combustion and their toxicity?

3   If not defined as flammable (inflammable), but combustible or burnable,
    does the material undergo decomposition or pyrolysis; how hazardous
    are the products?

4   What is the vapour density compared to air at STP?

5   What acute or chronic exposure effects have been reported from in-
    halation and skin absorption?

6   What bioelimination mechanism is known (*i.e.* blood, urine, hair, *etc.*)?
    Should this be monitored by a laboratory with specific expertise for
    such analyses?

7   What type of respiratory protection (gas mask, self-contained breathing
    apparatus, or air-supplied respirator) is needed, if any? Care in selection
    and use is vital.

8   Is skin contact (chemical burns or irritation), or skin penetration, a
    problem? Is proper and adequate eye protection worn at all times? Are
    gloves and other protective clothing properly selected and used if
    needed?

9   Should the experiment be conducted within a properly operating and
    clean fume cupboard with sufficient air velocity and protection for the
    chemist and others in case a runaway reaction, fire, or explosion should
    occur?

10 What first aid treatment is known? Do the medical doctors, nurses, and
    paramedic personnel available fully understand the effects of the specific
    chemicals? Is medical consultation available? Does the MSDS supply
    this information? Is the MSDS available from more than one source so
    comparisons can be made?

11 What analytical instrumentation is available for determining injurious
    concentrations of the gas or vapours in air, assuming that Permissible
    Exposure Limits (PELs) or Threshold Limit Values (TLVs), Time
    Weighted Average (TWAs), or Short Term Exposure Limit (TELs) have
    been recommended by the American Conference of Governmental

Industrial Hygienists. The Occupational Safety and Health Permissible Exposure Limits (PELs) are the official limits for enforcement purposes. Many chemicals have been tested for carcinogenic, mutagenic, or teratogenic effects, but many have not. (Since 1984, Occupational Exposure Limits (OELs) for UK are set, published, and revised annually by the Health & Safety Executive, as a Guidance Note.)

12 What fire control agent is effective on the material, and how will spills or residues be cleaned up and ultimately disposed of properly?

13 Will the waste from the cleanup be considered hazardous? If so, who will handle it?

3 The OSHA Chemical Laboratory Standard

The recognition that chemicals, even in relatively small quantities handled on a short-term basis, can present a hazard to the health and safety of workers (including chemists, engineers, technicians, emergency responders to emergency conditions, and paramedics, as well as others who may occupy the same or adjacent buildings) has long been an item of discussion. One opinion is that since chemists and other scientists and laboratory personnel have superior education, experience, and 'know-how', they should be exempt from 'rules, regulations, and other restraining thoughts'. An opposing view is that many chemists and laboratory workers in industry and academic workplaces have received only nominal training and supervision regarding safety. Information regarding serious or even fatal effects of acute and chronic exposures to benzene, carbon tetrachloride, and numerous other common chemicals have been largely ignored for years even after the medical and epidemiological data were generally available and published. Time is often significant; when vinyl chloride monomer was the subject of much debate and rule-making in 1974, attention was directed primarily to angiosarcoma, as a result of high-level exposures in the cleaning and handling of tanks. At that time, little consideration was given to the chronic low-level exposures. Detailed effects of both low-level and high-level exposures (causing the rare chronic liver disease) were finally published by the Public Health Service in June 1990.

The Occupational Safety and Health Act (PL 91–596, 29 December 1970) marked the first serious effort of the federal government to bring the personal injuries and illnesses arising from inadequate and incomplete attention to health and safety under formal supervision and control. During the past 20 years, it has had a positive effect on safety and occupational health. At first resistance to these measures was on the basis of cost—health and safety were considered unnecessary expenses and most losses were covered by insurance or other means. However, as the enforcement and legal aspects of the Act became more widely known, and since the fines for non-compliance have mounted, OSHA has had a positive effect on the 14 000 000 industrial locations in the United States. Other federal government laws and regulations and other agencies, such as the Environmental Protection Agency, the Food and Drug Administration, the Consumer Products Safety Commission, the Bureau of Alcohol,

Tobacco, and Firearms, the Department of Agriculture, and others have slowly raised the level of understanding but the American public is still relatively uninformed regarding the benefits, the hazards, and prudent control of chemicals and radiation.

An important, but largely overlooked aspect of chemical safety is the remedial action underway or proposed by the Environmental Protection Agency on contaminated chemical waste disposal sites. Many sites have been remedied to the point where the site 'does not pose a significant threat to public health or to the environment'. The remaining sites have not been so classified.

With this as an introduction, it is hardly surprising that chemical laboratories handling, using, and disposing of hazardous materials would eventually be the subject of rule-making. From this has evolved OSHA's Occupational Exposures to Hazardous Chemicals in Laboratories; Final Rule, 29.CFR Part 1910.1450(d)(2)(i) to (d)(x), published in the Federal Register, **55**, No. 21, 31 January 1990, with the mandate it was to be fully in compliance in covered laboratories by 31 January 1991. (Not all laboratories are affected.)

Basically, the rule cites and reinforces regulations which OSHA has applied for years with respect to general regulations for the standards, facilities, and classification of chemicals. It includes permissible exposure limits, exposure monitoring, and a specific chemical hygiene plan, all of which require information and training, medical consultation, hazard identification, and monitoring, use of protective equipment, record keeping, and references.

As defined in the rule, 'laboratory' means a facility where the laboratory use of hazardous chemicals occur. It is a workplace where relatively small quantities of hazardous chemicals are used on a non-production basis. Laboratory scale means work with substances in which the containers used for reactions, transfers, and other handling of substances are designed to be easily and safely manipulated by one person, and excludes those workplaces whose function is to produce commercial quantities of materials. This is a generic laboratory standard. Laboratories generally have many hazardous chemicals present to which exposures are intermittent rather than a few substances to which there are routine exposures. The consideration is that, in a laboratory using good practices, the development of risk assessment for the hundreds of chemicals is not required. Because the working conditions and exposures are of a different nature than those in general industry, the hazards should be regulated in a different way. It is recognized that many companies, academic institutions and government agencies have devised detailed guidelines for the handling of hazardous chemicals. Carcinogens, suspected carcinogens, and known or suspected mutagens and teratogens have been listed in many of these guides. Potential fire hazards have received much attention as well. Cryogenic fluids, when used in the laboratory, should receive special considerations. Both containment, adequate venting, and proper respect for the extremely cold temperatures, should be noted. Liquid hydrogen, for example, expands 777-fold in passing from a liquid to a gas, has a boiling point of $-252.7\,°C$, and a flammable or explosive range of 4.1 to 74.2% by volume in air. It is also interesting that liquid hydrogen is the only cryogen with a reverse Thompson–

Joule effect (heats instead of cools when it expands through a small orifice). Liquid nitrogen, the cryogen more widely used, can be contaminated by exposure to air, creating a mixture which, unlike nitrogen alone, may be an oxidizer. Skin burns from all cryogens deserve prompt medical attention as do thermal burns and other exposures to toxic or hazardous materials. Chlorine dioxide, usually made on site, is unusually hazardous. Widely used in the pulp and paper industry, where its properties are well known, extreme care should be taken with this powerful oxidizer and toxic compound, even on a laboratory scale.

In summary, laboratory hazards can be anticipated in many, but not all situations. However, a complete literature search of the reaction and its components, and care in handling, using, and disposing of excessive chemicals and waste products, along with proper and adequate protective clothing, adequate venting, and the following of recognized practices of good housekeeping and personal hygiene, will reduce greatly the probability and severity of injuries, especially when such precautions are backed up with experienced medical personnel who recognize the chemicals and their effects on humans.

Subject Index

This index is designed to enable readers to make a quick reference to topics in the book, with the exception of the coloured alphabetically compiled section on Hazardous Substances for which there is a separate index.

Hazardous Substances

H. H. HENNING AND S. G. LUXON

This section consists of monographs describing briefly the hazardous properties and effects upon the human body of approximately 1400 flammable, explosive, corrosive, and/or toxic substances commonly used in chemical laboratories. It also recommends first aid and fire fighting procedures in the event of an accident and suggests methods of dealing with spillages of these substances.

Some of the entries in the fourth edition of the Handbook have been omitted, either because the substances are no longer in common use or to make space for the inclusion of other more hazardous substances. This does not mean to imply that any chemical not included is therefore not hazardous in some way. It is prudent to regard every chemical used in the laboratory as having some hazardous property, especially if it is used carelessly, imprudently, or contrary to any accompanying instructions. Even water should be handled carefully as accidental contact with such substances as acetylides or phosphides could give rise to dangerous concentrations of flammable and/or toxic gases.

Also included in this section are all those substances dealt with in the first five volumes of the Chemical Safety Data Sheets, published by the Royal Society of Chemistry, together with all those substances, except for those not normally found in a laboratory environment, which are currently listed in the booklet EH40/92 Occupational Exposure Limits 1992 published by the Health and Safety Executive.

For the new entries much of the information in this section has been collected from a variety of reference books currently available in the library of the Royal Society of Chemistry at Burlington House, London. At the same time the opportunity has been taken to update and revise the existing entries.

Reference books which have been found to be particularly valuable in this context are:

(i) 'The Sigma-Aldrich Library of Chemical Safety Data', 2nd Edn., ed. R. E. Lenga, Sigma-Aldrich Corp., Milwaukee, Wisconsin, 1987.
(ii) 'Compendium of Safety Data Sheets for Research and Industrial Chemicals', ed. L. H. Keith and D. B. Waters, VCH Weinheim, Germany.

(iii) 'Dangerous Properties of Industrial Materials', 7th Edn., ed. N. I. Sax and R. J. Lewis, Van Nostrand Reinhold, New York, 1989.

(iv) 'Patty's Industrial Hygiene and Toxicology', ed. G. D. Clayton and F. E. Clayton, John Wiley and Sons, New York.

Heading

Each entry starts with the name of the chemical which may or may not be the recommended IUPAC name. All the names and many of the synonyms have been included in the 'Index of Chemicals' at the end of this section. No attempt has been made to complicate the work by including all possible synonyms and trade names and accordingly only accepted common names have been included. Also listed in the index are the CAS numbers so that it would be possible for any user, when seeking additional information, to ensure that the chemicals referred to are identical.

Physical Description

Where relevant information has been available, a physical description of the substance is included together with details of the melting point and/or boiling point. The solubility in, or reaction with water, is also given because of its importance in first aid, fire fighting, and dealing with spillages.

Risk and Safety Phrases

The risk and safety phrases have been taken from the Authorized Lists appended to the Classification, Packaging, and Labelling Regulations 1984 and subsequent updating Statutory Instruments. The chemicals are classified on the basis of health risks and are defined as follows:

- *Very toxic*. A substance which if it is inhaled, ingested, or penetrates the skin, may involve extremely serious acute or chronic health risks and even death.
- *Toxic*. A substance which if it is inhaled, ingested, or penetrates the skin, may involve serious acute or chronic health risks.
- *Harmful*. A substance which if it is inhaled, ingested, or penetrates the skin, may involve limited health risks.
- *Corrosive*. A substance which may on contact with living tissues destroy them.
- *Irritant*. A non-corrosive substance which, through immediate, prolonged, or repeated contact with the skin or mucous membranes, can cause inflammation.

Difficulties with the warnings used for flammable substances in the UK have not yet been completely resolved. Although it is widely agreed that 'flammable' is a better word than 'inflammable' to describe materials that burn readily in air, 'inflammable' remains the word used in common parlance and in earlier legislation, *e.g.* The Petroleum (Consolidation) Act 1928. The 1984 UK legisla-

Criteria for the Classification of Substances as Very Toxic, Toxic, or Harmful

'Very toxic', 'toxic', or 'harmful' means in relation to a dangerous substance not classified in accordance with Schedules 3, 4, or 5, that the substance has a toxicity falling within the range set out in the table below for that category.

Category	LD_{50} absorbed orally in rat mg kg^{-1}	LD_{50} absorbed percutaneously in rat or rabbit mg kg^{-1}	LC_{50} absorbed by inhalation in rat mg l^{-1} (4 hrs)
Very toxic	$\leqslant 25$	$\leqslant 15$	$\leqslant 0.5$
Toxic	> 25 to 200	> 50 to 400	> 0.5 to 2
Harmful	> 200 to 2000	> 400 to 2000	> 2 to 20

Note: If facts show that for the purposes of classification it is inadvisable to use the LD_{50} or LC_{50} values as a principal basis because the substance produces other effects, the substance shall be classified according to the magnitude of these effects.

tion seeks some reconciliation by specifying that hazardous substances correctly classified and labelled under its provisions will be deemed to comply with the requirements of the 1928 Act. The UK and European legislative wordings for degrees of flammability have been adopted in these monographs, as have the ranges of flash point defining those degrees. Thus, 'flammable' substances are those with a flash point between 21 °C and 55 °C, 'highly flammable' substances have flash points between 0 °C and 21 °C while 'extremely flammable' substances have flash points below 0 °C and a boiling point below 35 °C. According to the Highly Flammable Liquids and Liquefied Gases Regulations 1972 liquids with a flash point below 32 °C were designated as 'highly flammable' in the UK. Again, the 1984 legislation specifies that correctly classified and labelled substances will be deemed to comply with the requirements of the 1972 Regulations.

The safety phrases and risk phrases, cannot be regarded as comprehensive as the choice of phrases must of necessity be very selective. However Section 6(4)(c) of the Health and Safety at Work Act 1974 places on the supplier of a chemical the duty 'to take such steps as are necessary to secure that there will be available in connection with the use of the substance at work adequate information about the results of any relevant tests which have been carried out on or in connection with the substance and about any conditions necessary to ensure that it will be safe and without risks to health when properly used'. Where appropriate care must be taken to ensure that this additional information is obtained from the supplier and passed down through the management structure to the people actually handling the chemicals. Such information can be of maximum benefit only it it is allied to a solid basis of sound working practice for which adequate training and supervision are essential

Limit Values

This section gives the maximum exposure limits (MEL) and the occupational exposure standards (OES), both short-term and long-term, as published by the

Health and Safety Executive in EH40/92. They are defined as follows:

- a MEL is the maximum concentration of an airborne substance, averaged over a reference period (usually 8 hours long-term, or 10 minutes short-term) to which employees may be exposed by inhalation under any circumstances and is specified, together with the appropriate reference period, in Schedule 1 of COSHH;
- an OES is the concentration of an airborne substance, averaged over a reference period (usually 8 hours long-term or 10 minutes short-term), at which, according to current knowledge, there is no evidence that it is likely to be injurious to employees if they are exposed by inhalation, day after day to that concentration, and which is specified in a list approved by the Health and Safety Commission.

Similar limits and standards have been set by other bodies, *e.g.* threshold limit values (TLV) by the American Conference of Governmental Industrial Hygienists and maximum arbeitplatz konzentrations (MAK) by the German authorities. These values provide very useful guidance in the absence of published UK figures.

Toxic Effects

As has been indicated earlier, the risk phrases highlight the main toxic effects. Information amplifying these phrases or dealing with effects not highlighted is included in this section where it appears to be important. Additional information should always be obtained from the manufacturer or supplier of any chemical. It is his duty to provide this information under Section 6 of the Health and Safety at Work Act 1974.

Again it must be emphasized that the absence of any information on the toxicity of a particular substance does not mean that it is not toxic and all chemicals should be handled with care at all times. For corrosive and irritant substances, the parameters for which have been given earlier, advice on protective clothing and handling procedures can be obtained from the Health and Safety Executive.

Carcinogenic Substances

One possible effect of chronic exposure to various chemicals is that of the induction of cancer (carcinogenesis). Much effort has been devoted to understanding and evaluating this particular effect but there are many unresolved difficulties in the way of interpreting the results of animal experimentation and extrapolating these results to humans. Much information and guidance on interpretation is to be found in the series of monographs on the Evaluation of Carcinogenic Risk of Chemicals to Man published by the International Agency for Research on Cancer. It is wise to adopt the general policy in any teaching laboratory of not using any chemicals that are known or suspected to cause cancer in humans. In other specialized laboratories where such chemicals must be used for specific purposes the utmost care, allied to rigorous training and discipline, must be taken. The specific requirements of the Control of Substances Hazardous to Health Regulations 1989 and the appropriate Codes of Practice on Carcinogens must always be observed. For many of the individual entries in this section there is a comment on carcinogenesis but as advances in this field are proceeding apace, where appropriate additional information should always be obtained from the supplier or from the Health and Safety Executive.

Hazardous Reactions

In some cases where a hazardous reaction has been reported either in the preparation or in the use of a particular chemical this has been noted. Many of the references have been taken from the 'Handbook of Reactive Chemical Hazards' by L. Bretherick to which reference should be made for a more detailed assessment.

First Aid

The principles and practice of first aid treatment for chemical casualties have been presented in detail in Chapter 8. These have been summarized very briefly below and this summary provides a quick and easy reminder of the basic steps to be taken. It must be stressed that this information is essentially *first aid*, it is not a substitute for skilled medical attention.

General

In every case ensure that first aid treatment can be given without danger to yourself or others present.

Put on such protective clothing as may be necessary.

Remove the patient from the area of contamination. This may necessitate removing the patient's clothing to prevent further or secondary contamination.

If there is a risk of absorption by more than one route, the priority should be to deal with the exposure route which will most quickly cause harm.

After treatment keep the patient warm and quiet and provide comfort and reassurance as required.

If in doubt seek medical assistance.

Chemical on the Skin

If possible take the casualty to the nearest source of clean water.

Wash the affected parts of the body with plenty of cold or lukewarm running water. Soap my be useful but do not use any medicaments or any other chemicals to neutralize the contaminant. Continue washing for at least 10 minutes, longer if necessary, to remove all traces of chemical contaminant.

Dry the skin gently. Do not attempt to remove any clothing still stuck to the skin after cleansing with water.

Chemical in the Eyes

Flood the eyes with plenty of cool, clean water. Do not delay and do not use any medicaments or any other chemicals to neturalize the contaminant.

Ensure that the eyeball and the inside of the eyelids are properly bathed by gently prising open the eyelids during washing. Also make sure that the contaminated water runs off the face away from the eyes.

Bandage the eyes firmly, provide constant reassurance for the patient, and ensure that the patient is seen by a doctor.

Inhalation of Chemical

Ensure that you are properly protected and then enter the area to remove the casualty to fresh air.

If the casualty is unconscious but breathing, place on one side in the recovery position.

If breathing has stopped apply artificial respiration by the mouth-to-mouth or mouth-to-nose method or give oxygen by mask.

Remove any contaminated clothing and wash off any chemicals from the skin and eyes.

Ingestion of Chemicals

If the casualty is unconscious but breathing, place on one side in the recovery position.

If breathing has stopped apply artificial respiration by the mouth-to-mouth method or give oxygen by mask.

Remove any contaminated clothing and wash off any chemicals from the skin and eyes.

Encourage the patient to rinse out the mouth several times but do not induce vomiting nor give anything to drink if the patient finds it difficult to swallow.

Special Treatments

If the chemical involved is a cyanide or nitrile, hydrofluoric acid or a fluoride, or phenolic in character then special treatments will be indicated in addition to the general treatment summarized above.

Fire Hazard

The problem of fire hazards has been dealt with fully in Chapter 4. The short note in each individual monograph gives, where available or applicable;

   (i)   flashpoint—generally, but not exclusively, closed cup unless otherwise stated;
   (ii)   explosive limits in terms of the range of percentage flammable component in a mixture with air that presents risk of ignition;

(iii)   autoignition temperature, which is the minimum temperature required to initiate or maintain self-sustained combustion independent of the source of heat.

These figures have been taken from a variety of reference books and, particularly for flash point, slightly different figures can be found.

Most of the entries conclude with a recommendation for the types of extinguishing agent suitable for fighting a fire involving the substance concerned. Some recommendations taken from the fourth edition of the handbook were extensively based on the Fire Service Circulars issued by the Home Office. Additional information, particularly for the new entries, has been taken largely from 'The Sigma-Aldrich Library of Chemical Safety Data', ed. R. E. Lenga and the Fire Protection Guide on Hazardous Materials issued by the National Fire Protection Agency.

Spillage and Disposal

In this edition the individual entries do not include detailed methods of dealing with spillages and the subsequent disposal of the material collected, except for very specific advice in a few special cases. This is because for any incident there will be considerable differences in the action taken due firstly to the size of the spillage together with the flammability and toxicity of the substance(s) concerned and secondly on the availability of facilities for disposal of the material collected and on differences in local authority by-laws regarding tipping, running to waste, *etc.*

Instead it has been decided to give general guidance, set out below, on the handling and disposal of spillages so that each individual laboratory can formulate its own instructions for dealing with such incidents. It is expected that the staff will have been fully trained in emergency procedures including all aspects of the collection and disposal of any substances involved as there may be very little time to read the instructions in the event of a major emergency.

The following general rules can be applied to the extent necessary to deal with any spillage without danger to the operators involved or to the environment. These should be supplemented with advice from the chemical suppliers and from the relevant local authorities.

(i)   Eliminate all sources of heat and ignition for all flammable materials and also for those which form more toxic substances on exposure to heat. Care should be taken with flammable vapours and gases which are heavier than air as these can travel considerable distances and then flash back on reaching a distant source of ignition.

(ii)   Wear face shield or goggles, gloves, and, where appropriate to the situation, waterproof boots and protective jacket and trousers.

(iii)   Wear suitable breathing apparatus or have it available as appropriate, where the extent of the spillage and the toxicity of the material so requires.

(iv)   Ensure that all other personnel either leave the affected area or, if they remain to help, are also adequately and suitable protected.

(v)   Liquid spills can be treated in several ways:

   (a) If small, absorb on paper towels and evaporate in a fume cupboard or remove outside for evaporation or subsequent disposal by another method.

   (b) If large, absorb on sand or vermiculite, shovel into a covered container, and remove outside.

   (c) Any spillage up to about 2 ½ litres can be treated by adding a non-flammable dispersing agent and working it into an emulsion with brush and water. About 1 volume dispersing agent is needed for every 2 volumes of flammable water-insoluble liquid spilt (less with non-flammable water-insoluble liquids) together with 10 volumes of water. The emulsion can then be run to waste with large quantities of water. With these proportions there is no danger of a flammable vapour mixture developing in the drainage system. Advice should be obtained from the water authority before this method of disposal is used.

(vi)   Subsequent disposal of small amounts of material not run to waste can be dealt with by evaporation in the open air where no nuisance can be created. Larger amounts should be disposed of by burial in a licensed site or by incineration in an approved incinerator. In certain cases it may be necessary before disposal to neutralize the material or to treat it chemically to reduce the chances of a hazardous situation developing later. Detailed advice on chemical treatments for a number of substances can be found in 'Destruction of Hazardous Chemicals in the Laboratory' by G. Lunn and E. B. Sansome.

(vii) Most solid materials should be swept up dry or mixed with dry sand before being swept up and placed in buckets for removal and subsequent disposal. In some cases it may be possible to preserve the collected material for re-use or recycling. As many solid materials are metal salts, which may not be destroyed by dilution with water or by burning or burial and which may therefore remain a toxic hazard to the environment, specialist advice should be sought and the local authorities consulted before drawing up detailed plans for dealing with such spillages.

(viii) Following removal of the material from the site of the spillage, the area should be ventilated to remove any residual vapour, and/or washed with water and soap or detergent to remove any traces of material.

(ix)   Any contaminated personal or protective clothing should be thoroughly cleaned to remove all traces of contaminant. In some cases however it may be necessary to discard contaminated clothing.

(x)   Finally, each incident should be thoroughly investigated by the laboratory management to evaluate the cause of the spillage with a view to preventing further similar incidents and also to ensure that the instructions for handling such incidents are satisfactory.

1. Abietates

Hazardous reactions   Finely divided abietates of several metals are subject to spontaneous heating and ignition.

2. Acetaldehyde

Colourless, fuming liquid with pungent fruity odour; m.p. -123.5 ˚C; b.p. 20.2 ˚C; miscible with water.

RISKS
Extremely flammable – Irritating to eyes and respiratory system – Possible risk of irreversible effects (R12, R36/37, R40)

SAFETY PRECAUTIONS
Keep away from sources of ignition – No Smoking – Take precautionary measures against static discharges – Wear protective clothing and gloves (S16, S33, S36/37)

Limit values   OES short-term 150 p.p.m. (270 mg m^{-3}) under review; long-term 100 p.p.m. (180 mg m^{-3}) under review.

Toxic effects   The vapour and liquid are both extremely irritant. Repeated inhalation of vapour can cause delirium, hallucinations, loss of intelligence, *etc.* as in chronic alcoholism. Animal carcinogen.

Hazardous reactions   Extremely reactive with several classes of inorganic and organic chemicals.

First aid   Standard treatment for exposure by all routes (see pages 108-122).

Fire hazard   Flash point 38 ˚C; Explosive limits 4-60%; Autoignition temperature 185 ˚C. Extinguish fire with dry powder, carbon dioxide, or alcohol foam.

Spillage disposal   See general section.

RSC *Chemical Safety Data Sheets* Vol. 3, No. 1, 1990 gives extended coverage.

3. Acetamide

Moist, white, crystalline solid with a mousy odour; m.p. 79-81 ˚C; b.p. 222 ˚C; soluble in water.

Toxic effects   Irritant to skin, eyes, upper respiratory tract, and mucous membranes. Harmful if inhaled, ingested, or absorbed through the skin. Experimental carcinogen.

First aid   Standard treatment for exposure by all routes (see pages 108-122).

Fire hazard   Extinguish fires with water spray, carbon dioxide, dry chemical powder, or alcohol or polymer foam.

Spillage disposal   See general section.

4. Acetanilide

White, crystalline solid; m.p. 114-115 ˚C; b.p. 304-305 ˚C; sparingly soluble in cold water, soluble in hot water.

Toxic effects Irritant to the skin, eyes, upper respiratory tract, and mucous membranes. Harmful by inhalation, ingestion, or skin absorption. May cause cyanosis.

First aid Standard treatment for exposure by all routes (see pages 108-122).

Fire hazard Flash point 173 ˚C (open cup); Autoignition temperature 538 ˚C. Extinguish fires with water spray, carbon dioxide, dry chemical powder, or alcohol or polymer foam.

Spillage disposal See general section.

5. Acetic acid

Colourless liquid with pungent acrid odour; m.p. 17 ˚C; b.p. 118 ˚C; glacial acetic acid freezes to a crystalline solid in cool weather, miscible with water.

RISKS
Flammable – Causes severe burns (R10, R35)

SAFETY PRECAUTIONS
Keep out of reach of children – Do not breathe fumes – In case of contact with eyes, rinse immediately with plenty of water and seek medical advice (S2, S23, S26)

Limit values OES short-term 15 p.p.m. (37 mg m^{-3}); long-term 10 p.p.m. (25 mg m^{-3}).

Toxic effects The vapour and liquid are extremely irritant to all tissues.

Hazardous reactions Causes exothermic polymerization of acetaldehyde; violent or explosive reactions with some oxidants.

First aid Standard treatment for exposure by all routes (see pages 108-122).

Fire hazard Flash point 39 ˚C (closed cup); Explosive limits 5.4-16%; Autoignition temperature 465 ˚C. Extinguish fire with water spray, dry powder, carbon dioxide, or alcohol foam.

Spillage disposal See general section.

RSC *Chemical Safety Data Sheets* Vol 3, No. 2, 1990 gives extended coverage.

6. Acetic anhydride

Colourless liquid with strong acrid odour; m.p. -73 °C; b.p. 139 °C; reacts slowly with cold water to form acetic acid, but rapidly if hot and acid present.

RISKS
Flammable – Causes burns (R10, R34)

SAFETY PRECAUTIONS
In case of contact with eyes, rinse immediately with plenty of water and seek medical advice (S26)

Limit values   OES short-term 5 p.p.m. (20 mg m^{-3}).

Toxic effects   The vapour and liquid are severely irritant to all tissues.

Hazardous reactions   Causes exothermic polymerization of acetaldehyde; violent or explosive reactions with some oxidants.

First aid   Standard treatment for exposure by all routes (see pages 108-122).

Fire hazard   Flash point 49 °C (closed cup); Explosive limits 3-10%; Autoignition temperature 385 °C. Extinguish fire with water spray, dry powder, carbon dioxide, or alcohol foam.

Spillage disposal   See general section.

RSC *Chemical Safety Data Sheets* Vol. 3, No. 2, 1990 gives extended coverage.

7. Acetone

Colourless, volatile liquid with a sweetish odour; m.p. -94.6 °C; b.p. 56.4 °C; miscible with water.

RISKS
Highly flammable (R11)

SAFETY PRECAUTIONS
Keep container in a well ventilated place – Keep away from sources of ignition – No Smoking – Do not breathe vapour – Take precautionary measures against static discharges (S9, S16, S23, S33)

Limit values   OES short-term 1500 p.p.m. (3560 mg m^{-3}); long-term 750 p.p.m. (1780 mg m^{-3}).

Toxic effects   Inhalation of vapour may cause dizziness, narcosis, and coma. The liquid irritates the eyes and may cause severe damage. If swallowed may cause gastric irritation, narcosis, and coma.

Hazardous reactions   Vigorously oxidized by air in the presence of several inorganic oxidants.

First aid   Standard treatment for exposure by all routes (see pages 108-122).

Fire hazard   Flash point -19 °C (closed cup); Explosive limits 2.6-12.8%; Autoignition temperature 538 °C. Extinguish fire with dry powder, carbon dioxide, or alcohol-resistant foam.

Spillage disposal   See general section.

RSC *Chemical Safety Data Sheets* Vol. 1, No. 1, 1988 gives extended coverage.

8. Acetonitrile

Colourless, flammable, volatile liquid with ethereal odour and sweetish taste; m.p. -45 °C; b.p. 81.1 °C; miscible with water.

RISKS
Highly flammable – Toxic by inhalation, in contact with skin, and if swallowed (R11, R23/24/25)

SAFETY PRECAUTIONS
Keep away from sources of ignition – No Smoking – Take off immediately all contaminated clothing – If you feel unwell, seek medical advice (show label where possible) (S16, S27, S44)

Limit values   OES short-term 60 p.p.m. (105 mg m^{-3}); long-term 40 p.p.m. (70 mg m^{-3}).

Toxic effects   Inhalation of vapour may cause fatigue, nausea, diarrhoea, and abdominal pain; in severe cases there may be delirium, convulsions, paralysis, and coma. Evidence is lacking on the effects of skin absorption and ingestion, but these may be similar to those resulting from inhalation.

Hazardous reactions   Violent or explosive reactions with some inorganic compounds.

First aid   Special treatment for cyanides.

Fire hazard   Flash point 6 °C; Explosive limits 4.4-16%; Autoignition temperature 524 °C. Extinguish fire with dry powder, carbon dioxide, or water spray. Cyanide fumes will be produced from thermal decomposition.

Spillage disposal   See general section.

RSC *Chemical Safety Data Sheets* Vol. 1, No. 2, 1988 gives extended coverage.

9. Acetophenone

Yellow-tinted liquid or crystalline solid with a persistent flowery odour; m.p. 19-20 °C; b.p. 202 °C; slightly soluble in water.

Toxic effects   Vapour or mist is irritant to the skin, eyes, and mucous membranes. May be harmful by inhalation, ingestion, or absorption. Narcotic and hypnotic.

First aid   Standard treatment for exposure by all routes (see pages 108-122).

Fire hazard   Flash point 82 °C (closed cup); Autoignition temperature 471 °C. Extinguish fires with water spray, carbon dioxide, dry chemical powder, or alcohol or polymer foam.

Spillage disposal   See general section.

10. Acetyl bromide

Colourless to yellow liquid; m.p. -96 ˚C; b.p. 74-77 ˚C; decomposed by water with formation of hydrobromic acid and acetic acid.

Toxic effects   The vapour and liquid are severely irritant to all tissues.

Hazardous   Violent reaction with water and hydroxylic compounds.
  reactions

First aid   Standard treatment for exposure by all routes (see pages 108-122).

Spillage   See general section.
  disposal

11. Acetyl chloride

Colourless, fuming, volatile liquid with a pungent odour; m.p. -112 ˚C; b.p. 51 ˚C.; rapidly decomposed by water with formation of hydrochloric acid and acetic acid.

RISKS
Highly flammable – Reacts violently with water – Causes burns (R11, R14, R34)

SAFETY PRECAUTIONS
Keep container in a well ventilated place – Keep away from sources of ignition – No Smoking – In case of contact with eyes, rinse immediately with plenty of water and seek medical advice (S9, S16, S26)

Toxic effects   The vapour and liquid are severely irritant to all tissues.

Hazardous   Violent decomposition in preparation from phosphorus trichloride and acetic
  reactions   acid; violent reactions with water, dimethyl sulfoxide.

First aid   Standard treatment for exposure by all routes (see pages 108-122).

Fire hazard   Flash point 4 ˚C; Autoignition temperature 390 ˚C. Extinguish fire with water
                          spray, foam, dry powder, or carbon dioxide.

Spillage   See general section.
  disposal

12. Acetylene

Colourless gas; commercial gas has garlic-like odour due to impurities; b.p. -83 ˚C; slightly soluble in water.

RISKS
Heating may cause an explosion – Explosive with or without contact with air – Extremely flammable (R5, R6, R12)

SAFETY PRECAUTIONS
Keep container in a well ventilated place – Keep away from sources of ignition – No Smoking – Take precautionary measures against static discharges (S9, S16, S33)

Toxic effects   Inhalation of gas may cause dizziness, headache, and nausea. Mixed with
                          oxygen it can have narcotic properties but it is primarily an asphyxiant.

Hazardous reactions   The endothermic gas explodes, alone or mixed with air, under various circumstances; explosive reactions also with a variety of inorganic compounds.

First aid   Standard treatment for exposure by all routes (see pages 108-122).

Fire hazard   Explosive limits 3-82%; Autoignition temperature 335 °C. As the gas is supplied in a cylinder, turning off the valve will reduce any fire involving it; if possible cylinders should be removed quickly from an area in which a fire has developed.

13. Acetylenebis(triethyltin)

Hazardous reactions   Explodes on standing at room temperature.

14. Acetylides

Hazardous reactions   Most of these compounds explode readily and are very sensitive to shock, friction, and heat. As they contain no oxygen or nitrogen they explode as a result of the large amount of heat instantaneously generated and the explosion produces no gas. As a class they react vigorously with a variety of inorganic compounds. Most of them react violently with water producing flammable acetylene gas.

RSC *Chemical Safety Data Sheets* Vol. 2, No. 3, 1989 gives extended coverage of aluminium carbide.

15. Acetyl nitrate

Colourless, fuming liquid; b.p. 22 °C at 70 mm Hg.

Toxic effects   Has corrosive effects on the eyes.

Hazardous reactions   Unstable and liable to explode.

First aid   Standard treatment for exposure by all routes (see pages 108-122).

Spillage disposal   See general section.

16. Acetyl nitrite

Colourless liquid.

Hazardous reactions   Unstable and liable to explode.

17. Acetylsalicylic acid

Colourless, crystalline solid; m.p. 135 ˚C (rapid heating); b.p. 140 ˚C; slightly soluble in water.

Limit values	OES long-term 5 mg m^{-3}.
Toxic effects	Irritant to skin, eyes, upper respiratory tract, and mucous membranes. Can be absorbed through the skin. Experimental teratogen.
First aid	Standard treatment for exposure by all routes (see pages 108-122).
Fire hazard	Extinguish fires with water spray, carbon dioxide, dry chemical powder, or alcohol or polymer foam.
Spillage disposal	See general section.

18. Acrolein

Colourless to yellow, volatile liquid with pungent, choking odour; m.p. -87.7 ˚C; b.p. 52.5 ˚C; soluble in water.

RISKS
Highly flammable – Toxic by inhalation – Irritating to eyes, respiratory system, and skin (R11, R23, R36/37/38)

SAFETY PRECAUTIONS
Do not empty into drains – Take precautionary measures against static discharges – If you feel unwell, seek medical advice (show label where possible) (S29, S33, S44)

Limit values	OES short-term 0.3 p.p.m. (0.8 mg m^{-3}); long-term 0.1 p.p.m. (0.25 mg m^{-3}).
Toxic effects	The vapour and liquid are irritant to all tissues. Inhalation of the vapour may cause unconsciousness. Poisoning can result from absorption of the liquid through the skin.
Hazardous reactions	Liable to polymerize violently, especially in contact with strong acid or basic catalysts.
First aid	Standard treatment for exposure by all routes (see pages 108-122).
Fire hazard	Flash point -18 ˚C (open cup); Explosive limits 3-31%; Autoignition temperature 234 ˚C. Extinguish fire with dry powder, carbon dioxide, or alcohol foam.
Spillage disposal	See general section.

RSC *Chemical Safety Data Sheets* Vol. 3, No. 4, 1990, gives extended coverage.

19. Acrylamide

White, crystalline, odourless solid. May polymerize with violence on melting; m.p. 85 °C; soluble in water.

RISKS
Toxic by inhalation, in contact with skin, and if swallowed – Danger of cumulative effects (R23/24/25, R33)

SAFETY PRECAUTIONS
Take off immediately all contaminated clothing – If you feel unwell, seek medical advice (show label where possible) (S27, S44)

Limit values	MEL long-term 0.3 mg m^{-3}.
Toxic effects	Irritates the skin, eyes, and upper respiratory tract. Expected, from animal experiments, to affect the central nervous system as a result of skin absorption. Experimental carcinogen.
First aid	Standard treatment for exposure by all routes (see pages 108-122).
Fire hazard	Flash point 138 °C; Autoignition temperature 424 °C. Extinguish fires with water spray, carbon dioxide, or dry chemical powder.
Spillage disposal	See general section.

RSC *Chemical Safety Data Sheets* Vol. 4a, No. 2, 1991, gives extended coverage.

20. Acrylic acid

Colourless solid and liquid with acrid odour; m.p. 13.5 °C; b.p. 141 °C; miscible with water.

RISKS
Flammable – Causes burns (R10, R34)

SAFETY PRECAUTIONS
In case of contact with eyes, rinse immediately with plenty of water and seek medical advice – Wear suitable protective clothing (S26, S36)

Limit values	OES short-term 20 p.p.m. (60 mg m^{-3}); long-term 10 p.p.m. (30 mg m^{-3}).
Toxic effects	The vapour and liquid are irritant to all tissues.
Hazardous reactions	Liable to polymerize violently.
First aid	Standard treatment for exposure by all routes (see pages 108-122).
Fire hazard	Flash point 68 °C (open cup); Explosive limits 2.9%-8%; Autoignition temperature 360 °C. Extinguish fire with dry powder, alcohol foam, or carbon dioxide.
Spillage disposal	See general section.

RSC *Chemical Safety Data Sheets* Vol. 3. No. 5, 1990 gives extended coverage.

21. Acrylonitrile

Colourless, volatile liquid with mild pungent odour; m.p. -83.5 ˚C; b.p. 77 ˚C; soluble in water

RISKS
May cause cancer – Highly flammable – Toxic by inhalation, in contact with skin, and if swallowed – Irritating to skin (R45, R11, R23/24/25, R38)

SAFETY PRECAUTIONS
Avoid exposure – Obtain special instructions before use – Keep away from sources of ignition – No Smoking – Take off immediately all contaminated clothing – If you feel unwell, seek medical advice (show label where possible) (S53, S16, S27, S44)

Limit values	MEL long-term 2 p.p.m. (4 mg m^{-3}).
Toxic effects	Vapour may cause dizziness, nausea, and unconsciousness. The liquid may cause blistering, dermatitis, and acute effects by absorption. Poisonous if taken by mouth. Action similar to cyanides.
Hazardous reactions	Violent or explosive polymerization promoted by a variety of reagents including strong acids and bases.
First aid	Special treatment for cyanides.
Fire hazard	Flash point 0 ˚C (open cup); Explosive limits 3-17%; Autoignition temperature 482 ˚C. Extinguish fire with dry powder, carbon dioxide, or alcohol foam.
Spillage disposal	See general section. Large spills need specialist help.

RSC *Chemical Safety Data Sheets* Vol. 3, No. 6, 1990 gives extended coverage.

22. Adipic acid

White powder; m.p. 152 ˚C; b.p. 337.5 ˚C; slightly soluble in water.

RISKS
Irritating to eyes (R36)

Toxic effects	Harmful if inhaled or swallowed. Irritant to eyes and possibly skin.
First aid	Standard treatment for exposure by all routes (see pages 108-122).
Fire hazard	Flash point 196 ˚C; Autoignition temperature 420 ˚C. Extinguish fires with water spray, carbon dioxide, dry chemical powder, or alcohol or polymer foam.
Spillage disposal	See general section.

23. Allyl alcohol

Colourless, mobile liquid with a pungent, mustard-like smell; m.p. -129 °C; b.p. 96.9 °C; miscible with water.

RISKS
Highly flammable – Very toxic by inhalation – Irritating to eyes, respiratory system, and skin (R11, R26, R36/37/38)

SAFETY PRECAUTIONS
Keep away from sources of ignition – No Smoking – Wear eye/face protection – In case of accident or if you feel unwell, seek medical advice immediately (show label where possible) (S16, S39, S45)

Limit values   OES short-term 4 p.p.m. (10 mg m^{-3}); long-term 2 p.p.m. (5 mg m^{-3}).

Toxic effects   The vapour and liquid are irritant to all tissues.

Hazardous   Sulfuric acid may cause it to polymerize explosively. Violent reactions have
   reactions   occurred with other chemicals.

First aid   Standard treatment for exposure by all routes (see pages 108-122).

Fire hazard   Flash point 21.1 °C (closed cup); Explosive limits 2.5%-18%; Autoignition
                         temperature 378 °C. Extinguish fire with dry powder, carbon dioxide, or
                         alcohol foam.

Spillage   See general section.
   disposal

RSC *Chemical Safety Data Sheets* Vol. 4a, No. 3, 1991 gives extended coverage.

24. Allylamine

Colourless to light yellow liquid, with a sharp smell and burning taste; m.p. -88.2 °C; b.p. 56.5 °C; soluble in water.

RISKS
Highly flammable – Toxic by inhalation, in contact with skin, and if swallowed (R11, R23/24/25)

SAFETY PRECAUTIONS
Keep container in a well ventilated place – Keep away from sources of ignition – No Smoking – Avoid contact with skin and eyes – If you feel unwell, seek medical advice (show label where possible) (S9, S16, S24/25, S44)

Toxic effects   Severe irritant to skin, eyes, and upper respiratory tract.

First aid   Standard treatment for exposure by all routes (see pages 108-122), with
                         particular attention to the possibility of convulsions, shock and/or lung
                         congestion occurring.

Fire hazard   Flash point -29 °C (open cup); Explosive limits 2.2%-22%; Autoignition
                         temperature 374 °C. Extinguish fires with carbon dioxide, dry chemical
                         powder, or alcohol foam.

Spillage disposal	See general section.

RSC *Chemical Safety Data Sheets* Vol. 4a, No. 4, 1991 gives extended coverage.

25. Allyl benzenesulfonate

Hazardous reactions	Explosion of vacuum distillation residue.

26. Allyl bromide

Colourless liquid with an unpleseant smell; m.p. -119 °C; b.p. 71 °C; almost insoluble in water.

Toxic effects	The vapour and liquid are irritant to all tissues.
Hazardous reactions	Contact with some inorganic compounds may cause violent exothermic polymerization.
First aid	Standard treatment for exposure by all routes (see pages 108-122).
Fire hazard	Flash point -1 °C; Explosive limits 4-7%; Autoignition temperature 295 °C. Extinguish fire with water spray, foam, dry powder, or carbon dioxide.
Spillage disposal	See general section.

27. Allyl chloride

Colourless mobile liquid with unpleasant pungent smell; m.p. -134.5 °C; b.p. 44.6 °C; immiscible with water.

RISKS
Highly flammable – Very toxic by inhalation (R11, R26)

SAFETY PRECAUTIONS
Keep away from sources of ignition – No Smoking – Do not empty into drains – Take precautionary measures against static discharges – In case of accident or if you feel unwell, seek medical advice immediately (show label where possible) (S16, S29, S33, S45)

Limit values	OES short-term 2 p.p.m. (6 mg m^{-3}) under review; long-term 1 p.p.m. (3 mg m^{-3}) under review.
Toxic effects	The vapour and liquid are irritant to all tissues. Inhalation of the vapour will cause severe irritation and may lead to headache, dizziness, and, in high concentrations, unconsciousness. Muscular pains may follow skin absorption.
Hazardous reactions	Contact with various chemicals may cause exothermic violent polymerization.
First aid	Standard treatment for exposure by all routes (see pages 108-122).

Fire hazard   Flash point -31.7 °C (closed cup); Explosive limits 3.28%-11.15%; Autoignition temperature 391.7 °C. Extinguish fire with alcohol foam, dry powder, or carbon dioxide.

Spillage   See general section.
disposal

RSC *Chemical Safety Data Sheets* Vol. 4a, No. 5, 1991 gives extended coverage.

28. Allyl ethyl ether

Colourless liquid; b.p. 65-66 °C.

Toxic effects   Irritant to eyes, skin, and upper respiratory tract.

**Hazardous   Explosion of peroxide at end of distillation.
reactions**

First aid   Standard treatment for exposure by all routes (see pages 108-122).

Fire hazard   Flash point -5 °C. Extinguish fires with carbon dioxide, dry powder, or alcohol or polymer foam extinguishers.

Spillage   See general section.
disposal

29. Allyl glycidyl ether

Colourless liquid; b.p. 154 °C; soluble in water.

RISKS
Harmful by inhalation – May cause sensitization by skin contact (R20, R43)

SAFETY PRECAUTIONS
Avoid contact with skin and eyes (S24/25)

Limit values   OES short-term 10 p.p.m. (44 mg m^{-3}); long-term 5 p.p.m. (22 mg m^{-3}).

Toxic effects   The vapour and liquid are irritant to all tissues. Inhalation of vapour has caused pulmonary oedema. Ingestion of the liquid may cause depression of the central nervous system and sensitization.

First aid   Standard treatment for exposure by all routes (see pages 108-122).

Spillage   See general section.
disposal

30. Allyl iodide

Yellowish to brown liquid with unpleasant pungent odour; m.p. -99 ˚C; b.p. 101-103 ˚C; immiscible with water

RISKS
Flammable – Causes burns (R10, R34)

SAFETY PRECAUTIONS
Keep container tightly closed – In case of contact with eyes, rinse immediately with plenty of water and seek medical advice (S7, S26)

Toxic effects   The vapour and liquid are irritant to all tissues.

First aid   Standard treatment for exposure by all routes (see pages 108-122).

Fire hazard   Flash point < 21 ˚C. Extinguish fire with water spray, foam, dry powder, or carbon dioxide.

Spillage   See general section.
  disposal

31. Allyl isothiocyanate

Colourless to pale yellow, oily liquid; m.p. -80 ˚C; b.p. 152 ˚C; sparingly soluble in water.

Toxic effects   Irritant to skin, eyes, upper respiratory tract, and mucous membranes. Mild allergen.

Hazardous   Explosion at end of reaction between allyl chloride and sodium thiocyanate in
  reactions   autoclave.

First aid   Standard treatment for exposure by all routes (see pages 108-122).

Fire hazard   Flash point 46 ˚C. Extinguish fire with carbon dioxide or dry chemical powder extinguisher.

Spillage   See general section.
  disposal

32. Aluminium

Silvery, ductile metal; m.p. 660 ˚C; b.p. 2450 ˚C; insoluble in water.

RISKS
Contact with water liberates highly flammable gases – Spontaneously flammable in air (R15, R17)

SAFETY PRECAUTIONS
Keep container tightly closed and dry – In case of fire, use chemical powder extinguisher (S7/8, S43)

Limit values   OES long-term 10 mg m^{-3} (total inhalable dust), 5 mg m^{-3} (respirable dust) proposed change.

Toxic effects   Aluminium dust is a respiratory irritant and may be harmful if inhaled. The dust is also irritant and harmful to the eyes. Very large oral doses of aluminium are reported to be toxic.

Hazardous   Violent or explosive reactions with numerous oxidants and other substances.
   reactions

First aid   Standard treatment for exposure by all routes (see pages 108-122).

Fire hazard   Explosive limits lel 40-50 mg l^{-1} air for powder approx. 14 mm diameter. To extinguish fires use dry chemical powder extinguishers. Do NOT use water.

Spillage   See general section. Elemental aluminium can be recovered for recycling.
   disposal

RSC *Chemical Safety Data Sheets* Vol. 2, No. 1, 1989 gives extended coverage.

33. Aluminium alkyls

Mainly colourless liquids with a characteristic musty odour; react, sometimes vigorously, with water.

RISKS
Reacts violently with water – Spontaneously flammable in air – Causes burns (R14, R17, R34)

SAFETY PRECAUTIONS
Keep away from sources of ignition – No Smoking – In case of fire, use dry chemical powder. Do not use water. (S16, S43)

Limit values   OES long-term 2 mg m^{-3}.

Toxic effects   Extremely destructive to all body tissues causing severe skin burns.

Hazardous   React vigorously or explosively with certain alcohols. Pyrophoric if exposed to
   reactions   air.

First aid   Standard treatment for exposure by all routes (see pages 108-122).

Fire hazard   Extinguish fires with dry chemical powder. Do NOT use water.

Spillage   See general section.
   disposal

34. Aluminium bromide anhydrous

White to yellow-red lumps; m.p. 97 ˚C; b.p. 256 ˚C; violently decomposed by water with evolution of hydrogen bromide.

Limit values   OES long-term 2 mg m^{-3}.

Toxic effects   The dust Is highly irritant to all tissues. Heat is produced on contact with moist skin and other tissues resulting in severe thermal and/or acid burns.

First aid   Standard treatment for exposure by all routes (see pages 108-122).

Fire hazard   Extinguish fires by using carbon dioxide, dry chemical, or foam. Do NOT use water.

Spillage   See general section.
   disposal

RSC *Chemical Safety Data Sheets* Vol. 2, No. 2, 1989 gives extended coverage.

35. Aluminium carbide

Yellow crystals; m.p. 2100 ˚C; b.p. 2200 ˚C with decomposition; decomposed by water producing methane and acetylene.

Limit values OES long-term 2 mg m^{-3}.

Toxic effects Irritating to skin, eyes, upper respiratory tract, and mucous membranes. May be harmful if inhaled, ingested, or absorbed through the skin.

Hazardous reactions Incandesces in contact with potassium permanganate and lead peroxide.

First aid Standard treatment for exposure by all routes (see pages 108-122).

Fire hazard Extinguish fires with dry chemical powder. Do NOT use water.

Spillage disposal See general section.

RSC *Chemical Safety Data Sheets* Vol. 2, No. 3, 1989 gives extended coverage.

36. Aluminium chlorate

Colourless, deliquescent crystals; m.p. Decomposes; very soluble in water.

Limit values OES long-term 2 mg m^{-3}.

Hazardous reactions Explosion when aqueous solution was evaporated.

First aid Standard treatment for exposure by all routes (see pages 108-122).

Fire hazard Dangerous in contact with flammable matter and also produces toxic fumes on decomposition. Use water spray to extinguish fires.

Spillage disposal See general section. Also sensitive to shock and friction.

37. Aluminium chloride anhydrous

Yellow or off-white, deliquescent pieces, granules, or powder; m.p. sublimes at 178 ˚C; violently decomposed by water with the formation of hydrogen chloride.

RISKS
Causes burns (R34)

SAFETY PRECAUTIONS
Keep container tightly closed and dry – After contact with skin, wash immediately with plenty of water (S7/8, S28)

Limit values OES long-term 2 mg m^{-3}.

Toxic effects The dust is highly irritant to all tissues. Heat is produced on contact with moist skin or other tissues resulting in severe thermal and acid burns.

Hazardous reactions	Violent reactions with water, and several other substances.
First aid	Standard treatment for exposure by all routes (see pages 108-122).
Fire hazard	Extinguish fires by using dry chemical powder, carbon dioxide, or alcohol foam. Do NOT use water.
Spillage disposal	See general section.

RSC *Chemical Safety Data Sheets* Vol. 2, No. 4, 1989 gives extended coverage.

38. Aluminium fluoride

Hexagonal crystals; m.p. sublimes at 1260 °C; b.p. 1537 °C; sparingly soluble in water.

Limit values	OES long-term 2 mg m^{-3}.
Toxic effects	Irritant to skin and eyes causing severe burns. Inhalation, ingestion, and skin absorption may give rise to signs of fluoride poisoning.
First aid	Standard treatment, with special reference to fluorides, for exposure by all routes.
Fire hazard	Use extinguishing agent appropriate to the surroundings.
Spillage disposal	See general section.

RSC *Chemical Safety Data Sheets* Vol. 2, No. 5, 1989 gives extended coverage.

39. Aluminium hydride

Colourless powder; decomposes on contact with water producing explosive hydrogen.

Limit values	OES long-term 2 mg m^{-3}.
Hazardous reactions	May explode spontaneously at ambient temperatures; violent decomposition in certain circumstances.
First aid	Standard treatment for exposure by all routes (see pages 108-122).
Spillage disposal	See general section.

40. Aluminium hydroxide

White, crystalline powder, balls, or granules; insoluble in water.

Limit values	OES long-term 2 mg m^{-3}.
Toxic effects	Irritant to skin and eyes. Fibrosis, emphysema, and pneumothorax have been reported.

Hazardous reactions	May explode violently in contact with chlorinated rubbers.
First aid	Standard treatment for exposure by all routes (see pages 108-122).
Fire hazard	Extinguish fires with carbon dioxide, dry chemical powder, or alcohol or polymer foam.
Spillage disposal	See general section.

RSC *Chemical Safety Data Sheets* Vol. 2, No. 6, 1989 gives extended coverage.

41. Aluminium iodide

White leaflets, which turn brown due to presence of free iodine; m.p. 191 ˚C; b.p. 382 ˚C; reacts violently with water.

Limit values	OES long-term 2 mg m^{-3}.
Toxic effects	Severely irritating and destructive to skin, eyes, upper respiratory tract, and mucous membranes.
First aid	Standard treatment for exposure by all routes (see pages 108-122).
Fire hazard	Extinguish fires with dry chemical powder. Do NOT use water.
Spillage disposal	See general section.

RSC *Chemical Safety Data Sheets* Vol. 2, No. 7, 1989 gives extended coverage.

42. Aluminium isopropoxide

White powder; m.p. 128-132 ˚C; may decompose on contact with water.

RISKS
Highly flammable (R11)

SAFETY PRECAUTIONS
Keep container dry – Keep away from sources of ignition – No Smoking (S8, S16)

Limit values	OES long-term 2 mg m^{-3}.
Toxic effects	Irritant and destructive to all tissues. Inhalation may be fatal.
First aid	Standard treatment for exposure by all routes (see pages 108-122).
Spillage disposal	See general section.

43. Aluminium nitrate

White crystals; m.p. 73 ˚C; b.p. decomposes at 135 ˚C; soluble in water.

Limit values	OES long-term 2 mg m^{-3}.
Toxic effects	Irritant to skin, eyes, and upper respiratory tract. Ingestion may result in dizziness, abdominal cramps, vomiting, diarrhoea, convulsions, and possibly unconsciousness.
Hazardous reactions	Powerful oxidizer reacting violently or explosively with a variety of substances.
First aid	Standard treatment for exposure by all routes (see pages 108-122).
Fire hazard	Extinguish fires with water spray.
Spillage disposal	See general section.

RSC *Chemical Safety Data Sheets* Vol. 2, No. 8, 1989 gives extended coverage.

44. Aluminium oxide

Solid; m.p. 2072 ˚C; b.p. 2977 ˚C; insoluble in water.

Limit values	OES long-term 10 mg m^{-3} (as total inhalable dust), 5 mg m^{-3} (as respirable dust) proposed change.
Toxic effects	Irritant to skin, eyes, upper respiratory tract, and mucous membranes.
Hazardous reactions	Reacts vigorously and sometimes explosively with vinyl acetate vapour, halocarbons (above 200 ˚C), oxygen difluoride, sodium nitrite, and chlorine trifluoride.
First aid	Standard treatment for exposure by all routes (see pages 108-122).
Fire hazard	Use an extinguishing agent appropriate to the surroundings.
Spillage disposal	See general section.

RSC *Chemical Safety Data Sheets* Vol. 2, No. 9, 1989 gives extended coverage.

45. Aluminium phosphide

Dark grey or dark yellow crystals; m.p. over 1000 ˚C.

RISKS
Contact with water liberates toxic, highly flammable gas – Very toxic if swallowed (R15/29, R28)

SAFETY PRECAUTIONS
Keep locked up and out of reach of children – Do not breathe dust – In case of fire, use carbon dioxide or dry powder – Do not use water – In case of accident or if you feel unwell, seek medical advice immediately (show label where possible) (S1/2, S22, S43, S45)

Limit values	OES long-term 2 mg m^{-3}.

Toxic effects	Irritant to skin and eyes. Highly toxic on inhalation leading to headaches, nausea, dispnoea, jaundice, and possible death following myocardial infiltration and pulmonary oedema. Also highly toxic on ingestion resulting in signs of cardiac, respiratory, liver, and kidney involvement.
First aid	Standard treatment for exposure by all routes (see pages 108-122).
Fire hazard	Extinguish fires with carbon dioxide or dry chemical powder. Do NOT use water.
Spillage disposal	See general section.

RSC *Chemical Safety Data Sheets* Vol. 2, No. 10, 1989 gives extended coverage.

46. Aluminium sulfate

White powder; m.p. decomposes at 770 ˚C; partially soluble in water.

Limit values	OES long-term 2 mg m^{-3}.
Toxic effects	Irritant to skin, eyes, and upper respiratory tract. On ingestion causes ulceration and necrosis to all the mucous membranes.
First aid	Standard treatment for exposure by all routes (see pages 108-122).
Fire hazard	Extinguish fires with water spray, carbon dioxide, dry chemical powder, or alcohol or chemical foam.
Spillage disposal	See general section.

RSC *Chemical Safety Data Sheets* Vol. 2, No. 11, 1989 gives extended coverage.

47. Aluminium tetrahydroborate

Liquid; m.p. -64.5 ˚C; b.p. 44.5 ˚C; reacts with water.

Limit values	OES long-term 2 mg m^{-3}.
Hazardous reactions	Vapour is spontaneously flammable in air and explodes in oxygen.
First aid	Standard treatment for exposure by all routes (see pages 108-122).
Fire hazard	Extinguish fires with carbon dioxide or dry chemical powder.

48. Aluminium triformate

Limit values	OES long-term 2 mg m^{-3}.
Hazardous reactions	Explosion when aqueous solution was being evaporated.

49. Amidosulfuric acid

White crystals; m.p. approx. 205 ˚C with decomposition; soluble in water.

Toxic effects   The dust or solution irritates the eyes. Prolonged contact with the skin may cause irritation.

First aid   Standard treatment for exposure by all routes (see pages 108-122).

Spillage disposal   See general section.

50. 4-Aminobiphenyl

Colourless to tan crystals; m.p. 53 ˚C; b.p. 302 ˚C; soluble in water.

RISKS
May cause cancer – Harmful if swallowed (R45, R22)

SAFETY PRECAUTIONS
Avoid exposure – Obtain special instructions before use – If you feel unwell, seek medical advice (show label where possible) (S53, S44)

Toxic effects   Human carcinogen, use prohibited in the United Kingdom under The Carcinogenic Substances Regulations 1967. Inhalation or absorption through the skin of the dust has been recognized as a cause of bladder tumours. It is not therefore considered appropriate to deal with the hazards more fully in this book.

First aid   Standard treatment for exposure by all routes (see pages 108-122).

Fire hazard   Flash point >230 ˚C; Autoignition temperature 450 ˚C. Extinguish fires with carbon dioxide, dry powder, or alcohol or polymer foam.

Spillage disposal   See general section.

51. Aminoguanidinium nitrate

White, crystalline flakes; m.p. 145-147 ˚C; soluble in water.

Toxic effects   Irritating to skin, eyes, upper respiratory tract, and mucous membranes.

Hazardous reactions   Violent explosion when aqueous solution was being evaporated on a steam bath.

First aid   Standard treatment for exposure by all routes (see pages 108-122).

Fire hazard   Use water spray to extinguish fires.

Spillage disposal   See general section.

52. 2-Amino-2-methylpropan-1-ol

Colourless, viscous liquid or crystalline mass; m.p. 30-31 ˚C; b.p. 165 ˚C; miscible with water.

RISKS
Irritating to eyes, respiratory system, and skin (R36/37/38)

SAFETY PRECAUTIONS
In case of contact with eyes, rinse immediately with plenty of water and seek medical advice (S26)

Toxic effects   Mild skin irritant and moderately toxic orally. It can cause burns to the eyes. Inhalation may be fatal as a result of spasm, inflammation, and oedema.

Hazardous reactions   2-Amino-2-methylpropan-1-ol emits toxic fumes of nitrogen oxides when heated to decomposition. It is a moderate fire hazard when exposed to heat or flame and can react with oxidizing materials. It attacks copper, brass, and aluminium, but not steel or iron.

First aid   Standard treatment for exposure by all routes (see pages 108-122).

Fire hazard   Flash point 67.2 ˚C (open cup). Extinguish fires with dry chemical powder, or alcohol foam.

Spillage disposal   See general section.

RSC *Chemical Safety Data Sheets* Vol. 3, No. 7, 1990 gives extended coverage.

53. 2-Aminophenol

White to brown, light-sensitive crystals; m.p. 173 ˚C; b.p. sublimes at 153 ˚C at 11 mm Hg; sparingly soluble in water.

RISKS
Harmful by inhalation, in contact with skin, and if swallowed (R20/21/22)

SAFETY PRECAUTIONS
Avoid contact with skin and eyes (S24/25)

Toxic effects   May cause dermatitis and cyanosis by skin absorption.

First aid   Standard treatment for exposure by all routes (see pages 108-122).

Spillage disposal   See general section.

54. 4-Aminophenol

Off-white, crystalline solid; m.p. 189-190 ˚C; b.p. 284 ˚C (decomp.); sparingly soluble in water.

RISKS
Harmful by inhalation, in contact with skin, and if swallowed (R20/21/22)

SAFETY PRECAUTIONS
After contact with skin, wash immediately with plenty of water (S28)

Toxic effects   Irritant to skin, eyes, upper respiratory tract, and mucous membranes. Entry into the body may cause cyanosis or allergic response. Onset of symptoms may be delayed. Experimental teratogen.

First aid   Standard treatment for exposure by all routes (see pages 108-122).

Fire hazard   Extinguish fires with water spray, carbon dioxide, dry chemical powder, or alcohol or polymer foam.

Spillage disposal   See general section.

55. 2-Aminopropane

Colourless, volatile liquid with an ammoniacal odour; m.p. -101.2 ˚C; b.p. 33-34 ˚C; miscible with water.

RISKS
Extremely flammable – Irritating to eyes, respiratory system, and skin (R12, R36/37/38)

SAFETY PRECAUTIONS
Keep away from sources of ignition – No Smoking – In case of contact with eyes, rinse immediately with plenty of water and seek medical advice – Do not empty into drains (S16, S26, S29)

Toxic effects   Extremely destructive to all mucous membranes. Inhalation causes extreme irritation which can lead to death.

First aid   Standard treatment for exposure by all routes (see pages 108-122).

Fire hazard   Flash point -2 ˚C (open cup); Explosive limits about 2-10%; Autoignition temperature 420 ˚C. Extinguish fires with water spray, carbon dioxide, dry chemical powder, or alcohol or polymer foam.

Spillage disposal   See general section.

56. 3-Aminopropionitrile

The free base is a liquid, the hydrochloride a crystalline solid. Both have amine odour; b.p. 185 ˚C (free base).

Toxic effects   Experimental teratogen.

Hazardous reactions	Stored material exploded after polymerization of yellow solid.
First aid	Standard treatment for exposure by all routes (see pages 108-122).
Fire hazard	Decomposes on heating to produce toxic fumes.

57. 2-Aminopyridine

White crystals or powder; m.p. 58 °C; soluble in water

Limit values	OES short-term 2 p.p.m. (8 mg m^{-3}); long-term 0.5 p.p.m. (2 mg m^{-3}).
Toxic effects	Headache, dizziness, flushing of skin, shortness of breath, nausea, collapse, convulsions (possibly fatal).
First aid	Standard treatment for exposure by all routes (see pages 108-122).
Spillage disposal	See general section.

58. 2-Aminothiazole

White to yellowish crystals; m.p. 90 °C; slightly soluble in water

Toxic effects	Moderately toxic by ingestion.
Hazardous reactions	Material ignited in drying oven; violent explosion when nitrated with nitric/sulfuric acids.
First aid	Standard treatment for exposure by all routes (see pages 108-122).
Fire hazard	Extinguish fires with carbon dioxide, dry powder, or alcohol or polymer foam.
Spillage disposal	See general section.

59. Amitrole

Off-white powder; m.p. 157-159 °C; soluble in water.

RISKS
Harmful if swallowed – Possible risk of irreversible effects – Danger of serious damage to health by prolonged exposure (R22, R40, R48)

SAFETY PRECAUTIONS
Wear suitable protective clothing – Wear suitable gloves (S36, S37)

Toxic effects	May cause irritation. Suspect human carcinogen.
First aid	Standard treatment for exposure by all routes (see pages 108-122).
Fire hazard	Extinguish fire with water spray, carbon dioxide, dry chemical powder, or alcohol or polymer foam.
Spillage disposal	See general section.

60. Ammonia

Colourless gas with characteristic pungent odour; m.p. -77.7 ˚C; b.p. -33 to -35 ˚C; very soluble in water.

RISKS
Flammable – Toxic by inhalation (R10, R23)

SAFETY PRECAUTIONS
Keep container tightly closed and in a well ventilated place – Keep away from sources of ignition – No Smoking – In case of insufficient ventilation, wear suitable respiratory equipment (S7/9, S16, S38)

Limit values	OES short-term 35 p.p.m. (24 mg m^{-3}); long-term 25 p.p.m. (17 mg m^{-3}).
Toxic effects	The gas irritates all parts of the respiratory system and the eyes.
Hazardous reactions	Mixtures with air have exploded; violent reactions with some inorganic compounds.
First aid	Standard treatment for exposure by all routes (see pages 108-122).
Fire hazard	Explosive limits 16-25%; Autoignition temperature 651 ˚C. As the gas is supplied in a cylinder, turning off the valve will reduce any fire involving it; if possible cylinders should be removed quickly from an area in which a fire has developed.
Spillage disposal	Surplus gas or leaking cylinder can be vented slowly into water-fed scrubbing tower or column, or into a fume cupboard served by such a tower.

RSC *Chemical Safety Data Sheets* Vol. 3, No. 8, 1990 gives extended coverage.

61. Ammonia (solutions)

Ammonia solution is commonly supplied to laboratories as a 35% solution in water (0.88 specific gravity). In warm weather this strong solution develops pressure in its bottle and the cap must be released with care. The 25% solution (0.90) is free from this problem.

RISKS
Causes burns – Irritating to eyes, respiratory system, and skin (R34, R36/37/38)

SAFETY PRECAUTIONS
Keep container tightly closed – In case of contact with eyes, rinse immediately with plenty of water and seek medical advice (S7, S26)

Toxic effects	The solution will cause severe internal burns if ingested.
First aid	Standard treatment for exposure by all routes (see pages 108-122).
Spillage disposal	See general section.

62. Ammonium amidosulfate

Colourless crystals; m.p. 132-135 °C; b.p. decomposes at 160 °C; soluble in water.

Limit values	OES short-term 20 mg m^{-3}; long-term 10 mg m^{-3}.
Toxic effects	Irritant to skin and eyes. Harmful by any route.
Hazardous reactions	Vigorous exothermic hydrolysis of 60% solution with acid.
First aid	Standard treatment for exposure by all routes (see pages 108-122).
Fire hazard	Extinguish fires with carbon dioxide, dry powder, or alcohol or polymer foam.
Spillage disposal	See general section.

63. Ammonium bromate

Colourless crystals; m.p. explodes on heating; very soluble in water.

Toxic effects	Can cause paralysis of the central nervous system.
Hazardous reactions	Very friction-sensitive, may explode spontaneously.
First aid	Standard treatment for exposure by all routes (see pages 108-122).
Fire hazard	Decomposes on heating to produce toxic fumes.
Spillage disposal	See general section.

64. Ammonium chlorate

White, crystalline powder; m.p. explodes at 100 °C; soluble in water.

Hazardous reactions	Occasionally explodes spontaneously, always above 100 °C; cold saturated solution may decompose explosively.
First aid	Standard treatment for exposure by all routes (see pages 108-122).
Fire hazard	Decomposes on heating to give toxic fumes.
Spillage disposal	See general section.

65. Ammonium chloride

White, crystalline solid; m.p. 340 ˚C (sublimes); b.p. 520 ˚C; soluble in water.

RISKS
Harmful if swallowed – Irritating to eyes (R22, R36)

SAFETY PRECAUTIONS
Do not breathe dust (S22)

Limit values   OES short-term 20 mg m^{-3} (as fume); long-term 10 mg m^{-3} (as fume).

Toxic effects   Irritant to skin and eyes. Harmful on inhalation, ingestion, and skin absorption.

Hazardous reactions   Reacts vigorously or explosively with certain inorganic compounds.

First aid   Standard treatment for exposure by all routes (see pages 108-122).

Fire hazard   Use extinguishant appropriate to the surroundings.

Spillage disposal   See general section.

66. Ammonium chromate

Yellow powder; m.p. 185 ˚C (decomp.); soluble in water.

Toxic effects   High concentrations are extremely irritant and destructive to skin, eyes, upper respiratory tract, and mucous membranes. Carcinogenic.

First aid   Standard treatment for exposure by all routes (see pages 108-122).

Fire hazard   Use extinguishant appropriate to the surroundings.

Spillage disposal   See general section.

67. Ammonium dichromate

Orange, crystalline solid; m.p. 170 ˚C with decomposition; soluble in water.

RISKS
Explosive when dry – Contact with combustible material may cause fire – Irritating to eyes, respiratory system and skin – May cause sensitization by skin contact (R1, R8, R36/37/38, R43)

SAFETY PRECAUTIONS
After contact with skin, wash immediately with plenty of water – This material and its container must be disposed of in a safe way (S28, S35)

Limit values   OES long-term 0.05 mg m^{-3} (as Cr) under review.

Toxic effects   Irritant to skin, eyes, upper respiratory tract, and mucous membranes. Can cause nasal and skin ulcers. Absorption can lead to stomach pains, vomiting, diarrhoea, liver, and kidney damage and can result in death. Carcinogenic.

First aid	Standard treatment for exposure by all routes (see pages 108-122).
Fire hazard	Extinguish fires with water spray. In close confinement the deflagrating salt can become explosive.
Spillage disposal	See general section.

68. Ammonium fluoride

Colourless crystals; m.p. 125.6 °C; very soluble in water.

RISKS
Toxic by inhalation, in contact with skin, and if swallowed (R23/24/25)

SAFETY PRECAUTIONS
Keep locked up and out of reach of children – In case of contact with eyes, rinse immediately with plenty of water and seek medical advice – If you feel unwell, seek medical advice (show label where possible) (S1/2, S26, S44)

Limit values	OES long-term 2.5 mg m^{-3} (as F).
Toxic effects	Irritant to skin, eyes, and respiratory tract. Toxic on inhalation and ingestion.
First aid	Standard treatment for exposure by all routes (see pages 108-122), with special reference to fluorides.
Fire hazard	Extinguish fires with water spray.
Spillage disposal	See general section.

69. Ammonium iodate

Colourless crystals; m.p. decomposes at 150 °C; slightly soluble in water.

Hazardous reactions	Decomposed violently on touching with scoop.
First aid	Standard treatment for exposure by all routes (see pages 108-122).
Fire hazard	Decomposes on heating to produce toxic fumes.
Spillage disposal	See general section.

70. Ammonium nitrate

Colourless crystals; m.p. 169 °C; b.p. decomposes at about 210 °C; soluble in water.

| **Toxic effects** | Irritant to skin, eyes, upper respiratory tract, and mucous membranes. Prolonged exposure may produce gastro-intestinal upsets and blood disorders. |

Hazardous reactions	Reactions with several powdered metals and other inorganic substances.
First aid	Standard treatment for exposure by all routes (see pages 108-122).
Fire hazard	Mixtures of ammonium nitrate and combustible materials are readily ignited; mixtures with finely divided combustible materials can react explosively. Extinguish fire with water spray.
Spillage disposal	See general section. Ensure that site of spillage is thoroughly washed down to eliminate future fire risks.

71. Ammonium permanganate

Crystalline powder; m.p. explodes at 69 ˚C; slightly soluble in water.

Toxic effects	Damage to the central nervous system and the pulmonary system may result from inhalation of the fume or dust. Chronic manganese poisoning is a clearly characterized disease.
Hazardous reactions	Friction sensitive when dry and explodes at 60 ˚C.
First aid	Standard treatment for exposure by all routes (see pages 108-122).

72. Ammonium peroxodisulfate

White crystals; m.p. 120 ˚C with decomposition; soluble in water.

Hazardous reactions	Mixture with water and powdered aluminium may explode; mixture with sodium peroxide explodes on grinding in mortar.
Fire hazard	Mixtures with combustible materials are readily ignited. Extinguish fire with water spray.
Spillage disposal	See general section.

73. Ammonium picrate

Red or yellow crystals; m.p. decomposes on melting and explodes at 423 ˚C; slightly soluble in water.

Toxic effects	Moderately irritating to skin, eyes, and mucous membranes.
Hazardous reactions	Explodes on heating or impact.
First aid	Standard treatment for exposure by all routes (see pages 108-122).
Spillage disposal	See general section. But the material is explosive when dry.

74. Ammonium sulfide solution

Yellow liquid with offensive odour.

Toxic effects   Inhalation of vapour causes headache, giddiness, and loss of energy sometime after exposure; higher concentrations can cause unconsciousness. The liquid seriously irritates eyes, skin, and gastro-intestinal tract.

Hazardous reactions   Contact with acid liberates poisonous hydrogen sulfide.

First aid   Standard treatment for exposure by all routes (see pages 108-122).

Spillage disposal   See general section.

75. n-Amyl alcohol and s-amyl alcohol

Colourless liquids with pungent odours; m.p. (n-) -79 °C; b.p. (n-) 138 °C and (s-) 119 °C; both are slightly soluble in water.

RISKS
Flammable – Harmful by inhalation (R10, R20)

SAFETY PRECAUTIONS
Avoid contact with skin and eyes (S24/25)

Toxic effects   The alcohols are irritating to the skin, eyes, upper respiratory tract, and the gastro-intestinal tract. Systemic effects on livers and kidneys can follow inhalation, ingestion, or skin absorption.

Hazardous reactions   Explosion of a Raney nickel catalyst in the presence of amyl alcohol has been recorded. The decomposition of amyl alcohol in the presence of hydrogen trisulfide is extremely violent.

First aid   Standard treatment for exposure by all routes (see pages 108-122).

Fire hazard   Flash point (n-) 32 °C and (s-) 40 °C.; Explosive limits 1-10%; Autoignition temperature 280 °C. Extinguish fire with water spray, alcohol-resistant foam, dry powder, or carbon dioxide.

Spillage disposal   See general section.

RSC *Chemical Safety Data Sheets* Vol. 1, No. 4, 1988 gives extended coverage on n-amyl alcohol.

76. Amyl alcohol (commercial)

Commercial amyl alcohol is derived from several sources, each being a different mixture of isomers. The main constituent is usually 3-methyl-1-butanol. With the exception of solid 2,2-dimethyl-1-propanol all the other isomers are colourless liquids with a mild odour; b.p. between 101 °C and 131 °C; all are slightly soluble in water.

RISKS
Flammable – Harmful by inhalation (R10, R20)

SAFETY PRECAUTIONS
Avoid contact with skin and eyes (S24/25)

Limit values   OES short-term 125 p.p.m. (450 mg m^{-3}) (3-methyl-1-butanol); long-term 100 p.p.m. (360 mg m^{-3})(3-methyl-1-butanol).

Toxic effects   All the isomers are irritant to skin, eyes, upper respiratory tract, and mucous membranes. All are harmful following inhalation or ingestion.

First aid   Standard treatment for exposure by all routes (see pages 108-122).

Fire hazard   Flash point 45 °C; Explosive limits lel 1.2%; Autoignition temperature 343 °C (figures for 3-methyl-1-butanol). Figures for other isomers are similar. Extinguish fires with water spray, carbon dioxide, dry chemical powder, or alcohol foam.

Spillage disposal   See general section.

77. Amyl nitrite

Colourless or pale yellow, highly volatile liquid with pungent fruity odour; advisable to store in refrigerator; b.p. 99 °C; immiscible with water.

Toxic effects   If inhaled may cause headache, flushing of face, weakness, and collapse. If swallowed, similar effects may be expected.

First aid   Standard treatment for exposure by all routes (see pages 108-122).

Fire hazard   Flash point 10 °C; Autoignition temperature 209 °C. Extinguish fire with dry powder or carbon dioxide.

Spillage disposal   See general section.

78. Aniline

Colourless liquid with characteristic smell; darkens on exposure to air and light; m.p. -6.2 ˚C; b.p. 184.4 ˚C; slightly soluble in water.

RISKS
Toxic by inhalation, in contact with skin, and if swallowed – Danger of cumulative effects (R23/24/25, R33)

SAFETY PRECAUTIONS
After contact with skin, wash immediately with plenty of water – Wear protective clothing and gloves – If you feel unwell, seek medical advice (show label where possible) (S28, S36/37, S44)

Limit values	OES short-term 5 p.p.m. (20 mg m^{-3}) under review; long-term 2 p.p.m. (10 mg m^{-3}) under review.
Toxic effects	Inhalation of the vapour, absorption through the skin, or ingestion causes headache, drowsiness, cyanosis, mental confusion, and, in severe cases, convulsions. The liquid is dangerous to the eyes. There are chronic effects following prolonged exposure to the vapour or slight skin exposure over a period.
Hazardous reactions	Vigorously oxidized by a number of oxidants.
First aid	Standard treatment for exposure by all routes (see pages 108-122).
Fire hazard	Flash point 70-76 ˚C (closed cup); Explosive limits lel 1.3%; Autoignition temperature 700 ˚C. Extinguish fires with water spray, carbon dioxide, dry chemical powder, or foam.
Spillage disposal	See general section.

RSC *Chemical Safety Data Sheets* Vol. 4a, No. 8, 1991 gives extended coverage.

---

79. Anilinium salts

The commoner anilinium salts – the chloride and sulfate – are colourless to greyish-brown in colour; soluble in water.

RISKS
Toxic by inhalation, in contact with skin, and if swallowed – Danger of cumulative effects (R23/24/25, R33)

SAFETY PRECAUTIONS
After contact with skin, wash immediately with plenty of water – Wear protective clothing and gloves – If you feel unwell, seek medical advice (show label where possible) (S28, S36/37, S44)

Toxic effects	The acute and chronic effects of aniline poisoning are described under Aniline. Skin absorption does not occur so readily with the salts or their solutions as with aniline itself, but they are dangerous to the eyes, partly because of the acidity, and cause intense irritation. If taken by mouth, the effects of aniline poisoning (headache, drowsiness, cyanosis) will be apparent.

First aid   Standard treatment for exposure by all routes (see pages 108-122).

Spillage   See general section.
  disposal

80. Anisidines

o-Anisidine is a pale yellow to orange liquid, *p*-Anisidine is a pale yellow to brown solid; m.p. (*o*-) 5.2 ˚C; (*p*-) 59 ˚C; b.p. (*o*-) 225 ˚C; (*p*-) 243 ˚C; both *o*- and *p*-anisidine are insoluble in water.

RISKS
Very toxic by inhalation, in contact with skin, and if swallowed – Danger of cumulative effects (R26/27/28, R33)

SAFETY PRECAUTIONS
After contact with skin, wash immediately with plenty of water – Wear protective clothing and gloves – In case of accident or if you feel unwell, seek medical advice immediately (show label where possible) (S28, S36/37, S45)

Limit values   OES long-term 0.1 p.p.m. (0.5 mg m^{-3}).

Toxic effects These are similar to those of aniline poisoning, *i.e.* headache, drowsiness, and cyanosis. *o*-Anisidine is an irritant to skin and upper respiratory tract.

First aid   Standard treatment for exposure by all routes (see pages 108-122).

Fire hazard   Flash point 107 ˚C (closed cup). Extinguish fires with water spray, carbon dioxide, dry chemical powder, or alcohol or polymer foam.

Spillage   See general section.
  disposal

RSC *Chemical Safety Data Sheets* Vol. 4a, No. 9, 1991 gives extended coverage.

81. Anisole

Colourless liquid with a distinctive odour; m.p. -37.3 ˚C; b.p. 153.8 ˚C; insoluble in water.

Toxic effects May be irritant or harmful if absorbed into the body.

First aid   Standard treatment for exposure by all routes (see pages 108-122).

Fire hazard   Flash point 43 ˚C; Autoignition temperature 475 ˚C. Extinguish fires with water spray, carbon dioxide, dry chemical powder, or alcohol or polymer foam.

Spillage   See general section.
  disposal

82. Anthracene

Light yellow, shiny flakes; m.p. 218 ˚C; b.p. 342 ˚C; insoluble in water.

Toxic effects Irritant to skin, eyes, upper respiratory tract, or mucous membranes. Prolonged contact can cause pigmentation of the skin and cancer.

Hazardous reactions	Violent explosion with fluorine even at -210 ˚C.
First aid	Standard treatment for exposure by all routes (see pages 108-122).
Fire hazard	Flash point 121 ˚C; Explosive limits lel 0.6%; Autoignition temperature 538 ˚C. Extinguish fires with water spray, carbon dioxide, dry chemical powder, or alcohol or polymer foam.
Spillage disposal	See general section.

83. Anthraquinone

Dull yellow powder; m.p. 286 ˚C (sublimes); b.p. 377 ˚C; sparingly soluble in water.

Toxic effects	Irritant to skin and eyes. May cause allergic skin reaction.
First aid	Standard treatment for exposure by all routes (see pages 108-122).
Fire hazard	Flash point 185 ˚C. Extinguish fires with water spray, carbon dioxide, dry chemical powder, or alcohol or polymer foam.
Spillage disposal	See general section.

84. Antimony

Silvery-white, lustrous, brittle metal; m.p. 630.7 ˚C; b.p. 1635 ˚C; insoluble in water.

RISKS
Harmful by inhalation and in contact with skin (R20/21)

SAFETY PRECAUTIONS
Do not breathe dust (S22)

Limit values	OES long-term 0.5 mg m^{-3} (as Sb).
Toxic effects	Irritant to skin, eyes, and upper respiratory tract. Ingestion may result in irritation of the gastro-intestinal tract, vomiting, diarrhoea, and eventually death from respiratory or circulatory failure.
Hazardous reactions	Reacts violently or explosively with a variety of substances.
First aid	Standard treatment for exposure by all routes (see pages 108-122).
Fire hazard	Extinguish fires with water spray.
Spillage disposal	See general section.

RSC *Chemical Safety Data Sheets* Vol. 2, No. 12, 1989 gives extended coverage.

85. Antimony hydride

Colourless gas with an unpleasant smell; m.p. -88 °C; b.p. -18.4 °C; slightly soluble in water.

RISKS
Harmful by inhalation and if swallowed (R20/22)

SAFETY PRECAUTIONS
Do not breathe dust (S22)

Limit values   OES short-term 0.1 p.p.m. (0.5 mg m^{-3}); long-term 0.3 p.p.m. (1.5 mg m^{-3}).

Toxic effects   Mild irritant to skin and eyes. Inhalation may result in headaches, nausea, vomiting, weakness, irregular breathing and pulse, and anaemia.

Hazardous reactions   Reacts explosively with chlorine, chlorine water, concentrated nitric acid, ozone, or ammonia (on heating).

First aid   Standard treatment for exposure by all routes (see pages 108-122).

Fire hazard   In fire conditions, cool cylinders with water or remove them to a place of safety.

Spillage disposal   Dispose of via a licensed contractor.

RSC *Chemical Safety Data Sheets* Vol. 2, No. 13, 1989 gives extended coverage.

86. Antimony(III) nitride

Orange powder; m.p. decomposes on heating; decomposed by water.

Limit values   OES long-term 0.5 mg m^{-3} (as Sb).

Hazardous reactions   Impure material explodes mildly on heating in air or on contact with water or dilute acids.

First aid   Standard treatment for exposure by all routes (see pages 108-122).

Spillage disposal   See general section.

87. Antimony pentachloride

Red-yellow, oily liquid with an offensive smell; m.p. 2.8 °C; b.p. 140 °C; decomposed by water.

RISKS
Causes burns – Irritating to respiratory system (R34, R37)

SAFETY PRECAUTIONS
In case of contact with eyes, rinse immediately with plenty of water and seek medical advice (S26)

Limit values   OES long-term 0.5 mg m^{-3} (as Sb).

Toxic effects	Severely irritating to skin, eyes, and respiratory tract. Toxic if ingested. Mutagenic.
First aid	Standard treatment for exposure by all routes (see pages 108-122).
Fire hazard	Use a fire extinguishing agent, but NOT water, appropriate to the surroundings.
Spillage disposal	See general section.

RSC *Chemical Safety Data Sheets* Vol. 2, No. 14, 1989 gives extended coverage.

88. Antimony pentafluoride

Colourless, oily liquid; m.p. 7.0 °C; b.p. 149.5 °C; reacts violently with water.

RISKS
Harmful by inhalation and if swallowed (R20/22)

SAFETY PRECAUTIONS
Do not breathe dust (S22)

Limit values	OES long-term 0.5 mg m^{-3} (as Sb).
Toxic effects	Irritant to skin, eyes, upper respiratory tract, and mucous membranes. Ingestion may lead to hypotension, coma, and convulsions.
First aid	Standard treatment for exposure by all routes (see pages 108-122), with special reference to fluorides.
Fire hazard	Extinguish fires with carbon dioxide or dry chemical powder.
Spillage disposal	See general section.

RSC *Chemical Safety Data Sheets* Vol. 2, No. 15, 1989 gives extended coverage.

89. Antimony trichloride

Colourless, crystalline solid; m.p. 73.4 °C; b.p. 222.6 °C; soluble in water.

RISKS
Causes burns – Irritating to respiratory system (R34, R37)

SAFETY PRECAUTIONS
In case of contact with eyes, rinse immediately with plenty of water and seek medical advice (S26)

Limit values	OES long-term 0.5 mg m^{-3} (as Sb).
Toxic effects	Irritating and corrosive to skin, eyes, upper respiratory tract, and mucous membranes. Inhalation and ingestion can lead to considerable tissue damage. Mutagenic.
First aid	Standard treatment for exposure by all routes (see pages 108-122).

Fire hazard   Use an extinguishing agent, but NOT water, appropriate to the surroundings.

Spillage   See general section.
  disposal

RSC *Chemical Safety Data Sheets* Vol. 2, No. 16, 1989 gives extended coverage.

90. Antimony trifluoride

Colourless, deliquescent crystals; m.p. 292 ˚C; b.p. 376 ˚C; reacts violently with water.

RISKS
Toxic by inhalation, in contact with skin, and if swallowed (R23/24/25)

SAFETY PRECAUTIONS
Keep container tightly closed – In case of contact with eyes, rinse immediately with plenty of water and seek medical advice – If you feel unwell, seek medical advice (show label where possible) (S7, S26, S44)

Limit values   OES long-term 0.5 mg m^{-3} (as Sb).

Toxic effects   Irritant to skin, eyes, and upper respiratory tract. Ingestion may result in vomiting, diarrhoea, weakness, coma, and convulsions.

First aid   Standard treatment for exposure by all routes (see pages 108-122), with special reference to fluorides.

Fire hazard   Use an extinguishing agent, but NOT water, appropriate to the surroundings.

Spillage   See general section.
  disposal

RSC *Chemical Safety Data Sheets* Vol. 2, No. 17, 1989 gives extended coverage.

91. Antimony trioxide

White crystals; m.p. 573 ˚C; b.p. 656 ˚C; slightly soluble in water.

Limit values   OES long-term 0.5 mg m^{-3} (as Sb) under review.

Toxic effects   Irritant to skin, eyes, and upper respiratory tract. Ingestion may cause swellings of the lining of the mouth and gastro-intestinal disorders. Mutagenic. Suspect human carcinogen.

Hazardous   Reacts explosively with chlorinated rubber on heating and with perchloric acid
  reactions   when hot.

First aid   Standard treatment for exposure by all routes (see pages 108-122).

Fire hazard   Extinguish fires with water fog, carbon dioxide, dry chemical powder, or foam.

Spillage   See general section.
  disposal

RSC *Chemical Safety Data Sheets* Vol. 2, No. 18, 1989 gives extended coverage.

92. Antimony trisulfide

Grey-black, crystalline powder; m.p. 550 ˚C; b.p. 1150 ˚C; practically insoluble in water.

Limit values OES long-term 0.5 mg m⁻³ (as Sb) under review.

Toxic effects Irritant to skin and eyes. Harmful if swallowed. Sudden deaths from chronic heart disease have been reported following prolonged inhalation.

Hazardous reactions Reacts vigorously or explosively with a variety of substances.

First aid Standard treatment for exposure by all routes (see pages 108-122).

Fire hazard Extinguish fires with carbon dioxide, dry chemical powder, or alcohol or polymer foam.

Spillage disposal See general section.

RSC *Chemical Safety Data Sheets* Vol. 2, No. 19, 1989 gives extended coverage.

93. Arsenic and compounds

Most arsenic compounds are colourless powders or crystals – they include arsenites and arsenates of many metals; syrupy arsenic acid and arsenic trichloride are liquids.

RISKS
Toxic by inhalation and if swallowed (R23/25)

SAFETY PRECAUTIONS
Keep locked up and out of reach of children – When using do not eat, drink, or smoke – After contact with skin, wash immediately with plenty of water – If you feel unwell, seek medical advice (show label where possible) (S1/2, S20/21, S28, S44)

Limit values MEL long-term 0.1 mg m⁻³ (as As).

Toxic effects All compounds must be considered to be extremely poisonous. The inhalation of dust or fume irritates the mucous membranes and leads to arsenical poisoning. Certain compounds, especially the trichloride and arsenic acid, irritate the eyes and skin, and absorption causes poisoning. If swallowed arsenic compounds irritate the stomach severely and affect the heart, liver, and kidneys; nervousness, thirst, vomiting, diarrhoea, cyanosis, and collapse may be symptoms. The inhalation of small concentrations of dust or fume over a long period will cause poisoning; skin contact over a long period may cause ulceration. The metal itself has not been recognized as a noteworthy hazard although it can be irritant to skin, eyes, and upper respiratory tract.

First aid Standard treatment for exposure by all routes (see pages 108-122).

Fire hazard Most of the compounds are not combustible and an extinguishant appropriate to the surroundings can be used.

Spillage disposal The disposal of these in any quantity must be considered carefully in the light of local conditions and regulations. See general section.

RSC *Chemical Safety Data Sheets* Vol. 2, Nos. 43, 80, 97 & 98, 1989 and Vol. 4a, Nos. 10-13, 1991 give extended coverage.

94. Arsine

Colourless gas with a disagreeable garlic smell; m.p. -116.3 ˚C; b.p. -62.5 ˚C; slightly soluble in water.

RISKS
Toxic by inhalation and if swallowed (R23/25)

SAFETY PRECAUTIONS
Keep locked up and out of reach of children – When using do not eat, drink, or smoke – After contact with skin, wash immediately with plenty of water – If you feel unwell, seek medical advice (show label where possible) (S1/2, S20/21, S28, S44)

Limit values	OES long-term 0.05 p.p.m. (0.2 mg m^{-3}).
Toxic effects	A few inhalations may be fatal, death resulting from anoxia or pulmonary oedema. Symptoms of poisoning include headache, weakness, vertigo, and nausea. Damage is caused to kidneys and liver.
Hazardous reactions	The endothermic compound is capable of detonation. Reacts vigorously with certain chemicals.
First aid	Standard treatment for exposure by all routes (see pages 108-122). Prompt medical attention is vital.
Fire hazard	Explosive limits 3.9-77.8.
Spillage disposal	Leaking laboratory cylinders should be removed to open space and allowed to discharge as slowly as possible. Wear self-contained breathing apparatus throughout.

RSC *Chemical Safety Data Sheets* Vol. 4a, No. 14, 1991 gives extended coverage.

95. Asbestos

Toxic effects	Human carcinogen, use of all forms is controlled in the United Kingdom by the Control of Asbestos at Work Regulations 1987.
First aid	Standard treatment for exposure by all routes (see pages 108-122).

96. Auramine

Yellow flakes or powder; insoluble in water.

Toxic effects	Absorption through the skin may result in dermatitis and burns, nausea, and vomiting. In the United Kingdom the manufacture (not use) of this substance is controlled by the Carcinogenic Substances Regulations 1967.
First aid	Standard treatment for exposure by all routes (see pages 108-122).
Spillage disposal	See general section. Arrange for disposal by specialist contractor.

97. Azides and azido- compounds

Many compounds of both organic and inorganic derivation which contain the azide function are unstable or explosive under appropriate conditions of initiation.

Metal azides are usually explosive being sensitive to heat, friction, or impact. Contact with most acids will usually liberate hydrazoic acid, itself an explosive and highly toxic low boiling liquid. Prolonged contact with halogenated solvents can lead to the slow formation of explosive organic azides.

Organic azides are usually heat- or shock-sensitive compounds of varying degrees of stability, sensitivity to traces of strong acids and metallic salts which may catalyse explosive decomposition. Acyl azides of low molecular weight should not be isolated from solution as the concentrated solutions are dangerously explosive.

The presence of more than one azide group, particularly if on the same carbon atom, greatly reduces the stability. Unsaturated azides containing one or more double or triple bonds and one or more azide groups are particularly dangerous and must be handled with extreme care.

As a class the azides are of variable toxicity. Many of them are toxic with the main effects on the body being changes in blood pressure and inhibition of enzyme action.

RSC *Chemical Safety Data Sheets* Vol. 2, No. 99, 1989 gives extended coverage of sodium azide.

98. Aziridine

Mobile colourless liquid with ammoniacal odour; b.p. 55-56 °C; infinitely soluble in water.

RISKS
Highly flammable – Very toxic by inhalation, in contact with skin, and if swallowed – Possible risk of irreversible effects (R11, R26/27/28, R40)

SAFETY PRECAUTIONS
Keep container in a well ventilated place – Do not empty into drains – Wear suitable protective clothing – In case of accident or if you feel unwell, seek medical advice immediately (show label where possible) (S9, S29, S36, S45)

Limit values	OES long-term 0.5 p.p.m. (1 mg m^{-3}) under review.
Toxic effects	Experimental carcinogen and mutagen. Potent skin irritant and vesicant. Can cause corneal damage. Powerful emitic if swallowed.
Hazardous reactions	Erroneous published preparative procedure-liable to polymerize explosively. Forms explosive derivatives.
First aid	Standard treatment for exposure by all routes (see pages 108-122).
Fire hazard	Flash point -11 °C (closed cup); Explosive limits 3.6-46%; Autoignition temperature 322 °C. Extinguish fires with carbon dioxide, dry chemical powder, or alcohol foam. Dangerous fire and explosion hazard.
Spillage disposal	See general section.

RSC *Chemical Safety Data Sheets* Vol. 4a, No. 15, 1991 gives extended coverage.

99. Azodicarbonamide

Orange-red powder; b.p. decomposes above 195 ˚C; very slightly soluble in hot water.

Toxic effects   May be irritant to skin and eyes. May be harmful if swallowed. Irritation and sensitization follow inhalation.

First aid   Standard treatment for exposure by all routes (see pages 108-122).

Fire hazard   Extinguish fires with carbon dioxide, dry chemical powder, or alcohol or polymer foam.

Spillage disposal   See general section.

RSC *Chemical Safety Data Sheets* Vol. 3, No. 11, 1990 gives extended coverage.

100. Azoisobutyronitrile

Hazardous reactions   Decomposes when heated; explosive decomposition when technical material being recrystallized from acetone.

First aid   Standard treatment for exposure by all routes (see pages 108-122)

Spillage disposal   See general section.

101. Azo-*N*-nitroformamidine

Hazardous reactions   Decomposes explosively at 165 ˚C.

102. Barium

Silvery-white, slightly lustrous metal; m.p. 725 ˚C; b.p. 1640 ˚C; decomposes in water.

Toxic effects   Irritant to skin and eyes. Harmful by ingestion giving rise to nausea, vomiting, diarrhoea, muscular weakness, and possibly convulsions.

Hazardous reactions   Finely divided barium metal may ignite or explode in contact with water or in certain conditions with chlorinated hydrocarbons.

First aid   Standard treatment for exposure by all routes (see pages 108-122).

Fire hazard   Extinguish fires with dry chemical powder. Do NOT use water.

Spillage disposal   See general section. Large spills may require specialist help.

RSC *Chemical Safety Data Sheets* Vol. 2, No. 20, 1989 gives extended coverage.

103. Barium compounds

Most of the common barium compounds are white or colourless, crystalline solids; solubility in water varies.

RISKS
Harmful by inhalation and if swallowed (R20/22)

SAFETY PRECAUTIONS
After contact with skin, wash immediately with plenty of water (S28)

Limit values   OES long-term 0.5 mg m^{-3} (as Ba for soluble compounds).

Toxic effects   Skin and eye irritation varies with the fluoride and oxide being particularly dangerous. Soluble salts are rapidly absorbed from the gastro-intestinal tract and are very toxic. All the dusts irritate the upper respiratory tract and may cause baritosis, a benign pneumoconiosis.

Hazardous reactions   The oxide interacts vigorously with water and may be a potential fire and ignition hazard. The bromate, chlorate, and nitrate will decompose violently or explosively on heating.

First aid   Standard treatment for exposure by all routes (see pages 108-122), with special reference to fluorides.

Fire hazard   Use extinguishant appropriate to the surroundings.

Spillage disposal   See general section.

RSC *Chemical Safety Data Sheets* Vol. 2, Nos. 21-24 and 26, 1989 give extended coverage.

104. Barium peroxide

Greyish white, heavy powder; m.p. 450 ˚C; insoluble in water, but slowly decomposes in it.

RISKS
Contact with combustible material may cause fire – Harmful by inhalation and if swallowed (R8, R20/22)

SAFETY PRECAUTIONS ●
Keep away from food, drink and animal feeding stuffs – Keep out of reach of children (S13, S2)

Limit values   OES long-term 0.5 mg m^{-3} (as Ba).

Toxic effects   Irritant to all tissues.

Hazardous reactions   May ignite hydrogen sulfide, hydroxylamine, and organic materials especially in presence of water, mixtures with powdered metals ignite, explosions with acetic anhydride.

First aid   Standard treatment for exposure by all routes (see pages 108-122).

Fire hazard   Use flooding amounts of water in early stages of fire. Smother with suitable dry powder.

Spillage disposal   See general section.

RSC *Chemical Safety Data Sheets* Vol. 2, No. 25, 1989, gives extended coverage.

105. Barium sulfate

White or yellowish heavy powder; m.p. 1580 ˚C; insoluble in water.

Limit values   OES long-term 2 mg m^{-3} (as respirable dust).

Toxic effects   May cause local irritation of the skin and eyes. Very poorly absorbed from the gastro-intestinal tract. Inhalation can cause baritosis, a benign pneumoconiosis.

First aid   Standard treatment for exposure by all routes (see pages 108-122).

Fire hazard   Use extinguishant appropriate to the surroundings.

Spillage   See general section.
   disposal

RSC *Chemical Safety Data Sheets* Vol. 2, No. 26, 1989 gives extended coverage.

106. Benzaldehyde

Colourless, strongly refractive liquid.; m.p. -26 ˚C; b.p. 179 ˚C; slightly soluble in water.

RISKS
Harmful if swallowed (R22)

SAFETY PRECAUTIONS
Avoid contact with skin (S24)

Toxic effects   A central nervous system depressant in small doses and narcotic in high concentrations. Toxic by ingestion and skin absorption. May cause dermatitis on skin contact.

Hazardous   Violent oxidation by 90% performic acid.
   reactions

First aid   Standard treatment for exposure by all routes (see pages 108-122).

Fire hazard   Flash point 62 ˚C; Autoignition temperature 192 ˚C. Extinguish fires with water, dry powder, or alcohol foam.

Spillage   See general section.
   disposal

107. Benzene

Colourless, volatile liquid with characteristic odour; m.p. 5.5 ˚C; b.p. 80.1 ˚C; very slightly soluble in water.

RISKS
May cause cancer – Highly flammable – Toxic by inhalation, in contact with skin, and if swallowed – Danger of serious damage to health by prolonged exposure (R45, R11, R23/24/25, R48)

SAFETY PRECAUTIONS
Avoid exposure – Obtain special instructions before use – Keep away from sources of ignition – No Smoking – Do not empty into drains – If you feel unwell, seek medical advice (show label where possible) (S53, S16, S29, S44)

Limit values   MEL long-term 5 p.p.m. (16 mg m^{-3}).

Toxic effects   Inhalation of the vapour causes dizziness, headache, and excitement; high concentrations may cause unconsciousness and death. Repeated inhalation of low concentrations over a considerable period may cause severe, even fatal, blood disease. The vapour and liquid irritate the eyes and mucous membranes. The liquid is absorbed through the skin and poisoning may result from this. Extremely poisonous if taken by mouth.

Hazardous reactions   Complex with silver perchlorate exploded on crushing in mortar; certain mixtures with 04% of nitric acid are highly sensitive to detonation, mixture with liquid oxygen is explosive; benzene solution of rubber exploded when ozonized; reacts vigorously or explosively with other oxidants, interhalogens, and uranium hexafluoride.

First aid   Standard treatment for exposure by all routes (see pages 108-122).

Fire hazard   Flash point -11 ˚C; Explosive limits 1.4 – 8.0%; Autoignition temperature 560 ˚C. Extinguish fire with foam or dry powder. Fires involving benzene produce large amounts of sooty smoke.

Spillage disposal   See general section.

RSC *Chemical Safety Data Sheets* Vol. 1, No. 5, 1988 gives extended coverage.

108. Benzenediazonium salts

Hazardous reactions   In general these salts are unstable and are sensitive to friction, heat, and shock. They can react violently or explosively with a variety of substances.

First aid   Standard treatment for exposure by all routes (see pages 108-122).

109. Benzenesulfonyl chloride

Colourless to brown liquid; b.p. 251 ˚C with decomposition; reacts with water to form benzenesulfonic acid and hydrochloric acid.

Toxic effects   Irritates the eyes severely and causes skin burns. Causes severe internal irritation if taken by mouth.

Hazardous reactions   Spontaneous decomposition in storage, violent reaction with dimethyl sulfoxide.

First aid   Standard treatment for exposure by all routes (see pages 108-122).

Spillage disposal   See general section.

110. Benzenethiol

Colourless liquid with repulsive pungent smell; m.p. -14.8 ˚C; b.p. 168 ˚C; immiscible with water.

Limit values   OES long-term 0.5 p.p.m. (2 mg m^{-3}).

Toxic effects	Inhalation may result in headache and dizziness. Skin contact may lead to severe dermatitis.
Hazardous reactions	Violent explosion during preparation from benzenediazonium chloride.
First aid	Standard treatment for exposure by all routes (see pages 108-122).
Spillage disposal	See general section.

111. Benzidine and salts

The use of these compounds in the United Kingdom is now prohibited under The Carcinogenic Substances Regulations 1967. It is therefore not considered appropriate to deal with these hazards any further in this book.

RISKS
May cause cancer – Harmful if swallowed (R45, R22)

SAFETY PRECAUTIONS
Avoid exposure – Obtain special instructions before use – If you feel unwell, seek medical advice (show label where possible) (S53, S44)

112. Benzonitrile

Colourless liquid; m.p. -13 °C; b.p. 191 °C; sparingly soluble in water.

RISKS
Harmful in contact with skin and if swallowed (R21/22)

SAFETY PRECAUTIONS
Do not breathe vapour (S23)

Toxic effects	Harmful by inhalation, skin and eye contact, and ingestion.
First aid	Standard treatment for exposure by all routes (see pages 108-122).
Spillage disposal	See general section.

113. Benzophenone-3,3',4, 4'-tetracarboxylic dianhydride

Light yellow powder; m.p. 224-226 °C.

RISKS
Irritating to eyes and respiratory system (R36/37)

SAFETY PRECAUTIONS
Avoid contact with eyes (S25)

Toxic effects	May be irritant to skin, eyes, and upper respiratory tract. May be harmful following inhalation, ingestion, or skin absorption.
First aid	Standard treatment for exposure by all routes (see pages 108-122).
Fire hazard	Extinguish fires with water spray, carbon dioxide, dry chemical powder, or alcohol or polymer foam.
Spillage disposal	See general section.

RSC *Chemical Safety Data Sheets* Vol. 3, No. 12, 1990 gives extended coverage.

114. Benzo[*a*]pyrene

Yellow powder; m.p. 179 ˚C; b.p. 495 ˚C.

RISKS
May cause cancer – May cause heritable genetic damage – May cause birth defects (R45, R46, R47)

SAFETY PRECAUTIONS
Avoid exposureb – Obtain special instructions before use – If you feel unwell, seek medical advice (show label where possible) (S53, S44)

Toxic effects	Suspected human carcinogen. Irritant to skin and eyes.
First aid	Standard treatment for exposure by all routes (see pages 108-122).
Fire hazard	Extinguish fires with water spray, carbon dioxide, dry chemical powder, or alcohol or polymer foam.
Spillage disposal	See general section.

115. *p*-Benzoquinone

Yellow crystals with a characteristic irritating smell; m.p. 115.7 ˚C; b.p. sublimes; slightly soluble in water.

RISKS
Toxic by inhalation and if swallowed – Irritating to eyes, respiratory system, and skin (R23/25, R36/37/38)

SAFETY PRECAUTIONS
In case of contact with eyes, rinse immediately with plenty of water and seek medical advice – After contact with skin, wash immediately with plenty of water – If you feel unwell, seek medical advice (show label where possible) (S26, S28, S44)

Limit values	OES short-term 0.3 p.p.m. (1.2 mg m^{-3}); long-term 0.1 p.p.m. (0.4 mg m^{-3}).
Toxic effects	The dust irritates the respiratory system severely. Skin or eye contact is very irritating and can cause severe local damage. Highly irritant and dangerous if taken by mouth.
Hazardous reactions	Drums of moist material self-heated and decomposed.

First aid	Standard treatment for exposure by all routes (see pages 108-122).
Fire hazard	Autoignition temperature 560 ˚C.
Spillage disposal	See general section.

116. Benzotriazole

Colourless to tan needle like crystals; m.p. 98.5 ˚C; b.p. 204 ˚C at 15 mm Hg; sparingly soluble in water.

Toxic effects	Moderately toxic by ingestion and intraperitoneal routes. Severe eye and skin irritant.
Hazardous reactions	Large batch exothermally decomposed and then detonated during distillation at 160 ˚C/2.5 mbar.
First aid	Standard treatment for exposure by all routes (see pages 108-122).
Fire hazard	Water spray may be used to extinguish fires.
Spillage disposal	See general section.

117. Benzoyl chloride

Colourless, fuming liquid with pungent smell; m.p. 0 ˚C; b.p. 197 ˚C; reacts with water forming benzoic acid and hydrochloric acid.

RISKS
Causes burns (R34)

SAFETY PRECAUTIONS
In case of contact with eyes, rinse immediately with plenty of water and seek medical advice (S26)

Toxic effects	The vapour irritates the respiratory system and eyes. The liquid is very irritating to the skin, eyes, and internal organs.
Hazardous reactions	Violent reaction with dimethyl sulfoxide.
First aid	Standard treatment for exposure by all routes (see pages 108-122).
Spillage disposal	See general section.

118. Benzoyl nitrate

| **Hazardous reactions** | Unstable liquid which explodes on rapid heating and may also explode on exposure to light or contact with moist cellulose. |

119. Benzvalene

Hazardous reactions   Exploded violently when scratched.

120. Benzyl acetate

Colourless liquid; m.p. -51.5 ˚C; b.p. 215.5 ˚C; sparingly soluble in water.

Toxic effects   Irritant to skin, eyes, upper respiratory tract, and mucous membranes. Can cause decrease in blood pressure and depth of respiration, also increase in heart beat. Experimental carcinogen.

First aid   Standard treatment for exposure by all routes (see pages 108-122).

Fire hazard   Flash point 102 ˚C; Autoignition temperature 462 ˚C. Extinguish fires with water spray, carbon dioxide, dry chemical powder, or alcohol or polymer foam.

Spillage disposal   See general section.

121. Benzyl alcohol

Colourless liquid, faint aromatic odour; m.p. -15 ˚C; b.p. 205 ˚C; slightly soluble in water.

RISKS
Harmful by inhalation and if swallowed (R20/22)

SAFETY PRECAUTIONS
In case of contact with eyes, rinse immediately with plenty of water and seek medical advice (S26)

Toxic effects   Irritant to skin, eyes, upper respiratory tract, and mucous membranes. Can be absorbed through the skin.

Hazardous reactions   Benzyl alcohol contaminated with 1.4% hydrogen bromide and 1.1% dissolved iron(ii) polymerizes exothermally above 100 ˚C. Mixture with 58% sulfuric acid decomposes explosively at about 180 ˚C.

First aid   Standard treatment for exposure by all routes (see pages 108-122).

Fire hazard   Flash point 100 ˚C (closed cup); Autoignition temperature 435 ˚C. Extinguish fires with water spray, carbon dioxide, dry chemical powder, or alcohol or polymer foam.

Spillage disposal   See general section.

122. Benzylamine

Colourless liquid; b.p. 185 ˚C; miscible with water.

RISKS
Causes burns (R34)

SAFETY PRECAUTIONS
In case of contact with eyes, rinse immediately with plenty of water and seek medical advice (S26)

Toxic effects   Irritant to skin, eyes, and mucous membranes.

First aid   Standard treatment for exposure by all routes (see pages 108-122).

Spillage   See general section.
disposal

123. Benzyl bromide

Colourless to brown-yellow liquid with acrid smell; b.p. 198 ˚C; immiscible with water.

RISKS
Irritating to eyes, respiratory system, and skin (R36/37/38)

SAFETY PRECAUTIONS
Wear eye/face protection (S39)

Toxic effects   The vapour irritates the respiratory system, eyes, and skin. The liquid causes skin burns. Assumed to cause severe internal irritation and damage if taken by mouth.

Hazardous   Material drying over molecular sieve polymerized, evolving hydrogen
reactions   bromide.

First aid   Standard treatment for exposure by all routes (see pages 108-122).

Spillage   See general section.
disposal

124. Benzylbutyl phthalate

Colourless, viscous liquid; m.p. -35 ˚C; b.p. 370 ˚C; slightly soluble in water.

Limit values   OES long-term 5 mg m^{-3}.

Toxic effects   Irritant to skin, eyes, upper respiratory tract, and mucous membranes. Harmful by inhalation, ingestion, or skin absorption. Experimental carcinogen.

First aid   Standard treatment for exposure by all routes (see pages 108-122).

Fire hazard   Flash point 199 ˚C (closed cup). Extinguish fires with water spray, carbon dioxide, dry chemical powder, or alcohol or polymer foam.

Spillage   See general section.
disposal

125. Benzyl chloride

Colourless to brown-yellow liquid with acrid smell; b.p. 179 °C; immiscible with water.

RISKS
Irritating to eyes, respiratory system, and skin (R36/37/38)

SAFETY PRECAUTIONS
Wear eye/face protection (S39)

Limit values OES long-term 1 p.p.m. (5 mg m^{-3}) under review.

Toxic effects The vapour irritates the respiratory system, eyes, and skin. The liquid causes skin burns. Assumed to cause severe internal irritation and damage if taken by mouth.

Hazardous reactions Absence of sufficient base to prevent acidity developing led to violent reaction or explosions.

First aid Standard treatment for exposure by all routes (see pages 108-122).

Spillage disposal See general section.

126. Benzyl chloroformate

Colourless to yellow, fuming, oily liquid; decomposes slowly at room temperature; b.p. 103 °C at 29 mbar (2.9x10^2 Pa); insoluble in water.

RISKS
Causes burns – Irritating to respiratory system (R34, R37)

SAFETY PRECAUTIONS
In case of contact with eyes, rinse immediately with plenty of water and seek medical advice (S26)

Toxic effects Vapour causes severe irritation of eyes and respiratory system. Liquid blisters the skin and severely damages the eyes.

First aid Standard treatment for exposure by all routes (see pages 108-122).

Spillage disposal See general section.

127. Benzylidene chloride

Colourless liquid; m.p. -16.1 °C; b.p. 205 °C; immiscible with water.

RISKS
Irritating to eyes, respiratory system, and skin (R36/37/38)

SAFETY PRECAUTIONS
Wear eye/face protection (S39)

Toxic effects The vapour irritates the respiratory system and eyes. The liquid irritates the eyes and skin. If taken by mouth, internal irritation and damage may follow.

| First aid | Standard treatment for exposure by all routes (see pages 108-122). |
| Spillage disposal | See general section. |

128. Benzylidyne chloride

Colourless to yellow, fuming liquid; b.p. 214 ˚C; immiscible with water.

Toxic effects	The vapour irritates all parts of the respiratory system. The vapour and liquid irritate the eyes and skin. If taken by mouth, internal irritation and damage must be assumed.
First aid	Standard treatment for exposure by all routes (see pages 108-122).
Spillage disposal	See general section.

129. Benzylidyne fluoride

Colourless liquid with an aromatic odour; b.p. 101 ˚C; immiscible with water.

Toxic effects	Animal experiments indicate the risk of central nervous system depression through inhalation, absorption, or ingestion.
First aid	Standard treatment for exposure by all routes (see pages 108-122).
Fire hazard	Flash point 12 ˚C. Extinguish fire with water spray, foam, dry powder, or carbon dioxide.
Spillage disposal	See general section.

130. Benzyloxyacetylene

| Hazardous reactions | Explosion if heated above 60 ˚C during vacuum distillation. |

131. Benzyltriethylammonium permanganate

| Hazardous reactions | The quaternary oxidant, previously reported as stable, is explosive under unexpectedly mild conditions. Related compounds are similarly unpredictable. |
| First aid | Standard treatment for exposure by all routes (see pages 108-122). |

132. Beryllium

Greyish-white, hard, light metal; m.p. 1278 ˚C; b.p. 2970 ˚C; insoluble in water.

RISKS
Very toxic by inhalation and in contact with skin – Irritating to respiratory system – Danger of very serious irreversible effects (R26/27, R37, R39)

SAFETY PRECAUTIONS
In case of contact with eyes, rinse immediately with plenty of water and seek medical advice – After contact with skin, wash immediately with plenty of water – In case of accident or if you feel unwell, seek medical advice immediately (show label where possible) (S26, S28, S45)

Limit values   OES long-term 0.002 mg m^{-3} under review.

Toxic effects   Beryllium is absorbed slowly from the lungs into the blood and is then transported to the skeletal system, liver, and kidneys. Acute effects include damage to the respiratory system. Poorly absorbed from the gastro-intestinal tract. Human carcinogen.

Hazardous reactions   Heavy impact flashes with mixtures of powdered metal and certain chlorinated solvents; incandescent reaction when heated with phosphorus.

First aid   Standard treatment for exposure by all routes (see pages 108-122).

Fire hazard   Extinguish fires with suitable dry powder. Do NOT use water.

Spillage disposal   See general section. Exposed personnel should bathe carefully and all contaminated clothing and equipment dealt with separately.

RSC *Chemical Safety Data Sheets* Vol. 2, No. 27, 1989, gives extended coverage.

133. Beryllium compounds

White powder or crystals; the chloride, fluoride, and sulfate are soluble in water, the oxide insoluble.

RISKS
Very toxic by inhalation and in contact with skin – Irritating to respiratory system – Danger of very serious irreversible effects (R26/27, R37, R39)

SAFETY PRECAUTIONS
In case of contact with eyes, rinse immediately with plenty of water and seek medical advice – After contact with skin, wash immediately with plenty of water – In case of accident or if you feel unwell, seek medical advice immediately (show label where possible) (S26, S28, S45)

Toxic effects   Particles penetrating the skin through wounds and abrasions may cause local damage difficult to heal. Symptoms of poisoning, indicated by respiratory troubles or cyanosis, may develop within a week or after a latent period of even several years.

First aid   Standard treatment for exposure by all routes (see pages 108-122).

Fire hazard   Extinguish fire by using extinguishant appropriate to the surroundings, but do NOT use water on beryllium chloride.

Spillage disposal The disposal of these in even small quantities must be considered carefully in the light of local conditions and regulations. Large amounts should involve either recycling or an approved contractor. See general section.

RSC *Chemical Safety Data Sheets* Vol. 2, Nos. 28-31, 1989, give extended coverage.

134. Biphenyl

White crystals or powder, faint characteristic odour; m.p. 70 ˚C; b.p. 256 ˚C; insoluble in water.

Limit values OES short-term 0.6 p.p.m. (4 mg m^{-3}); long-term 0.2 p.p.m. (1.5 mg m^{-3}).

Toxic effects May also cause irritation of respiratory tract and eyes, nausea, and depressed appetite.

First aid Standard treatment for exposure by all routes (see pages 108-122).

Spillage disposal See general section.

135. Bis(4-aminophenyl)methane

Light brown solid; sparingly soluble in water.

Limit values OES short-term 0.5 p.p.m. (4 mg m^{-3}) under review; long-term 0.1 p.p.m. (0.8 mg m^{-3}) under review.

Toxic effects Inhalation and/or ingestion of dust leads to liver damage.

First aid Standard treatment for exposure by all routes (see pages 108-122).

Spillage disposal See general section.

136. Bis(2-chloroethyl) ether

Colourless liquid with sweet pleasant odour; m.p. -51.9 ˚C; b.p. 178.5 ˚C; practically insoluble in water.

RISKS
Flammable – Very toxic by inhalation, in contact with skin, and if swallowed – Possible risk of irreversible effects (R10, R26/27/28, R40)

SAFETY PRECAUTIONS
Keep container tightly closed and in a well ventilated place – Take off immediately all contaminated clothing – In case of insufficient ventilation, wear suitable respiratory equipment – In case of accident or if you feel unwell, seek medical advice immediately (show label where possible) (S7/9, S27, S38, S45)

Toxic effects The vapour irritates the respiratory system and high concentrations may result in lung damage after a latent period of some hours. The vapour and liquid irritate the eyes.

Hazardous reactions   Liable to form explosive peroxides on exposure to air and light.

First aid   Standard treatment for exposure by all routes (see pages 108-122).

Fire hazard   Flash point 55 °C (closed cup); Explosive limits lel 2.7%; Autoignition temperature 369 °C. Extinguish fire with alcohol-resistant foam, dry powder, or carbon dioxide.

Spillage disposal   See general section.

RSC *Chemical Safety Data Sheets* Vol. 1. No. 25, 1988 gives extended coverage.

137. Bis(chloromethyl) ether

Colourless liquid with a suffocating odour; b.p. 106 °C; decomposed by water.

RISKS
May cause cancer – Flammable – Harmful if swallowed – Toxic in contact with skin – Very toxic by inhalation (R45, R10, R22, R24, R26)

SAFETY PRECAUTIONS
Avoid exposure – Obtain special instructions before use – In case of accident or if you feel unwell, seek medical advice immediately (show label where possible) (S53, S45)

Limit values   OES long-term 0.001 p.p.m. under review.

Toxic effects   The vapour irritates the respiratory system and eyes, the primary effects being probably due to rapid formation of hydrochloric acid and formaldehyde. A high incidence of lung cancer has been observed. Assumed to be toxic and corrosive if swallowed.

First aid   Standard treatment for exposure by all routes (see pages 108-122).

Spillage disposal   See general section.

138. Bis[4-(2,3-epoxypropoxy) phenyl]propane

Solid; m.p. 43 °C.

RISKS
Irritating to eyes and skin – May cause sensitization by skin contact (R36/38, R43)

SAFETY PRECAUTIONS
After contact with skin, wash immediately with plenty of water – Wear suitable gloves and eye/face protection (S28, S37/39)

Toxic effects   Mildly toxic by ingestion. Experimental carcinogen.

First aid   Standard treatment for exposure by all routes (see pages 108-122).

RSC *Chemical Safety Data Sheets* Vol. 3, No. 13, 1990 gives extended coverage.

139. Bis(2,3-epoxypropyl) ether

Colourless liquid with irritant odour; b.p. 260 ˚C.

Limit values   OES long-term 0.1 p.p.m. (0.6 mg m^{-3}).

Toxic effects   Irritant to skin, eyes, and mucous membranes. Inhalation can be fatal due to chemical pneumonitis and pulmonary oedema. Experimental skin carcinogen. Chronic exposure can cause bone marrow depression.

First aid   Standard treatment for exposure by all routes (see pages 108-122).

Fire hazard   Flash point 64 ˚C. Extinguish large fires with water spray or foam. Additionally carbon dioxide or dry chemical powder can be used on small fires.

**Spillage
disposal**   See general section.

140. Bis(2-ethylhexyl)phthalate

Colourless liquid; m.p. -50 ˚C; b.p. 384 ˚C; insoluble in water.

Limit values   OES short-term 10 mg m^{-3}; long-term 5 mg m^{-3}.

Toxic effects   Irritant to skin, eyes, upper respiratory tract, and mucous membranes. Absorption can lead to nausea and central nervous system depression.

First aid   Standard treatment for exposure by all routes (see pages 108-122).

Fire hazard   Flash point 207 ˚C; Explosive limits lel 0.3%. Extinguish fires with carbon dioxide, dry chemical powder, or alcohol or polymer foam. Water spray may also be used but may cause frothing.

**Spillage
disposal**   See general section.

141. Bis(isocyanatophenyl)methane

Yellow crystals or fused solid with irritating smell; m.p. 37 ˚C; b.p. 194-199 ˚C; hydrolysed by water.

Limit values   MEL short-term 0.07 mg m^{-3} (as NCO-); long-term 0.02 mg m^{-3} (as NCO-).

Toxic effects   The vapour and dust are irritating to the eyes and respiratory system. The solid or molten material irritates the eyes and skin and must be considered poisonous if taken by mouth. An allergic sensitizer.

**Hazardous
reactions**   By storing at 5 ˚C a tendency to form polymeric solids is reduced to a minimum.

First aid   Standard treatment for exposure by all routes (see pages 108-122).

Fire hazard   Flash point 218 ˚C.

**Spillage
disposal**   See general section.

142. Bis(methoxyethyl) ether

Colourless liquid; m.p. -68 °C; b.p. 162 °C; miscible with water.

Toxic effects	Irritant to skin and eyes. Ingestion can result in nausea, cramps, and weakness leading to coma.
Hazardous reactions	Forms peroxides on exposure to air and light, explosive reactions of peroxidized solvent with aluminium hydrides.
First aid	Standard treatment for exposure by all routes (see pages 108-122).
Fire hazard	Flash point 70 °C. Alcohol foam may be used to extinguish fires.

143. Bismuth

Lustrous, brittle, crystalline solid with a pink tinge; m.p. 271.4 °C; b.p. 1420-1560 °C; insoluble in water.

Toxic effects	May be irritating to the respiratory tract, eyes, and skin. Low oral toxicity due to its poor absorption from the bowel.
Hazardous reactions	The finely divided metal reacts violently or explosively with certain inorganic compounds.
First aid	Standard treatment for exposure by all routes (see pages 108-122).
Fire hazard	To extinguish fire use extinguishant appropriate to the surroundings.
Spillage disposal	See general section. Spills should be swept into a bag and held for re-use or recycling.

RSC *Chemical Safety Data Sheets* Vol. 2, No. 32, 1989 gives extended coverage.

144. Bismuth oxychloride

White, lustrous, crystalline powder; m.p. at red heat with decomposition; practically insoluble in water.

Toxic effects	Does not appear to present any major biological hazards.
First aid	Standard treatment for exposure by all routes (see pages 108-122).
Fire hazard	Extinguish fires with carbon dioxide, dry chemical powder, or alcohol or polymer foam.
Spillage disposal	See general section.

RSC *Chemical Safety Data Sheets* Vol. 2, No. 33, 1989 gives extended coverage.

145. Bismuth pentafluoride

Crystalline powder; m.p. sublimes at 120 °C; reacts vigorously with water.

Toxic effects	Irritant to skin, eyes, upper respiratory tract, and gastro-intestinal tissues.
Hazardous reactions	Reacts vigorously with water and may ignite.

146. Bismuth telluride

Grey, hygroscopic crystals; m.p. 43 °C; b.p. 210 °C; reacts violently with water.

Hazardous reactions	Strongly exothermal reaction with ammonia. Incandescent reaction with warm red or white phosphorus.

159. Bromine

Dark reddish-brown, fuming liquid; b.p. 59 °C; slightly soluble in water.

RISKS
Very toxic by inhalation – Causes severe burns (R26, R35)

SAFETY PRECAUTIONS
Keep container tightly closed and in a well ventilated place – In case of contact with eyes, rinse immediately with plenty of water and seek medical advice (S7/9, S26)

Limit values	OES short-term 0.3 p.p.m. (2 mg m^{-3}); long-term 0.1 p.p.m. (0.7 mg m^{-3}).
Toxic effects	The vapour irritates all parts of the respiratory system, the eyes, and mucous membranes. The liquid burns the skin and eyes, and would cause internal damage following ingestion.
Hazardous reactions	Bromine reacts with varying degrees of violence with a large number of compounds and elements.
First aid	Standard treatment for exposure by all routes (see pages 108-122).
Spillage disposal	See general section.

160. Bromine pentafluoride

Colourless to pale yellow fuming liquid with pungent odour; m.p. -61.3 °C; b.p. 40 °C; reacts vigorously and possibly explosively with water.

Limit values	OES short-term 0.3 p.p.m. (2 mg m^{-3}); long-term 0.1 p.p.m. (0.7 mg m^{-3}).
Toxic effects	This vapour is severely irritant to skin, eyes, and respiratory system. The liquid causes severe burns to all tissues.
Hazardous reactions	The liquid reacts violently with many organic compounds and some inorganic compounds. It explodes or ignites in contact with hydrogen-containing

Substances 147–158

We regret that substances 147–158 appear out of sequence.
*For the continuation of entry **159. Bromine pentafluoride** please see p. 247.*

147. Bismuth trioxide

Yellow powder; m.p. 824 °C; almost insoluble in water.

Toxic effects	May be irritant to skin, eyes, and respiratory tract. Prolonged ingestion may give rise to mental, muscular, and motor co-ordination disorders.
First aid	Standard treatment for exposure by all routes (see pages 108-122).
Fire hazard	Extinguish fires with water, carbon dioxide, dry chemical powder, or alcohol or polymer foam.
Spillage disposal	See general section.

RSC *Chemical Safety Data Sheets* Vol. 2, No. 35, 1989 gives extended coverage.

148. Bis(trichloroacetyl) peroxide

Hazardous reactions	Explodes on standing at room temperature.

149. Bis(trifluoroacetyl) peroxide

Hazardous reactions	Explodes on standing at room temperature.

150. Borane – tetrahydrofuran complex

Colourless liquid; reacts violently with water.

Toxic effects	Vapour or mist are irritant to skin, eyes, upper respiratory tract, or mucous membranes. Exposure may produce headaches, nausea, and vomiting.
Hazardous reactions	May decompose in storage with liberation of hydrogen and bursting of bottle.
Fire hazard	Flash point –17 °C. Extinguish fires with dry powder extinguisher.
Spillage disposal	See general section.

151. Borazine

Colourless liquid; m.p. −58 ˚C; b.p. 53 ˚C; reacts with water producing toxic and flammable boron hydrides.

Toxic effects Irritant to skin, eyes, upper respiratory tract, and mucous membranes.

Hazardous reactions Sealed ampoules exploded in daylight.

152. Boran-2-one

Colourless crystalline solid with distinctive odour; m.p. 180 ˚C; b.p. 204 ˚C; insoluble in water.

Limit values OES short-term 3 p.p.m. (18 mg m^{-3}); long-term 2 p.p.m. (12 mg m^{-3}).

Toxic effects Irritant to skin, eyes, and upper respiratory tract. Absorption into the body can cause dizziness, excitation, and convulsions.

First aid Standard treatment for exposure by all routes (see pages 108-122).

Fire hazard Flash point 65.5 ˚C (closed cup); Autoignition temperature 466 ˚C. Extinguish fires with water spray, dry chemical powder, or foam extinguishing agents.

Spillage disposal See general section.

153. Boron

Yellow or brown powder; m.p. 2300 ˚C; b.p. 2550 ˚C; insoluble in water.

Toxic effects Toxic by ingestion and inhalation. Irritant to skin, eyes, upper respiratory tract, and mucous membranes.

Hazardous reactions Ignites in chlorine or fluorine at ambient temperature. Reacts violently or explosively with oxidants. Many of the violent reactions of boron with reagents other than hot powerful oxidants now attributed to previous use of impure boron samples.

First aid Standard treatment for exposure by all routes (see pages 108-122).

Fire hazard Extinguish fires with carbon dioxide or dry powder.

Spillage disposal See general section.

154. Boron oxide

White, crystalline powder; m.p. 450 ˚C; b.p. 1860 ˚C; slightly soluble in water.

Limit values OES short-term 20 mg m^{-3}; long-term 10 mg m^{-3}.

Toxic effects Irritant to eyes, upper respiratory tract, and mucous membranes. Harmful if inhaled or swallowed.

First aid   Standard treatment for exposure by all routes (see pages 108-122).

Fire hazard   Use extinguishant appropriate to surroundings.

Spillage   See general section.
disposal

155. Boron tribromide

Colourless, fuming liquid with a pungent odour; m.p. −46 ˚C; b.p. 91.3 ˚C; reacts violently with water.

RISKS
Reacts violently with water − Very toxic by inhalation and if swallowed − Causes severe burns (R14, R26/28, R35)

SAFETY PRECAUTIONS
Keep container in a well ventilated place − In case of contact with eyes, rinse immediately with plenty of water and seek medical advice − After contact with skin, wash immediately with plenty of water − Wear suitable protective clothing − In case of accident or if you feel unwell, seek medical advice immediately (show label where possible) (S9, S26, S28, S36, S45)

Limit values   OES short-term 1 p.p.m. (10 mg m^{-3}).

Toxic effects   The vapour irritates all parts of the respiratory system and eyes. The liquid burns the skin and eyes. If taken by mouth, there would be severe internal burning.

Hazardous   Reacts violently when poured into an excess of water and explosively when
reactions   water is poured into it. Mixture with sodium metal explodes on impact.

First aid   Standard treatment for exposure by all routes (see pages 108-122).

Spillage   See general section.
disposal

156. Boron trichloride

Colourless, fuming liquid or vapour with pungent, irritating odour; m.p. −107 ˚C; b.p. 12.5 ˚C; reacts rapidly with water forming boric and hydrochloric acids.

RISKS
Reacts violently with water − Very toxic by inhalation and if swallowed − Causes burns (R14, R26/28, R34)

SAFETY PRECAUTIONS
Keep container in a well ventilated place − In case of contact with eyes, rinse immediately with plenty of water and seek medical advice − After contact with skin, wash immediately with plenty of water − Wear suitable protective clothing − In case of accident or if you feel unwell, seek medical advice immediately (show label where possible) (S9, S26, S28, S36, S45)

Toxic effects   The gas irritates the eyes, skin, and respiratory system. The liquid irritates and burns the skin and eyes. If taken by mouth there would be severe internal burning.

Hazardous reactions	Reacts violently with aniline and phosphine.
First aid	Standard treatment for exposure by all routes (see pages 108-122).
Spillage disposal	See general section.

157. Boron trifluoride

Colourless, fuming gas with a pungent, suffocating odour; m.p. −127.1 ˚C; b.p. −100.4 ˚C; decomposes in water forming fluoroboric and boric acids.

RISKS
Reacts violently with water – Very toxic by inhalation – Causes severe burns (R14, R26, R35)

SAFETY PRECAUTIONS
Keep container in a well ventilated place – In case of contact with eyes, rinse immediately with plenty of water and seek medical advice – After contact with skin, wash immediately with plenty of water – Wear suitable protective clothing – In case of accident or if you feel unwell, seek medical advice immediately (show label where possible) (S9, S26, S28, S36, S45)

Limit values	OES short-term 1 p.p.m. (3 mg m^{-3}).
Hazardous reactions	Reacts with hot alkali or alkaline earth (not Mg) metals with incandescence.
First aid	Standard treatment for exposure by all routes (see pages 108-122).
Spillage disposal	Surplus gas or leaking cylinders can be vented slowly into water-fed scrubbing tower or column, or into a fume cupboard served by such a tower.

158. Boron triiodide

Colourless, hygroscopic crystals; m.p. 43 ˚C; b.p. 210 ˚C; reacts violently with water.

| **Hazardous reactions** | Strongly exothermal reaction with ammonia. Incandescent reaction with warm red or white phosphorus. |

*For substance **159. Bromine** and the opening of the entry for substance **160. Bromine pentafluoride** please see p. 246.*

materials. Reacts violently and may ignite with a very wide variety of substances.

First aid	Standard treatment for exposure by all routes (see pages 108-122).
Fire hazard	Extinguish fires with dry chemical powder. Do NOT use water.
Spillage disposal	See general section.

161. Bromine trifluoride

Colourless to grey-yellow, fuming liquid with pungent choking smell; m.p. 8.8 °C; b.p. 127 °C; reacts vigorously with water.

Toxic effects	The vapours severely irritate and may burn the eyes, skin, and respiratory system. The liquids burn all human tissue and cause severe damage.
Hazardous reactions	Violent or explosive reactions occur with a wide variety of organic and inorganic materials.
First aid	Standard treatment for exposure by all routes (see pages 108-122).
Spillage disposal	See general section.

162. Bromoacetamide

White, crystalline powder; m.p. 70-80 °C; soluble in water.

Toxic effects	Irritant to skin, eyes, upper respiratory tract, and mucous membranes. May be harmful on inhalation, ingestion, or skin absorption.
Hazardous reactions	Decomposes rapidly when hot in presence of moisture and light.
First aid	Standard treatment for exposure by all routes (see pages 108-122).
Fire hazard	Extinguish fires with carbon dioxide, dry powder, or alcohol or polymer foam.
Spillage disposal	See general section.

163. Bromoacetic acid

Colourless to pale brown solid; m.p. 50 °C; soluble in water.

RISKS
Toxic by inhalation, in contact with skin, and if swallowed – Causes severe burns (R23/24/25, R35)

SAFETY PRECAUTIONS
Wear suitable protective clothing, gloves, and eye/face protection – If you feel unwell, seek medical advice (show label where possible) (S36/37/39, S44)

Toxic effects	Contact of the solid or solution with the eyes causes severe burns. The effect on the skin is not immediate and blisters may not appear for 12 hours or more

after contact. Can be assumed to cause severe internal irritation and damage if taken by mouth.

First aid   Standard treatment for exposure by all routes (see pages 108-122).

Spillage   See general section.
disposal

164. Bromoacetone oxime

Hazardous   Explodes during distillation.
reactions

First aid   Standard treatment for exposure by all routes (see pages 108-122).

165. Bromoacetylene

Gas; b.p. -2 °C.

Hazardous   Unstable – may burn or explode on contact with air.
reactions

166. Bromoaziridine

Hazardous   Unstable – explodes during or shortly after distillation.
reactions

167. Bromobenzene

Colourless liquid with an aromatic smell; b.p. 156 °C; immiscible with water.

RISKS
Flammable – Irritating to skin (R10, R38)

Toxic effects   Little is known about the toxic properties, but its relationship with benzene suggests caution in handling. The vapour may be narcotic in high concentrations. It should be assumed to be poisonous through skin absorption and if taken by mouth.

Hazardous   May react violently with sodium.
reactions

First aid   Standard treatment for exposure by all routes (see pages 108-122).

Fire hazard   Flash point 51 °C; Autoignition temperature 566 °C. Extinguish fire with dry powder or carbon dioxide.

Spillage   See general section.
disposal

168. Bromochloromethane

Colourless liquid with sweetish odour; b.p. 69 °C; insoluble in water.

Limit values OES short-term 250 p.p.m. (1300 mg m^{-3}); long-term 200 p.p.m. (1050 mg m^{-3}).

Toxic effects The vapour irritates the respiratory system and the eyes. The liquid irritates the eyes, skin, and gastro-intestinal tract.

First aid Standard treatment for exposure by all routes (see pages 108-122).

Spillage disposal See general section.

169. Bromodimethylaniline

White to off-white powder; m.p. 55 °C; b.p. 264 °C; insoluble in water.

Toxic effects Irritant to skin, eyes, upper respiratory tract, and mucous membranes.

Hazardous reactions Exploded during vacuum distillation.

First aid Standard treatment for exposure by all routes (see pages 108-122).

Fire hazard Flash point >110 °C. Extinguish fires using water spray, carbon dioxide, dry powder, or alcohol or polymer foam.

Spillage disposal See general section.

170. Bromoethane

Colourless, volatile liquid with ethereal odour; b.p. 38 °C; sparingly soluble in water.

RISKS
Harmful by inhalation, in contact with skin, and if swallowed (R20/21/22)

SAFETY PRECAUTIONS
After contact with skin, wash immediately with plenty of water (S28)

Limit values OES short-term 250 p.p.m. (1110 mg m^{-3}); long-term 200 p.p.m. (890 mg m^{-3}).

Toxic effects The vapour irritates the respiratory system and has anaesthetic and narcotic effects. The liquid irritates the eyes and is poisonous if taken by mouth, causing damage to the kidneys.

Hazardous reactions In preparation from ethanol and bromine.

First aid Standard treatment for exposure by all routes (see pages 108-122).

Spillage disposal See general section.

171. Bromoethylfuran

Hazardous   Very unstable – will explode violently.
reactions

First aid   Standard treatment for exposure by all routes (see pages 108-122).

172. Bromoform

Heavy, colourless liquid with smell like chloroform; m.p. 6-7 ˚C; b.p. 150 ˚C; only slightly soluble in water.

RISKS
Toxic by inhalation – Irritating to eyes and skin (R23, R36/38)

SAFETY PRECAUTIONS
After contact with skin, wash immediately with plenty of water – If you feel unwell, seek medical advice (show label where possible) (S28, S44)

Limit values   OES long-term 0.5 p.p.m. (5 mg m^{-3}).

Toxic effects The vapour is lachrymatory and irritates the respiratory system. The liquid irritates the skin. Ingestion can cause respiratory difficulties, tremors, and loss of consciousness.

Hazardous   Reacts violently with acetone if catalysed by powdered bases, even in
reactions   presence of diluting solvents. Mixtures with alkali metals are shock sensitive, even when the stimulus is far away.

First aid   Standard treatment for exposure by all routes (see pages 108-122).

Spillage   See general section.
disposal

RSC *Chemical Safety Data Sheets* Vol. 1, No. 6, 1988 gives extended coverage.

173. Bromomethane

Colourless liquid or gas with faint chloroform-like odour; m.p. -93.66 ˚C; b.p. 3.56 ˚C; liquid forms a crystalline hydrate with cold water.

RISKS
Very toxic by inhalation (R26)

SAFETY PRECAUTIONS
Keep locked up and out of reach of children – Keep container tightly closed and in a well ventilated place – Avoid contact with skin and eyes (S1/2, S7/9, S24/25)

Limit values   OES short-term 15 p.p.m. (60 mg m^{-3}); long-term 5 p.p.m. (20 mg m^{-3}).

Toxic effects Short exposures to high concentrations of vapour cause headache, dizziness, nausea, vomiting, and weakness; this may be followed by mental excitement, convulsions, and even acute mania. The longer inhalation of lower concentrations may lead to bronchitis and pneumonia. Both the vapour and liquid cause severe damage to the eyes. The liquid burns the skin, blisters appearing several hours after contact; itching and reddening of the skin may precede this. Assumed to be very poisonous if taken by the mouth.

Hazardous reactions Forms pyrophoric Grignard-type compounds with zinc, aluminium, and magnesium; delayed explosion in reaction with dimethyl sulfoxide.

Fire hazard Explosive limits 13.5%-14.5%; Autoignition temperature 536.7 °C.

Spillage disposal Leaking cylinder should be placed in a well ventilated fume cupboard and vented slowly until discharged.

174. 1-Bromopropane

Liquid; m.p. -110 °C; b.p. 70.9 °C; sparingly soluble in water.

RISKS
Highly flammable – Very toxic by inhalation, in contact with skin, and if swallowed (R11, R26/27/28)

SAFETY PRECAUTIONS
Keep container tightly closed and in a well ventilated place – Do not empty into drains – In case of accident or if you feel unwell, seek medical advice immediately (show label where possible) (S7/9, S29, S45)

Toxic effects Irritant to skin, eyes, and upper respiratory tract.

First aid Standard treatment for exposure by all routes (see pages 108-122), with particular attention to the possible occurrence of convulsions, shock, and/or lung congestion.

Fire hazard Flash point <22 °C (closed cup); Explosive limits lel 4.6%; Autoignition temperature 490 °C. Extinguish fires with carbon dioxide, dry chemical powder, or alcohol or polymer foam.

Spillage disposal See general section.

RSC *Chemical Safety Data Sheets* Vol. 4a, No. 23, 1991 gives extended coverage.

175. Bromopropyne

Almost colourless liquid with a sharp odour; m.p. -61 °C; b.p. 88-90 °C.

Toxic effects Vapour and mist are irritant to skin, eyes, and upper respiratory tract. Also extremely toxic by ingestion.

Hazardous reactions Classed as extremely shock-sensitive. Also danger of explosion in contact with copper, high-copper alloys, mercury, and silver.

First aid Standard treatment for exposure by all routes (see pages 108-122).

Fire hazard Flash point 65 °C (closed cup). Extinguish fires with foam, dry chemical, or carbon dioxide extinguishers. Emits highly toxic fumes on heating to decomposition.

Spillage disposal See general section.

176. Bromosuccinimide

White to pale buff, crystalline solid smelling faintly of bromine; m.p. 177-181 ˚C with decomposition.

Toxic effects   Irritates or burns the skin, eyes, or respiratory system. Strongly irritant if taken by mouth.

Hazardous reactions Reacts violently with aniline, diallyl sulfide, and hydrazine hydrate.

First aid   Standard treatment for exposure by all routes (see pages 108-122).

Spillage disposal   See general section.

177. Bromotetramethylguanidine

Hazardous reactions Unstable and explodes if heated above 50 ˚C.

First aid   Standard treatment for exposure by all routes (see pages 108-122).

178. Bromotrifluoromethane

Colourless gas; m.p. -168 ˚C; b.p. -58 ˚C.

Limit values   OES short-term 1200 p.p.m. (7300 mg m^{-3}); long-term 1000 p.p.m. (6100 mg m^{-3}).

Toxic effects   Irritant to eyes, upper respiratory tract, and mucous membranes. Inhalation of high concentrations can cause dizziness, disorientation, narcosis, nausea, and vomiting.

First aid   Standard treatment for exposure by all routes (see pages 108-122).

Fire hazard   Extinguish fires with water spray, carbon dioxide, dry chemical powder, or alcohol or polymer foam.

Spillage disposal   See general section.

179. Buta-1,3-diene

Colourless gas; m.p. -108.9 ˚C; b.p. -4.7 ˚C; insoluble in water.

RISKS
Extremely flammable liquefied gas (R13)

SAFETY PRECAUTIONS
Keep container in a well ventilated place – Keep away from sources of ignition – No Smoking – Take precautionary measures against static discharges (S9, S16, S33)

Limit values   MEL long-term 10 p.p.m. (22 mg m^{-3}).

Toxic effects	The gas has narcotic effects in high concentration and can irritate the skin.
Hazardous reactions	May explode when heated under pressure. Peroxides formed on long contact with air are explosive but may also initiate polymerization. Violent or explosive reactions with some chemicals.
First aid	Standard treatment for exposure by all routes (see pages 108-122).
Fire hazard	Flash point -7 ˚C; Explosive limits 2-11.5%; Autoignition temperature 429 ˚C. The gas is supplied in a cylinder and turning off the valve will reduce any fire involving it; if possible cylinders should be removed quickly from an area in which a fire has developed.
Spillage disposal	Surplus gas or leaking cylinder can be vented slowly to air in a fume cupboard or a safe area or the gas burnt-off in a suitable burner.

180. Butadiene diepoxide

Colourless liquid; m.p. 4 ˚C (DL-form); -19 ˚C (*meso-*); b.p. 142 ˚C (*meso-*); 138 ˚C (mixture); 144 ˚C (DL-form); miscible with water.

RISKS
Toxic by inhalation, in contact with skin, and if swallowed – Irritating to eyes, respiratory system and skin – Possible risk of irreversible effects – May cause sensitization by inhalation and skin contact (R23/24/25, R36/37/38, R40, R42/43)

SAFETY PRECAUTIONS
Do not breathe vapour – Avoid contact with skin – If you feel unwell, seek medical advice (show label where possible) (S23, S24, S44)

Toxic effects	In aminal studies caused carcinogenesis, reduction in spermatogenesis, leukopaenia, and lymphopaenia.
First aid	Standard treatment for exposure by all routes (see pages 108-122).
Fire hazard	Use extinguishant appropriate to the surroundings.
Spillage disposal	See general section.

RSC *Chemical Safety Data Sheets* Vol. 4a, No. 24, 1991 gives extended coverage.

181. Butadiyne

Liquid; m.p. -36.4 ˚C; b.p. 10.3 ˚C.

Hazardous reactions	Polymerizes rapidly above 0 ˚C and is a gas above 10 ˚C. Potentially very explosive.

182. Butane

Colourless gas; m.p. -138 ˚C; b.p. -0.5 ˚C; sparingly soluble in water.

RISKS
Extremely flammable liquefied gas (R13)

SAFETY PRECAUTIONS
Keep container in a well ventilated place – Keep away from sources of ignition – No Smoking – Take precautionary measures against static discharges (S9, S16, S33)

Limit values   OES short-term 750 p.p.m. (1780 mg m^{-3}); long-term 600 p.p.m. (1430 mg m^{-3}).

Toxic effects   The gas has an anaesthetic effect but is not toxic.

First aid   Standard treatment for exposure by all routes (see pages 108-122).

Fire hazard   Flash point -60 ˚C; Explosive limits 1.9-8.5%; Autoignition temperature 405 ˚C. The gas is supplied in a cylinder and turning off the valve will reduce any fire involving it; if possible, cylinders should be removed quickly from an area in which a fire has developed.

Spillage disposal   Surplus gas or leaking cylinder can be vented slowly to air in a fume cupboard or a safe area or the gas burnt-off in a suitable burner.

183. Butane-2,3-dione monoxime

Solid; m.p. 77-78 ˚C; b.p. 185-186 ˚C.

Hazardous reactions   Has exploded during vacuum distillation.

First aid   Standard treatment for exposure by all routes (see pages 108-122).

184. Butan-1-ol

Colourless liquid with rancid odour; m.p. -88.9 ˚C; b.p. 118 ˚C; soluble in water.

RISKS
Flammable – Harmful by inhalation (R10, R20)

SAFETY PRECAUTIONS
Keep away from sources of ignition – No Smoking (S16)

Limit values   OES short-term 50 p.p.m. (150 mg m^{-3}).

Toxic effects   Butanol is readily absorbed through the skin and gastro-intestinal tract. It is irritating to the eyes and skin. Vapour is irritating to the respiratory tract and may cause central nervous system depression.

Hazardous reactions   Reacts with finely divided chromium trioxide causing ignition. Attacks aluminium and reacts with oxidizing agents.

First aid   Standard treatment for exposure by all routes (see pages 108-122).

| **Fire hazard** | Flash point 29 ˚C (open cup); Explosive limits 1.4-11.2%; Autoignition temperature 345 ˚C. Extinguish fires with dry chemical powder, carbon dioxide, or alcohol-resistant foam. |
| **Spillage disposal** | See general section. |

RSC *Chemical Safety Data Sheets* Vol. 1, No. 7, 1988 gives extended coverage.

185. Butan-2-ol

Colourless liquid; m.p. -114.7 ˚C; b.p. 99.5 ˚C; slightly soluble in water.

RISKS
Flammable – Harmful by inhalation (R10, R20)

SAFETY PRECAUTIONS
Keep away from sources of ignition – No Smoking (S16)

Limit values	OES short-term 150 p.p.m. (450 mg m^{-3}); long-term 100 p.p.m. (300 mg m^{-3}).
Toxic effects	The liquid irritates the eyes, skin, and gastro-intestinal tract. May have irritant and narcotic action if taken by mouth.
Hazardous reactions	Liable to form explosive peroxides on exposure to air and light which should be decomposed before the ether is distilled to small volume.
First aid	Standard treatment for exposure by all routes (see pages 108-122).
Fire hazard	Flash point 24 ˚C; Explosive limits 1.4-11%; Autoignition temperature 406 ˚C. Extinguish fire with water spray, dry powder, or carbon dioxide.
Spillage disposal	See general section.

186. Butan-2-one

Colourless liquid with irritating odour like acetone; m.p. -85.9 ˚C; b.p. 80 ˚C; soluble in water.

RISKS
Highly flammable – Irritating to eyes and respiratory system (R11, R36/37)

SAFETY PRECAUTIONS
Keep container in a well ventilated place – Keep away from sources of ignition – No Smoking – Avoid contact with eyes – Take precautionary measures against static discharges (S9, S16, S25, S33)

Limit values	OES short-term 300 p.p.m. (885 mg m^{-3}); long-term 200 p.p.m. (590 mg m^{-3}).
Toxic effects	Inhalation of vapour may cause dizziness, headache, and nausea. The liquid irritates the eyes and may cause severe damage. If swallowed may cause gastric irritation and narcosis. Weak teratogen.
Hazardous reactions	Vigorous reaction with chloroform in presence of bases. Explosive peroxides formed by action of H_2O_2/HNO_3.
First aid	Standard treatment for exposure by all routes (see pages 108-122).

Fire hazard   Flash point 6 ˚C; Explosive limits 1.9-11%; Autoignition temperature 516 ˚C. Extinguish fire with water spray, dry powder, carbon dioxide, or alcohol-resistant foam.

Spillage   See general section.
  disposal

RSC *Chemical Safety Data Sheets* Vol. 1, No. 72, 1988 gives extended coverage.

187. Butan-2-one peroxide

Colourless liquid; b.p. 118 ˚C with decomposition; soluble in water.

Toxic effects   Corrosive and irritant to skin, eyes, and mucous membranes. Absorption into the body can cause dyspnoea, stupor, CNS depression, gastritis, and oesophagitis.

Hazardous   May ignite or explode if mixed with readily oxidizable, organic, or flammable
  reactions   materials.

First aid   Standard treatment for exposure by all routes (see pages 108-122).

Fire hazard   Flash point 82 ˚C.

Spillage   See general section.
  disposal

188. Butenes

Colourless gases; b.p. all boil between -6 ˚C and 4 ˚C; all are insoluble in water.

RISKS
Extremely flammable liquefied gas (R13)

SAFETY PRECAUTIONS
Keep container in a well ventilated place – Keep away from sources of ignition – No Smoking – Take precautionary measures against static discharges (S9, S16, S33)

Toxic effects   The butenes are generally regarded as simple asphyxiants with some anaesthetic properties.

First aid   Standard treatment for exposure by all routes (see pages 108-122).

Fire hazard   Flash point all below -7 ˚C; Explosive limits 1.6-9.7%; Autoignition temperature all between 230 ˚C and 390 ˚C. The gases are supplied in cylinders and turning off the valve will reduce any fire involving them; if possible cylinders should be removed quickly from an area in which a fire has developed.

Spillage   Surplus gas or leaking cylinder can be vented slowly to air in a fume
  disposal   cupboard or a safe open area or the gas burnt in a suitable burner.

189. But-1-en-3-yne

Gas with acetylene-like odour; b.p. 2-3 ˚C.

Hazardous reactions   Forms explosive compounds on contact with air or silver nitrate, explodes on heating under pressure.

Fire hazard   Flash point <-5 ˚C. Emits toxic fumes on heating to decomposition.

190. Butoxyacetylene

Liquid; b.p. 106-108 ˚C.

Hazardous reactions   Explodes on heating in sealed tubes.

Fire hazard   Emits acrid fumes on heating to decomposition.

191. Butyl acetate

Yellowish liquid with rancid odour.; m.p. -73 ˚C; b.p. 126 ˚C; slightly soluble in water.

RISKS
Flammable (R10)

Limit values   OES short-term 200 p.p.m. (950 mg m^{-3}); long-term 150 p.p.m. (710 mg m^{-3}).

Toxic effects   The vapour may irritate the respiratory system and cause headache and nausea. The liquid will irritate the eyes and may irritate the skin. If taken by mouth, the liquid will cause irritation and act as a depressant of the central nervous system.

First aid   Standard treatment for exposure by all routes (see pages 108-122).

Fire hazard   Flash point 29 ˚C (open cup); Explosive limits 1.4-7.5%; Autoignition temperature 421 ˚C. Extinguish fire with foam, dry powder, or carbon dioxide.

Spillage disposal   See general section.

RSC *Chemical Safety Data Sheets* Vol. 1, No. 8, 1988 gives extended coverage.

192. s-Butyl acetate

Colourless liquid; m.p. -99 ˚C; b.p. 111-112 ˚C; insoluble in water.

RISKS
Highly flammable (R11)

SAFETY PRECAUTIONS
Keep away from sources of ignition – No Smoking – Do not breathe vapour – Do not empty into drains – Take precautionary measures against static discharges (S16, S23, S29, S33)

Limit values   OES short-term 250 p.p.m. (1190 mg m^{-3}); long-term 200 p.p.m. (950 mg m^{-3}).

Toxic effects	Irritant to skin, eyes, upper respiratory tract, and mucous membranes. Harmful by inhalation, ingestion, and skin absorption.
First aid	Standard treatment for exposure by all routes (see pages 108-122).
Fire hazard	Flash point 16 °C; Explosive limits 1.3-7.5%. Extinguish fires with water spray, carbon dioxide, dry chemical powder, or alcohol or polymer foam.
Spillage disposal	See general section.

193. t-Butyl acetate

Colourless liquid; b.p. 97 °C; insoluble in water.

RISKS
Highly flammable (R11)

SAFETY PRECAUTIONS
Keep away from sources of ignition – No Smoking – Do not breathe vapour – Do not empty into drains – Take precautionary measures against static discharges (S16, S23, S29, S33)

Limit values	OES short-term 250 p.p.m. (1190 mg m^{-3}); long-term 200 p.p.m. (950 mg m^{-3}).
Toxic effects	Irritant to skin, eyes, upper respiratory tract, and mucous membranes. Harmful by inhalation, ingestion, or skin absorption.
First aid	Standard treatment for exposure by all routes (see pages 108-122).
Fire hazard	Flash point 15 °C (closed cup). Extinguish fires with water spray, carbon dioxide, dry chemical powder, or alcohol or polymer foam.
Spillage disposal	See general section.

194. Butyl acrylate

Colourless liquid; b.p. 145 °C; immiscible with water.

RISKS
Irritating to eyes, respiratory system, and skin – May cause sensitization by skin contact (R36/37/38, R43)

SAFETY PRECAUTIONS
Keep container in a well ventilated place (S9)

Limit values	OES long-term 10 p.p.m. (55 mg m^{-3}).
Toxic effects	Assumed to be poisonous if taken by mouth.
First aid	Standard treatment for exposure by all routes (see pages 108-122).
Fire hazard	Flash point 49 °C. Extinguish fire with foam, dry powder, or carbon dioxide.
Spillage disposal	See general section.

195. Butylamines

Colourless liquids with ammoniacal odour; b.p. all boil between 45 °C and 78 °C; miscible with water.

RISKS
Highly flammable – Irritating to eyes, respiratory system, and skin (R11, R36/37/38)

SAFETY PRECAUTIONS
Keep away from sources of ignition – No Smoking – In case of contact with eyes, rinse immediately with plenty of water and seek medical advice – Do not empty into drains (S16, S26, S29)

Limit values   OES short-term 5 p.p.m. (15 mg m^{-3}) (n-isomer).

Toxic effects   Assumed to be very irritant and poisonous if taken by mouth.

First aid   Standard treatment for exposure by all routes (see pages 108-122).

Fire hazard   Flash point -12 °C(n-), -9 °C(iso-); Explosive limits 1.7-9.8%(n-), 1.7-9.8%(t-); Autoignition temperature 312 °C(n-),378 °C (iso). Extinguish fire with water spray, dry powder, or carbon dioxide.

Spillage disposal   See general section.

196. 2-t-Butylaminoethyl methacrylate

Liquid; b.p. 100-105 °C.

RISKS
Irritating to eyes and skin – May cause sensitization by skin contact (R36/38, R43)

SAFETY PRECAUTIONS
In case of contact with eyes, rinse immediately with plenty of water and seek medical advice (S26)

Toxic effects   Skin and eye irritant.

First aid   Standard treatment for exposure by all routes (see pages 108-122).

Fire hazard   Flash point 11 °C (open cup). Extinguish fires with water spray, dry chemical powder, or alcohol foam.

Spillage disposal   See general section.

RSC *Chemical Safety Data Sheets* Vol. 3, No. 14, 1990 gives extended coverage.

197. Butyl chloroformate

Clear, colourless liquid; b.p. 140-145 °C; insoluble in and decomposed by water.

RISKS
Flammable – Toxic by inhalation – Causes burns (R10, R23, R34)

SAFETY PRECAUTIONS
In case of contact with eyes, rinse immediately with plenty of water and seek medical advice –
Wear suitable protective clothing – If you feel unwell, seek medical advice (show label where
possible) (S26, S36, S44)

Limit values   OES long-term 1 p.p.m. (5.6 mg m^{-3}).

Toxic effects   Extremely irritant and destructive to skin, eyes, upper respiratory tract, and
                     mucous membranes. Inhalation may be fatal following spasm, inflammation,
                     chemical pneumonitis, and pulmonary oedema. Can also be fatal following
                     ingestion or skin absorption.

First aid   Standard treatment for exposure by all routes (see pages 108-122).

Fire hazard   Flash point 46 °C (closed cup). Extinguish fires with carbon dioxide, dry
                     chemical powder, or alcohol or polymer foam.

Spillage   See general section.
disposal

RSC *Chemical Safety Data Sheets* Vol. 4a, No. 25, 1991 gives extended coverage.

198. Butyl diazoacetate

Hazardous   Ignites on prolonged exposure to air; hydrolysis may be explosive.
reactions

Fire hazard   Emits toxic fumes on heating to decomposition.

199. Butyldichloroborane

Liquid; b.p. 106-108 °C.

Hazardous   Ignites on prolonged exposure to air; hydrolysis may be explosive.
reactions

200. Butyl 2,3-epoxypropyl ether

Colourless liquid; b.p. 164-166 ˚C; slightly soluble in water.

RISKS
Harmful by inhalation – May cause sensitization by skin contact (R20, R43)

SAFETY PRECAUTIONS
Avoid contact with skin and eyes (S24/25)

Limit values   OES long-term 25 p.p.m. (135 mg m^{-3}).

Toxic effects   Also harmful by ingestion or skin absorption. Skin sensitizer and central nervous system depressant.

First aid   Standard treatment for exposure by all routes (see pages 108-122).

Fire hazard   Extinguish fires with water spray, carbon dioxide, dry chemical powder, or alcohol or polymer foam.

Spillage   See general section.
disposal

201. Butyl ethyl ketone

Colourless liquid with ketone-type odour; m.p. -36.7 ˚C; b.p. 148 ˚C; insoluble in water.

RISKS
Flammable – Harmful by inhalation – Very toxic by inhalation (R10, R20, R26)

SAFETY PRECAUTIONS
Avoid contact with skin (S24)

Limit values   OES short-term 75 p.p.m. (345 mg m^{-3}); long-term 50 p.p.m. (230 mg m^{3}).

Toxic effects   Irritant to skin, eyes, upper respiratory tract, and all mucous membranes. Also harmful by ingestion or skin absorption.

First aid   Standard treatment for exposure by all routes (see pages 108-122).

Fire hazard   Flash point 46.1 ˚C (open cup). Extinguish fires with water spray, carbon dioxide, dry chemical powder, or alcohol or polymer foam.

Spillage   See general section.
disposal

202. t-Butyl hydroperoxide

Colourless liquid stable below 75 ˚C; slightly soluble in water.

Toxic effects   The liquid irritates the eyes and skin. Assumed to be toxic if taken by mouth.

Hazardous   Liable to explode when distilled.
reactions

First aid   Standard treatment for exposure by all routes (see pages 108-122).

Fire hazard   Flash point 27 ˚C. Extinguish fire with water spray, dry powder, or carbon dioxide.

Spillage   See general section.
disposal

203. Butyl hypochlorite

Pale yellow liquid with an irritating odour; b.p. 77-78 ˚C.

Toxic effects   Irritant to skin, eyes, and mucous membranes.

Hazardous   Ampoules liable to burst unless stored cool and in dark.
reactions

Fire hazard   Emits toxic fumes on heating to decomposition.

204. Butyl lactate

Colourless liquid; m.p. -28 ˚C; b.p. 185-187 ˚C; slightly soluble in water.

Limit values   OES long-term 5 p.p.m. (25 mg m^{-3}).

Toxic effects   Irritant to skin, eyes, and all mucous membranes. Can be absorbed through the skin.

First aid   Standard treatment for exposure by all routes (see pages 108-122).

Fire hazard   Flash point 69 ˚C. Extinguish fires with water spray, carbon dioxide, dry chemical powder, or alcohol or polymer foam.

Spillage   See general section.
disposal

205. Butyl mercaptan

Colourless liquid; m.p. -116 ˚C; b.p. 98 ˚C; slightly soluble in water.

Toxic effects   Irritant to skin, eyes, upper respiratory tract, and mucous membranes. Exposure can cause nausea, headache, and vomiting.

First aid   Standard treatment for exposure by all routes (see pages 108-122).

Fire hazard   Flash point 12 ˚C (closed cup). Extinguish fires with carbon dioxide, dry chemical powder, or alcohol or polymer foam.

Spillage   See general section.
disposal

206. Butyl methacrylate

Colourless, mobile liquid, normally supplied containing a small amount of stabilizing agent (*e.g.* 0.01% quinol); b.p. 163 °C; slightly soluble in water.

RISKS
Irritating to eyes, respiratory system, and skin – May cause sensitization by skin contact (R36/37/38, R43)

Toxic effects	The vapour irritates the eyes and respiratory system. The liquid irritates the eyes and may irritate the skin. Considered moderately toxic if taken by mouth.
First aid	Standard treatment for exposure by all routes (see pages 108-122).
Fire hazard	Flash point 52 °C. Extinguish fire with foam, dry powder, or carbon dioxide.
Spillage disposal	See general section.

207. Butyl peracetate

Clean colourless liquid; insoluble in water.

Toxic effects	Moderately toxic by inhalation and ingestion.
Hazardous reactions	Explodes violently when rapidly heated.
First aid	Standard treatment for exposure by all routes (see pages 108-122).
Fire hazard	Flash point <27 °C. Emits acrid smoke and fumes on heating to decomposition. Water spray may be used to extinguish fires.
Spillage disposal	See general section.

208. Butyl perbenzoate

Colourless to slightly yellow liquid with mild aromatic odour; m.p. 8 °C; b.p. 112 °C with decomposition; insoluble in water.

Toxic effects	Irritant to skin, eyes, and mucous membranes. Toxic by ingestion.
Hazardous reactions	Exploded during interrupted vacuum distillation. May react violently with some organic materials.
First aid	Standard treatment for exposure by all routes (see pages 108-122).
Fire hazard	Flash point 19 °C. Emits acrid smoke and fumes on heating to decomposition. Extinguish fires with water spray, carbon dioxide, dry chemical powder, or alcohol or polymer foam.
Spillage disposal	See general section.

209. 2-s-Butylphenol

Viscous, colourless liquid; m.p. 16 ˚C; b.p. 226-228 ˚C at 25 mm Hg.

Limit values OES long-term 5 p.p.m. (30 mg m^{-3}).

Toxic effects Extremely destructive to tissues of skin, eyes, upper respiratory tract, and mucous membranes. Inhalation may be fatal following spasm, pneumonitis, and pulmonary oedema.

First aid Standard treatment for exposure by all routes (see pages 108-122).

Fire hazard Flash point 107.2 ˚C. Extinguish fires with water spray, carbon dioxide, dry chemical powder, or alcohol or polymer foam.

Spillage disposal See general section.

210. But-1-yne

Colourless liquid and gas; b.p. 8.1 ˚C; insoluble in water.

Toxic effects The toxicity has not been fully investigated. It probably has some anaesthetic activity and can act as a simple asphyxiant.

First aid Standard treatment for exposure by all routes (see pages 108-122).

Fire hazard The gas is supplied in a cylinder and turning off the valve will reduce any fire involving it. If possible cylinders should be removed quickly from an area in which a fire has developed.

Spillage disposal Surplus gas or a leaking cylinder can be vented slowly to air in a fume cupboard or a safe open area or the gas burnt-off in a suitable burner.

211. Butynedinitrile

Hazardous reactions Potentially explosive in pure state or in concentrated solutions.

First aid Standard treatment for exposure by all routes (see pages 108-122), with reference to treatment for cyanides.

Fire hazard Emits toxic fumes on heating to decomposition.

212. Butyne-1,4-diol

Straw or amber coloured crystals; m.p. 57.5 ˚C; b.p. 194 ˚C at 100 mm Hg.

RISKS
Toxic if swallowed – Causes burns (R25, R34)

SAFETY PRECAUTIONS
Do not breathe dust – Wear suitable protective clothing – If you feel unwell, seek medical advice (show label where possible) (S22, S36, S44)

Toxic effects	Irritant to skin, eyes, upper respiratory tract, and mucous membranes. Prolonged or repeated exposure may give rise to allergic reactions or narcotic effects and central nervous system depression.
Hazardous reactions	Explodes on distillation in presence of traces of alkali or alkaline earth hydroxides or halides.
First aid	Standard treatment for exposure by all routes (see pages 108-122).
Fire hazard	Flash point 152 ˚C (open cup). Extinguish fires using water spray, carbon dioxide, dry powder, or alcohol or polymer foam.
Spillage disposal	See general section.

213. Butynethiol

Hazardous reactions	Exposure to air results in formation of polymer which may explode on heating.
Fire hazard	Emits toxic fumes on heating to decomposition.

214. Butyraldehyde

Colourless liquid; m.p. -99 ˚C; b.p. 76 ˚C; slightly soluble in water.

RISKS
Highly flammable (R11)

SAFETY PRECAUTIONS
Keep container in a well ventilated place – Do not empty into drains – Take precautionary measures against static discharges (S9, S29, S33)

Toxic effects	The vapour may irritate the eyes and respiratory system. The liquid will irritate the eyes and may irritate the skin. Assumed to be irritant and possibly narcotic if swallowed.
First aid	Standard treatment for exposure by all routes (see pages 108-122).
Fire hazard	Flash point -6.7 ˚C; Autoignition temperature 230 ˚C. Extinguish fire with water spray, foam, dry powder, or carbon dioxide.
Spillage disposal	See general section.

215. Butyraldehyde oxime

Liquid; m.p. -29.5 ˚C; b.p. 152 ˚C.

RISKS
Harmful if swallowed – Toxic in contact with skin – Irritating to eyes (R22, R24, R36)

SAFETY PRECAUTIONS
Do not breathe vapour – Wear suitable protective clothing – If you feel unwell, seek medical advice (show label where possible) (S23, S36, S44)

Hazardous reactions	Large batch exploded violently during vacuum distillation caused by metal-catalysed Beckmann rearrangement.
First aid	Standard treatment for exposure by all routes (see pages 108-122).
Fire hazard	Flash point 58 ˚C (closed cup). Alcohol foam may be used to extingiush fires.
Spillage disposal	See general section.

216. Butyric acid

Colourless, oily liquid with very pungent smell; b.p. 163.5 ˚C; miscible with water.

RISKS
Causes burns (R34)

SAFETY PRECAUTIONS
In case of contact with eyes, rinse immediately with plenty of water and seek medical advice – Wear suitable protective clothing (S26, S36)

Toxic effects	Irritates or burns skin and eyes.
First aid	Standard treatment for exposure by all routes (see pages 108-122).
Spillage disposal	See general section.

217. Butyric and isobutyric anhydrides

Colourless liquids with pungent odour; b.p. (n-) 200 ˚C; (iso-) 182 ˚C; react with water to form the corresponding acids.

Toxic effects	The liquids burn the eyes and may burn the skin. They are irritant and corrosive if taken by mouth.
First aid	Standard treatment for exposure by all routes (see pages 108-122).
Spillage disposal	See general section.

218. Butyronitrile

Colourless llquid; m.p. -112.6 °C; b.p. 117 °C; immiscible with water but tends to break down to cyanide by hydrolysis.

RISKS
Flammable – Toxic by inhalation, in contact with skin, and if swallowed (R10, R23/24/25)

SAFETY PRECAUTIONS
If you feel unwell, seek medical advice (show label where possible) (S44)

Toxic effects Rats exposed to its vapour rapidly develop weakness, laboured breathing, and convulsions which usually result in death. Cases of poisoning seem to warrant the same urgent attention as those caused by hydrogen cyanide.

First aid Special treatment as for hydrogen cyanide.

Fire hazard Flash point 26 °C (open cup). Extinguish fire with dry powder or carbon dioxide.

Spillage disposal See general section.

219. Butyryl nitrate

Hazardous reactions Detonates on heating.

Fire hazard Emits toxic fumes on heating to decomposition.

220. Cadmium

Silver-white, malleable metal; m.p. 320.9 °C; b.p. 765 °C; insoluble in water.

RISKS
Harmful by inhalation, in contact with skin, and if swallowed (R20/21/22)

SAFETY PRECAUTIONS
Do not breathe dust (S22)

Limit values MEL long-term 0.05 mg m^{-3} (as Cd).

Toxic effects Irritant to skin and eyes. May produce allergic reactions. May be carcinogenic.

Hazardous reactions The powdered metal reacts violently or explosively with fused ammonium nitrate; it reacts vigorously with selenium or tellurium on warming.

First aid Standard treatment for exposure by all routes (see pages 108-122).

Fire hazard Use extinguishing agent appropriate to the surroundings.

Spillage disposal See general section.

RSC *Chemical Safety Data Sheets* Vol. 4a, No. 28, 1991 gives extended coverage.

221. Cadmium chloride

Hexagonal, colourless crystals; m.p. 568 °C; b.p. 960 °C; soluble in water.

RISKS
May cause cancer – Toxic by inhalation and if swallowed – Danger of serious damage to health by prolonged exposure (R45, R23/25, R48)

SAFETY PRECAUTIONS
Avoid exposure – Obtain special instructions before use – If you feel unwell, seek medical advice (show label where possible) (S53, S44)

Limit values	MEL long-term 0.05 mg m^{-3} (as Cd).
Toxic effects	Eye irritant. Probable human carcinogen.
Hazardous reactions	Can react violently with certain inorganic compounds.
First aid	Standard treatment for exposure by all routes (see pages 108-122), with particular attention to the possible occurrence of lung congestion.
Fire hazard	Use extinguishant appropriate to the surroundings.
Spillage disposal	See general section.

RSC *Chemical Safety Data Sheets* Vol. 4a, No. 29, 1991 gives extended coverage.

222. Cadmium cyanide

White, crystalline precipitate; m.p. 200 °C with decomposition; soluble in water.

RISKS
Very toxic by inhalation, in contact with skin, and if swallowed – Contact with acids liberates very toxic gas – Danger of cumulative effects – Possible risk of irreversible effects (R26/27/28, R32, R33, R40)

SAFETY PRECAUTIONS
Keep locked up and out of reach of children – Keep container tightly closed – After contact with skin, wash immediately with plenty of water – Do not empty into drains – In case of accident or if you feel unwell, seek medical advice immediately (show label where possible) (S1/2, S7, S28, S29, S45)

Limit values	MEL long-term 0.05 mg m^{-3} (as Cd).
Toxic effects	Irritant to the eyes. Probable human carcinogen.
Hazardous reactions	Tendency to instability under certain initiation conditions.
First aid	Standard treatment for exposure by all routes (see pages 108-122).
Fire hazard	Water may be used to extinguish fires.
Spillage disposal	See general section. Large spills need specialist help.

RSC *Chemical Safety Data Sheets* Vol. 4a, No. 30, 1991 gives extended coverage.

223. Cadmium diamide

Solid; m.p. decomposes at 120 ˚C.

RISKS
Harmful by inhalation, in contact with skin, and if swallowed (R20/21/22)

SAFETY PRECAUTIONS
Do not breathe dust (S22)

Limit values	MEL long-term 0.05 mg m^{-3} (as Cd).
Hazardous reactions	May explode when heated rapidly. Reacts violently with water.
First aid	Standard treatment for exposure by all routes (see pages 108-122).
Spillage disposal	See general section.

224. Cadmium fluoride

White, cubic crystals; m.p. 1100 ˚C; b.p. 1758 ˚C; soluble in water.

RISKS
Toxic by inhalation and if swallowed – Danger of cumulative effects – Possible risk of irreversible effects (R23/25, R33, R40)

SAFETY PRECAUTIONS
Do not breathe dust – If you feel unwell, seek medical advice (show label where possible) (S22, S44)

Limit values	MEL long-term 0.05 mg m^{-3} (as Cd).
Toxic effects	Irritant to skin, eyes, and upper respiratory tract. Can be fatal following inhalation or ingestion. Probable human carcinogen.
First aid	Standard treatment for exposure by all routes (see pages 108-122).
Fire hazard	Use extinguishant appropriate to the surroundings.
Spillage disposal	See general section.

RSC *Chemical Safety Data Sheets* Vol. 4a, No. 31, 1991 gives extended coverage.

225. Cadmium nitrate

White crystals; m.p. 350 ˚C; soluble in water.

RISKS
Harmful by inhalation, in contact with skin, and if swallowed (R20/21/22)

SAFETY PRECAUTIONS
Do not breathe dust (S22)

Limit values	MEL long-term 0.05 mg m^{-3} (as Cd).

Toxic effects   Severe skin and eye irritant. Destructive to membranes of the upper respiratory tract which can lead to inflammation, oedema, and death. Probable human carcinogen.

First aid   Standard treatment for exposure by all routes (see pages 108-122).

Fire hazard   Use extinguishant appropriate to the surroundings.

Spillage disposal   See general section.

RSC *Chemical Safety Data Sheets* Vol. 4a, No. 32, 1991 gives extended coverage.

226. Cadmium oxide

Amorphous (1) or cubic brown crystals (2); m.p. (1) decomposes 900-1000 ˚C, (2) sublimes 1559 ˚C; b.p. >1500 ˚C; insoluble in water.

RISKS
Toxic by inhalation and if swallowed – Danger of cumulative effects – Possible risk of irreversible effects (R23/25, R33, R40)

SAFETY PRECAUTIONS
Do not breathe dust – If you feel unwell, seek medical advice (show label where possible) (S22, S44)

Limit values   MEL short-term fume 0.05 mg m^{-3} (as Cd); long-term 0.05 mg m^{-3} (as Cd).

Toxic effects   Irritant to skin and eyes. Inhalation of fumes causes respiratory distress which progresses and become fatal. Long-term exposure to low concentrations can cause damage to kidneys and lungs. Probable human carcinogen.

First aid   Standard treatment for exposure by all routes (see pages 108-122), with special attention to the possible occurrence of lung congestion.

Fire hazard   Use extinguishant appropriate to the surroundings.

Spillage disposal   See general section.

RSC *Chemical Safety Data Sheets* Vol. 4a, No. 33, 1991 gives extended coverage.

227. Cadmium propionate

RISKS
Harmful by inhalation, in contact with skin, and if swallowed (R20/21/22)

SAFETY PRECAUTIONS
Do not breathe dust (S22)

Limit values   MEL long-term 0.05 mg m^{-3} (as Cd).

Hazardous reactions   The salt exploded during drying in oven.

First aid   Standard treatment for exposure by all routes (see pages 108-122).

Spillage disposal   See general section.

228. Cadmium selenide

Grey powder; m.p. above 1250 ˚C.

RISKS
Harmful by inhalation, in contact with skin, and if swallowed (R20/21/22)

SAFETY PRECAUTIONS
Do not breathe dust (S22)

Limit values   MEL long-term 0.05 mg m^{-3} (as Cd).

Hazardous reactions   Mixtures of powdered metal and selenium may explode.

First aid   Standard treatment for exposure by all routes (see pages 108-122).

Fire hazard   Use extinguishing medium appropriate to surroundings.

Spillage disposal   See general section.

229. Cadmium sulfate

White, rhombic crystals; m.p. 1000 ˚C; soluble in water.

RISKS
Harmful by inhalation, in contact with skin, and if swallowed (R20/21/22)

SAFETY PRECAUTIONS
Do not breathe dust (S22)

Limit values   MEL long-term 0.05 mg m^{-3} (as Cd).

Toxic effects   Irritant to eyes, upper respiratory tract, and gastro-intestinal membranes. Long-term exposure can lead to kidney damage. Possible human carcinogen.

First aid   Standard treatment for exposure by all routes (see pages 108-122).

Fire hazard   Use extinguishant appropriate to the surroundings.

Spillage disposal   See general section.

RSC *Chemical Safety Data Sheets* Vol. 4a, No. 34, 1991 gives extended coverage.

230. Caesium and compounds

Hazardous reactions   Ignites immediately in air, oxygen, and on contact with water, violent or incandescent reactions with halogens, phosphorus, and sulfur.

231. Caesium carbonate

Colourless, crystalline solid, very deliquescent.; m.p. 610 ˚C; extremely soluble in water.

Toxic effects	Irritant to skin, eyes, and upper respiratory tract.
First aid	Standard treatment for exposure by all routes (see pages 108-122).
Fire hazard	Extinguish fires with water spray, carbon dioxide, dry chemical powder, or alcohol or polymer foam.
Spillage disposal	See general section.

RSC *Chemical Safety Data Sheets* Vol. 2, No. 37, 1989 gives extended coverage.

232. Caesium chloride

Colourless, deliquescent cubic crystals.; m.p. 646 ˚C; b.p. 1303 ˚C; very soluble in water.

Toxic effects	Moderately toxic on ingestion and irritant to skin, eyes, and respiratory system. Has mutagenic activity.
First aid	Standard treatment for exposure by all routes (see pages 108-122).
Fire hazard	Extinguish fires with water spray, carbon dioxide, dry chemical powder, or alcohol or polymer foam.
Spillage disposal	See general section.

RSC *Chemical Safety Data Sheets* Vol. 2, No. 38, 1989 gives extended coverage.

233. Caesium fluoride

Extremely hygroscopic, colourless, crystalline solid; m.p. 682-703 ˚C; b.p. 1253 ˚C; extremely soluble in water.

Limit values	OES long-term 2.5 mg m^{-3} (as F).
Toxic effects	Very destructive to tissues of the eyes and upper respiratory tract. Rapidly absorbed from the gastro-intestinal system and ingestion of large quantities could be fatal.
First aid	Standard treatment for exposure by all routes (see pages 108-122), with special reference to fluorides.
Fire hazard	Use extinguishing agent suitable for the surroundings.
Spillage disposal	See general section.

RSC *Chemical Safety Data Sheets* Vol. 2, No. 39, 1989 gives extended coverage.

234. Caesium hydroxide

Colourless or yellowish, very deliquescent crystals; m.p. 272.3 °C; very soluble in water.

Limit values   OES long-term 2 mg m^{-3}.

Toxic effects   Highly irritant and corrosive to eyes and upper respiratory tract.

First aid   Standard treatment for exposure by all routes (see pages 108-122).

Fire hazard   Use dry chemical powder to extinguish fires.

Spillage disposal   See general section.

RSC *Chemical Safety Data Sheets* Vol. 2, No. 40, 1989 gives extended coverage.

235. Caesium nitrate

Glittering, hexagonal prisms; m.p. 414 °C; decomposes on further heating; soluble in water.

Toxic effects   Irritant to skin, eyes, upper respiratory tract, and all mucous membranes.

First aid   Standard treatment for exposure by all routes (see pages 108-122).

Fire hazard   Extinguish fires with water spray.

Spillage disposal   See general section.

RSC *Chemical Safety Data Sheets* Vol. 2, No. 41, 1989 gives extended coverage.

236. Calcium

Lustrous, silver white, soft, ductile metal which tarnishes to a bluish-grey colour on exposure to moist air; m.p. 842 °C; b.p. 1484 °C; reacts with water.

RISKS
Contact with water liberates highly flammable gases (R15)

SAFETY PRECAUTIONS
Keep container dry – Avoid contact with skin and eyes – In case of fire, use powder – Never use water (S8, S24/25, S43)

Toxic effects   Reaction with moisture on skin, eyes, and internal tissues may cause irritation and burning.

Hazardous reactions   Pyrophoric when finely divided; reacts violently or explosively with a variety of chemicals; reaction with water or dilute acids may be violent.

First aid   Standard treatment for exposure by all routes (see pages 108-122).

Fire hazard   Extinguish fires with dry graphite, soda ash, powdered sodium chloride, or appropriate metal fire extinguishing dry powder. Do NOT use water.

Spillage disposal   See general section.

RSC *Chemical Safety Data Sheets* Vol. 2, No. 42, 1989 gives extended coverage.

237. Calcium bis-2-iodylbenzoate

Hazardous   Overdried formulated granules exploded.
reactions

238. Calcium carbide

Grey solid; m.p. 2300 °C; reacts with water producing acetylene.

RISKS
Contact with water liberates highly flammable gases (R15)

SAFETY PRECAUTIONS
Keep container dry – In case of fire, use class D fire extinguisher – Never use water (S8, S43)

Toxic effects   Extremely destructive to all tissues. Inhalation may result in coughing, laryngitis, shortness of breath, headaches, nausea, and vomiting; may eventually be fatal.

Hazardous   Incandesces with PbF_2 at room temperature, with HCl on warming, with Mg
reactions   when heated in air, with Cl_2, Br_2, and I_2 at temperatures over 245 °C; very vigorous reaction with boiling methanol; forms highly sensitive explosive with silver nitrate solution; a mixture with sodium peroxide is explosive.

First aid   Standard treatment for exposure by all routes (see pages 108-122).

Fire hazard   Use class D fire extinguishing materials only. Do NOT use water.

Spillage   See general section.
disposal

239. Calcium carbonate

White powder or lumps; m.p. 825 °C; almost insoluble in water.

Limit values   OES long-term 10 mg m^{-3} (total inhalable dust), 5 mg m^{-3} (respirable dust).

Toxic effects   Irritant to skin and eyes. Slight nasal irritation and sneezing on inhalation.

First aid   Standard treatment for exposure by all routes (see pages 108-122).

Fire hazard   Use extinguishant applicable to the surroundings.

Spillage   See general section.
disposal

240. Calcium chromate

Yellow crystals; barely soluble in water.

RISKS
May cause cancer – Harmful if swallowed (R45, R22)

SAFETY PRECAUTIONS
Avoid exposure – Obtain special instructions before use – If you feel unwell, seek medical advice (show label where possible) (S53, S44)

Limit values   MEL long-term 0.05 mg m^{-3} (as Cr).

Toxic effects   Irritant and highly corrosive to skin, eyes, upper respiratory tract, and all mucous membranes. Animal carcinogen.

First aid   Standard treatment for exposure by all routes (see pages 108-122).

Fire hazard   Use extinguishant appropriate to the surroundings.

Spillage disposal   See general section.

RSC *Chemical Safety Data Sheets* Vol. 2, No. 44, 1989 gives extended coverage.

241. Calcium cyanamide

Colourless crystals, although commercial grade may be a greyish-black powder; m.p. 1300 °C; decomposed by water.

Limit values   OES short-term 1 mg m^{-3}; long-term 0.5 mg m^{-3}.

Toxic effects   Inhalation of dust can cause severe irritation of the mucous membranes. It irritates the eyes and can cause conjunctivitis, and will also burn the skin. If taken by mouth, there is severe internal irritation and damage.

Hazardous reactions   Commercial grades containing up to 2% calcium carbide may liberate acetylene and cause explosions on contact with water.

First aid   Standard treatment for exposure by all routes (see pages 108-122).

Fire hazard   Extinguish fires with a dry agent.

Spillage disposal   See general section.

RSC *Chemical Safety Data Sheets* Vol. 2, No. 45, 1989 gives extended coverage.

242. Calcium dihydride

White, crystalline solid; m.p. decomposes at about 600 °C; decomposed by water.

RISKS
Contact with water liberates highly flammable gases (R15)

SAFETY PRECAUTIONS
Keep container tightly closed and dry – Avoid contact with skin and eyes – In case of fire, use extinguishant appropriate to the surroundings. Do NOT use water (S7/8, S24/25, S43)

Hazardous reactions   Mixtures with various bromates, chlorates, perchlorates, and other compounds explode on grinding.

Spillage disposal   See general section.

243. Calcium disilicide

Solid; insoluble in water.

Hazardous reactions   Explodes when milled in carbon tetrachloride; ignites in close contact with alkali metal fluorides; mixture with potassium or sodium nitrate ignites readily.

244. Calcium hypochlorite

White or off-white powder, smelling of chlorine; m.p. decomposes on heating at 100 °C; absorbs water and is decomposed by it.

RISKS
Contact with combustible material may cause fire – Contact with acids liberates toxic gas – Causes burns (R8, R31, R34)

SAFETY PRECAUTIONS
Keep out of reach of children – In case of contact with eyes, rinse immediately with plenty of water and seek medical advice – In case of fire, use water spray (S2, S26, S43)

Toxic effects   Dust irritates the respiratory system, skin, eyes, and alimentary system. High concentrations are extremely destructive to all tissues.

Hazardous reactions   Pure compound is powerful oxidant, 'burns' with evolution of oxygen; violent reactions with hydroxy-compounds, nitromethane, sulfur compounds, and many combustibles.

First aid   Standard treatment for exposure by all routes (see pages 108-122).

Fire hazard   Extinguish fires with water spray.

Spillage disposal   See general section.

245. Calcium oxide

White, amorphous lumps and powder; m.p. 2572 ˚C; b.p. 2850 ˚C; reacts vigorously with water forming calcium hydroxide.

Limit values	OES long-term 2 mg m^{-3}.
Toxic effects	The dust irritates the skin, eyes, gastro-intestinal tract, and respiratory system.
Hazardous reactions	Incandesces in contact with liquid hydrogen fluoride; some mixtures with water develop enough heat to ignite combustible materials; glass bottles of the oxide may burst due to hydration expansion when the hydroxide is formed.
First aid	Standard treatment for exposure by all routes (see pages 108-122).
Fire hazard	Avoid water unless it is necessary for the other burning materials, in which case flood with water to absorb the heat generated.
Spillage disposal	See general section.

RSC *Chemical Safety Data Sheets* Vol. 2, No. 48, 1989 gives extended coverage.

246. Calcium peroxide

Pearly tetragonal crystals or a yellowish powder; m.p. decomposes at 275 ˚C; slightly soluble in cold water, decomposed by hot water.

Hazardous reactions	Grinding with oxidizable materials may cause fire.
First aid	Standard treatment for exposure by all routes (see pages 108-122).

247. Calcium peroxodisulfate

Hazardous reactions	Shock-sensitive; explodes violently.
Spillage disposal	See general section.

248. Calcium silicate

White powder; m.p. 1540 ˚C; insoluble in water but reacts on prolonged contact.

Limit values	OES long-term 10 mg m^{-3} (total inhalable dust), 5 mg m^{-3} (respirable dust).
Toxic effects	Irritant to upper respiratory tract.
First aid	Standard treatment for exposure by all routes (see pages 108-122).

249. Calcium silicide

Insoluble in cold water, decomposed by hot water.

Hazardous reactions	Reacts vigorously with acid; the silanes evolved ignite.
First aid	Standard treatment for exposure by all routes (see pages 108-122).
Spillage disposal	See general section.

250. Calcium sulfate

Colourless crystals or a white powder; m.p. 1450 °C (anhydrite); sparingly soluble in water.

Toxic effects	Toxicity unknown.
Hazardous reactions	Reduced violently or explosively by aluminium powder. Contact with diazomethane vapour may result in detonation.
First aid	Standard treatment for exposure by all routes (see pages 108-122).
Spillage disposal	See general section.

251. Calcium sulfide

White, crystalline solid; m.p. decomposes on heating; decomposed by water producing hydrogen sulfide.

RISKS
Contact with acids liberates toxic gas – Irritating to eyes, respiratory system, and skin (R31, R36/37/38)

SAFETY PRECAUTIONS
After contact with skin, wash immediately with plenty of water (S28)

Hazardous reactions	Reacts vigorously with chromyl chloride, lead dioxide; explodes with potassium chlorate and potassium nitrate.
First aid	Standard treatment for exposure by all routes (see pages 108-122).
Fire hazard	Use extinguishant appropriate to the surroundings.
Spillage disposal	See general section.

252. Caprolactam

White, crystalline leaflets; m.p. 69 °C; b.p. 100 °C at 3 mm Hg; highly soluble in water and hygroscopic.

Limit values	OES short-term dust 3 mg m^{-3}, vapour 10 p.p.m. (40 mg m^{-3}); long-term dust 1 mg m^{-3}, vapour 5 p.p.m. (20 mg m^{-3}).

Toxic effects	Irritant to skin, eyes, upper respiratory tract, and mucous membranes. Toxic on inhalation or ingestion.
First aid	Standard treatment for exposure by all routes (see pages 108-122).
Fire hazard	Flash point 125 °C (open cup). Extinguish small fires with dry powder or carbon dioxide. Large fires are best treated with alcohol-resistant foam. Produces toxic fumes on decomposition.
Spillage disposal	See general section.

RSC *Chemical Safety Data Sheets*, Vol. 1, No. 10, 1988 gives extended coverage.

253. Carbides

Hazardous reactions	Most of these compounds explode readily and are very sensitive to shock, friction, and heat. As they contain no oxygen or nitrogen they explode as a result of the large amount of heat instantaneously generated and the explosion produces no gas. As a class they react vigorously with a variety of organic compounds. Most of them react violently with water producing flammable acetylene gas.
Spillage disposal	Keep cool and do NOT store near metal powders.

254. Carbon

Black powder; m.p. 3652-3697 °C; b.p. 4200 °C approx; insoluble in water.

Limit values	OES short-term 7 mg m^{-3} (as carbon black); long-term 3.5 mg m^{-3} (as carbon black).
Hazardous reactions	Activated carbon is a potential fire hazard; contamination with drying oils or oxidizing agents may ignite it spontaneously; numerous oxidants in intimate contact with carbon may cause ignition or explosion.
First aid	Standard treatment for exposure by all routes (see pages 108-122).
Spillage disposal	See general section.

255. Carbon dioxide

Colourless, odourless gas; m.p. -56.6 °C at 5.2 atm. pressure; b.p. sublimes at -78.5 °C; soluble in water.

Limit values	OES short-term 15000 p.p.m. (27000 mg m^{-3}); long-term 5000 p.p.m. (9000 mg m^{-3}).
Toxic effects	Low to medium concentrations can cause respiratory stimulation and affect the blood circulation. High concentrations lead to increased symptoms and can cause unconsciousness and death.

Hazardous reactions	May react violently with various metal oxides or reducing metals; mixtures with alkali metals explode if shocked.
First aid	Standard treatment for exposure by all routes (see pages 108-122).
Fire hazard	Use water to keep cylinders cool, but move them away if possible.

256. Carbon disulfide

Colourless to yellow liquid, with unpleasant odour; m.p. -111 ˚C; b.p. 46 ˚C; slightly soluble in water.

RISKS
Extremely flammable – Very toxic by inhalation (R12, R26)

SAFETY PRECAUTIONS
Take off immediately all contaminated clothing – Do not empty into drains – Take precautionary measures against static discharges – In case of fire, use dry chemical or carbon dioxide extinguishers – In case of accident or if you feel unwell, seek medical advice immediately (show label where possible) (S27, S29, S33, S43, S45)

Limit values	MEL long-term 10 p.p.m. (30 mg m^{-3}).
Toxic effects	When inhaled high concentrations produce narcotic effects and may result in unconsciousness. Repeated inhalation of the vapour over a period may cause severe damage to the nervous system, including failure of vision, mental disturbance, and paralysis. The liquid and vapour irritate the eyes. The liquid is poisonous if taken by mouth or absorbed through the skin.
Hazardous reactions	Many fires and explosions have been caused by the ignition of the vapour from liquid poured down laboratory sinks, and ignition was caused by high intensity flash illumination; violent or explosive reactions occur with a variety of substances.
First aid	Standard treatment for exposure by all routes (see pages 108-122).
Fire hazard	Flash point -30 ˚C; Explosive limits 1-44%; Autoignition temperature 100 ˚C. Extinguish fire with dry powder or carbon dioxide extinguishers.
Spillage disposal	See general section.

RSC *Chemical Safety Data Sheets* Vol. 1, No. 11, 1988 gives extended coverage.

257. Carbon monoxide

Colourless, odourless gas; m.p. -207 ˚C; b.p. -191.3 ˚C; slightly soluble in water

RISKS
Extremely flammable – Toxic by inhalation (R12, R23)

SAFETY PRECAUTIONS
Keep container tightly closed – Keep away from sources of ignition – No Smoking (S7, S16)

Limit values	OES short-term 300 p.p.m. (330 mg m^{-3}); long-term 50 p.p.m. (55 mg m^{-3}).
Toxic effects	Causes unconsciousness due to anoxia resulting from the combination of carbon monoxide with haemoglobin. Gas in lower concentrations causes

	headache, throbbing of temples, nausea, followed possibly by collapse. Chronic effects include headache, nausea, and weakness.
Hazardous reactions	May react explosively with a variety of compounds.
Fire hazard	As the gas is supplied in a cylinder, turning off the valve will reduce any fire involving it; if possible cylinders should be removed quickly from an area in which a fire has developed.
Spillage disposal	Surplus gas or a leaking cylinder can be vented slowly to the air in a safe open area or burnt-off in a suitable gas burner.

258. Carbon tetrabromide

White, crystalline solid; m.p. 48.4 °C (α-form); 90.1 °C (β-form); b.p. 189.5 °C; insoluble in water.

Limit values	OES short-term 0.3 p.p.m. (4 mg m^{-3}); long-term 0.1 p.p.m. (1.4 mg m^{-3}).
Toxic effects	The vapour is narcotic in high concentrations. Assumed that the solid is poisonous by skin absorption and if taken by mouth.
Hazardous reactions	Will react vigorously or explosively with alkali metals.
First aid	Standard treatment for exposure by all routes (see pages 108-122).
Fire hazard	Use water spray, carbon dioxide, dry powder, or alcohol or polymer foam.
Spillage disposal	See general section.

259. Carbon tetrachloride

Heavy, colourless liquid with a characteristic odour; m.p. -22.6 °C; b.p. 77 °C; immiscible with water.

RISKS
Very toxic by inhalation and in contact with skin (R26/27)

SAFETY PRECAUTIONS
Keep out of reach of children – In case of insufficient ventilation, wear suitable respiratory equipment – In case of accident or if you feel unwell, seek medical advice immediately (show label where possible) (S2, S38, S45)

| **Limit values** | OES long-term 2 p.p.m. (12.6 mg m^{-3}) proposed change. |
| **Toxic effects** | Inhalation of vapour can cause headache, mental confusion, depression, fatigue, loss of appetite, nausea, vomiting, and coma, these symptoms sometimes taking many hours to appear. The vapour and liquid irritate the eyes. It causes internal irritation, nausea, and vomiting if taken by mouth and is readily absorbed through the gastro-intestinal tract. Dermatitis may follow repeated contact with the liquid. Experimental carcinogen, moderate teratogen. |

Hazardous reactions	May react violently or explosively with a wide variety of substances.
First aid	Standard treatment for exposure by all routes (see pages 108-122).
Fire hazard	The material is non-flammable but may generate phosgene and other toxic gases if heated.
Spillage disposal	See general section.

RSC *Chemical Safety Data Sheets* Vol. 1, No. 12, 1988 gives extended coverage.

260. Carbonyl difluoride

Colourless gas with pungent odour; m.p. -114 ˚C; b.p. -83.1 ˚C; decomposed by water.

Toxic effects	Highly irritant to tissues. Long-term effects due to the release of fluoride ions in contact with moist body tissues.
First aid	Standard treatment for exposure by all routes (see pages 108-122), with particular reference to fluorides.

261. Catechol

Colourless, crystalline powder; m.p. 105 ˚C; b.p. 245 ˚C with decomposition; soluble in water.

Limit values	OES long-term 5 p.p.m. (20 mg m^{-3}).
Toxic effects	Irritates the eyes severely, causing burns. Irritates the skin and causes poisoning by absorption. Assumed to be irritant and poisonous if taken by mouth.
Hazardous reactions	Explodes on contact with concentrated nitric acid.
First aid	Standard treatment for exposure by all routes (see pages 108-122).
Fire hazard	Flash point 127 ˚C. Extinguish fires with water, carbon dioxide, or dry powder.
Spillage disposal	See general section.

262. Cellulose

White or off-white powder; insoluble in water.

Limit values	OES short-term 20 mg m^{-3} (as total inhalable dust); long-term 10 mg m^{-3} (as total inhalable dust), 5mg m^{-3} (as respirable dust).
Hazardous reactions	Reacts vigorously with calcium oxide and oxidants such as bleaching powder, perchlorates, perchloric acid, sodium chlorate, fluorine, nitric acid, sodium nitrate, nitrite, and peroxide.
First aid	Standard treatment for exposure by all routes (see pages 108-122).

Fire hazard	Use water spray, carbon dioxide, dry powder, or alcohol or polymer foam to extinguish fires.
Spillage disposal	See general section.

263. Cellulose nitrate

White or yellowish-white amorphous powder or matted filaments (usually supplied moistened with alcohol); insoluble in water.

RISKS
Highly flammable (R11)

SAFETY PRECAUTIONS
Keep away from sources of ignition – No Smoking – Take precautionary measures against static discharges – Wear suitable gloves and eye/face protection (S16, S33, S37/39)

Toxic effects	Mildly toxic.
Hazardous reactions	The hazards of cellulose intake centre on its flammability and explosive potential.
First aid	Standard treatment for exposure by all routes (see pages 108-122).
Fire hazard	Autoignition temperature 160-170 ˚C.
Spillage disposal	See general section.

264. Cerium

Grey, metallic granules; m.p. 804 ˚C; b.p. 2930 ˚C; insoluble in water but slowly oxidized by water.

Toxic effects	Irritant to skin, eyes, upper respiratory tract, and mucous membranes.
Hazardous reactions	Ignites and burns brightly at 160 ˚C, burns and reacts violently with a variety of substances.
First aid	Standard treatment for exposure by all routes (see pages 108-122).
Fire hazard	Use class D extinguishing materials only. Do NOT use water.
Spillage disposal	See general section.

265. Cerium nitride

Solid; reacts with water producing ammonia and hydrogen.

Toxic effects	Generally of low toxicity.
Hazardous reactions	Contact with limited amount of water or dilute acid causes rapid incandescence with ignition.

First aid   Standard treatment for exposure by all routes (see pages 108-122).

Spillage   See general section.
  disposal

266. Chloral

Colourless, oily liquid with pungent, irritating odour; b.p. 98 ˚C; soluble in water forming chloral hydrate.

Toxic effects   The vapour irritates the respiratory system and eyes. The liquid irritates the eyes and skin. If taken by mouth it will show the effects of chloral hydrate, namely nausea, vomiting, coldness of extremities, and unconsciousness.

First aid   Standard treatment for exposure by all routes (see pages 108-122).

Spillage   See general section.
  disposal

267. Chloral hydrate

Colourless crystals with a slightly acrid smell and a slightly bitter, caustic taste; m.p. 52 ˚C; b.p. 95.7 ˚C; soluble in water.

RISKS
Toxic if swallowed – Irritating to eyes and skin (R25, R36/38)

SAFETY PRECAUTIONS
Avoid contact with eyes – If you feel unwell, seek medical advice (show label where possible) (S25, S44)

Toxic effects   If taken by mouth it may cause nausea, vomiting, coldness of extremities, and unconsciousness.

First aid   Standard treatment for exposure by all routes (see pages 108-122).

Spillage   See general section.
  disposal

268. Chloramine

Hazardous   Hazardous preparation, stable in ethereal solution, but solvent-free material
  reactions   decomposes violently or explosively.

269. Chloric acid

Colourless liquid; b.p. decomposes at 40 ˚C.

Toxic effects   Strongly irritant to all tissues.

Hazardous reactions	Aqueous solution explodes if evaporated too far; it ignites filter paper; reactions with other oxidizable substances are similar to those of chlorates.
First aid	Standard treatment for exposure by all routes (see pages 108-122).
Spillage disposal	See general section.

270. Chlorinated biphenyls

Vary from mobile liquids through viscous liquids to white, crystalline solids; commercial material usually contain 42% or 54% chlorine; insoluble in water.

Limit values	OES short-term (42% chlorine) 2 mg m^{-3}, (54% chlorine) 1 mg m^{-3} under review; long-term (42% chlorine) 1 mg m^{-3}, (54% chlorine) 0.5 mg m^{-3} under review.
Toxic effects	Irritant to skin, eyes, and upper respiratory tract. Inhalation, ingestion, and skin absorption can cause systemic effects. Skin contact can cause chloracne. Suspected carcinogen.
First aid	Standard treatment for exposure by all routes (see pages 108-122).
Fire hazard	Flash point (42%) 176-180 °C (open cup); (54%) 222 °C.
Spillage disposal	The preferred method of disposal is high temperature incineration.

271. Chlorinated diphenyl oxide

White to yellowish waxy semisolids; b.p. 2320-260 °C at 8 torr (hexachloro derivative).

Toxic effects	Prolonged skin contact may cause acneform dermatitis. Prolonged exposure to the vapour can cause liver damage.

272. Chlorine

Greenish-yellow gas with irritating odour; m.p. -101 °C; b.p. -33.6 °C; soluble in water.

RISKS
Toxic by inhalation – Irritating to eyes, respiratory system, and skin (R23, R36/37/38)

SAFETY PRECAUTIONS
Keep container tightly closed and in a well ventilated place – If you feel unwell, seek medical advice (show label where possible) (S7/9, S44)

Limit values	OES short-term 1 p.p.m. (3 mg m^{-3}); long-term 0.5 p.p.m. (1.5 mg m^{-3}).
Toxic effects	The gas causes severe lung irritation and damage. It also irritates the eyes and can cause conjunctivitis. In high concentrations the gas irritates the skin.
Hazardous reactions	Numerous reports of violent or explosive reactions with a wide variety of elements and compounds.
Fire hazard	Use water spray to keep cylinders cool or remove them from vicinity of.

RSC *Chemical Safety Data Sheets*, Vol. 4a, No. 37, 1991 gives extended coverage.

273. Chlorine dioxide

Red-yellow gas or orange/red crystals with pungent odour; m.p. -59 °C; b.p. 9.9 °C at 731 mm Hg with explosion; readily soluble in water.

Limit values	OES short-term 0.3 p.p.m. (0.9 mg m^{-3}); long-term 0.1 p.p.m. (0.3 mg m^{-3}).
Toxic effects	Moderately toxic by inhalation.
Hazardous reactions	Explodes violently under slightest provocation in a variety of situations.
First aid	Standard treatment for exposure by all routes (see pages 108-122).
Spillage disposal	See general section.

274. Chlorine fluoride

Colourless gas or yellow liquid with sweet odour; b.p. 11.8 °C; explodes on contact with water.

Toxic effects	Highly irritant and causes severe burns.
Hazardous reactions	Powerful oxidant reacting violently with a wide range of materials.
First aid	Standard treatment for exposure by all routes (see pages 108-122).
Spillage disposal	See general section.

275. Chlorine pentafluoride

Hazardous reactions	Very vigorous reaction with water and anhydrous nitric acid.
First aid	Standard treatment for exposure by all routes (see pages 108-122).

276. Chlorine perchlorate

Hazardous reactions	Shock-sensitive and liable to explode, violent reactions with haloalkanes and haloalkenes.
First aid	Standard treatment for exposure by all routes (see pages 108-122).

277. Chlorine trifluoride

Colourless gas or yellow-green liquid with somewhat sweet but highly irritant smell; b.p. 11.75 °C; reacts violently with water.

Limit values	OES short-term 0.1 p.p.m. (0.4 mg m^{-3}).

Toxic effects	The vapour severely irritates the eyes, skin, and respiratory system, and may cause burns. The liquid severely burns all human tissue.
Hazardous reactions	Reacts violently or explosively with a wide variety of elements and compounds.
First aid	Standard treatment for exposure by all routes (see pages 108-122).
Fire hazard	Extinguish fires with water spray or dry chemical powder.
Spillage disposal	See general section.

278. Chloroacetaldehyde

Clear, colourless liquid with pungent odour; m.p. -16.3 ˚C; b.p. 85 ˚C at 748 mm Hg.

Limit values	OES short-term 1 p.p.m. (3 mg m^{-3}).
Toxic effects	Aqueous solutions are very destructive to all tissues.
First aid	Standard treatment for exposure by all routes (see pages 108-122).
Fire hazard	Flash point 88 ˚C.
Spillage disposal	See general section.

279. 2-Chloroacetamide

White, crystalline powder; m.p. 120 ˚C; b.p. 225 ˚C with decomposition; moderately soluble in water.

Toxic effects	High concentrations are destructive to tissues of the skin, eyes, upper respiratory tract, and mucous membranes.
First aid	Standard treatment for exposure by all routes (see pages 108-122).
Fire hazard	Use water spray, carbon dioxide, dry powder, or alcohol or polymer foam to extinguish fires.
Spillage disposal	See general section.

280. N-Chloroacetamide

Crystalline solid; m.p. 110 ˚C; very soluble in water.

Hazardous reactions	Exploded during desiccation of solid and during concentration of chloroform solution.

281. Chloroacetic acid

Colourless to pale brown deliquescent crystals; m.p. 61-63 ˚C (commercial grade); b.p. 189.4 ˚C; very soluble in water.

Toxic effects	The solid or its solutions severely irritate or burn the eyes and also produce severe skin burns which may only be apparent several hours after contact. Assumed to cause severe internal irritation and damage if taken by mouth.
First aid	Standard treatment for exposure by all routes (see pages 108-122).
Fire hazard	Large fires can be treated with water spray. For small fires use carbon dioxide, dry powder, or water spray.
Spillage disposal	See general section.

282. Chloroacetone

Colourless liquid with pungent odour; darkens on exposure to light; m.p. -44.5 ˚C; b.p. 121 ˚C; soluble in water.

Toxic effects	Irritant and destructive to tissues of skin, eyes, upper respiratory tract, and mucous membranes. May be fatal following inhalation, ingestion, or skin absorption.
Hazardous reactions	May polymerize explosively on storage.
First aid	Standard treatment for exposure by all routes (see pages 108-122).
Fire hazard	Flash point 27 ˚C. Extinguish fires using carbon dioxide, dry powder, or alcohol or polymer foam extinguishants.
Spillage disposal	See general section.

283. Chloroacetonitrile

Clear, colourless liquid with a pungent smell; b.p. 126-127 ˚C; reacts with water.

RISKS
Toxic by inhalation, in contact with skin, and if swallowed (R23/24/25)

SAFETY PRECAUTIONS
If you feel unwell, seek medical advice (show label where possible) (S44)

Toxic effects	Irritant to all mucous membranes.
First aid	Standard treatment for exposure by all routes (see pages 108-122).
Fire hazard	Flash point 56 ˚C; Explosive limits lel 1.0%. Emits toxic fumes on heating to decomposition.
Spillage disposal	See general section.

284. Chloroacetyl chloride

Colourless to pale yellow liquid; b.p. 106 ˚C; reacts with water.

RISKS
Causes burns – Irritating to respiratory system (R34, R37)

SAFETY PRECAUTIONS
Keep container in a well ventilated place – In case of contact with eyes, rinse immediately with plenty of water and seek medical advice (S9, S26)

Limit values   OES long-term 0.05 p.p.m. (0.2 mg m^{-3}) under review.

Toxic effects   The vapour severely irritates all parts of the respiratory system, skin, and eyes. The liquid may produce blisters after contact. Assumed to cause severe internal irritation and damage if taken by mouth.

First aid   Standard treatment for exposure by all routes (see pages 108-122).

Spillage disposal   See general section.

285. Chloroacetylene

Gas; b.p. -30 ˚C.

Hazardous reactions   Endothermic: may burn or explode in contact with air.

286. Chloroanilines

The *o*- and *m*-chloroanilines are colourless to yellow or brown liquids with characteristic sweet odour; *p*-chloroaniline is an almost colourless crystalline solid or powder; m.p. *o*- 0 ˚C; *m*- -10 ˚C; *p*-70 ˚C; b.p. *o*- 209 ˚C; *m*- 229 ˚C; *p*- 230 ˚C; all are insoluble in water.

RISKS
Toxic by inhalation, in contact with skin, and if swallowed – Danger of cumulative effects (R23/24/25, R33)

SAFETY PRECAUTIONS
After contact with skin, wash immediately with plenty of water – Wear protective clothing and gloves – If you feel unwell, seek medical advice (show label where possible) (S28, S36/37, S44)

Toxic effects   The inhalation of vapour and absorption through the skin may cause cyanosis, and damage to the liver and kidneys. Assumed that similar poisoning will result from ingestion.

Hazardous reactions   Subject to exothermic decomposition during high-temperature distillation.

First aid   Standard treatment for exposure by all routes (see pages 108-122).

Fire hazard   Flash point *o*- 97 ˚C; *m*- 123 ˚C; *p*- above 104 ˚C. Use water spray, carbon
                     dioxide, dry powder, or alcohol or polymer foam to extinguish fires.

Spillage   See general section.
disposal

287. 1-Chloroaziridine

Hazardous   Liable to explode on long storage.
reactions

288. 2-Chlorobenzaldehyde

Colourless liquid or crystals; m.p. 11 ˚C; b.p. 208 ˚C; very slightly soluble in water.

RISKS
Causes burns (R34)

SAFETY PRECAUTIONS
In case of contact with eyes, rinse immediately with plenty of water and seek medical advice
(S26)

Toxic effects   Contact with skin or eyes may cause irritation or burns. There may be severe
                     irritation and damage if the substance is swallowed.

First aid   Standard treatment for exposure by all routes (see pages 108-122).

Fire hazard   Flash point 87 ˚C. Use water spray, carbon dioxide, dry powder, or alcohol or
                     polymer foam to extinguish fires

Spillage   See general section.   •
disposal

289. Chlorobenzene

Clear, colourless liquid with a faint, not unpleasant, almond-like odour; b.p. 132 ˚C; immiscible
with water.

RISKS
Flammable – Harmful by inhalation (R10, R20)

SAFETY PRECAUTIONS
Avoid contact with skin and eyes (S24/25)

Limit values   OES long-term 50 p.p.m. (230 mg m^{-3}).

Toxic effects   The vapour may cause drowsiness and unconsciousness. The liquid irritates
                     the skin and eyes, and may cause stupor and unconsciousness after a few
                     hours if taken by mouth.

Hazardous   Explodes with finely divided sodium.
reactions

First aid   Standard treatment for exposure by all routes (see pages 108-122).

Fire hazard	Flash point 29 ˚C; Explosive limits 1.3-7%; Autoignition temperature 593 ˚C. Extinguish fire with foam, dry powder, or carbon dioxide.
Spillage disposal	See general section.

RSC *Chemical Safety Data Sheets* Vol. 1, No. 13, 1988 gives extended coverage.

290. 1-Chlorobenzotriazole

Hazardous reactions	May ignite spontaneously.

291. 2-Chlorobuta-1,3-diene

Colourless liquid with pungent odour; b.p. 59.4 ˚C; slightly soluble in water.

RISKS
Extremely flammable – Harmful by inhalation (R12, R20)

SAFETY PRECAUTIONS
Keep container in a well ventilated place – Keep away from sources of ignition – No Smoking – Do not empty into drains – Take precautionary measures against static discharges (S9, S16, S29, S33)

Limit values	OES long-term 10 p.p.m. (36 mg m^{-3}).
Toxic effects	Inhalation causes depression of the central nervous system and injury to lungs, liver, and kidneys. Eye contact can cause conjunctivitis and necrosis of the cornea.
Hazardous reactions	Autoxidizes very rapidly to produce unstable peroxide which will catalyse exothermic polymerization of monomer.
First aid	Standard treatment for exposure by all routes (see pages 108-122).
Fire hazard	Flash point -20 ˚C; Explosive limits 4.0-20.0%. Emits toxic fumes on heating to decomposition.
Spillage disposal	See general section.

292. 1-Chlorobutan-2-one

Liquid; b.p. 137.5 ˚C.

Hazardous reactions	Bottle of stabilized material exploded spontaneously.
First aid	Standard treatment for exposure by all routes (see pages 108-122).

293. 1-Chlorobut-1-en-3-one

Hazardous reactions   Liable to explode soon after preparation.

First aid   Standard treatment for exposure by all routes (see pages 108-122).

294. 3-Chlorocyclopentene

Oily liquid; b.p. 25-31 ˚C at 30 mm Hg.

Hazardous reactions   Explosive decomposition after brief storage.

First aid   Standard treatment for exposure by all routes (see pages 108-122).

295. Chlorodifluoromethane

Gas; m.p. -146 ˚C; b.p. -40.8 ˚C.

Limit values   OES long-term 1000 p.p.m. (3500 mg m^{-3}) proposed change.

Toxic effects   Asphyxiant in high concentrations.

Hazardous reactions   Reacts exothermically with aluminium.

First aid   Standard treatment for exposure by all routes (see pages 108-122).

Fire hazard   Autoignition temperature 632 ˚C.

296. 4-Chloro-2,6-dinitroaniline

Orange-yellow, crystalline solid; m.p. 146 ˚C.

Hazardous reactions   Dangerous decomposition set in during large-scale manufacture from 4-chloro-2-nitroaniline. Explosion occurred during large-scale diazotization of the amine.

First aid   Standard treatment for exposure by all routes (see pages 108-122).

297. 1-Chloro-2,4-dinitrobenzene

Pale yellow crystals; m.p. α 53.4 °C, β 43 °C, χ 27 °C; b.p. 315 °C; insoluble in water.

RISKS
Toxic by inhalation, in contact with skin, and if swallowed – Danger of cumulative effects (R23/24/25, R33)

SAFETY PRECAUTIONS
After contact with skin, wash immediately with plenty of water – Wear suitable gloves – If you feel unwell, seek medical advice (show label where possible) (S28, S37, S44)

Toxic effects	The dust, or vapour from the molten compound, irritates the respiratory system. It also irritates the skin and may cause dermatitis. Cyanosis and liver injury may follow inhalation or skin absorption. Assumed to be poisonous if taken by mouth.
Hazardous reactions	Has been used as an explosive; reaction with ammonia under pressure may be violent.
First aid	Standard treatment for exposure by all routes (see pages 108-122).
Fire hazard	Flash point 194 °C (closed cup); Explosive limits 2-22%.
Spillage disposal	See general section.

298. 1-Chloro-2, 3-epoxypropane

Colourless liquid with irritant chloroform-like odour; m.p. -57 °C; b.p. 118 °C; immiscible with water.

RISKS
May cause cancer – Flammable – Toxic by inhalation, in contact with skin, and if swallowed – Causes burns – May cause sensitization by skin contact (R45, R10, R23/24/25, R34, R43)

SAFETY PRECAUTIONS
Avoid exposure – Obtain special instructions before use – Keep container in a well ventilated place – If you feel unwell, seek medical advice (show label where possible) (S53, S9, S44)

Limit values	OES short-term 5 p.p.m. (20 mg m^{-3}) under review; long-term 2 p.p.m. (8 mg m^{-3}) under review.
Toxic effects	The vapour irritates the respiratory system and in severe cases can cause respiratory paralysis. Prolonged exposure to low concentrations may cause conjunctivitis, chronic weariness, and stomach upset. The vapour and liquid irritate the eyes and may cause conjunctivitis. Poisonous by skin absorption and ingestion. Skin blistering and severe pain may develop after latent period, possibly with sensitization and dermatitis.
Hazardous reactions	Reacts violently with aniline, isopropylamine, sulfuric acid, and trichloroethylene; polymerizes exothermically.
First aid	Standard treatment for exposure by all routes (see pages 108-122).

Fire hazard   Flash point 31 ˚C (open cup); Explosive limits 3.8-21%; Autoignition temperature 411 ˚C. Extinguish fire with water spray, dry powder, or carbon dioxide.

Spillage disposal   See general section.

RSC *Chemical Safety Data Sheets* Vol. 1, No. 39, 1988 gives extended coverage.

299. Chloroethane

Colourless gas and liquid with pungent, ethereal odour; b.p. 12.4 ˚C; sparingly soluble in water.

Limit values   OES short-term 1250 p.p.m. (3250 mg m⁻³); long-term 1000 p.p.m. (2600 mg m⁻³).

Toxic effects   The vapour is mildly irritating to the mucous membranes; at high concentrations it is narcotic.

Hazardous reactions   Mixture with potassium is shock-sensitive.

First aid   Standard treatment for exposure by all routes (see pages 108-122).

Fire hazard   Flash point -50 ˚C; Explosive limits 3.6-15.4%; Autoignition temperature 519 ˚C. As the gas is supplied in a cylinder, turning off the valve will reduce any fire involving it; if possible, cylinders should be removed quickly from an area in which a fire has developed.

Spillage disposal   Surplus gas or leaking cylinder can be vented slowly to air in a safe open area or the gas burnt-off in a suitable burner.

300. 2-Chloroethanol

Colourless liquid with a faint ethereal smell; m.p. -69 ˚C; b.p. 128.8 ˚C; miscible with water.

RISKS
Very toxic by inhalation, in contact with skin, and if swallowed (R26/27/28)

SAFETY PRECAUTIONS
Keep container tightly closed and in a well ventilated place – After contact with skin, wash immediately with plenty of water – In case of accident or if you feel unwell, seek medical advice immediately (show label where possible) (S7/9, S28, S45)

Limit values   OES short-term 1 p.p.m. (3 mg m⁻³).

Toxic effects   The vapour causes nausea, headaches, vomiting, stupefaction, and unconsciousness. It irritates the mucous membranes. The liquid is rapidly absorbed by the skin, producing similar effects to inhalation. Assumed to be extremely poisonous if taken by mouth.

First aid   Standard treatment for exposure by all routes (see pages 108-122).

Fire hazard   Flash point 60 ˚C (open cup); Explosive limits 4.9-15.9%; Autoignition temperature 425 ˚C.

Spillage disposal   See general section.

301. 2-Chloroethylamine

Oily liquid.

Hazardous reactions	May polymerize explosively.
First aid	Standard treatment for exposure by all routes (see pages 108-122).

302. Chlorofluoroalkanes

The commonly available compounds in this series are colourless gases or low-boiling liquids supplied in cylinders.

Toxic effects	They are all relatively innocuous gases, their toxicity being of the same order as nitrogen or carbon dioxide.
First aid	Standard treatment for exposure by all routes (see pages 108-122).
Fire hazard	Explosive limits 9-14.8%; Autoignition temperature 632 °C. Chlorodifluoroethane is the only one that is flammable. As the gas is supplied in a cylinder, turning off the valve will reduce any fire involving it; cylinders should be removed quickly from an area in which a fire has developed.
Spillage disposal	Surplus gas or leaking cylinder can be vented slowly in air in a safe open area or gas burnt-off in a suitable burner.

303. Chloroform

Colourless volatile liquid with a characteristic odour; m.p. -63.5 °C; b.p. 61 °C; very slightly soluble in water.

RISKS
Harmful by inhalation and if swallowed – Irritating to skin – Possible risk of irreversible effects – Danger of serious damage to health by prolonged exposure (R20/22, R38, R40, R48)

SAFETY PRECAUTIONS
Wear protective clothing and gloves (S36/37)

Limit values	OES long term 2 p.p.m. (9.8 mg m^{-3}) proposed change.
Toxic effects	The vapour has anaesthetic properties, causing drowsiness, giddiness, headache, nausea, vomiting, and unconsciousness. The vapour and liquid irritate the eyes causing conjunctivitis. The liquid is poisonous if taken by mouth. Animal carcinogen, suspected human carcinogen.
Hazardous reactions	Vigorous reaction with a variety of compounds.
First aid	Standard treatment for exposure by all routes (see pages 108-122).
Fire hazard	Liberates phosgene when heated to a high temperature.
Spillage disposal	See general section.

RSC *Chemical Safety Data Sheets* Vol. 1, No. 14, 1988 gives extended coverage.

304. Chlorogermane

Hazardous   Reacts with ammonia to form explosive product.
reactions

305. 3-Chloro-2-hydroxypropionitrile

Hazardous   May explode during vacuum distillation.
reactions

First aid   Standard treatment for exposure by all routes (see pages 108-122).

306. 3-Chloro-2-hydroxypropyl perchlorate

Hazardous   Explodes violently on heating.
reactions

First aid   Standard treatment for exposure by all routes (see pages 108-122).

307. Chloromethane

Colourless gas; b.p. -24 ˚C; soluble in water.

RISKS
Extremely flammable liquefied gas – Harmful by inhalation – Possible risk of irreversible effects – Danger of serious damage to health by prolonged exposure (R13, R20, R40, R48)

SAFETY PRECAUTIONS
Keep container in a well ventilated place – Keep away from sources of ignition – No Smoking – Take precautionary measures against static discharges (S9, S16, S33)

Limit values   OES short-term 100 p.p.m. (210 mg m^{-3}); long-term 50 p.p.m. (105 mg m^{-3}).

Toxic effects   Inhalation of vapour may cause dizziness, drowsiness, nausea, stomach pains, visual disturbances, mental confusion, and unconsciousness; heavy exposure can be fatal and some symptoms may be delayed.

Hazardous   Ignites or explodes on contact with a variety of substances.
reactions

First aid   Standard treatment for exposure by all routes (see pages 108-122).

Fire hazard   Flash point below 0 ˚C; Explosive limits 10.7-17.4%; Autoignition temperature 632 ˚C. As the gas is supplied in a cylinder, turning off the valve will reduce any fire involving it; if possible cylinders should be removed quickly from an area in which a fire has developed.

Spillage   Surplus gas or leaking cylinder can be vented slowly to air in a safe open
disposal   area or burnt-off in a suitable burner.

308. 2-Chloromethylfuran

Liquid.

Hazardous reactions   Liable to explode violently owing to polymerization or decomposition.

First aid   Standard treatment for exposure by all routes (see pages 108-122).

309. Chloromethyl methyl ether

Colourless liquid with suffocating smell; m.p. -103.5 °C; b.p. 59 °C; decomposed by water.

RISKS
May cause cancer – Highly flammable – Harmful by inhalation, in contact with skin, and if swallowed (R45, R11, R20/21/22)

SAFETY PRECAUTIONS
Avoid exposure – Obtain special instructions before use – Keep container in a well ventilated place – Keep away from sources of ignition – No Smoking – If you feel unwell, seek medical advice (show label where possible) (S53, S9, S16, S44)

Toxic effects   Vapour irritates the respiratory system and eyes, primary effects probably being due to fairly rapid formation of hydrochloric acid, formaldehyde, and methanol. A high incidence of lung cancer has been observed.

First aid   Standard treatment for exposure by all routes (see pages 108-122).

Fire hazard   Flash point 15 °C.

Spillage disposal   See general section.

310. 4-Chloro-2-methylphenol

Brown, crystalline solid; m.p. 45-48 °C; b.p. 220-225 °C.

RISKS
Harmful in contact with skin and if swallowed – Irritating to skin (R21/22, R38)

SAFETY PRECAUTIONS
In case of contact with eyes, rinse immediately with plenty of water and seek medical advice – After contact with skin, wash immediately with plenty of water (S26, S28)

Toxic effects   Effects vary from mild irritation to severe destruction of tissue dependent on intensity and duration of exposure.

Hazardous reactions   Vigorous reaction followed by explosion when large quantity was left in contact with concentrated solution of sodium hydroxide.

First aid   Standard treatment for exposure by all routes (see pages 108-122).

Fire hazard   Flash point above 110 °C. Extinguish fire with water spray.

Spillage disposal   See general section.

311. 2-Chloromethylthiophene

Hazardous   Unstable and gradually decomposes; closed containers may explode.
reactions

312. Chloronitroanilines

2-Chloro-4-nitro- and 4-chloro-2-nitroanilines are yellow to brownish-yellow powders or crystals; m.p. 2,4-107 °C; 4,2- 116 °C; insoluble in water.

RISKS
Very toxic by inhalation and in contact with skin – Danger of cumulative effects (R26/27, R33)

SAFETY PRECAUTIONS
After contact with skin, wash immediately with plenty of water – Wear protective clothing and gloves – In case of accident or if you feel unwell, seek medical advice immediately (show label where possible) (S28, S36/37, S45)

Toxic effects   Absorption through the skin may cause dermatitis, cyanosis, and damage to the liver and kidneys. Assumed that similar poisoning will result from ingestion. Onset of symptoms may be delayed.

First aid   Standard treatment for exposure by all routes (see pages 108-122).

Fire hazard   Use water spray, carbon dioxide, dry powder, or alcohol or polymer foam to extinguish fires.

Spillage   See general section.
disposal

313. Chloronitrobenzenes

Light yellow crystals with a sweet smell; m.p. 83 °C; b.p. *o-* 245-5 °C; *m-* 236 °C; *p-* 242 °C; insoluble in water.

RISKS
Toxic by inhalation, in contact with skin, and if swallowed – Danger of cumulative effects (R23/24/25, R33)

SAFETY PRECAUTIONS
After contact with skin, wash immediately with plenty of water – Wear suitable gloves – If you feel unwell, seek medical advice (show label where possible) (S28, S37, S44)

Limit values   OES short-term 2 mg m^{-3}; long-term 1 mg m^{-3}.

Toxic effects   The dust or vapour from the molten compounds irritates the respiratory system. Contact with the skin may cause dermatitis. Cyanosis and liver injury may follow inhalation or skin absorption. Assumed to be poisonous if taken by mouth.

Hazardous   Runaway reaction of *o*-chloronitrobenzene with ammonia under pressure
reactions   involved 6 simultaneous fault conditions. *p*-Chloronitrobenzene reacted violently and finally explosively when added to a solution of sodium methoxide in methanol (B556-557).

First aid   Standard treatment for exposure by all routes (see pages 108-122).

Fire hazard   Flash point 127 °C closed cup. Use water spray, carbon dioxide, dry chemical powder, or alcohol or polymer foam to extinguish fires.

Spillage   See general section.
disposal

314. Chloronitromethane

Liquid; b.p. 122-123 °C; moderately soluble in water.

Hazardous   Product of chlorination of nitromethane decomposed explosively during
reactions   vacuum distillation.

First aid   Standard treatment for exposure by all routes (see pages 108-122).

315. Chloronitrotoluenes

2,4- Pale yellow, crystalline solid; 4,3- golden yellow liquid; m.p. 2,4- 58-61 °C; b.p. 2,4- 360 °C; 4,3-260 °C at 745 mm Hg.

Toxic effects   Irritant to skin, eyes, upper respiratory tract, and mucous membranes.

Hazardous   Residue from vacuum distillation of crude 2-chloro-4-nitrotoluene or isomers
reactions   exploded; exothermic reaction of 4-chloro-3-nitrotoluene with copper cyanide in pyridine exploded.

First aid   Standard treatment for exposure by all routes (see pages 108-122).

Fire hazard   Flash point 4,3- above 110 °C. Use water spray, carbon dioxide, dry powder, or alcohol or polymer foam to extinguish fires.

Spillage   See general section.
disposal

316. 1-Chloropentafluoroethane

Colourless gas; m.p. -77 °C; b.p. -39 °C.

Limit values   OES long-term 1000 p.p.m. (6320 mg m^{-3}).

Toxic effects   Inhalation in high concentrations can cause dizziness, inco-ordination, narcosis, nausea, central nervous system depression, and unconsciousness.

First aid   Standard treatment for exposure by all routes (see pages 108-122).

Fire hazard   Use water spray to keep containers cool or move them out of the affected area.

Spillage   See general section.
disposal

317. Chlorophenols

o-Chlorophenol is a colourless to pale brown liquid; m-chlorophenol is a colourless to yellow solid; p-chlorophenol is a colourless to pale brown, crystalline solid all with a phenolic (carbolic) odour; m.p. (o-) 9.0 °C; (m-) 33.0 °C; (p-) 41-43 °C; b.p. (o-) 174 °C; (m-) 214 °C; (p-) 217-220 °C; o-Chlorophenol is soluble in water, the others sparingly so.

RISKS
Harmful by inhalation, in contact with skin, and if swallowed (R20/21/22)

SAFETY PRECAUTIONS
Keep out of reach of children – After contact with skin, wash immediately with plenty of water (S2, S28)

Toxic effects	The vapours when inhaled irritate the respiratory system. In contact with the eyes they cause irritation or burning. They irritate and may burn the skin and must be assumed to be very poisonous and irritant if taken by mouth.
First aid	Standard treatment for exposure by all routes (see pages 108-122).
Fire hazard	Flash point (o-) 63.0 °C; (p-) 121 °C. Use carbon dioxide, dry powder, or alcohol or polymer foam to extinguish fires.
Spillage disposal	See general section.

318. Chlorophenyllithium

Solid.

Hazardous reactions	Absence of solvent or presence of traces of oxygen in reaction atmosphere may cause explosion.
First aid	Standard treatment for exposure by all routes (see pages 108-122).

319. Chloropropanes

Colourless, volatile liquids; b.p. 1-isomer, 47 °C; 2-isomer 35 °C; immiscible with water.

RISKS
Highly flammable – Harmful by inhalation, in contact with skin, and if swallowed (R11, R20/21/22)

SAFETY PRECAUTIONS
Keep container in a well ventilated place – Do not empty into drains (S9, S29)

Toxic effects	Inhalation of 1-chloropropane irritates the respiratory system: high concentrations of both isomers cause narcosis. The liquids irritate the eyes. Assumed to be poisonous if taken by mouth.
First aid	Standard treatment for exposure by all routes (see pages 108-122).

Fire hazard   Flash point below -18 ˚C for 1-isomer; -32 ˚C for 2-isomer; Explosive limits
                       2.6-11.1% for 1-isomer; 2.8-10.7% for 2-isomer; Autoignition temperature
                       520 ˚C for 1-isomer; 592 ˚C for 2-isomer. Extinguish fire with water spray,
                       foam, dry powder, or carbon dioxide.

Spillage   See general section.
   disposal

320. 3-Chloropropiolonitrile

Crystalline solid; m.p. 42-42.5 ˚C.

Hazardous   Ignites spontaneously in air. Explosion hazard when heated in nearly closed
   reactions   vessel.

First aid   Standard treatment by all routes with special reference to cyanides.

321. Chloropropionyl chloride

Colourless liquid with acrid odour; b.p. 143 ˚C; reacts with water forming chloropropionic acid
and hydrogen chloride.

Toxic effects   The vapour irritates the eyes and respiratory system severely. The liquid
                       burns the eyes and skin and causes severe internal irritation and damage if
                       taken by mouth.

First aid   Standard treatment for exposure by all routes (see pages 108-122).

Fire hazard   Flash point 61 ˚C. Extinguish fires with carbon dioxide or dry powder. Do
                       NOT use water.

Spillage   See general section.
   disposal

322. 3-Chloropropyne

Liquid; b.p. 65 ˚C.

Hazardous   Violent reaction with ammonia under pressure followed by explosion.
   reactions

First aid   Standard treatment for exposure by all routes (see pages 108-122).

Fire hazard   Flash point <15 ˚C.

323. *o*-Chlorostyrene

Yellow liquid; m.p. -63.1 ˚C; b.p. 58.60 ˚C at 7 mm Hg; insoluble in water.

Toxic effects   Irritant to skin and eyes. Harmful if inhaled or swallowed.

First aid	Standard treatment for exposure by all routes (see pages 108-122).
Fire hazard	Flash point 58 ˚C. Extinguish fires with water spray, carbon dioxide, dry chemical powder, or alcohol or polymer foam.
Spillage disposal	See general section.

324. *N*-Chlorosuccinimide

White powder; m.p. 150-152 ˚C.

Toxic effects	Irritant and extremely destructive to all tissues. Inhalation may be fatal.
Hazardous reactions	Violent or explosive reaction with aliphatic alcohols, benzylamine, hydrazine hydrate.
Fire hazard	Use water spray, carbon dioxide, dry powder, or alcohol or polymer foam to extinguish fires.
Spillage disposal	See general section.

325. Chlorosulfonic acid

Colourless to brown fuming liquid; b.p. 151 ˚C; decomposes with explosive violence when mixed with water.

RISKS
Reacts violently with water – Causes severe burns – Irritating to respiratory system (R14, R35, R37)

SAFETY PRECAUTIONS
In case of contact with eyes, rinse immediately with plenty of water and seek medical advice (S26)

Limit values	OES long-term 1 mg m^{-3}.
Toxic effects	The fumes are very irritant to the lungs and mucous membranes and irritate the eyes severely. The liquid burns the skin and eyes. If taken by mouth there would be severe local and internal corrosive effects.
Hazardous reactions	Powerful oxidizing agent.
First aid	Standard treatment for exposure by all routes (see pages 108-122).
Spillage disposal	See general section.

326. *N*-Chlorotetramethylguanidine

| **Hazardous reactions** | Unstable; explodes if heated above 50 ˚C. |
| **First aid** | Standard treatment for exposure by all routes (see pages 108-122). |

327. Chlorotoluenes

All three isomers are liquids at normal temperatures; b.p. (*o-*) 159 ˚C; (*m-*) 162 ˚C; (*p-*) 162 ˚C; all insoluble in water.

RISKS
Harmful by inhalation (R20)

SAFETY PRECAUTIONS
Avoid contact with skin – Keep out of reach of children (S24, S2)

Limit values   OES long-term 50 p.p.m. (250 mg m^{-3}) for 2-chlorotoluene.

Toxic effects   The vapour of the chlorotoluenes is considered potentially toxic in moderate concentrations and is known to be narcotic at high concentrations in the case of the 3-isomer.

First aid   Standard treatment for exposure by all routes (see pages 108-122).

Fire hazard   Flash point 47-50 ˚C. Extinguish fire with foam, dry powder, or carbon dioxide.

**Spillage
disposal**   See general section.

328. 2-Chloro-6-(trichloromethyl)pyridine

Solid; m.p. 62.5-62.9 ˚C; b.p. 136-137.5 ˚C at 11 mm Hg; sparingly soluble in water.

Limit values   OES short-term 20 mg m^{-3}; long-term 10 mg m^{-3}.

Toxic effects   Toxic by ingestion and skin contact.

First aid   Standard treatment for exposure by all routes (see pages 108-122).

**Spillage
disposal**   See general section.

329. Chloryl hypofluorite

**Hazardous
reactions**   Explosive.

Fire hazard   Emits toxic fumes on heating to decomposition.

330. Chloryl perchlorate

**Hazardous
reactions**   Very powerful oxidant; reacts violently or explosively with ethanol, stop-cock grease, wood, and organic matter generally. Liable to explode on heating or contact with water, thionyl chloride, even at -70 ˚C.

First aid   Standard treatment for exposure by all routes (see pages 108-122).

Fire hazard   Emits toxic fumes on heating to decomposition.

331. Chromium

Greyish metal; m.p. 1903 ±10 ˚C; b.p. 2642 ˚C; insoluble in water.

Limit values   OES long-term 0.5 mg m^{-3}.

Toxic effects   Irritant to skin, eyes, upper respiratory tract, and mucous membranes. Prolonged exposure can cause systemic damage. Possible carcinogen.

First aid   Standard treatment for exposure by all routes (see pages 108-122).

Fire hazard   Use extinguishant appropriate to surroundings.

Spillage disposal   See general section.

332. Chromium salts

Generally lightly coloured solids but a few are coloured liquids; solubility in water varies.

Limit values   OES long-term 0.5 mg m^{-3} (Cr^{II} and Cr^{III}), MEL 0.05 mg m^{-3} (Cr^{VI}).

Toxic effects   All are irritant to skin, eyes, upper respiratory tract, and mucous membranes; some are extremely destructive to these tissues. Some may cause allergic skin reactions. All are toxic if inhaled or ingested.

First aid   Standard treatment for exposure by all routes (see pages 108-122).

Fire hazard   Water spray may be used on some of the salts, but in the absence of specific confirmation it would be prudent to use carbon dioxide, dry chemical powder, or alcohol or polymer foam.

Spillage disposal   See general section.

333. Chromium trioxide

Dark red crystalline masses or flakes; m.p. 196 ˚C; b.p. 250 ˚C with decomposition; very soluble in water.

RISKS
Contact with combustible material may cause fire – Causes severe burns – May cause sensitization by skin contact (R8, R35, R43)

SAFETY PRECAUTIONS
After contact with skin, wash immediately with plenty of water (S28)

Limit values   OES long-term 0.05 mg m^{-3}.

Toxic effects   The dust irritates all parts of the respiratory system. The solid and its solutions cause severe burns to the eyes and skin. Frequent exposure of skin to the material may result in ulceration. If taken by mouth there would be severe internal irritation and damage.

Hazardous reactions   Very powerful oxidant; violent or explosive reactions with a wide variety of substances.

First aid	Standard treatment for exposure by all routes (see pages 108-122).
Fire hazard	Use water to extinguish fires. Take care against possibility of steam explosion.
Spillage disposal	See general section.

RSC *Chemical Safety Data Sheets* Vol. 3, No. 15, 1990 gives extended coverage.

334. Chromyl acetate

Limit values	MEL long-term 0.05 mg m^{-3} (as Cr).
Hazardous reactions	Explosion occurred when being prepared from chromium trioxide and acetic anhydride.
First aid	Standard treatment for exposure by all routes (see pages 108-122).

335. Chromyl chloride

Red fuming liquid with pungent musty odour; b.p. 117 ˚C; decomposed vigorously by water.

RISKS
Contact with combustible material may cause fire – Causes severe burns (R8, R35)

SAFETY PRECAUTIONS
Keep container tightly closed and dry – Do not breathe dust – After contact with skin, wash immediately with plenty of water (S7/8, S22, S28)

Limit values	MEL long-term 0.05 mg m^{-3} (as Cr).
Toxic effects	The vapour irritates all parts of the respiratory system and the eyes. The liquid burns the skin and eyes. Frequent exposure of skin to the material may result in ulceration. If taken by mouth there would be severe local and internal irritation and damage.
Hazardous reactions	Explosions may occur with a variety of substances. Can ignite organic matter on contact.
First aid	Standard treatment for exposure by all routes (see pages 108-122).
Spillage disposal	See general section.

336. Chromyl nitrate

Limit values	MEL long-term 0.05 mg m^{-3} (as Cr).
Hazardous reactions	Powerful oxidant and nitrating agent which ignites many hydrocarbons and organic solvents on contact.
First aid	Standard treatment for exposure by all routes (see pages 108-122).
Fire hazard	Emits toxic fumes on heating to decomposition.

337. Chromyl perchlorate

Limit values   MEL long-term 0.05 mg m^{-3} (as Cr).

Hazardous reactions   Explodes violently above 80 °C; ignites organic solvents.

338. Cinnamaldehyde

Light yellow, oily liquid; m.p. -7.5 °C; b.p. 253 °C with decomposition; slightly soluble in water.

Toxic effects   Irritant to skin, eyes, upper respiratory tract, and mucous membranes.

Hazardous reactions   Rags soaked in caustic soda solution and the aldehyde ignited in waste bin.

First aid   Standard treatment for exposure by all routes (see pages 108-122).

Fire hazard   Flash point 71 °C. Use water spray, carbon dioxide, dry powder, or alcohol or polymer foam to extinguish fires.

Spillage disposal   See general section.

339. Cobalt metal

Silvery metal; m.p. 1495 °C; b.p. 2880 °C; insoluble in water.

Limit values   OES long-term 0.1 mg m^{-3} under review.

Toxic effects   Irritant to skin, eyes, upper respiratory tract, and mucous membranes. Suspected carcinogen.

Hazardous reactions   Finely divided cobalt is pyrophoric in air.

First aid   Standard treatment for exposure by all routes (see pages 108-122).

Fire hazard   Use extinguishant appropriate to immediate surroundings.

Spillage disposal   See general section.

340. Cobalt nitride

Limit values   OES long-term 0.1 mg m^{-3} (as Co) under review.

Hazardous reactions   A pyrophoric powder.

Fire hazard   Emits toxic fumes on heating to decomposition.

341. Cobalt(II) perchlorate hydrates

Limit values   OES long-term 0.1 mg m^{-3} (as Co) under review.

Hazardous reactions   Partial dehydration of the stable hexahydrate to a trihydrate led to explosion.

342. Cobalt trifluoride

Brownish, crystalline solid; decomposed by water.

Limit values   OES long-term 0.1 mg m^{-3} (as Co) under review.

Hazardous reactions   Fluorinating agent: reacts violently with hydrocarbons and water.

Fire hazard   Emits toxic fumes on heating to decomposition.

343. Copper

Red brown granules when powdered; m.p. 1083 °C; b.p. 2595 °C; insoluble in water.

Limit values   OES short-term 2 mg m^{-3} (dusts and mists) (as Cu); long-term 0.2 mg m^{-3} (fume), 1 mg m^{-3} (dusts and mists) (as Cu).

Toxic effects   Irritant to skin, eyes, upper respiratory tract, and mucous membranes. Exposure can cause lung damage, stomach pains, diarrhoea, and blood changes. Experimental tumourigen and teratogen.

Hazardous reactions   Mixtures of finely divided copper with chlorates or iodates explode on friction, shock, or heating. Violent reactions with a wide variety of substances.

First aid   Standard treatment for exposure by all routes (see pages 108-122).

Fire hazard   Use dry chemical powder. Do NOT use water.

Spillage disposal   See general section.

344. Copper compounds

Blue or greenish-blue crystals or powders.

Limit values   OES short-term 2 mg m^{-3} (dusts and mists) (as Cu); long-term 1 mg m^{-3} (dusts and mists) (as Cu).

Toxic effects   The dust irritates the mucous membranes. The dust and solutions of salts irritate the eyes. Ingestion may cause violent vomiting and diarrhoea with intense abdominal pain and collapse.

Hazardous reactions   Some of the compounds such as hydride, nitride, and phosphinate may decompose explosively on heating.

First aid   Standard treatment for exposure by all routes (see pages 108-122).

Spillage disposal   See general section.

345. Cresols

m-Cresol is a colourless, yellowish to brown-yellow, or pinkish crystalline solid or liquid with a sweet tarry smell; *o*- and *p*-cresols are colourless crystals with a phenolic odour; m.p. *o*- 30.9 °C; *m*- 11-12 °C; *p*- 36 °C; b.p. *o*- 191.2 °C; *m*- 202 °C; *p*- 202.5 °C; all are sparingly soluble in water.

RISKS
Toxic in contact with skin and if swallowed – Causes burns (R24/25, R34)

SAFETY PRECAUTIONS
Keep out of reach of children – After contact with skin, wash immediately with plenty of water – If you feel unwell, seek medical advice (show label where possible) (S2, S28, S44)

Limit values   OES long-term 5 p.p.m. (22 mg m^{-3}) for all isomers.

Toxic effects   The vapours from heated cresols are irritant to the respiratory system. Cresols burn the eyes severely and irritate or burn the skin. Considerable absorption through the skin may give effects similar to those caused by ingestion, namely headache, dizziness, nausea, vomiting, stomach pain, exhaustion, and possibly coma. Repeated inhalation or absorption of small amounts may cause damage to liver or kidneys; repeated contact with skin may cause dermatitis.

Hazardous reactions   Explodes at about 90 °C.

First aid   Standard treatment for exposure by all routes (see pages 108-122).

Fire hazard   Flash point *o*- 81 °C; *m*- and *p*- 86 °C; Explosive limits lel *o*- 1.4%; *m*- 1.19%; *p*- 1.1%; Autoignition temperature *o*- 599 °C; *m*- 560 °C; *p*- 671 °C. Use water spray, carbon dioxide, dry powder, or alcohol or polymer foam to extinguish fires.

Spillage disposal   See general section.

346. Cresyl glycidyl ether

Colourless to light yellow liquid; b.p. 259 °C; sparingly soluble in water.

RISKS
Irritating to skin (R38)

SAFETY PRECAUTIONS
In case of contact with eyes, rinse immediately with plenty of water and seek medical advice – After contact with skin, wash immediately with plenty of water (S26, S28)

Toxic effects   Toxic if inhaled, moderately toxic if ingested, and may cause conjunctivitis and dermatitis.

First aid   Standard treatment for exposure by all routes (see pages 108-122).

Fire hazard   Flash point 93.3 °C (closed cup). Extinguish fires with water, dry chemical powder, or alcohol or polymer foam.

Spillage disposal   See general section.

RSC *Chemical Safety Data Sheets* Vol. 3, No. 16, 1990 gives extended coverage.

347. Crotonaldehyde

Colourless liquid with a pungent, suffocating odour; b.p. 104 ˚C; immiscible with water.

RISKS
Highly flammable – Toxic by inhalation – Irritating to eyes, respiratory system, and skin (R11, R23, R36/37/38)

SAFETY PRECAUTIONS
Do not empty into drains – Take precautionary measures against static discharges – If you feel unwell, seek medical advice (show label where possible) (S29, S33, S44)

Limit values OES short-term 6 p.p.m. (18 mg m^{-3}) under review; long-term 2 p.p.m. (6 mg m^{-3}) under review.

Toxic effects Assumed to be irritant and poisonous if taken by mouth.

Hazardous reactions Exploded when heated with butadiene in autoclave at 180 ˚C; explodes on contact with concentrated nitric acid.

First aid Standard treatment for exposure by all routes (see pages 108-122).

Fire hazard Flash point 13 ˚C; Explosive limits 2.1-15.5%; Autoignition temperature 207 ˚C. Extinguish fire with water spray, foam, dry powder, or carbon dioxide.

Spillage disposal See general section.

348. Cumene

Colourless liquid; m.p. -96 ˚C; b.p. 152 ˚C; insoluble in water.

RISKS
Flammable – Irritating to respiratory system (R10, R37)

Limit values OES short-term 75 p.p.m. (370 mg m^{-3}) proposed change; long-term 25 p.p.m. (120 mg m^{-3}) proposed change.

Toxic effects The vapour irritates the respiratory system and the eyes, possibly causing conjunctivitis. The liquid irritates the skin and, by absorption and slow elimination, may cause serious poisoning. Assumed to be harmful if taken by mouth.

Hazardous reactions Violent reactions have been reported with nitric acid, oleum, and chlorosulfonic acids. Reactions with oxidizing agents may lead to explosive hazards.

First aid Standard treatment for exposure by all routes (see pages 108-122).

Fire hazard Flash point 44 ˚C; Explosive limits 0.9-6.5%; Autoignition temperature 425 ˚C. Extinguish fire with foam, dry powder, or carbon dioxide.

Spillage disposal See general section.

RSC *Chemical Safety Data Sheets* Vol. 1, No. 15, 1988 gives extended coverage.

349. Cyanamide

Deliquescent crystals; m.p. 45 ˚C; b.p. 260 ˚C.

RISKS
Toxic if swallowed – Irritating to eyes and skin – May cause sensitization by skin contact (R25, R36/38, R43)

SAFETY PRECAUTIONS
Keep in a cool place – Do not breathe dust – Wear suitable protective clothing – If you feel unwell, seek medical advice (show label where possible) (S3, S22, S36, S44)

Limit values	OES long-term 2 mg m^{-3}.
Hazardous reactions	Acids or alkalis and moisture speed decomposition which may become violent above 49 ˚C; explosive polymerization may occur on evaporating aqueous solutions to dryness or storing unstabilized material.
Fire hazard	Flash point 140.5 ˚C (closed cup).

350. Cyanides

Colourless crystals or powders which are soluble in water and react with acids to generate hydrogen cyanide.

Limit values	OES long-term 5 mg m^{-3} (as -CN).
Toxic effects	Cyanides and their solutions, and hydrogen cyanide liberated from these by the action of acids, are extremely poisonous. Both the cyanide solutions and the gas can be absorbed through the skin. Whatever the route of absorption, severe poisoning may result. The early warning symptoms of poisoning are general weakness and heaviness of the arms and legs, increased difficulty in breathing, headache, dizziness, nausea, and vomiting. These may be rapidly followed by pallor, unconsicousness, cessation of breathing, and death.
First aid	Exposure by any route requires special treatment for soluble cyanides.
Spillage disposal	See general section.

RSC *Chemical Safety Data Sheets* Vol. 2, Nos. 46, 84, and 101, 1989 give extended coverage.

351. Cyanogen

Colourless gas with almond-like odour; b.p. -21 ˚C; soluble in water.

RISKS
Highly flammable – Toxic by inhalation (R11, R23)

SAFETY PRECAUTIONS
Do not breathe vapour – If you feel unwell, seek medical advice (show label where possible) (S23, S44)

Limit values	OES long-term 10 p.p.m. (20 mg m^{-3}).

Toxic effects	The gas irritates the respiratory system, leading to headache, dizziness, rapid pulse, nausea, vomiting, unconsciousness, convulsions, and death, depending upon exposure. The gas irritates the eyes causing lachrymation.
Hazardous reactions	Reaction of the endothermic compound with various oxidants may be explosive.
First aid	Standard treatment for exposure by all routes (see pages 108-122).
Fire hazard	Explosive limits 6-32%. As the gas is supplied in a cylinder, turning off the valve will reduce any fire involving it, if possible, cylinders should be removed quickly from an area in which a fire has developed.
Spillage disposal	Surplus gas or leaking cylinder can be vented slowly in a water-fed scrubbing tower or column, or into a fume cupboard served by such a tower.

352. Cyanogen chloride

Colourless liquid or gas; m.p. -6.5 ˚C; b.p. 13.1 ˚C; soluble in water.

Limit values	OES short-term 0.3 p.p.m. (0.6 mg m^{-3}).
Toxic effects	Highly irritant to the upper respiratory tract with high concentrations being quickly fatal.
Hazardous reactions	Crude cyanogen chloride prepared from hydrogen cyanide and chlorine may trimerize violently to cyanuric chloride.
First aid	Standard treatment for exposure by all routes (see pages 108-122), with special reference to cyanides.

353. Cyanogen *N,N*-dioxide

Hazardous reactions	Decomposes at -45 ˚C under vacuum before exploding.

354. Cyclohexadiene

Colourless liquid; m.p. -98 ˚C; b.p. 79 ˚C; insoluble in water.

Toxic effects	Harmful by inhalation, ingestion, and skin absorption. Irritant to skin and eyes.
Hazardous reactions	Slowly forms explosive peroxide on contact with air.
First aid	Standard treatment for exposure by all routes (see pages 108-122).
Fire hazard	Flash point 26 ˚C. Use carbon dioxide, dry powder, or alcohol or polymer foam to extinguish fires.
Spillage disposal	See general section.

355. Cyclohexane

Colourless, mobile liquid with a sweetish odour when pure but a pungent odour when impure; m.p. 6.5 °C; b.p. 81 °C; very slightly soluble in water.

RISKS
Highly flammable (R11)

SAFETY PRECAUTIONS
Keep container in a well ventilated place – Keep away from sources of ignition – No Smoking – Take precautionary measures against static discharges (S9, S16, S33)

Limit values	OES short-term 300 p.p.m. (1030 mg m^{-3}) proposed change; long-term 100 p.p.m. (340 mg m^{-3}) proposed change.
Toxic effects	Irritates the eyes, skin, and respiratory system; the inhalation of high concentrations may cause narcosis. Assumed to be irritant and narcotic if taken by mouth.
Hazardous reactions	Addition of liquid dinitrogen tetraoxide to hot cyclohexane caused explosion.
First aid	Standard treatment for exposure by all routes (see pages 108-122).
Fire hazard	Flash point -18 °C; Explosive limits 1.3-8%; Autoignition temperature 260 °C. Extinguish fire with foam, dry powder, or carbon dioxide.
Spillage disposal	See general section.

RSC *Chemical Safety Data Sheets* Vol. 1, No. 16, 1988 gives extended coverage.

356. Cyclohexane-1,2-dione

Bright yellow solid; m.p. 35-38 °C.

Toxic effects	Irritant. May be harmful by all routes of exposure.
Hazardous reactions	Explosion when being prepared by oxidation of cyclohexanol with nitric acid.
First aid	Standard treatment for exposure by all routes (see pages 108-122).
Fire hazard	Flash point 84 °C. Use water spray, carbon dioxide, dry powder, or alcohol or polymer foam to extinguish fires.
Spillage disposal	See general section.

357. Cyclohexanol

Colourless hygroscopic crystals or viscous liquid with camphor or menthol-like odour; m.p. 24 °C; b.p. 161 °C; fairly soluble in water.

RISKS
Harmful by inhalation and if swallowed – Irritating to respiratory system and skin (R20/22, R37/38)

SAFETY PRECAUTIONS
Avoid contact with skin and eyes (S24/25)

Limit values	OES long-term 50 p.p.m. (200 mg m^{-3}).
Toxic effects	The vapour may irritate the eyes, skin, and respiratory system. The liquid irritates the eyes and may cause conjunctivitis and more serious damage. The liquid irritates the skin and major absorption may lead to tremors and kidney or liver damage. Irritant and damaging to the alimentary system if taken by mouth.
Hazardous reactions	Explosion with nitric acid.
First aid	Standard treatment for exposure by all routes (see pages 108-122).
Fire hazard	Flash point 68 °C (closed cup); Explosive limits lel 2.4%; Autoignition temperature 300 °C. Extinguish fires with dry powder, alcohol-resistant foam, or carbon dioxide.
Spillage disposal	See general section.

RSC *Chemical Safety Data Sheets* Vol. 1, No. 17, 1988 gives extended coverage.

358. Cyclohexanone

Colourless oily liquid with odour somewhat similar to that of acetone; m.p. -16.4 °C; b.p. 156 °C; soluble in water.

RISKS
Flammable – Harmful by inhalation (R10, R20)

SAFETY PRECAUTIONS
Avoid contact with eyes (S25)

Limit values	OES short-term 100 p.p.m. (400 mg m^{-3}); long term 25 p.p.m. (100 mg m^{-3}).
Toxic effects	The vapour may irritate the eyes, skin, and respiratory system. The liquid irritates the eyes and may cause corneal damage. The liquid may irritate the skin. Irritant and damaging to the alimentary system if taken by mouth.
Hazardous reactions	Forms explosive peroxide with hydrogen peroxide; may explode when added to nitric acid at about 75 °C.
First aid	Standard treatment for exposure by all routes (see pages 108-122).
Fire hazard	Flash point 44 °C; Explosive limits 1.1-9.4%; Autoignition temperature 420 °C. Extinguish fire with water spray, foam, dry powder, or carbon dioxide.
Spillage disposal	See general section.

RSC *Chemical Safety Data Sheets* Vol. 1, No. 18, 1988 gives extended coverage.

359. Cyclohexanone oxime

Crystalline solid; m.p. 89 °C; b.p. 204 °C with slight decomposition; soluble in water.

Hazardous reactions	Beckmann rearrangement with oleum to give caprolactam failed to start and led to explosion.
First aid	Standard treatment for exposure by all routes (see pages 108-122).

360. Cyclohexene

Colourless liquid; b.p. 83 °C; insoluble in water.

Limit values	OES long-term 300 p.p.m. (1015 mg m^{-3}).
Toxic effects	The vapour irritates the respiratory system and may irritate the eyes and skin. Assumed to be irritant if taken by mouth.
First aid	Standard treatment for exposure by all routes (see pages 108-122).
Fire hazard	Flash point -60 °C. Extinguish fire with foam, dry powder, or carbon dioxide.
Spillage disposal	See general section.

361. Cyclohexylamine

Colourless liquid with an ammoniacal, fishy odour; m.p. -17.7 °C; b.p. 134 °C; soluble in water.

RISKS
Flammable – Harmful in contact with skin and if swallowed – Causes burns (R10, R21/22, R34)

SAFETY PRECAUTIONS
Wear suitable protective clothing, gloves, and eye/face protection (S36/37/39)

Limit values	OES long-term 10 p.p.m. (40 mg m^{-3}).
Toxic effects	The vapour may irritate the eyes and respiratory system, causing difficulty in breathing. The liquid can burn the eyes and skin; absorption through the skin may cause nausea and vomiting. Assumed to be poisonous if taken by mouth.
First aid	Standard treatment for exposure by all routes (see pages 108-122).
Fire hazard	Flash point 31 °C; Explosive limits 1.5-9.4%; Autoignition temperature 293 °C. Extinguish fire with foam, dry powder, or carbon dioxide. Emits toxic fumes when heated to decomposition.
Spillage disposal	See general section.

RSC *Chemical Safety Data Sheets* Vol. 3, No. 17, 1990 gives extended coverage.

362. Cyclopenta-1,3-diene

Colourless liquid; m.p. -85 °C; b.p. 42 °C; insoluble in water.

Toxic effects   The main toxic effects arise from inhalation which leads to depression of central nervous system, liver damage, and a benzene-like effect upon the blood and narcosis. Chronic effects include headache, abdominal pains, jaundice, and anaemia. An acute effect may be narcosis.

Hazardous reactions   Dimerization is exothermic and may cause rupture of closed, uncooled containers; heat sensitive explosive peroxides formed on exposure to oxygen.

First aid   Standard treatment for exposure by all routes (see pages 108-122).

Fire hazard   Flash point 25 °C.

Spillage disposal   See general section.

363. Cyclopentadienylsodium

Hazardous reactions   Pyrophoric in air.

364. Cyclopentane

Colourless, mobile liquid; b.p. 49.3 °C; insoluble in water.

RISKS
Highly flammable (R11)

SAFETY PRECAUTIONS
Keep container in a well ventilated place – Keep away from sources of ignition – No Smoking – Do not empty into drains – Take precautionary measures against static discharges (S9, S16, S29, S33)

Toxic effects   May act as mild narcotic in high concentrations.

First aid   Standard treatment for exposure by all routes (see pages 108-122).

Fire hazard   Flash point -6.7 °C. Extinguish fire with foam, dry powder, or carbon dioxide.

Spillage disposal   See general section.

365. Cyclopentanone

Colourless, oily liquid; m.p. -51 °C; b.p. 131 °C.

RISKS
Flammable – Irritating to eyes and skin (R10, R36/38)

SAFETY PRECAUTIONS
Do not breathe our (S23)

Toxic effects   May be harmful by all routes of exposure.

| **Hazardous reactions** | Mixtures with nitric acid and hydrogen peroxide react vigorously and may become explosive. |

First aid   Standard treatment for exposure by all routes (see pages 108-122).

Fire hazard   Flash point 30 ˚C. Use carbon dioxide, dry powder, or alcohol or polymer foam to extinguish fires.

Spillage disposal   See general section.

366. Cyclopentanone oxime

Off-white, crystalline powder; m.p. 53-55 ˚C.

Toxic effects   May be harmful by all routes of exposure.

Hazardous reactions   Two instances of violet Beckmann rearrangement.

First aid   Standard treatment for exposure by all routes (see pages 108-122).

Fire hazard   Flash point 92 ˚C. Use water spray, carbon dioxide, dry powder, or alcohol or polymer foam to extinguish fires.

Spillage disposal   See general section.

367. Cyclopropane

Colourless gas with smell like that of petroleum spirit; b.p. -33 ˚C; fairly soluble in water.

RISKS
Extremely flammable liquefied gas (R13)

SAFETY PRECAUTIONS
Keep container in a well ventilated place – Keep away from sources of ignition – No Smoking – Take precautionary measures against static discharges (S9, S16, S33)

Toxic effects   The gas is an anaesthetic and is employed for this purpose.

First aid   Remove from exposure, rest, and keep warm.

Fire hazard   Explosive limits 2.4-10.4%; Autoignition temperature 498 ˚C. As the gas is supplied in a cylinder, turning off the valve will reduce any fire involving it; if possible, cylinders should be removed quickly from an area in which a fire has developed.

Spillage disposal   Surplus gas or a leaking cylinder can be vented slowly to air in a safe open area or burnt-off in a suitable gas burner.

368. *p*-Cymene

Colourless liquid; m.p. -68 ˚C; b.p. 177 ˚C; insoluble in water.

Toxic effects   Irritant to skin and eyes. Swiftly absorbed through the skin. Toxic effects to the blood and nervous system have been observed following inhalation and ingestion.

First aid   Standard treatment for exposure by all routes (see pages 108-122).

Fire hazard   Flash point 47 ˚C; Explosive limits 0.7-5.6%; Autoignition temperature 436 ˚C. Extinguish fires with carbon dioxide, dry chemical powder, or foam.

Spillage disposal   See general section.

RSC *Chemical Safety Data Sheets* Vol. 1, No. 19, 1988 gives extended coverage.

369. Decaborane

White powder; m.p. 99-100 ˚C.

Toxic effects   Irritant to upper respiratory tract and mucous membranes. If inhaled, ingested, or absorbed through the skin can cause nausea, dizziness, headache, damage to liver and kidneys, and can be fatal.

Hazardous reactions   Forms impact-sensitive mixtures with ethers and halocarbons; ignites in oxygen at 100 ˚C.

First aid   Standard treatment for exposure by all routes (see pages 108-122).

Fire hazard   Extinguish fires with carbon dioxide or dry chemical powder extinguishers.

Spillage disposal   See general section.

370. Decalin

Liquids; m.p. (*cis-*) -43 ˚C; (*trans-*) -31.5 ˚C; b.p. (*cis-*) 194.6 ˚C; (*trans-*) 185.5 ˚C; insoluble in water.

Toxic effects   Irritant to mucous membranes. Repeated or prolonged skin contact can cause dermatitis. On inhalation it has anaesthetic and central nervous system depressant activity.

First aid   Standard treatment for exposure by all routes (see pages 108-122).

Fire hazard   Flash point 58 ˚C.

371. Diacetyl peroxide

Crystalline solid with very pungent odour; m.p. 30 ˚C; aqueous solution slowly evolves oxygen.

Toxic effects   Irritant to skin, eyes, and upper respiratory tract.

Hazardous reactions   When dry is a shock-sensitive explosive; use in ethereal solution.

First aid   Standard treatment for exposure by all routes (see pages 108-122).

Fire hazard   Emits acrid fumes on heating to decomposition. Extinguish fires with water spray.

Spillage disposal   See general section.

372. Diallyl ether

Liquid with odour of radishes; b.p. 94.3 ˚C.

Toxic effects	Poisonous by ingestion.
Hazardous reactions	Hazardous preparation, peroxidizes readily in air and sunlight to explosive peroxide.
Fire hazard	Flash point -6 ˚C (open cup). Use alcohol foam to extinguish fires. Emits acrid fumes when heated to decomposition.

373. Diallyl phosphite

Hazardous reactions	Liable to explode during distillation.

374. Diallyl phthalate

Colourless, oily liquid; m.p. -70 ˚C; b.p. 157 ˚C; insoluble in water.

RISKS
Harmful if swallowed (R22)

SAFETY PRECAUTIONS
Avoid contact with skin and eyes (S24/25)

Limit values	OES long-term 5 mg m^{-3}.
Toxic effects	The liquid irritates the eyes and skin and can cause internal disorders by continued skin absorption. Assumed to be harmful if taken by mouth.
First aid	Standard treatment for exposure by all routes (see pages 108-122).
Fire hazard	Flash point 110 ˚C. Use water spray, carbon dioxide, dry chemical powder, or alcohol or polymer foam to extinguish fires.
Spillage disposal	See general section.

375. Diallyl sulfate

Unpleasant smelling liquid.

Hazardous reactions	Exploded during distillation.

376. 1,2-Diaminoethane

Clear, colourless liquid with ammoniacal odour; b.p. 117 ˚C; miscible with water.

RISKS
Flammable – Harmful in contact with skin and if swallowed – Causes burns – May cause sensitization by skin contact (R10, R21/22, R34, R43)

SAFETY PRECAUTIONS
Keep container in a well ventilated place – In case of contact with eyes, rinse immediately with plenty of water and seek medical advice – Wear suitable protective clothing, gloves, and eye/face protection (S9, S26, S36/37/39)

Limit values	OES long-term 10 p.p.m. (25 mg m^{-3}).
Toxic effects	The vapour irritates the respiratory system. The liquid and vapour cause irritation of skin and eyes. Repeated inhalation of vapour or skin contact may cause sensitization of skin or respiratory system. If swallowed, may cause digestive disturbance and possible damage to the kidneys.
Hazardous reactions	May ignite on contact with cellulose nitrate; dangerous reactions with nitromethane and di-isopropyl peroxydicarbonate.
First aid	Standard treatment for exposure by all routes (see pages 108-122).
Fire hazard	Flash point 43 ˚C (open cup); Explosive limits 4.2-14.4%; Autoignition temperature 385 ˚C. Extinguish fire with alcohol-resistant foam, dry powder, or carbon dioxide.
Spillage disposal	See general section.

RSC *Chemical Safety Data Sheets* Vol. 1, No. 44, 1988 and Vol. 3, No. 24, 1990 give extended coverage.

377. Diaminohexane

Colourless leaflets; m.p. 42 ˚C; b.p. 205 ˚C; freely soluble in water.

Toxic effects	Strongly irritant and corrosive to all tissues. Harmful by inhalation, ingestion, and skin absorption.
First aid	Standard treatment for exposure by all routes (see pages 108-122).
Fire hazard	Flash point 71 ˚C; Explosive limits 0.7-6.3%. Use water spray, carbon dioxide, dry chemical powder, or alcohol or polymer foam to extinguish fires.
Spillage disposal	See general section.

378. Dianilinium dichromate

Hazardous reactions	Unstable on storage.

379. *o*-Dianisidine and its dihydrochloride

o-Dianisidine is a colourless to grey-mauve powder. Its dihydrochloride is a colourless to grey powder; m.p. 131.5 ˚C; the free base is insoluble in water;the dihydrochloride is sparingly soluble.

RISKS
May cause cancer – Harmful if swallowed (R45, R22)

SAFETY PRECAUTIONS
Avoid exposure-obtain special instructions before use – If you feel unwell, seek medical advice (show label where possible) (S53, S44)

Toxic effects	The dust irritates the nose severely, causing sneezing. Solutions of the dihydrochloride irritate the eyes. The effects of ingestion are not recorded. There is evidence that *o*-dianisidine, through continued absorption, can cause cancer of the bladder. The use of *o*-dianisidine and its salts is controlled in the United Kindgom by The Carcinogenic Substances Regulations 1967.
Hazardous reactions	Standard treatment for exposure by all routes (see pages 108-122)
Fire hazard	Flash point 206 ˚C. Use water spray, carbon dioxide, dry chemical powder, or alcohol or polymer foam to extinguish fires.
Spillage disposal	See general section.

380. Diazoacetonitrile

Hazardous reactions	Liable to explode through friction.

381. Diazocyclohexanone

Hazardous reactions	May explode on heating.

382. Diazocyclopentadiene

Yellow liquid; m.p. -23 ˚C.

Hazardous reactions	Exploded violently during vacuum distillation.

383. Diazoindene

Dark red oil or orange crystals; m.p. 23-25 ˚C.

Hazardous reactions   Distillation residue may explode.

384. Diazomalononitrile

Hazardous reactions   May explode at 75 ˚C.

First aid   Standard treatment for exposure by all routes (see pages 108-122), with special reference to cyanides.

385. Diazomethane

A yellow gas generally employed for organic synthesis in chloroform or ethereal solution and prepared as part of the process; m.p. -145 ˚C; b.p. -23 ˚C.

Limit values   OES long-term 0.2 p.p.m. (0.4 mg m^{-3}) under review.

Toxic effects   Inhalation may result in chest discomfort, headache, weakness, and, in severe cases, collapse. Experimental carcinogen.

Hazardous reactions   Gaseous diazomethane may explode on ground glass surfaces and when heated to about 100 ˚C; concentrated solutions may also explode especially if impurities are present; explosions occur on contact with alkali metals and the exothermic reaction with calcium sulfate is also dangerous.

First aid   Standard treatment for exposure by all routes (see pages 108-122).

386. Diazophenylacetophenone

Hazardous reactions   May explode if heated above 40 ˚C.

387. Dibenzoyl peroxide

White, granular crystals, normally supplied moistened with about 30% water; m.p. 103-105 ˚C; insoluble in water.

RISKS
Extreme risk of explosion by shock, friction, fire, or other sources of ignition – Irritating to eyes, respiratory system, and skin (R3, R36/37/38)

SAFETY PRECAUTIONS
Keep container tightly closed, in a cool well ventilated place – Keep away from heat – Take off immediately all contaminated clothing – Avoid shock and friction – Wear suitable gloves and eye/face protection (S3/7/9, S14, S27, S34, S37/39)

Limit values   OES long-term 5 mg m^{-3}.

Toxic effects	Irritant to skin, eyes, and upper respiratory tract.
Hazardous reactions	Dry material burns and is sensitive to heat (explodes above mp), shock, friction or contact with combustible materials; has exploded on heating with ethylene and carbon tetrachloride under pressure; explosions have resulted from contact with dimethylaniline or dimethyl sulfide; ignition occurs with methyl methacrylate; explosion with aniline; recrystallization from hot chloroform leads to explosions.
First aid	Standard treatment for exposure by all routes (see pages 108-122).
Fire hazard	Extinguish fires with water spray.
Spillage disposal	See general section. Large quantities may need specialist advice.

388. Dibenzylamine

Colourless, oily liquid with ammoniacal odour; b.p. 300 ˚C with decomposition; immiscible with water.

SAFETY PRECAUTIONS
Avoid contact with skin and eyes (S24/25)

Toxic effects	The liquid irritates and burns the skin and eyes. Corrosive and highly irritant if taken by mouth.
First aid	Standard treatment for exposure by all routes (see pages 108-122).
Spillage disposal	See general section.

389. Dibenzyl ether

Liquid; b.p. 289 ˚C; insoluble in water.

Hazardous reactions	Exploded with aluminium dichloride hydride diethyl etherate.
Fire hazard	Flash point 135 ˚C. Extinguish fires with water spray – frothing may occur.

390. Dibenzyl phosphite

Pale yellow liquid; b.p. 110-120 ˚C at 0.01 mm Hg.

Toxic effects	Toxicological properties not thoroughly investigated but may be harmful on inhalation, ingestion, or skin absorption.
Hazardous reactions	Decomposes at 160 ˚C.
First aid	Standard treatment for exposure by all routes (see pages 108-122).
Fire hazard	Flash point above 110 ˚C. Use water spray, carbon dioxide, dry chemical powder, or alcohol or polymer foam to extinguish fires.
Spillage disposal	See general section.

391. Diborane

Colourless gas with sickly sweet odour; m.p. -165 °C; b.p. -92.5 °C; decomposed by water.

Limit values   OES long-term 0.1 p.p.m. (0.1 mg m^{-3}).

Toxic effects   Inhalation can cause pulmonary oedema, tightness of the chest, chills, fever, pericardial pain, nausea, shivering, and drowsiness.

Hazardous   Usually ignites in air and delayed ignition may be followed by violent
reactions   explosions; reacts explosively with chlorine and forms explosive compound with dimethylsulfoxide; reacts violently with halocarbon liquids; explosive reaction with tetravinyllead, and with nitrogen trifluoride in liquid phase.

First aid   Standard treatment for exposure by all routes (see pages 108-122).

Fire hazard   Flash point -90 °C; Autoignition temperature 38-52 °C. Use water spray to keep cylinders cool or remove them from the vacinity.

392. Diboron tetrachloride

Liquid; m.p. -92.6 °C; b.p. 65.5 °C.

Hazardous   Exposure to air may cause explosion.
reactions

393. Diboron tetrafluoride

Hazardous   Extremely explosive in presence of oxygen; violent reaction or ignition with
reactions   metal oxides (B68-69).

394. Dibromoacetylene

Heavy liquid with unpleasant odour; m.p. -25 to -23 °C; b.p. about 76 °C; insoluble in water.

Hazardous   Ignites in air and explodes on heating.
reactions

395. Dibromobenzoquinone chloroimide

Hazardous   Liable to decompose explosively on heating.
reactions

396. Dibromodifluoromethane

Colourless liquid; m.p. -142 ˚C; b.p. 22-23 ˚C.

Limit values   OES short-term 150 p.p.m. (1290 mg m^{-3}); long-term 100 p.p.m. (860 mg m^{-3}).

Toxic effects   Inhalation can cause lung irritation, chest pain, and oedema which can prove fatal. Prolonged exposure by any route can cause liver damage.

First aid   Standard treatment for exposure by all routes (see pages 108-122).

Fire hazard   Extinguish fires with water spray, carbon dioxide, dry chemical powder, or alcohol or polymer foam.

Spillage disposal   See general section.

397. 1,2-Dibromoethane

Colourless liquid with sweetish chloroform-like odour, turns brown on exposure to light; b.p. 131 ˚C; slightly soluble in water.

RISKS
May cause cancer – Toxic by inhalation, in contact with skin, and if swallowed – Irritating to eyes, respiratory system and skin (R45, R23/24/25, R36/37/38)

SAFETY PRECAUTIONS
Avoid exposure – Obtain special instructions before use – If you feel unwell, seek medical advice (show label where possible) (S53, S44)

Limit values   MEL long-term 0.5 p.p.m. (4 mg m^{-3}).

Toxic effects   The vapour irritates the respiratory system and may have narcotic action. The vapour and liquid irritate the eyes. The liquid irritates the skin and may cause dermatitis. Poisonous by skin absorption and ingestion, effects being nausea, vomiting, pain, and jaundice resulting from liver and kidney damage. Suspected carcinogen.

Hazardous reactions   Reaction with magnesium may become violent. Highly reactive with many other substances.

First aid   Standard treatment for exposure by all routes (see pages 108-122).

Fire hazard   Extinguish fires with water spray, carbon dioxide, dry chemical powder, or alcohol or polymer foam.

Spillage disposal   See general section.

RSC *Chemical Safety Data Sheets* Vol. 4a, No. 71, 1991 gives extended coverage.

398. Dibromomethane

Colourless liquid; m.p. -52 °C; b.p. 96-98 °C.

RISKS
Harmful by inhalation (R20)

SAFETY PRECAUTIONS
Avoid contact with skin (S24)

Hazardous reactions	Forms shock-sensitive explosive with potassium.
First aid	Standard treatment for exposure by all routes (see pages 108-122).
Fire hazard	Use water spray, carbon dioxide, dry chemical powder, or alcohol or polymer foam to extinguish fires.
Spillage disposal	See general section.

399. 2-Dibutylaminoethanol

Clear colourless liquid; b.p. 230 °C; slightly soluble in water.

Toxic effects	Severely irritant to skin and eyes.
First aid	Standard treatment for exposure by all routes (see pages 108-122).
Fire hazard	Flash point 104 °C. Extinguish fires with carbon dioxide or dry chemical powder.
Spillage disposal	See general section.

400. Di-t-butyl chromate

Red, crystalline solid; m.p. -5 °C to 0 °C; hydrolysed by water.

Limit values	OES long-term 0.5 mg m^{-3} (as Cr(II)).
Hazardous reactions	Addition of t-butanol to chromium trioxide resulted in explosion, explosive oxidation of valencene.

401. 2,6-Di-t-butyl-*p*-cresol

Slightly yellow, crystalline solid; m.p. 70 °C; b.p. 265 °C.

Toxic effects	Irritant to skin, eyes, upper respiratory tract, and mucous membranes. Harmful by inhalation, ingestion, or skin absorption.
First aid	Standard treatment for exposure by all routes (see pages 108-122).

Fire hazard   Flash point 127 ˚C; Autoignition temperature 470 ˚C. Extinguish fires with carbon dioxide, dry chemical powder, or alcohol or polymer foam.

Spillage   See general section.
   disposal

402. Dibutyl ether

Colourless liquid; b.p. 142 ˚C; insoluble in water.

RISKS
Flammable – Irritating to eyes, respiratory system, and skin (R10, R36/37/38)

Toxic effects   The liquid irritates the eyes and is considered to present a slight hazard by skin absorption.

Hazardous   On exposure to light and air may form explosive peroxides which should be
   reactions   decomposed before distillation to small volume.

First aid   Standard treatment for exposure by all routes (see pages 108-122).

Fire hazard   Flash point 25 ˚C; Explosive limits 1.5-7.6%; Autoignition temperature 194 ˚C. Extinguish fire with dry powder or carbon dioxide.

Spillage   See general section.
   disposal

403. Dibutyl hydrogen phosphate

Colourless to brown odourless liquid; insoluble in water.

Limit values   OES short-term 2 p.p.m. (10 mg m^{-3}); long-term 1 p.p.m. (5 mg m^{-3}).

Toxic effects Irritant to skin, eyes, and upper respiratory tract. Inhalation can cause cough, shortness of breath and headache.

404. Di-t-butylnitrophenol

Yellow, crystalline solid; m.p. 157.5 ˚C with decomposition.

Hazardous   Exploded on heating to 100 ˚C.
   reactions

405. Di-t-butyl peroxide

Liquid; m.p. -40 °C; sparingly soluble in water.

RISKS
Highly flammable – Irritating to respiratory system and skin (R11, R37/38)

SAFETY PRECAUTIONS
Keep container tightly closed, in a cool well ventilated place – Keep away from heat – Take off immediately all contaminated clothing – Wear suitable gloves and eye/face protection (S3/7/9, S14, S27, S37/39)

Toxic effects	Irritant to skin, eyes, and upper respiratory tract.
Hazardous reactions	Addition of t-butanol to mixtures of hydrogen peroxide and sulfuric acid may result in explosions. The peroxide is thermally unstable.
First aid	Standard treatment for exposure by all routes (see pages 108-122).
Fire hazard	Extinguish fires with water spray.
Spillage disposal	See general section.

406. Dibutyl phthalate

Viscous, colourless liquid; m.p. -35 °C; b.p. 340 °C; very slightly soluble in water.

Limit values	OES short-term 10 mg m^{-3}; long-term 5 mg m^{-3}.
Toxic effects	Irritating to skin, eyes, and upper respiratory tract. Exposure can cause nausea, dizziness, and headaches.
Hazardous reactions	Mixture with liquid chlorine in S.S. bomb reacted explosively at 118 °C.
First aid	Standard treatment for exposure by all routes (see pages 108-122).
Fire hazard	Flash point 171 °C; Autoignition temperature 399 °C. Use carbon dioxide or dry chemical powder to extinguish fires. Water spray and foam are effective but may cause frothing.
Spillage disposal	See general section.

407. Dichlorine oxide

Liquid.

Hazardous reactions	The liquid explodes on pouring, the gas on heating or sparking; explodes with many oxidizable materials.

408. Dichlorine trioxide

Gas.

Hazardous   Vapour explodes well below 0 ˚C.
  reactions

409. Dichloroacetic acid

Colourless liquid with pungent odour; b.p. 194 ˚C; miscible with water.

RISKS
Causes severe burns (R35)

SAFETY PRECAUTIONS
In case of contact with eyes, rinse immediately with plenty of water and seek medical advice
(S26)

Toxic effects   The vapour irritates the eyes and respiratory system. The liquid burns the
                      skin and eyes. Assumed to cause severe irritation and damage if taken by
                      mouth.

First aid   Standard treatment for exposure by all routes (see pages 108-122).

Spillage   See general section.
  disposal

410. Dichloroacetyl chloride

Colourless, fuming liquid with acrid penetrating odour; b.p. 107 ˚C; reacts with water forming
dichloroacetic and hydrochloric acids.

RISKS
Causes severe burns (R35)

SAFETY PRECAUTIONS
Keep container in a well ventilated place – In case of contact with eyes, rinse immediately with
plenty of water and seek medical advice (S9, S26)

Toxic effects   The vapour irritates the eyes and respiratory system. The liquid burns the
                      skin and eyes. Assumed to cause severe irritation and damage if taken by
                      mouth.

First aid   Standard treatment for exposure by all routes (see pages 108-122).

Spillage   See general section.
  disposal

411. Dichloroacetylene

Volatile liquid; m.p. -66 ˚C to -64 ˚C; b.p. 33 ˚C.

Limit values   OES short-term 0.1 p.p.m. (0.4 mg m^{-3}).

Toxic effects   Toxic by inhalation affecting the CNS. Symptoms also include disabling nausea and intense jaw pain. Experimental carcinogen.

Hazardous reactions   Heat-sensitive explosive gas which ignites in contact with air.

First aid   Standard treatment for exposure by all routes (see pages 108-122).

412. 3,4-Dichloroaniline

White crystals; m.p. 71-72 ˚C; b.p. 272 ˚C; practically insoluble in water.

RISKS
Toxic by inhalation, in contact with skin, and if swallowed – Danger of cumulative effects (R23/24/25, R33)

SAFETY PRECAUTIONS
After contact with skin, wash immediately with plenty of water – Wear protective clothing and gloves – If you feel unwell, seek medical advice (show label where possible) (S28, S36/37, S44)

Toxic effects   Irritant to skin, eyes, and upper respiratory tract. May be fatal following inhalation due to inflammation, oedema, and chemical pneumonitis.

Hazardous reactions   Explosions have occurred during preparation.

First aid   Standard treatment for exposure by all routes (see pages 108-122).

Fire hazard   Flash point 166 ˚C; Explosive limits 2.8% (153 ˚C)-7.2% (179 ˚C); Autoignition temperature 509 ˚F. Extinguish fires with water spray, carbon dioxide, dry chemical powder, or alcohol or polymer foam.

Spillage disposal   See general section.

RSC *Chemical Safety Data Sheets* Vol. 4a, No. 49, 1991 gives extended coverage.

413. Dichlorobenzenes

o-Isomer is colourless liquid with a pleasant aromatic smell; *p*-isomer consists of colourless, volatile crystals with characteristic disinfectant smell; m.p. (*o*-) -17.5 ˚C; (*p*-) 53 ˚C; b.p. (*o*-) 180 ˚C; (*p*-) 174 ˚C; both isomers are insoluble in water.

RISKS
Harmful if swallowed (R22)

SAFETY PRECAUTIONS
Keep out of reach of children – Avoid contact with skin and eyes (S2, S24/25)

Limit values   OES short-term (1,2-) 50 p.p.m. (300 mg m^{-3}) (1,4-) 110 p.p.m. (675 mg m^{-3}); long-term (1,2-) 75 p.p.m. (450 mg m^{-3}).

Toxic effects	Inhalation of vapours may cause drowsiness and irritation of nose; both isomers irritate the eyes; the *o*-isomer is more irritating to the skin and may cause dermatitis; long exposure to either isomer may result in damage to liver and kidneys.
First aid	Standard treatment for exposure by all routes (see pages 108-122).
Fire hazard	Flash point (*o*-) 65 ˚C; Explosive limits (*o*-) 2.2-9.2%; Autoignition temperature (*o*-) 648 ˚C. Extinguish fires with dry chemical, foam, or carbon dioxide.
Spillage disposal	See general section.

RSC *Chemical Safety Data Sheets* Vol. 1, No. 21, 1988 gives extended coverage.

414. Dichlorobenzidine and dihydrochloride

Dichlorobenzidine and the hydrochloride are colourless or pale crystalline solids; dichlorobenzidine is scarcely soluble in water and the hydrochloride slightly soluble.

RISKS
May cause cancer – Harmful in contact with skin – May cause sensitization by skin contact (R45, R21, R43)

SAFETY PRECAUTIONS
Avoid exposure – Obtain special instructions before use – If you feel unwell, seek medical advice (show label where possible) (S53, S44)

Toxic effects	The dust irritates the eyes and skin and may lead to allergic reactions. There is evidence that 3, 3'-dichlorobenzidine through continued absorption may cause bladder tumours.
First aid	Standard treatment for exposure by all routes (see pages 108-122).
Spillage disposal	See general section.

415. 1,4-Dichlorobut-2-yne

Yellow liquid; b.p. 165-168 ˚C.

Toxic effects	Causes severe irritation and destruction of tissues of skin, eyes, upper respiratory tract, and mucous membranes. Harmful if inhaled, ingested, or absorbed through the skin.
Hazardous reactions	A preferred method using dichloromethane as diluent is indicated; a distillation residue exploded.
First aid	Standard treatment for exposure by all routes (see pages 108-122).
Fire hazard	Flash point 160 ˚C. Use water spray, carbon dioxide, dry chemical powder, or alcohol or polymer foam to extinguish fires.
Spillage disposal	See general section.

416. Dichlorodifluoromethane

Colourless, odourless gas; m.p. -160 ˚C; b.p. -28 ˚C.

Limit values	OES short-term 1250 p.p.m. (6200 mg m^{-3}); long-term 1000 p.p.m. (4950 mg m^{-3}).
Toxic effects	Low general toxicity but exposure to high concentrations can cause disorientation, nausea, vomiting, narcosis, cardiac dysrhythmias, hypotension, and death.
Hazardous reactions	May react exothermically with aluminium. Magnesium dust ignited at 400 ˚C in vapour and suspension exploded violently on sparking.
First aid	Standard treatment for exposure by all routes (see pages 108-122).
Fire hazard	Use water spray to keep cylinder cool or move it out of contact with fire.
Spillage disposal	See general section.

417. 1,3-Dichloro-5,5-dimethylimidazolidindione

Colourless powder; m.p. 132 ˚C; b.p. 212 ˚C with decomposition; sparingly soluble in water.

Limit values	OES short-term 0.4 mg m^{-3}; long-term 0.2 mg m^{-3}.
Toxic effects	Irritant to skin, eyes, upper respiratory tract, and mucous membranes.
Hazardous reactions	Violent explosion with xylene.
First aid	Standard treatment for exposure by all routes (see pages 108-122).
Fire hazard	Use carbon dioxide or dry chemical powder to extinguish fires.
Spillage disposal	See general section.

418. 1,1-Dichloroethane

Colourless liquid with a chloroform-like odour; m.p. -97 ˚C; b.p. 57.3 ˚C; slightly soluble in water.

RISKS
Extremely flammable – Harmful by inhalation (R12, R20)

SAFETY PRECAUTIONS
Keep container tightly closed – Keep away from sources of ignition – No Smoking – Do not empty into drains – Take precautionary measures against static discharges (S7, S16, S29, S33)

Limit values	OES short-term 400 p.p.m. (1620 mg m^{-3}); long-term 200 p.p.m. (810 mg m^{-3}).
Toxic effects	In high concentrations, the vapour irritates the eyes and respiratory system; it may also cause drowsiness, headache, vomiting, and mental confusion.

Continued exposure to low concentrations may result in dizziness, nausea, and abdominal pain, and there may be damage to the eyes and liver. The liquid may cause serious damage to the eyes. Poisonous if taken by mouth. The liquid irritates the skin. Dermatitis may follow repeated contact.

Hazardous reactions   Mixtures with dinitrogen tetroxide or potassium are explosive when subjected to shock; reaction with aluminium powder may be violent or explosive; mixtures with nitric acid easily detonate.

First aid   Standard treatment for exposure by all routes (see pages 108-122).

Fire hazard   Flash point 14 °C (open cup); Explosive limits 5.6-11.4%; Autoignition temperature 493 °C. Extinguish fire with water spray, foam, dry powder, or carbon dioxide.

Spillage disposal   See general section.

RSC *Chemical Safety Data Sheets* Vol. 1, No. 23, 1988 gives extended coverage.

419. 1,2-Dichloroethane

Colourless liquid with a chloroform-like odour; m.p. 0 °C; b.p. 84 °C; slightly soluble in water.

RISKS
May cause cancer – Highly flammable – Harmful if swallowed – Irritating to eyes, respiratory system and skin (R45, R11, R22, R36/37/38)

SAFETY PRECAUTIONS
Avoid exposure-obtain special instructions before use – Keep away from sources of ignition – No Smoking – Do not empty into drains – If you feel unwell, seek medical advice (show label where possible) (S53, S16, S29, S44)

Limit values   OES short-term 15 p.p.m. (60 mg m^{-3}) under review; long-term 10 p.p.m. (40 mg m^{-3}) under review.

Toxic effects   In high concentrations, the vapour irritates the eyes and respiratory system; it may also cause drowsiness, headache, vomiting, and mental confusion. Continued exposure to low concentrations may result in dizziness, nausea, and abdominal pain, and there may be damage to the eyes and liver. The liquid may cause serious damage to the eyes. Poisonous if taken by mouth. The liquid irritates the skin. Dermatitis may follow repeated contact.

Hazardous reactions   Mixtures with dinitrogen tetroxide or potassium are explosive when subjected to shock; reaction with aluminium powder may be violent or explosive; mixtures with nitric acid easily detonate.

First aid   Standard treatment for exposure by all routes (see pages 108-122).

Fire hazard   Flash point 13 °C; Explosive limits 6.2-15.9%; Autoignition temperature 413 °C. Extinguish fire with water spray, foam, dry powder, or carbon dioxide.

Spillage disposal   See general section.

RSC *Chemical Safety Data Sheets* Vol. 1, No. 23, 1988 gives extended coverage.

420. 2,2-Dichloroethylamine

Liquid.

Hazardous reactions   Violent explosion when ethereal solution was being evaporated under vacuum.

421. 1,1-Dichloroethylene

Colourless, volatile liquid with chloroform-like odour; b.p. 32 ˚C; almost insoluble in water.

RISKS
Extremely flammable – Harmful by inhalation – Possible risk of irreversible effects (R12, R20, R40)

SAFETY PRECAUTIONS
Keep container tightly closed – Keep away from sources of ignition – No Smoking – Do not empty into drains (S7, S16, S29)

Limit values   MEL long-term 10 p.p.m. (40 mg m^{-3}).

Toxic effects   Inhalation of vapour may cause drowsiness and anaesthesia, repeated exposure may lead to liver and kidney damage. Liquid irritates skin and eyes, and must be assumed to be toxic on ingestion.

Hazardous reactions   Rapidly absorbs oxygen forming a violently explosive peroxide; reaction products with ozone are particularly dangerous; reaction under pressure with chlorotrifluoroethylene may develop into explosive polymerization.

First aid   Standard treatment for exposure by all routes (see pages 108-122).

Fire hazard   Flash point -15 ˚C (open cup); Explosive limits 5.6-11.4%; Autoignition temperature 458 ˚C. Extinguish fires with water spray, foam, dry powder, or carbon dioxide.

Spillage disposal   See general section.

422. 1,2-Dichloroethylene

Colourless, volatile liquid with a slight chloroform-like odour; m.p. -80.5 ˚C; b.p. 59 ˚C; slightly soluble in water.

RISKS
Highly flammable – Harmful by inhalation (R11, R20)

SAFETY PRECAUTIONS
Keep container tightly closed – Keep away from sources of ignition – No Smoking – Do not empty into drains (S7, S16, S29)

Limit values   OES short-term 250 p.p.m. (1000 mg m^{-3}); long-term 200 p.p.m. (790 mg m^{-3}).

Toxic effects   The vapour irritates the eyes and mucous membranes; in high concentrations it may cause drowsiness and unconsciousness. Continued exposure to low concentrations of vapour may cause drowsiness and digestive disturbance.

Hazardous reactions	Contact with solid caustic alkalies or their concentrated solutions will form chloroacetylene which ignites in air; forms explosive mixtures with dinitrogen tetroxide.
First aid	Standard treatment for exposure by all routes (see pages 108-122).
Fire hazard	Flash point 2 °C; Explosive limits 9.7-12.8%; Autoignition temperature 460 °C. Extinguish fire with foam, dry powder, or carbon dioxide.
Spillage disposal	See general section.

RSC *Chemical Safety Data Sheets* Vol. 1, No. 24, 1988 gives extended coverage.

423. Dichlorofluoromethane

Colourless gas; m.p. -135 °C; b.p. 8.9 °C; insoluble in water.

Limit values	OES long-term 10 p.p.m. (40 mg m^{-3}).
Toxic effects	Can cause rapid suffocation. Exposure to high concentrations of vapour may cause light headiness, nausea, vomiting, narcosis, cardiac dysrhythmias, and death.
First aid	Standard treatment for exposure by all routes (see pages 108-122).
Fire hazard	Use water spray to keep cylinder cool or move it out of the danger area.

424. 1,6-Dichloro-2,4-hexadiyne

Liquid.

Hazardous reactions	Extremely shock-sensitive explosive.

425. Dichloromethane

Colourless, volatile liquid with chloroform-like odour; m.p. -96.7 °C; b.p. 40 °C; immiscible with water.

RISKS
Harmful by inhalation (R20)

SAFETY PRECAUTIONS
Avoid contact with skin (S24)

Limit values	MEL short-term 250 p.p.m. (870 mg m^{-3}) under review; long-term 100 p.p.m. (350 mg m^{-3}).
Toxic effects	The vapour irritates the eyes and respiratory system and may cause headache and nausea; high concentrations may result in cyanosis and unconsciousness. The liquid irritates the eyes. Mildly toxic if taken by mouth.

Hazardous reactions	Solution of dinitrogen pentaoxide in dichloromethane liable to explode; violent explosive reactions with several other substances.
First aid	Standard treatment for exposure by all routes (see pages 108-122).
Fire hazard	Explosive limits 12-18%; Autoignition temperature 600 ˚C. Use water spray to keep containers cool. Toxic and irritant fumes evolved.
Spillage disposal	See general section.

RSC *Chemical Safety Data Sheets* Vol. 1, No. 26, 1988 gives extended coverage.

426. *N,N*-Dichloromethylamine

Yellow liquid; b.p. 59-60 ˚C.

Hazardous reactions	Exploded on warming with water or on distillation over calcium hypochlorite; exploded on contact with solid sodium sulfide.

427. 2,2'-Dichloro-4,4'-methylenedianiline (MOCA)

Light brown pellets; m.p. 99-107 ˚C; sparingly soluble in water.

RISKS
May cause cancer – Harmful if swallowed (R45, R22)

SAFETY PRECAUTIONS
Avoid exposure – Obtain special instructions before use – If you feel unwell, seek medical advice (show label where possible) (S53, S44)

Limit values	MEL long-term 0.005 mg m^{-3}.
Toxic effects	Irritant to skin, eyes, and upper respiratory tract. Can be absorbed through the skin. Can cause methaemoglobinaemia. Suspect carcinogen.
First aid	Standard treatment for exposure by all routes (see pages 108-122).
Fire hazard	Use extinguishant appropriate to the surroundings.
Spillage disposal	See general section.

RSC *Chemical Safety Data Sheets* Vol. 4a, No. 50, 1991 gives extended coverage.

428. 1,4-Dichloro-5-nitrobenzene

Colourless crystals; m.p. 33 ˚C; insoluble in water.

Toxic effects	These are not well documented but it is assumed that, in common with other substituted benzene compounds of this type, it is irritant to the eyes, skin, and

respiratory system in vapour form. The liquid or solid would irritate the skin, eyes, and respiratory system.

First aid   Standard treatment for exposure by all routes (see pages 108-122).

Spillage   See general section.
disposal

429. 1,1-Dichloro-1-nitroethane

Clear, colourless liquid; b.p. 124 °C; sparingly soluble in water.

RISKS
Toxic by inhalation, in contact with skin, and if swallowed (R23/24/25)

SAFETY PRECAUTIONS
In case of contact with eyes, rinse immediately with plenty of water and seek medical advice –
If you feel unwell, seek medical advice (show label where possible) (S26, S44)

Toxic effects   A strong irritant. Inhalation causes pulmonary oedema, congestion, and acute bronchitis. May also affect liver, kidneys, and circulatory system.

First aid   Standard treatment for exposure by all routes (see pages 108-122).

Fire hazard   Flash point 76 °C (open cup). Extinguish fires with water spray, carbon dioxide, or dry chemical powder.

Spillage   See general section.
disposal

RSC *Chemical Safety Data Sheets* Vol. 4a, No. 51, 1991 gives extended coverage.

430. 2,4-Dichlorophenol

Colourless crystals; m.p. 45 °C; b.p. 210 °C; almost insoluble in water.

RISKS
Harmful if swallowed – Irritating to eyes and skin (R22, R36/38)

SAFETY PRECAUTIONS
In case of contact with eyes, rinse immediately with plenty of water and seek medical advice –
After contact with skin, wash immediately with plenty of water (S26, S28)

Toxic effects   Causes severe irritation or burns in contact with the eyes and skin.

First aid   Standard treatment for exposure by all routes (see pages 108-122).

Fire hazard   Flash point 114 °C (open cup). Extinguish fires with water spray or alcohol foam.

Spillage   See general section.
disposal

431. 2,4-Dichlorophenoxyacetic acid

Off-white, crystalline powder; m.p. 136-140 ˚C; sparingly soluble in water.

RISKS
Harmful by inhalation, in contact with skin, and if swallowed (R20/21/22)

SAFETY PRECAUTIONS
Keep out of reach of children – Keep away from food, drink, and animal feeding stuffs (S2,, S13)

Limit values   OES short-term 20 mg m^{-3}; long-term 10 mg m^{-3}.

Toxic effects   Irritant to skin and eyes. Repeated exposure may give rise to damage to the liver or heart.

First aid   Standard treatment for exposure by all routes (see pages 108-122).

Fire hazard   Extinguish fires with water spray, carbon dioxide, dry chemical powder, or alcohol or polymer foam.

Spillage disposal   See general section.

432. 1,2-Dichloropropane

Colourless liquid with chloroform-like odour; m.p. -100 ˚C; b.p. 96.8 ˚C; sparingly soluble in water.

RISKS
Highly flammable – Harmful by inhalation (R11, R20)

SAFETY PRECAUTIONS
Keep container in a well ventilated place – Keep away from sources of ignition – No Smoking – Do not empty into drains – Take precautionary measures against static discharges (S9, S16, S29, S33)

Toxic effects   Irritant to skin and eyes. Chronic exposure causes depression of the central nervous system and injury to the liver and kidneys.

First aid   Standard treatment for exposure by all routes (see pages 108-122).

Fire hazard   Flash point 15 ˚C; Explosive limits 3.4-14.5%; Autoignition temperature 557 ˚C. Small fires can be dealt with by using carbon dioxide, dry chemical powder, or alcohol resistant foam. The last named is most useful for tackling large fires.

Spillage disposal   See general section.

RSC *Chemical Safety Data Sheets* Vol. 1, No. 27, 1988 gives extended coverage.

433. 1,3-Dichloro-2-propanol

Colourless liquid with ethereal odour; b.p. 174 ˚C; sparingly soluble in water.

RISKS
May cause cancer – Harmful in contact with skin – Toxic if swallowed (R45, R21, R25)

SAFETY PRECAUTIONS
Avoid exposure – Obtain special instructions before use – If you feel unwell, seek medical advice (show label where possible) (S53, S44)

Toxic effects	The vapour irritates the eyes and respiratory system. Inhalation may cause headache, vertigo, nausea, vomiting, and pulmonary oedema. The liquid irritates the skin and eyes. Nausea, vomiting, coma, and liver damage may result if it is taken by mouth.
First aid	Standard treatment for exposure by all routes (see pages 108-122).
Spillage disposal	See general section.

434. Dichloropropene

(1,1-) Colourless liquid; (1,3-) yellow liquid; b.p. (1,1-) 76-77 ˚C; (1,3-) 105-106 ˚C at 730 mm Hg.

RISKS
Highly flammable – Toxic if swallowed (R11, R25)

SAFETY PRECAUTIONS
Keep away from sources of ignition – No Smoking – Do not empty into drains – Take precautionary measures against static discharges – If you feel unwell, seek medical advice (show label where possible) (S16, S29, S33, S44)

Limit values	OES short-term 10 p.p.m. (50 mg m^{-3}); long-term 1 p.p.m. (5 mg m^{-3}).
Toxic effects	Both isomers are irritant to skin, eyes, upper respiratory tract, and mucous membranes. Harmful if inhaled, ingested, or absorbed through the skin.
First aid	Standard treatment for exposure by all routes (see pages 108-122).
Fire hazard	Flash point (1,1-) 0 ˚C;(1,3-) 27 ˚C. Extinguish fires with carbon dioxide, dry chemical powder, or alcohol or polymer foam.
Spillage disposal	See general section.

435. 2,2-Dichloropropionic acid

Colourless liquid; b.p. 90-92 ˚C at 14 mm Hg; soluble in water.

Toxic effects	Extremely destructive to tissues of the skin, eyes, upper respiratory tract, and mucous membranes. Inhalation may be fatal following spasm, chemical pnuemonitis, and pulmonary oedema.

First aid   Standard treatment for exposure by all routes (see pages 108-122).

Fire hazard   Flash point above 110 ˚C. Extinguish fires with carbon dioxide, dry chemical powder, or alcohol or polymer foam.

Spillage disposal   See general section.

436. Dichlorosilane

Colourless gas; m.p. -122 ˚C; b.p. 8.3 ˚C.

Toxic effects   Extremely destructive to tissues of the skin, eyes, upper respiratory tract, and mucous membranes. Inhalation may cause chemical pneumonitis, pulmonary oedema, and eventually death.

Hazardous reactions   In spite of 70% chlorine content, it may ignite spontaneously in air.

First aid   Standard treatment for exposure by all routes (see pages 108-122).

Fire hazard   Explosive limits 4.1-99.0%; Autoignition temperature 58 ˚C. Use water spray to keep cylinders cool or move away from the fire if there is no risk.

Spillage disposal   See general section.

437. Dichlorotetrafluoroethane

Colourless gas with sweetish chloroform-like odour; m.p. -94 ˚C; b.p. 3.8 ˚C; insoluble in water.

Limit values   OES short-term 1250 p.p.m. (8750 mg m^{-3}); long-term 1000 p.p.m. (7000 mg m^{-3}).

Toxic effects   Can cause rapid suffocation. Prolonged exposure can have a narcotic effect.

First aid   Standard treatment for exposure by all routes (see pages 108-122).

Fire hazard   Use water spray to keep cylinders cool or remove them from the danger area.

438. Dicrotonoyl peroxide

Hazardous reactions   Very shock-sensitive explosive.

439. Dicyclohexylamine

Clear, yellow liquid with fishy smell; m.p. 20 ˚C; b.p. 256 ˚C; slightly soluble in water.

RISKS
Harmful if swallowed – Causes burns (R22, R34)

SAFETY PRECAUTIONS
Wear suitable protective clothing, gloves, and eye/face protection (S36/37/39)

Toxic effects	Extremely irritant and destructive to tissues of the skin, eyes, upper respiratory tract, and gastro-intestinal tract. Inhalation may be fatal as a result of inflammation, chemical pneumonitis, and pulmonary oedema. Possible animal carcinogen.
First aid	Standard treatment for exposure by all routes (see pages 108-122).
Fire hazard	Flash point above 98.9 ˚C (open cup). Extinguish fires with water spray, carbon dioxide, dry chemical powder, or alcohol or polymer foam.
Spillage disposal	See general section.

RSC *Chemical Safety Data Sheets* Vol. 3, No. 18, 1990 gives extended coverage.

440. Dicyclohexylcarbodiimide

White powder; m.p. 34 ˚C; reacts with water to form the urea.

Toxic effects	The vapour irritates the eyes severely, and prolonged use of the compound may lead to sensitization and subsequent allergic reactions. Assumed to be harmful if taken by mouth, though probably fairly rapidly inactivated by water.
First aid	Standard treatment for exposure by all routes (see pages 108-122).
Spillage disposal	See general section.

441. Dicyclohexylcarbonyl peroxide

Oily liquid.

Hazardous reactions	In bulk, may explode without apparent reason.

442. Dicyclohexylmethane-4,4'-diisocyanate

Clear, colourless liquid; reacts with water.

RISKS
Toxic by inhalation – Irritating to eyes, respiratory system, and skin – May cause sensitization by inhalation and skin contact (R23, R36/37/38, R42/43)

SAFETY PRECAUTIONS
In case of contact with eyes, rinse immediately with plenty of water and seek medical advice – After contact with skin, wash immediately with plenty of water – In case of insufficient ventilation, wear suitable respiratory equipment – In case of accident or if you feel unwell, seek medical advice immediately (show label where possible) (S26, S28, S38, S45)

Toxic effects	Very little information is available. It is irritant to the skin and pulmonary system, and can cause allergic skin reactions.

First aid Standard treatment for exposure by all routes (see pages 108-122). Immediate treatment may be required if allergic asthma occurs.

Fire hazard Extinguish fires with water spray, carbon dioxide, dry chemical powder, or alcohol or polymer foam.

Spillage disposal See general section.

RSC *Chemical Safety Data Sheets* Vol. 4a, No. 52, 1991 gives extended coverage.

443. Dicyclohexyl phthalate

Colourless, crystalline solid; m.p. 66 ˚C; b.p. 220-228 ˚C at 4 mm Hg; insoluble in water.

Limit values OES long-term 5 mg m^{-3}.

Toxic effects Irritant to skin, eyes, upper respiratory tract, and mucous membranes.

First aid Standard treatment for exposure by all routes (see pages 108-122).

Fire hazard Flash point 107 ˚C. Extinguish fires using water spray, carbon dioxide, dry chemical powder, or alcohol or polymer foam.

Spillage disposal See general section.

444. Dicyclopentadiene

The commercial product that is commonly used is a colourless liquid with a camphor-like odour. The pure compound is in the form of colourless crystals; m.p. 33 ˚C (pure compound); b.p. about 167 ˚C (commercial product); insoluble in water.

Limit values OES long-term 5 p.p.m. (30 mg m^{-3}).

Toxic effects These are not well documented, but suppliers advise against the inhalation of vapour and contact with the skin.

First aid Standard treatment for exposure by all routes (see pages 108-122).

Fire hazard Flash point 35 ˚C. Extinguish fire with foam, dry powder, or carbon dioxide.

Spillage disposal See general section.

445. 1,1-Diethoxyethane

Colourless, volatile liquid with a pleasant odour; b.p. 102 ˚C; sparingly soluble in water.

RISKS
Highly flammable – Irritating to eyes and skin (R11, R36/38)

SAFETY PRECAUTIONS
Keep container in a well ventilated place – Keep away from sources of ignition – No Smoking – Take precautionary measures against static discharges (S9, S16, S33)

Toxic effects These are not well documented. High concentrations of vapour are liable to cause narcosis when inhaled.

Hazardous reactions	Liable to form explosive peroxides on exposure to light and air, which requires that these be decomposed before the ether is distilled to a small volume.
First aid	Standard treatment for exposure by all routes (see pages 108-122).
Fire hazard	Flash point -20 ˚C; Explosive limits 1.7-10.4%; Autoignition temperature 230 ˚C. Extinguish fire with water spray, foam, dry powder, or carbon dioxide.
Spillage disposal	See general section.

446. Diethylamine

Colourless liquid with an ammoniacal odour; b.p. 56 ˚C; miscible with water.

RISKS
Highly flammable – Irritating to eyes and respiratory system (R11, R36/37)

SAFETY PRECAUTIONS
Keep away from sources of ignition – No Smoking – In case of contact with eyes, rinse immediately with plenty of water and seek medical advice – Do not empty into drains (S16, S26, S29)

Limit values	OES short-term 25 p.p.m. (75 mg m^{-3}); long-term 10 p.p.m. (30 mg m^{-3}).
Toxic effects	The vapour irritates the eyes and respiratory system. The liquid irritates the skin and eyes. Assumed to be poisonous if taken by mouth.
First aid	Standard treatment for exposure by all routes (see pages 108-122).
Fire hazard	Flash point below -26 ˚C; Explosive limits 1.8-10.1%; Autoignition temperature 312 ˚C. Extinguish fire with water spray, foam, dry powder, or carbon dioxide.
Spillage disposal	See general section.

447. 2-Diethylaminoethanol

Colourless, hygroscopic liquid; b.p. 163 ˚C; miscible with water.

RISKS
Irritating to eyes, respiratory system, and skin (R36/37/38)

SAFETY PRECAUTIONS
After contact with skin, wash immediately with plenty of water (S28)

Toxic effects	The vapour irritates the eyes and respiratory system. The liquid injures the eyes and is absorbed by the skin which may be irritated. The liquid is moderately toxic if taken by mouth.
First aid	Standard treatment for exposure by all routes (see pages 108-122).
Fire hazard	Extinguish fires with alcohol-resistant foam.
Spillage disposal	See general section.

448. 3-(Diethylamino)propylamine

Colourless liquid with a fishy smell; b.p. 165-170 ˚C; miscible with water.

Toxic effects   Irritant and corrosive to tissues of skin, eyes, upper respiratory, and gastro-intestinal tracts. Inhalation may be fatal due to bronchial spasm, inflammation, and oedema.

First aid   Standard treatment for exposure by all routes (see pages 108-122).

Fire hazard   Flash point 58.8 ˚C (open cup). Extinguish fires with water spray, carbon dioxide, dry chemical powder, or foam extinguishants.

Spillage disposal   See general section.

RSC *Chemical Safety Data Sheets* Vol. 3, No. 20, 1990 gives extended coverage.

449. *N,N*-Diethylaniline

Colourless to yellow liquid; m.p. -38 ˚C; b.p. 215-216 ˚C; sparingly soluble in water.

RISKS
Toxic by inhalation, in contact with skin, and if swallowed – Danger of cumulative effects (R23/24/25, R33)

SAFETY PRECAUTIONS
After contact with skin, wash immediately with plenty of water – Wear suitable gloves – If you feel unwell, seek medical advice (show label where possible) (S28, S37, S44)

Toxic effects   Excessive breathing of the vapour, absorption through the skin, or ingestion can cause headaches, drowsiness, cyanosis, and, in severe cases, convulsions. It also has harmful effects on the eyes.

First aid   Standard treatment for exposure by all routes (see pages 108-122).

Fire hazard   Flash point 97 ˚C; Autoignition temperature 630 ˚C. Extinguish fires with water spray, carbon dioxide, dry chemical powder, or alcohol or polymer foam.

Spillage disposal   See general section.

RSC *Chemical Safety Data Sheets* Vol. 4a, No. 53, 1991 gives extended coverage.

450. Diethylarsine

Liquid; b.p. 105 ˚C.

Hazardous reactions   Inflames in air.

First aid   Standard treatment for exposure by all routes (see pages 108-122).

451. Diethyl azoformate

Hazardous reactions   Shock-sensitive explosive.

452. Diethylberyllium

Colourless liquid; m.p. -12 ˚C; b.p. 110 ˚C at 15 mm Hg; reacts violently with water.

Hazardous reactions   Ignites in air.

453. Diethylcadmium

Oily liquid with musty odour; m.p. -21 ˚C; b.p. 64 ˚C; decomposed by moisture.

Hazardous reactions   Liable to explode under different circumstances.

454. Diethyl carbonate

Colourless liquid with pleasant, ethereal smell; b.p. 126 ˚C; practically insoluble in water.

Toxic effects   The vapour irritates the eyes and respiratory system. The liquid irritates the eyes and is assumed to be irritant and harmful if taken by mouth.

First aid   Standard treatment for exposure by all routes (see pages 108-122).

Fire hazard   Flash point 25 ˚C. Extinguish fire with water spray, foam, dry powder, or carbon dioxide.

Spillage disposal   See general section.

455. Diethylene glycol

Clear, colourless, viscous liquid with practically no odour; m.p. -8 ˚C; b.p. 245.8 ˚C; miscible in water.

Limit values   OES long-term 23 p.p.m. (100 mg m^{-3}).

Toxic effects   Diethylene glycol is not irritant to tissues, but is is harmful following ingestion. Animal carcinogen.

First aid   Standard treatment for exposure by all routes (see pages 108-122).

Fire hazard   Flash point 124 ˚C (open cup); Explosive limits 1.7-10.6%; Autoignition temperature 229 ˚C. Small fires can be dealt with by use of carbon dioxide, dry chemical powder, or by flooding with water. Large fires are best controlled with alcohol-resistant foam.

Spillage   See general section.
   disposal

RSC *Chemical Safety Data Sheets* Vol. 1, No. 28, 1988 gives extended coverage.

456. Diethylene glycol diacrylate

Liquid; b.p. 94 ˚C at 0.225 mm Hg.

RISKS
Toxic in contact with skin – Irritating to eyes and skin – May cause sensitization by skin contact (R24, R36/38, R43)

SAFETY PRECAUTIONS
After contact with skin, wash immediately with plenty of water – Wear eye/face protection – If you feel unwell, seek medical advice (show label where possible) (S28, S39, S44)

Toxic effects   Main effects are on the skin and eyes.

First aid   Standard treatment for exposure by all routes (see pages 108-122), with particular attention to the possible occurrence of convulsions and/or lung congestion.

Fire hazard   Extinguish fires using water spray, carbon dioxide, dry chemical powder, or alcohol or polymer foam.

Spillage   See general section.
   disposal

RSC *Chemical Safety Data Sheets* Vol. 4a, No. 54, 1991 gives extended coverage.

457. Diethyl ether

Colourless, highly volatile liquid with characteristic odour; m.p. -116.2 ˚C; b.p. 34 ˚C; slightly soluble in water.

RISKS
Extremely flammable – May form explosive peroxides (R12, R19)

SAFETY PRECAUTIONS
Keep container in a well ventilated place – Keep away from sources of ignition – No Smoking – Do not empty into drains – Take precautionary measures against static discharges (S9, S16, S29, S33)

Limit values   OES short-term 500 p.p.m. (1500 mg m^{-3}); long-term 400 p.p.m. (1200 mg m^{-3}).

Toxic effects   Inhalation of vapour may cause drowsiness, dizziness, mental confusion, faintness, and, in high concentrations, unconsciousness. Ingestion may also produce these effects. Continued inhalation of low concentrations may cause loss of appetite, dizziness, fatigue, and nausea. Repeated inhalation or swallowing may lead to 'ether habit', with symptoms resembling chronic alcoholism.

Hazardous   Liable to explode following formation of peroxides; powerful oxidants also
   reactions   produce explosive reactions readily; reacts vigorously with sulfuryl chloride.

First aid	Standard treatment for exposure by all routes (see pages 108-122).
Fire hazard	Flash point -45 ˚C; Explosive limits 1.85-36%; Autoignition temperature 160 ˚C. Extinguish fire with dry powder, carbon dioxide, or alcohol-resistant foam.
Spillage disposal	See general section.

RSC *Chemical Safety Data Sheets* Vol. 1, No. 30, 1988 gives extended coverage.

458. Diethylmagnesium

Polymeric solid; m.p. 176 ˚C with decomposition.

Hazardous reactions	Water ignites solid or ethereal solution, carbon dioxide ignites solid.

459. Diethyl malonate

Colourless liquid; m.p. -50 ˚C; b.p. 199 ˚C; slightly soluble in water.

Toxic effects	Irritant to skin and eyes.
First aid	Standard treatment for exposure by all routes (see pages 108-122).
Fire hazard	Flash point 100 ˚C. Extinguish fires with carbon dioxide, dry chemical powder, or alcohol or polymer foam.
Spillage disposal	See general section.

460. Diethylmethylphosphine

Hazardous reactions	May ignite on long exposure to air.

461. Diethyl oxalate

Colourless liquid with aromatic odour; m.p. -41 ˚C; b.p. 185 ˚C; decomposes in water.

RISKS
Harmful if swallowed – Irritating to eyes (R22, R36)

SAFETY PRECAUTIONS
Do not breathe es (S23)

Toxic effects	High concentrations are extremely destructive to tissues in all parts of the body. Inhalation may cause burning sensation, coughing, headaches, nausea, and vomiting.

First aid Standard treatment for exposure by all routes (see pages 108-122).

Fire hazard Flash point 75 °C. Extinguish fires with carbon dioxide, dry chemical powder, or alcohol or polymer foam.

Spillage disposal See general section.

462. Diethyl peroxide

Liquid; b.p. 65 °C; very slightly soluble in water.

Hazardous reactions Explosive.

Fire hazard Explosive limits lel 2.3%.

463. *N,N*-Diethyl-*p*-phenylenediamine and salts

The free base is a reddish-brown liquid; the hydrochloride and sulfate are buff or grey crystalline powders. All darken on exposure to light and air; b.p. (free base) 261 °C; they are insoluble in water.

RISKS
Toxic if swallowed – Causes burns (R25, R34)

SAFETY PRECAUTIONS
In case of contact with eyes, rinse immediately with plenty of water and seek medical advice – Wear suitable protective clothing – If you feel unwell, seek medical advice (show label where possible) (S26, S36, S44)

Toxic effects These are not recorded but may be expected to be similar to those resulting from contact with phenylenediamine, namely eye and skin irritation; dermatitis and more serious eye injury may also result from major contact.

First aid Standard treatment for exposure by all routes (see pages 108-122).

Fire hazard Flash point above 110 °C. Use water spray, carbon dioxide, dry chemical powder, or alcohol or polymer foam to extinguish fires.

Spillage disposal See general section.

RSC *Chemical Safety Data Sheets* Vol. 4a, No. 55, 1991 gives extended coverage of the free base.

464. Diethylphosphine

Liquid; b.p. 85 °C.

Toxic effects Toxic by inhalation and ingestion

| **Hazardous reactions** | Readily ignites in air. |
| **Fire hazard** | Extinguish fires with carbon dioxide, dry chemical powder, or foam. |

465. Diethyl phosphite

Colourless liquid; b.p. 50-51 ˚C at 2 mm Hg.

Toxic effects	Irritant to skin, eyes, upper respiratory tract, and mucous membranes. Harmful by inhalation, ingestion, or skin absorption.
Hazardous reactions	Violent reaction with *p*-nitrophenol in absence of solvent.
First aid	Standard treatment for exposure by all routes (see pages 108-122).
Fire hazard	Flash point 90 ˚C. Use water spray, carbon dioxide, dry chemical powder, or alcohol or polymer foam to extinguish fires.
Spillage disposal	See general section.

466. Diethyl phthalate

Viscous colourless liquid; m.p. -3 ˚C; b.p. 298-299 ˚C; insoluble in water.

Limit values	OES short-term 10 mg m^{-3}; long-term 5 mg m^{-3}.
Toxic effects	Irritant to skin, eyes, upper respiratory tract, and mucous membranes. Also harmful if inhaled or absorbed through the skin.
First aid	Standard treatment for exposure by all routes (see pages 108-122).
Fire hazard	Flash point 160 ˚C; Autoignition temperature 455 ˚C. Extinguish fires with carbon dioxide or dry chemical powder. Water spray or foam can also be used but may cause frothing.
Spillage disposal	See general section.

467. Diethyl sulfate

Colourless liquid with faint ethereal odour; b.p. 209 ˚C with decomposition; insoluble in water.

RISKS
May cause cancer – May cause heritable genetic damage – Harmful by inhalation, in contact with skin and if swallowed – Causes burns (R45, R46, R20/21/22, R34)

SAFETY PRECAUTIONS
Avoid exposure – Obtain special instructions before use – In case of contact with eyes, rinse immediately with plenty of water and seek medical advice – If you feel unwell, seek medical advice (show label where possible) (S53, S26, S44)

Toxic effects   Extremely destructive to all tissues of the mucous membranes, upper respiratory tract, skin, and eyes. May be fatal if inhaled, ingested, or absorbed through the skin.

Hazardous reactions   Violent reactions with 2,7-dinitro-9-phenylphenanthridine/water and potassium t-butoxide.

First aid   Standard treatment for exposure by all routes (see pages 108-122).

Fire hazard   Extinguish fires with water spray or alcohol foam.

Spillage disposal   See general section.

468. Diethylzinc

Liquid; m.p. -28 ˚C; b.p. 118 ˚C.

Toxic effects   Little toxicological information available.

Hazardous reactions   Pyrophoric in air; violent reaction with a variety of substances.

Fire hazard   Extinguish fire with carbon dioxide or dry powder extinguishers.

469. Difluoramine

Hazardous reactions   A dangerous explosive.

470. Difluorodiazene

Hazardous reactions   Reacts explosively with hydrogen above 90 ˚C. Very unstable explosive.

471. Difluoroethylene

Colourless gas with faint ethereal odour; b.p. -83 ˚C; slightly soluble in water.

Toxic effects   The gas, according to current evidence, has no substantial toxicity but shows the asphyxiant properties of non-toxic gases such as nitrogen.

Hazardous reactions   Reaction with hydrochloric acid may become extremely violent.

Fire hazard   Explosive limits 5.5-21.3%. As the gas is supplied in a cylinder, turning off the valve will reduce any fire involving it; if possible, cylinders should be removed quickly from an area in which a fire has developed.

Spillage disposal   Surplus gas or a leaking cylinder should be vented to air in a safe area.

472. Di-2-furoyl peroxide

Hazardous   Explodes violently on heating and friction.
reactions

473. Dihexanoyl peroxide

Hazardous   Explodes at 85 ˚C.
reactions

474. Di-iodoacetylene

Colourless, rhombic crystals; m.p. 78-82 ˚C decomposes at 80-100 ˚C; slightly soluble in water.

Hazardous   Explodes on friction, impact, and on heating to 84 ˚C.
reactions

475. Di-iodoamine

Hazardous   Explosive.
reactions

476. Di-isobutyl ketone

Colourless liquid with a mild odour; b.p. 166 ˚C; insoluble in water.

RISKS
Flammable – Irritating to respiratory system (R10, R37)

SAFETY PRECAUTIONS
Avoid contact with skin (S24)

Limit values   OES long-term 25 p.p.m. (150 mg m^{-3}).

Toxic effects   The vapour may irritate the respiratory system and is narcotic in high concentrations. The liquid irritates the eyes and may irritate the skin. Low acute oral toxicity.

First aid   Standard treatment for exposure by all routes (see pages 108-122).

Fire hazard   Flash point 49 ˚C; Explosive limits 0.8-6.2%; Autoignition temperature 396 ˚C. Extinguish fires with dry powder, alcohol-resistant foam, or carbon dioxide.

Spillage   See general section.
disposal

RSC *Chemical Safety Data Sheets* Vol. 1, No. 31, 1988 gives extended coverage.

477. Di-isobutyryl peroxide

Toxic effects   Irritant to skin, eyes, and upper respiratory tract.

Hazardous   Explodes on standing at room temperature; solution in ether exploded during
reactions   evaporation.

First aid   Standard treatment for exposure by all routes (see pages 108-122).

478. 2,4-Di-isocyanatotoluene

Pale yellow liquid with sharp pungent smell; commonly contains about 20% of the 2,6-isomer;
m.p. 19.5-21.5 °C; b.p. 251 °C; reacts with water with evolution of carbon dioxide.

RISKS
*Very toxic by inhalation – Irritating to eyes, respiratory system, and skin – May cause
sensitization by inhalation (R26, R36/37/38, R42)*

SAFETY PRECAUTIONS
*In case of contact with eyes, rinse immediately with plenty of water and seek medical advice –
After contact with skin, wash immediately with plenty of water – In case of insufficient
ventilation, wear suitable respiratory equipment – In case of accident or if you feel unwell, seek
medical advice immediately (show label where possible) (S26, S28, S38, S45)*

Limit values   MEL short-term 0.07 mg m^{-3} (as -NCO); long-term 0.02 mg m^{-3} (as -NCO).

Toxic effects   The vapour irritates the respiratory system and may cause bronchial asthma.
The vapour and liquid are very irritating to the eyes. The liquid irritates the
skin and may cause severe dermatitis. Assumed to be highly irritant and
poisonous if taken by mouth.

Hazardous   May polymerize vigorously on contact with bases and acyl chlorides, or slowly
reactions   with gas evolution by diffusion of moisture into polythene containers.

First aid   Standard treatment for exposure by all routes (see pages 108-122). If allergic
asthma occurs immediate treatment will be necessary.

Fire hazard   Flash point 132 °C; Explosive limits 0.9-9.5%. Extinguish fires with carbon
dioxide or dry chemical powder.

Spillage   See general section.
disposal

RSC *Chemical Safety Data Sheets* Vol. 4b, No. 143, 1991 gives extended coverage.

479. Di-isodecyl phthalate

Clear viscous liquid; m.p. -50 °C; b.p. 250-257 °C; almost insoluble in water.

Limit values   OES long-term 5 mg m^{-3}.

Toxic effects   Irritant to skin and eyes. Ingestion may lead to nausea, vomiting, and
dizziness.

First aid   Standard treatment for exposure by all routes (see pages 108-122).

Fire hazard   Flash point 232 ˚C. Extinguish fires with water spray, carbon dioxide, dry chemical powder, or alcohol or polymer foam.

Spillage   See general section.
    disposal

480. Di-isononyl phthalate

Clear, colourless liquid; almost insoluble in water.

Limit values   OES long-term 5 mg m⁻³.

Toxic effects   Irritant to skin and eyes. Ingestion may cause nausea, headache, and dizziness.

First aid   Standard treatment for exposure by all routes (see pages 108-122).

Fire hazard   Flash point above 93 ˚C. Extinguish fires with water spray, carbon dioxide, dry chemical powder, or alcohol or polymer spray.

Spillage   See general section.
    disposal

481. Di-isooctyl phthalate

Liquid; m.p. -4 ˚C; b.p. 239 ˚C at 5 mm Hg; insoluble in water.

Limit values   OES long-term 5 mg m⁻³.

Toxic effects   Skin irritant. Mildly toxic by ingestion and skin contact.

First aid   Standard treatment for exposure by all routes (see pages 108-122).

Fire hazard   Flash point 210 ˚C. Extinguish fires with water spray, carbon dioxide, dry chemical powder, or alcohol or polymer foam.

Spillage   See general section.
    disposal

482. Di-isopropylamine

Colourless, strongly alkaline liquid; b.p. 84 ˚C; miscible with water.

RISKS
Highly flammable – Irritating to eyes, respiratory system, and skin (R11, R36/37/38)

SAFETY PRECAUTIONS
Keep container in a well ventilated place – Keep away from sources of ignition – No Smoking (S9, S16)

Limit values   OES long-term 5 p.p.m. (20 mg m⁻³).

Toxic effects   The vapour irritates eyes and respiratory system. The liquid irritates the skin and eyes and may cause burns to the eyes. Assumed to be irritant and poisonous if taken internally.

First aid	Standard treatment for exposure by all routes (see pages 108-122).
Fire hazard	Flash point -1 ˚C. Extinguish fire with dry powder or carbon dioxide.
Spillage disposal	See general section.

483. Di-isopropyl ether

Colourless liquid with ethereal odour; m.p. -60 ˚C; b.p. 69 ˚C; soluble in water.

RISKS
Highly flammable – May form explosive peroxides (R11, R19)

SAFETY PRECAUTIONS
Keep container in a well ventilated place – Keep away from sources of ignition – No Smoking – Take precautionary measures against static discharges (S9, S16, S33)

Limit values	OES short-term 310 p.p.m. (1320 mg m^{-3}); long-term 250 p.p.m. (1050 mg m^{-3}).
Toxic effects	The vapour irritates the respiratory system and eyes and inhalation may lead to headache, dizziness, nausea, vomiting, and narcosis. The liquid irritates the eyes causing conjunctivitis; it will defat the skin and may lead to dermatitis. If taken internally, it gives effects similar to those indicated for inhalation of the vapour.
Hazardous reactions	Formation of peroxides responsible for numerous explosions. Vigorous reaction with propionyl chloride.
First aid	Standard treatment for exposure by all routes (see pages 108-122).
Fire hazard	Flash point -28 ˚C; Explosive limits 1.4-7.9%; Autoignition temperature 443 ˚C. Extinguish fire with alcohol-resistant foam, dry powder, or carbon dioxide.
Spillage disposal	See general section.

RSC *Chemical Safety Data Sheets* Vol. 1, No. 32, 1988 gives extended coverage.

484. Di-isopropyl peroxydicarbonate

Colourless, crystalline solid; m.p. 8-10 ˚C; almost insoluble in water.

Toxic effects	Severe eye irritant. Moderately toxic by inhalation and skin contact.
Hazardous reactions	Undergoes self-accelerating decomposition when warmed above its melting point (10 ˚C) which may become dangerously violent; ignites with dimethylaniline.
Fire hazard	Emits acid fumes on decomposition.

485. Diketene

Colourless liquid with pungent odour; b.p. 127 ˚C; decomposed by water.

RISKS
Flammable – Harmful by inhalation (R10, R20)

SAFETY PRECAUTIONS
Keep in a cool place (S3)

Toxic effects	Vapour irritates the respiratory system and the eyes severely, causing lachrymation. The liquid irritates the skin and may cause burns. It is assumed to cause severe damage if taken internally.
Hazardous reactions	Violent polymerization is catalysed by acids or bases; residues decomposed on standing.
First aid	Standard treatment for exposure by all routes (see pages 108-122).
Fire hazard	Flash point 46 ˚C. Extinguish fire with water spray, dry powder, or carbon dioxide.
Spillage disposal	See general section.

486. Di-linear 79 phthalate

Liquid; insoluble in water.

Limit values	OES long-term 5 mg m^{-3}.
Toxic effects	Irritant to skin and eyes. Toxic by inhalation and ingestion.
First aid	Standard treatment for exposure by all routes (see pages 108-122).
Fire hazard	Extinguish fires with water spray, carbon dioxide, dry chemical powder, or alcohol or polymer foam.
Spillage disposal	See general section.

487. Dimercury dicyanide oxide

Limit values	OES short-term 0.15 mg m^{-3} (as Hg); long-term 0.05 mg m^{-3} (as Hg).
Hazardous reactions	Heat and impact sensitive explosive.

488. 1,2-Dimethoxyethane

Colourless liquid with sharp, ethereal odour; b.p. 85 ˚C; miscible with water.

RISKS
Flammable – May form explosive peroxides – Harmful by inhalation (R10, R19, R20)

SAFETY PRECAUTIONS
Avoid contact with skin and eyes (S24/25)

Toxic effects	These are not well documented apart from animal experiments which indicate that the vapour is irritant to the respiratory system and that skin and eyes are liable to be affected by contact with the liquid.
Hazardous reactions	Liable to form explosive peroxides on exposure to air and light which should be decomposed before the ether is distilled to a small volume.
First aid	Standard treatment for exposure by all routes (see pages 108-122).
Fire hazard	Flash point 4.5 ˚C. Extinguish fire with water spray, foam, dry powder, or carbon dioxide.
Spillage disposal	See general section.

489. Dimethoxymethane

Colourless, volatile liquid with mild sweet odour; b.p. 42 ˚C; soluble in water.

RISKS
Flammable – May form explosive peroxides – Harmful by inhalation (R10, R19, R20)

SAFETY PRECAUTIONS
Avoid contact with skin and eyes (S24/25)

Limit values	OES short-term 1250 p.p.m. (3880 mg m^{-3}); long-term 1000 p.p.m. (3100 mg m^{-3}).
Toxic effects	Considered to be of low toxicity though high concentrations may cause narcosis. It has produced injury to lungs, liver, kidneys, and heart in experiments on animals. It is irritating to the eyes and, on prolonged contact, to the skin. It is toxic by skin absorption.
First aid	Standard treatment for exposure by all routes (see pages 108-122).
Fire hazard	Flash point -18 ˚C (open cup); Explosive limits 2.2-13.8%; Autoignition temperature 237 ˚C. Extinguish fire with dry powder, alcohol foam, or carbon dioxide.
Spillage disposal	See general section.

RSC *Chemical Safety Data Sheets* Vol. 1, No. 67, 1988 gives extended coverage.

490. 2,2-Dimethoxypropane

Colourless liquid; m.p. -47 ˚C; b.p. 83 ˚C.

Toxic effects	Irritant to skin, eyes, upper respiratory tract, and mucous membranes. Harmful if inhaled or ingested.
Hazardous reactions	Violent explosions when dehydration of hydrated manganese and nickel perchlorates was attempted using the ether. As a *gem*-diether it is subject to cool flame behaviour and subsequent explosion.
First aid	Standard treatment for exposure by all routes (see pages 108-122).
Fire hazard	Flash point -11 ˚C; Explosive limits 6.0-31.0%. Extinguish fires with carbon dioxide, dry chemical powder, or alcohol or polymer foam. Do NOT use water.
Spillage disposal	See general section.

491. *N,N*-Dimethylacetamide

Colourless liquid, slight basic odour; m.p. -20 ˚C; b.p. 165 ˚C; miscible with water.

RISKS
Harmful by inhalation and in contact with skin – Irritating to eyes (R20/21, R36)

SAFETY PRECAUTIONS
In case of contact with eyes, rinse immediately with plenty of water and seek medical advice – After contact with skin, wash immediately with plenty of water – Wear suitable protective clothing (S26, S28, S36)

Limit values	OES short-term 15 p.p.m. (50 mg m^{-3}); long-term 10 p.p.m. (35 mg m^{-3}).
Toxic effects	Prolonged contact with the skin may lead to absorption and liver damage.
Hazardous reactions	Exothermic reactions with some chlorinated hydrocarbons, violent in presence of iron.
First aid	Standard treatment for exposure by all routes (see pages 108-122).
Fire hazard	Flash point 70 ˚C (open cup); Explosive limits 2-11.5%; Autoignition temperature 490 ˚C. Extinguish fires with water, dry chemical powder, alcohol-resistant foam, or carbon dioxide.
Spillage disposal	See general section.

RSC *Chemical Safety Data Sheets* Vol. 1, No. 33, 1988 gives extended coverage.

492. Dimethylamine and solutions

Colourless gas at ordinary temperatures. Commonly available in aqueous and ethanolic solution; m.p. -92 ˚C; b.p. 7 ˚C; readily soluble in water.

RISKS
Extremely flammable liquefied gas – Irritating to eyes and respiratory system (R13, R36/37)

SAFETY PRECAUTIONS
Keep away from sources of ignition – No Smoking – In case of contact with eyes, rinse immediately with plenty of water and seek medical advice – Do not empty into drains (S16, S26, S29)

Limit values	OES long-term 10 p.p.m. (18 mg m^{-3}).
Toxic effects	The vapour irritates the mucous membranes and respiratory system; in high concentrations it may affect the nervous system. The vapour and solutions irritate the eyes. The solutions may irritate the skin. Poisonous if taken by mouth. Experimental carcinogen in animal studies.
Hazardous reactions	The gas incandesces on contact with fluorine; causes maleic anhydride to decompose exothermically above 150 ˚C.
First aid	Standard treatment for exposure by all routes (see pages 108-122).
Fire hazard	Flash point -18 ˚C; Explosive limits 2.8-14.4%; Autoignition temperature 400 ˚C. (Gas) Extinguish fire with water spray, foam, dry powder, or carbon dioxide.
Spillage disposal	(Liquid and solutions) See general section.

RSC *Chemical Safety Data Sheets* Vol. 1, No. 34, 1988 gives extended coverage.

493. 2-Dimethylaminoethanol

Colourless liquid; b.p. 135 ˚C; miscible with water.

RISKS
Flammable – Irritating to eyes, respiratory system, and skin (R10, R36/37/38)

SAFETY PRECAUTIONS
After contact with skin, wash immediately with plenty of water (S28)

Toxic effects	Little is recorded about these.
First aid	Standard treatment for exposure by all routes (see pages 108-122).
Fire hazard	Flash point 41 ˚C (open cup). Extinguish fire with water spray, foam, dry powder, or carbon dioxide.
Spillage disposal	See general section.

494. 3-(Dimethylamino)propylamine

Colourless liquid; m.p. -100 ˚C; b.p. 133 ˚C; miscible with water.

Toxic effects	The vapour and liquid are irritant and corrosive to skin, eyes, upper respiratory, and gastro-intestinal tracts. May be absorbed through the skin. Inhalation may be fatal as a result of bronchial spasm, irritation, and oedema.
First aid	Standard treatment for exposure by all routes (see pages 108-122).
Fire hazard	Flash point 45 ˚C. Extinguish fires with water spray, carbon dioxide, dry chemical powder, or alcohol or polymer foam.
Spillage disposal	See general section.

RSC *Chemical Safety Data Sheets* Vol. 3, No. 21, 1990 gives extended coverage.

495. *N,N*-Dimethylaniline

Yellow to brownish oily liquid; m.p. 2.5 ˚C; b.p. 193.1 ˚C; sparingly soluble in water.

RISKS
Toxic by inhalation, in contact with skin, and if swallowed – Danger of cumulative effects (R23/24/25, R33)

SAFETY PRECAUTIONS
After contact with skin, wash immediately with plenty of water – Wear suitable gloves – If you feel unwell, seek medical advice (show label where possible) (S28, S37, S44)

Limit values	OES short-term 10 p.p.m. (50 mg m^{-3}); long-term 5 p.p.m. (25 mg m^{-3}).
Toxic effects	Excessive breathing of the vapour or absorption of the liquid through the skin can cause headache, drowsiness, cyanosis, mental confusion, and, in severe cases, convulsions. The liquid is dangerous to the eyes and the above effects can also be experienced if it is swallowed. Continued exposure to the vapour or slight skin exposure to the liquid over a period may affect the nervous system and the blood causing fatigue, loss of appetite, headache, and dizziness.
Hazardous reactions	Contact with a drop causes dibenzoyl peroxide or di-isopropyl peroxydicarbonate to explode.
First aid	Standard treatment for exposure by all routes (see pages 108-122).
Fire hazard	Flash point 62.7 ˚C (closed cup); Explosive limits lel 1.0%; Autoignition temperature 316 ˚C. Use carbon dioxide, dry chemical powder, or alcohol or polymer foam to extinguish fires.

RSC *Chemical Safety Data Sheets* Vol. 4a, No. 56, 1991 gives extended coverage.

496. Dimethylantimony chloride

Hazardous reactions	Ignites in air at 40 ˚C.

497. Dimethylarsine

Colourless liquid; b.p. 35.6 °C at 747 mm Hg.

Hazardous reactions	Inflames in air.
Fire hazard	Use water spray, carbon dioxide, dry chemical powder, or foam to extinguish fires.

498. 3,5-Dimethylbenzoic acid

White crystals; m.p. 172-174 °C.

Toxic effects	Irritant to skin, eyes, upper respiratory tract, and mucous membranes. May be harmful by inhalation, ingestion, or skin absorption.
Hazardous reactions	Explosion when mesitylene being oxidized with nitric acid in autoclave.
First aid	Standard treatment for exposure by all routes (see pages 108-122).
Fire hazard	Use water spray, carbon dioxide, dry chemical powder, or alcohol or polymer foam to extinguish fires.
Spillage disposal	See general section.

499. Dimethylberyllium

Crystalline solid; m.p. 270 °C with sublimation; reacts explosively with water.

Hazardous reactions	Ignites in moist air or in carbon dioxide.

500. 2,3-Dimethylbuta-1,3-diene

Colourless liquid; m.p. -76 °C; b.p. 68-69 °C.

Toxic effects	Irritant to upper respiratory tract and mucous membranes. Harmful if inhaled or ingested
Hazardous reactions	Polymeric peroxide autoxidation residue exploded violently on ignition.
First aid	Standard treatment for exposure by all routes (see pages 108-122).
Fire hazard	Flash point -22 °C. Use carbon dioxide, dry chemical powder, or alcohol or polymer foam to extinguish fires.
Spillage disposal	See general section.

501. 1,3-Dimethylbutyl acetate

Liquid; m.p. -63.8 °C; b.p. 146.3 °C; almost insoluble in water.

Limit values	OES short-term 100 p.p.m. (600 mg m^{-3}); long-term 50 p.p.m. (300 mg m^{-3}).
Toxic effects	May irritate body tissues. Toxic by inhalation, ingestion, or skin absorption.
First aid	Standard treatment for exposure by all routes (see pages 108-122).
Fire hazard	Flash point 45 °C. Extinguish fires with carbon dioxide, dry chemical powder, or alcohol or polymer foam.
Spillage disposal	See general section.

502. Dimethylcadmium

Liquid with musty odour; m.p. -4.2 °C; b.p. 105.5 °C.

Limit values	MEL long-term 0.05 mg m^{-3} (as Cd).
Hazardous reactions	Peroxide formed on exposure to air is explosive.

503. Dimethylcarbamoyl chloride

Colourless liquid; b.p. 165 °C; decomposed by water.

RISKS
May cause cancer – Harmful if swallowed – Toxic by inhalation – Irritating to eyes, respiratory system and skin (R45, R22, R23, R36/37/38)

SAFETY PRECAUTIONS
Avoid exposure – Obtain special instructions before use – If you feel unwell, seek medical advice (show label where possible) (S53, S44)

Toxic effects	The vapour irritates the eyes and respiratory system severely. The liquid burns the eyes and skin. Assumed to cause severe internal irritation and damage if taken by mouth.
First aid	Standard treatment for exposure by all routes (see pages 108-122).
Spillage disposal	See general section.

504. Dimethyl carbonate

Colourless liquid with pleasant odour; b.p. 90 °C; insoluble in water.

RISKS
Highly flammable – Harmful by inhalation, in contact with skin, and if swallowed (R11, R20/21/22)

SAFETY PRECAUTIONS
Keep container in a well ventilated place – Do not empty into drains (S9, S29)

Toxic effects	The vapour from the hot liquid irritates the eyes and respiratory system. Prolonged inhalation of vapour has resulted in liver damage in experimental animals. The liquid irritates the skin and eyes. Assumed to be poisonous if taken by mouth.
Hazardous reactions	Reacts vigorously or violently with a range of materials.
First aid	Standard treatment for exposure by all routes (see pages 108-122).
Fire hazard	Flash point 10 ˚C (open cup). Extinguish fire using water spray, foam, dry powder, or carbon dioxide.
Spillage disposal	See general section.

505. Dimethyldichlorosilane

Colourless, fuming liquid; b.p. 70 ˚C; reacts violently with water.

RISKS
Highly flammable – Irritating to eyes, respiratory system, and skin (R11, R36/37/38)

Toxic effects	The liquid will also irritate the gastro-intestinal system if ingested.
First aid	Standard treatment for exposure by all routes (see pages 108-122).
Fire hazard	Flash point -9 ˚C; Explosive limits 3.4-9.5%. Extinguish fire with dry powder, carbon dioxide, dry sand, or earth.
Spillage disposal	See general section.

506. Dimethyl ether

Colourless gas with slight ethereal odour; b.p. -25 ˚C; slightly soluble in water.

RISKS
Extremely flammable liquefied gas (R13)

SAFETY PRECAUTIONS
Keep container in a well ventilated place – Keep away from sources of ignition – No Smoking – Take precautionary measures against static discharges (S9, S16, S33)

Toxic effects	The gas is about one fourth as potent as diethyl ether as an anaesthetic but is not used for this purpose because of other toxic effects, notably rushing of blood through the head and sickness.
First aid	Standard treatment for exposure by all routes (see pages 108-122).
Fire hazard	Flash point -41 ˚C; Explosive limits 3.4-18%; Autoignition temperature 350 ˚C. As the gas is supplied in a cylinder, turning off the valve will reduce any fire involving it; if possible, cylinders should be removed quickly from an area in which a fire has developed.
Spillage disposal	Surplus gas or a leaking cylinder can be vented slowly to air in a safe area or the gas burnt-off in a suitable burner.

507. *N,N*-Dimethylethylamine

Colourless liquid; m.p. -140 ˚C; b.p. 37.5 ˚C.

Limit values   OES short-term 15 p.p.m. (45 mg m^{-3}); long-term 10 p.p.m. (30 mg m^{-3}).

Toxic effects   Extremely destructive to skin, eyes, upper respiratory tract, and mucous membranes. Inhalation may be fatal as a result of spasm, inflammation, chemical pnuemonitis, and pulmonary oedema.

First aid   Standard treatment for exposure by all routes (see pages 108-122).

Fire hazard   Flash point -36 ˚C. Extinguish fires with carbon dioxide, dry chemical powder, or alcohol or polymer foam.

Spillage disposal   See general section.

508. *N,N*-Dimethylformamide

Colourless liquid with faint amine-like odour; m.p. -60 ˚C; b.p. 153 ˚C; miscible with water.

RISKS
Harmful by inhalation and in contact with skin – Irritating to eyes (R20/21, R36)

SAFETY PRECAUTIONS
In case of contact with eyes, rinse immediately with plenty of water and seek medical advice – After contact with skin, wash immediately with plenty of water – Wear suitable protective clothing (S26, S28, S36)

Limit values   OES short-term 20 p.p.m. (60 mg m^{-3}); long-term 10 p.p.m. (30 mg m^{-3}).

Toxic effects   The vapour from the hot liquid irritates the eyes and respiratory system. The liquid irritates the skin and eyes. Toxic if taken by mouth. Prolonged inhalation of vapour has resulted in liver damage in experimental animals.

Hazardous reactions   Reacts vigorously or violently with a range of materials.

First aid   Standard treatment for exposure by all routes (see pages 108-122).

Fire hazard   Flash point 57 ˚C; Explosive limits 2.2-15.2%; Autoignition temperature 445 ˚C. Extinguish fires with dry chemical powder, alcohol-resistant foam, or carbon dioxide extinguishers.

Spillage disposal   See general section.

RSC *Chemical Safety Data Sheets* Vol. 1, No. 35, 1988 gives extended coverage.

509. Dimethyl glyoxime

White powder; m.p. 240-241 ˚C with decomposition; practically insoluble in water.

Toxic effects   Skin contact can give rise to irritation and adsorption.

First aid   Standard treatment for exposure by all routes (see pages 108-122).

Fire hazard Extinguish fires with water spray, carbon dioxide, dry chemical powder, or alcohol or polymer foam.

Spillage disposal See general section.

510. 1,1-Dimethylhydrazine

Colourless to yellow, hygroscopic liquid; b.p. 62.5 ˚C; very soluble in water.

RISKS
May cause cancer – Highly flammable – Toxic by inhalation and if swallowed – Causes burns (R45, R11, R23/25, R34)

SAFETY PRECAUTIONS
Avoid exposure – Obtain special instructions before use – Keep away from sources of ignition – No Smoking – Take precautionary measures against static discharges – If you feel unwell, seek medical advice (show label where possible) (S53, S16, S33, S44)

Toxic effects Extremely destructive to tissues of skin, eyes, upper respiratory tract, and mucous membranes. May be fatal if inhaled, ingested, or absorbed through the skin

Hazardous reactions Ignites violently with oxidants.

First aid Standard treatment for exposure by all routes (see pages 108-122).

Fire hazard Flash point 1 ˚C; Explosive limits 2.0-95.0%; Autoignition temperature 248 ˚C. Use water spray, carbon dioxide, dry chemical powder, or alcohol or polymer foam to extinguish fires.

Spillage disposal See general section.

511. Dimethylketen

Yellow liquid; m.p. -97.5 ˚C; b.p. 34 ˚C at 750 mm Hg; decomposed by water.

Hazardous reactions Forms extremely explosive peroxide when exposed to air.

Fire hazard Emits acrid fumes on heating to decomposition.

512. Dimethylmagnesium

Polymeric solid; m.p. stable to 220 ˚C; ignited by water.

Hazardous reactions Water ignites the solid or its ethereal solution.

513. Dimethylnitrosamine

Clear, yellow liquid with a faint characteristic odour; b.p. 151-153 ˚C; soluble in water.

RISKS
May cause cancer – Toxic if swallowed – Very toxic by inhalation – Danger of serious damage to health by prolonged exposure (R45, R25, R26, R48)

SAFETY PRECAUTIONS
Avoid exposure – Obtain special instructions before use – In case of accident or if you feel unwell, seek medical advice immediately (show label where possible) (S53, S45)

Toxic effects   Irritant to skin and eyes, in addition to above risks.

First aid   Standard treatment for exposure by all routes (see pages 108-122).

Fire hazard   Flash point 61 ˚C. Extinguish fires with carbon dioxide, dry chemical powder, or alcohol or polymer foam.

Spillage   See general section.
disposal

RSC *Chemical Safety Data Sheets* Vol. 4a, No. 57, 1991 gives extended coverage.

514. *N,N*-Dimethyl-*p*-nitrosoaniline

Green powder; m.p. 93 ˚C; insoluble in water.

Toxic effects   The dust irritates the respiratory system. The dust irritates the eyes and skin and may cause dermatitis. Irritant and poisonous if taken by mouth.

Hazardous   Delayed violent reaction with acetic anhydride.
reactions

First aid   Standard treatment for exposure by all routes (see pages 108-122).

Spillage   See general section.
disposal

515. 1,2-Dimethyl-1-nitrosohydrazine

Liquid.

Hazardous   Deflagrates on heating.
reactions

516. Dimethyl peroxide

Gas or liquid; b.p. 10 ˚C.

Toxic effects   Irritant to skin, eyes, and upper respiratory tract.

Hazardous reactions	Heat- and shock-sensitive explosive as liquid or vapour.
First aid	Standard treatment for exposure by all routes (see pages 108-122).

517. *N,N*-Dimethylphenylenediamine

Reddish-violet crystals; m.p. 53 ˚C (also stated as 41 ˚C); b.p. 262 ˚C; soluble in water.

RISKS
Toxic by inhalation, in contact with skin, and if swallowed (R23/24/25)

SAFETY PRECAUTIONS
After contact with skin, wash immediately with plenty of water – If you feel unwell, seek medical advice (show label where possible) (S28, S44)

Toxic effects	Irritant to skin, eyes, and upper respiratory tract. May be absorbed through the skin.
First aid	Standard treatment for exposure by all routes (see pages 108-122), with particular attention to the possible occurrence of convulsions and/or shock.
Fire hazard	Flash point 90 ˚C. Extinguish fires with water spray, carbon dioxide, dry chemical powder, or alcohol or polymer foam.
Spillage disposal	See general section.

RSC *Chemical Safety Data Sheets* Vol. 4a, No. 58, 1991 gives extended coverage.

518. 3,3-Dimethyl-1-phenyltriazene

Yellow liquid; b.p. 125-127 ˚C at 19 mm Hg.

Toxic effects	Harmful if inhaled, ingested, or absorbed through the skin. May cause eye and skin irritation.
Hazardous reactions	Exploded on attempted distillation at atmospheric pressure.
First aid	Standard treatment for exposure by all routes (see pages 108-122).
Fire hazard	Flash point 107 ˚C. Extinguish fires with water spray.
Spillage disposal	See general section.

519. Dimethylphosphine

Colourless liquid; b.p. 25 ˚C; insoluble in water.

Hazardous reactions	Readily ignites in air.

520. Dimethyl phthalate

Colourless liquid; m.p. 6 °C; b.p. 284 °C; sparingly soluble in water.

Limit values	OES short-term 10 mg m^{-3}; long-term 5 mg m^{-3}.
Toxic effects	Irritant to skin, eyes, upper respiratory tract, and mucous membranes. Harmful if inhaled, ingested, or absorbed through the skin. May cause reproductive disorders.
First aid	Standard treatment for exposure by all routes (see pages 108-122).
Fire hazard	Flash point 146 °C; Explosive limits 0.9-8.0%; Autoignition temperature 490 °C. Extinguish fires with carbon dioxide or dry chemical powder. Water spray or foam can also be used but may cause frothing.
Spillage disposal	See general section.

521. 2,2-Dimethylpropane

Colourless liquid or gas; b.p. 9.5 °C; insoluble in water.

RISKS
Extremely flammable liquefied gas (R13)

SAFETY PRECAUTIONS
Keep container in a well ventilated place – Keep away from sources of ignition – No Smoking – Take precautionary measures against static discharges (S9, S16, S33)

Toxic effects	This is classed as a simple asphyxiant, anaesthetic gas of low toxicity which may show irritant and narcotic effects in high concentrations.
First aid	Standard treatment for exposure by all routes (see pages 108-122).
Fire hazard	Flash point below -7 °C; Explosive limits 1.4-7.5%; Autoignition temperature 450 °C. As the gas is supplied in a cylinder, turning off the valve will reduce any fire involving it; if possible, cylinders should be removed from an area in which a fire has developed.
Spillage disposal	Surplus gas or leaking cylinder can be vented slowly to air in a safe area or the gas burnt-off in a suitable burner.

522. Dimethyl selenate

Hazardous reactions	Explodes at about 150 °C when distilled at atmospheric pressure.

523. Dimethyl sulfate

Colourless, odourless, oily liquid; m.p. -31.8 ˚C; b.p. 188 ˚C with decomposition; slightly soluble in water.

RISKS
May cause cancer – Toxic if swallowed – Very toxic by inhalation – Causes burns (R45, R25, R26, R34)

SAFETY PRECAUTIONS
Avoid exposure – Obtain special instructions before use – In case of contact with eyes, rinse immediately with plenty of water and seek medical advice – Take off immediately all contaminated clothing – In case of accident or if you feel unwell, seek medical advice immediately (show label where possible) (S53, S26, S27, S45)

Limit values	OES short-term 0.1 p.p.m. (0.5 mg m^{-3}) under review; long-term 0.1 p.p.m. (0.5 mg m^{-3}) under review.
Toxic effects	Vapour causes severe irritation of respiratory system, with possible severe lung injury after a latent period. Vapour and liquid irritate or burn the eyes severely after a latent period, resulting in temporary or permanent dimming of vision. The vapour or liquid may blister the skin and skin absorption may result in severe poisoning after a latent period. Extremely poisonous and irritant if taken by mouth.
Hazardous reactions	Reacts violently with concentrated aqueous ammonia; ignites with barium chlorite.
First aid	Standard treatment for exposure by all routes (see pages 108-122).
Fire hazard	Flash point 83.5 ˚C (open cup); Autoignition temperature 495 ˚C. Use water spray, carbon dioxide, dry chemical powder, or alcohol or polymer foam to extinguish fires.
Spillage disposal	See general section.

524. Dimethyl sulfoxide

Colourless, odourless, hygroscopic liquid with bitter taste; m.p. 18.5 ˚C; b.p. 189 ˚C; miscible with water.

Toxic effects	May cause redness, itching, and scaling of skin and damage to eyes. Absorbed readily by skin and volunteers have reported nausea, vomiting, cramps, chills, and drowsiness from applications.
Hazardous reactions	Subject to thermal decomposition just above b.p. Reacts violently or explosively with a wide variety of chemicals.
First aid	Standard treatment for exposure by all routes (see pages 108-122).
Fire hazard	Flash point 95 ˚C; Explosive limits 2.6-42%; Autoignition temperature 215 ˚C. Extinguish fires with carbon dioxide, dry chemical powder, or foam extinguishers.
Spillage disposal	See general section.

RSC *Chemical Safety Data Sheets* Vol. 1, No. 36, 1988 gives extended coverage.

525. *N,N-*Dimethyltoluidine

Pale yellow liquid, darkens on storage and in light; b.p. 211 ˚C; insoluble in water.

RISKS
Toxic by inhalation, in contact with skin, and if swallowed – Danger of cumulative effects (R23/24/25, R33)

SAFETY PRECAUTIONS
After contact with skin, wash immediately with plenty of water – Wear protective clothing and gloves – If you feel unwell, seek medical advice (show label where possible) (S28, S36/37, S44)

Toxic effects   Irritant to the eyes. May be harmful if inhaled, ingested, or absorbed through the skin.

First aid   Standard treatment for exposure by all routes (see pages 108-122), with particular attention to the possible occurrence of convulsions and/or shock.

Fire hazard   Flash point 83 ˚C (closed cup); Explosive limits 1.2%-7.0%. Extinguish fires with water spray, carbon dioxide, dry chemical powder, or alcohol or polymer foam.

Spillage disposal   See general section.

RSC *Chemical Safety Data Sheets* Vol. 4a, No. 60, 1991 gives extended coverage.

526. Dimethylzinc

Liquid; m.p. -42.5 ˚C; b.p. 46 ˚C; explosive reaction on contact with water.

Hazardous reactions   Ignites in air and explodes in oxygen.

527. Dinaphthoyl peroxide

White or pale yellow solid; m.p. 98 ˚C with decomposition.

Hazardous reactions   Explodes on friction.

528. 2,4-Dinitroaniline

Yellow granules or powder; m.p. 188 ˚C; insoluble in water.

RISKS
Very toxic by inhalation, in contact with skin, and if swallowed – Danger of cumulative effects
(R26/27/28, R33)

SAFETY PRECAUTIONS
After contact with skin, wash immediately with plenty of water – Wear protective clothing and
gloves – In case of accident or if you feel unwell, seek medical advice immediately (show label
where possible) (S28, S36/37, S45)

Toxic effects	Skin absorption is liable to cause dermatitis and cyanosis. Irritant to the skin, eyes, and upper respiratory tract.
Hazardous reactions	Hazardous preparation; a high fire and explosion risk with fast flame propagation.
First aid	Standard treatment for exposure by all routes (see pages 108-122).
Fire hazard	Use water spray, carbon dioxide, dry chemical powder, or alcohol or polymer foam to extinguish fires.
Spillage disposal	See general section.

529. 1,2-Dinitrobenzene

Pale yellow, crystalline powder; m.p. 117-119 ˚C; insoluble in water.

RISKS
Very toxic by inhalation, in contact with skin, and if swallowed – Danger of cumulative effects
(R26/27/28, R33)

SAFETY PRECAUTIONS
After contact with skin, wash immediately with plenty of water – Wear protective clothing and
gloves – In case of accident or if you feel unwell, seek medical advice immediately (show label
where possible) (S28, S36/37, S45)

Limit values	OES short-term 0.5 p.p.m. (3 mg m^{-3}); long-term 0.15 p.p.m. (1 mg m^{-3}).
Toxic effects	Absorption into the body leads to the formation of methaemoglobin which can cause cyanosis. May be fatal if ingested or absorbed through the skin. Onset of symptoms may be delayed for several hours.
Hazardous reactions	Mixtures with concentrated nitric acid possess high explosive properties.
First aid	Standard treatment for exposure by all routes (see pages 108-122).
Fire hazard	Use water spray, carbon dioxide, dry chemical powder, or alcohol or polymer foam to extinguish fires.
Spillage disposal	See general section.

530. 1,3-Dinitrobenzene

Colourless to yellow crystals; m.p. 80-90 °C; b.p. 297 °C; insoluble in water.

RISKS
Very toxic by inhalation, in contact with skin, and if swallowed – Danger of cumulative effects
(R26/27/28, R33)

SAFETY PRECAUTIONS
After contact with skin, wash immediately with plenty of water – Wear protective clothing and
gloves – In case of accident or if you feel unwell, seek medical advice immediately (show label
where possible) (S28, S36/37, S45)

Limit values   OES short-term 0.5 p.p.m. (3 mg m^{-3}); long-term 0.15 p.p.m. (1 mg m^{-3}).

Toxic effects   The vapour may cause headache, vertigo, and vomiting; in severe cases this
may be followed by exhaustion, cyanosis, drowsiness, and unconsciousness.
Contact will damage the eyes and skin absorption may lead to the above
symptoms. Poisonous if taken by mouth.

First aid   Standard treatment for exposure by all routes (see pages 108-122).

Fire hazard   Use water spray, carbon dioxide, dry chemical powder, or alcohol or polymer
foam to extinguish fires.

Spillage   See general section.
disposal

531. 2,4-Dinitrobenzenesulfenyl chloride

Solid; m.p. 95-96 °C.

Hazardous   Must not be overheated as it may explode.
reactions

532. Dinitrobutenes

Hazardous   Liable to violent decomposition or explosion when heated.
reactions

533. 4,6-Dinitro-*o*-cresol

Yellow crystals or powder; m.p. 85.8 °C; b.p. 312 °C; almost insoluble in water.

RISKS
Very toxic by inhalation, in contact with skin, and if swallowed – Danger of cumulative effects
(R26/27/28, R33)

SAFETY PRECAUTIONS
Keep locked up – Keep away from food, drink and animal feeding stuffs – After contact with
skin, wash immediately with plenty of water – In case of accident or if you feel unwell, seek
medical advice immediately (show label where possible) (S1, S13, S28, S45)

Limit values   OES short-term 0.6 mg m^{-3}; long-term 0.2 mg m^{-3}.

Toxic effects   The inhalation of dust may cause profuse sweating, fever, shortness of breath, and yellow coloration of skin of hands and feet. Similar symptoms may follow ingestion and absorption through the skin. Skin contact may cause dermatitis.

First aid   Standard treatment for exposure by all routes (see pages 108-122).

Fire hazard   Use water spray, carbon dioxide, dry chemical powder, or alcohol or polymer foam to extinguish fires.

Spillage   See general section.
 disposal

534. 1,2-Dinitroethane

Yellow liquid; m.p. -22.3 ˚C; b.p. 114 ˚C with explosion; insoluble in water.

Limit values   OES short-term 0.2 p.p.m. (1.2 mg m^{-3}); long-term 0.2 p.p.m. (1.2 mg m^{-3}).

Toxic effects   Absorbed through the skin, lungs, and gastro-intestinal tract. Acute exposure results in headache, nausea, vomiting, hypotension, and tachycardia.

First aid   Standard treatment for exposure by all routes (see pages 108-122).

535. Dinitrogen oxide

Colourless gas with mild sweetish taste; m.p. -102.5 ˚C; b.p. -89.5 ˚C; slightly soluble in water.

Toxic effects   Anaesthetic on inhalation.

Hazardous   Endothermic, with oxygen content 1.7 times that of air. Amorphous boron
 reactions   ignites when heated in the gas; a mixture of phosphine with excess of the oxide can be exploded by sparking; decomposes exothermally if heated locally; explosive mixtures formed with some substances.

First aid   Remove from exposure and give 100% oxygen if patient is unconscious.

536. Dinitrogen pentaoxide

White, crystalline solid; m.p. 30 ˚C; b.p. 47 ˚C with decomposition; soluble in water.

Hazardous   Alkali metals burn in gas, mercury and arsenic are vigorously oxidized;
 reactions   explodes with naphthalene, other organic materials react vigorously; reacts explosively with sulfur dichloride and sulfuryl chloride.

537. Dinitrogen tetroxide

A red-brown gas or yellow liquid; m.p. -9.3 ˚C; b.p. 20 ˚C; decomposes in water.

RISKS
Very toxic by inhalation – Irritating to respiratory system (R26, R37)

SAFETY PRECAUTIONS
Keep container tightly closed and in a well ventilated place – In case of contact with eyes, rinse immediately with plenty of water and seek medical advice – In case of accident or if you feel unwell, seek medical advice immediately (show label where possible) (S7/9, S26, S45)

Limit values	OES short-term 5 p.p.m. (9 mg m^{-3}) (as NO_2); long-term 3 p.p.m. (5 mg m^{-3}) (as NO_2).
Toxic effects	Although nitrogen dioxide has some irritant effect upon the respiratory system, its danger lies in the delay before its full effects upon the lungs are shown by feelings of weakness and coldness, headache, nausea, dizziness, abdominal pain, and cyanosis; in severe cases, convulsions and asphyxia may follow.
Hazardous reactions	Reacts violently or explosively with wide range of materials.
First aid	It is advisable in all cases where appreciable inhalation of nitrogen dioxide is believed to have occurred to obtain medical attention immediately, even if the person exposed is not complaining of discomfort. Removal from exposure, followed by rest and warmth are essential until under professional care. In severe cases administer oxygen through a face mask. If breathing has stopped apply artificial respiration.
Fire hazard	Use water spray to keep cylinders cool or move it out of the affected area.
Spillage disposal	Surplus gas or a leaking cylinder can be vented slowly into a water-fed scrubbing tower or column in a fume cupboard, or into a fume cupboard served with such a tower.

RSC *Chemical Safety Data Sheets* Vol. 4a, No.61, 1991 gives extended coverage.

538. 2,4-Dinitrophenol

Yellow crystals with bitter taste; m.p. 112 ˚C; sparingly soluble in water.

RISKS
Toxic by inhalation, in contact with skin, and if swallowed – Danger of cumulative effects (R23/24/25, R33)

SAFETY PRECAUTIONS
After contact with skin, wash immediately with plenty of water – Wear suitable gloves – If you feel unwell, seek medical advice (show label where possible) (S28, S37, S44)

Toxic effects	The inhalation of dust may cause profuse sweating, fever, shortness of breath, and yellow coloration of skin of hands and feet. Similar symptoms may follow ingestion and absorption through the skin. Skin contact may cause dermatitis.

First aid   Standard treatment for exposure by all routes (see pages 108-122).

Fire hazard   Use water spray, carbon dioxide, dry chemical powder, or alcohol or polymer foam to extinguish fires.

Spillage   See general section.
disposal

539. Dinitrophenylhydrazine

Red, crystalline powder; m.p. about 200 °C with decomposition; slightly soluble in water which is normally added in storage to reduce explosion risk.

Toxic effects Irritant to skin and eyes. Harmful if inhaled, ingested, or absorbed through the skin. Symptoms may be delayed by several hours. Absorption into the body gives rise to methaemoglobinaemia which can cause cyanosis.

First aid   Standard treatment for exposure by all routes (see pages 108-122).

Fire hazard   Use water spray, carbon dioxide, dry chemical powder, or alcohol or polymer foam to extinguish fires.

Spillage   See general section.
disposal

540. 1,2-Dinitropropane

Colourless liquid with disagreeable odour; m.p. -27.7 °C; b.p. 92 °C at 10 torr but decomposes above 121 °C; sparingly soluble in water.

Limit values OES short-term 0.2 p.p.m. (1.2 mg m^{-3}); long-term 0.2 p.p.m. (1.2 mg m^{-3}).

Toxic effects Irritant to skin and eyes. Systemic effects follow inhalation or ingestion.

First aid   Standard treatment for exposure by all routes (see pages 108-122).

541. 3,5-Dinitro-*o*-toluamide

Yellowish solid; m.p. 177 °C; very slightly soluble in water.

Hazardous   A batch of material exploded while being processed under certain conditions.
reactions   The finely powdered material is a significant dust explosion hazard.

First aid   Standard treatment for exposure by all routes (see pages 108-122).

542. 2,4-Dinitrotoluene

Yellow needles; m.p. 69.5 °C; b.p. 300 °C with slight decomposition; insoluble in water.

RISKS
Toxic by inhalation, in contact with skin, and if swallowed – Danger of cumulative effects (R23/24/25, R33)

SAFETY PRECAUTIONS
After contact with skin, wash immediately with plenty of water – Wear suitable gloves – If you feel unwell, seek medical advice (show label where possible) (S28, S37, S44)

Limit values	OES short-term 5 mg m^{-3} under review; long-term 1.5 mg m^{-3} under review.
Toxic effects	The vapour and crystals irritate the eyes. Contact with the skin may cause dermatitis. Onset of symptoms following absorption into the body may be delayed by several hours.
Hazardous reactions	Prolonged heating with alkali at over 150 °C may cause explosion.
First aid	Standard treatment for exposure by all routes (see pages 108-122).
Fire hazard	Flash point 207 °C (closed cup). Use water spray, carbon dioxide, dry chemical powder, or alcohol or polymer foam to extinguish fires.
Spillage disposal	See general section.

RSC *Chemical Safety Data Sheets* Vol. 4a, No. 63, 1991 gives extended coverage.

543. Dinonyl phthalate

Liquid; b.p. 413 °C; insoluble in water.

Limit values	OES long-term 5 mg m^{-3}.
Toxic effects	Irritant to skin and eyes. Toxic on inhalation or ingestion.
First aid	Standard treatment for exposure by all routes (see pages 108-122).
Fire hazard	Flash point 215 °C. Extinguish fires with water spray, carbon dioxide, dry chemical powder, or alcohol or polymer foam.
Spillage disposal	See general section.

544. 1,4-Dioxane

Colourless, almost odourless, liquid; m.p. 11.8 °C; b.p. 101 °C; miscible with water.

RISKS
Highly flammable – Irritating to eyes and respiratory system – Possible risk of irreversible effects (R11, R36/37, R40)

SAFETY PRECAUTIONS
Keep away from sources of ignition – No Smoking – Wear protective clothing and gloves (S16, S36/37)

Limit values	OES short-term 100 p.p.m. (360 mg m^{-3}); long-term 25 p.p.m. (90 mg m^{-3}).
Toxic effects	The vapour irritates nose and eyes and this may be followed by headache and drowsiness. High concentrations may also cause nausea and vomiting, while injury to the kidney and liver are possible. Shows some of the above effects when taken by mouth. It can be absorbed through the skin. Weak animal carcinogen.
Hazardous reactions	Forms explosive peroxides on exposure to air; reacts almost explosively with Raney nickel above 210 °C; addition complex with sulfur trioxide decomposes violently on storage.
First aid	Standard treatment for exposure by all routes (see pages 108-122).
Fire hazard	Flash point 12 °C; Explosive limits 2-22%; Autoignition temperature 180 °C. Extinguish fire with water spray, dry powder, or carbon dioxide.
Spillage disposal	See general section.

RSC *Chemical Safety Data Sheets* Vol. 1, No. 37, 1988 gives extended coverage.

545. Dioxolane

Colourless liquid; m.p. -95 °C; b.p. 74-75 °C; miscible with water.

RISKS
Highly flammable (R11)

SAFETY PRECAUTIONS
Keep away from sources of ignition – No Smoking (S16)

Toxic effects	Irritant to the eyes. Low acute toxicity.
First aid	Standard treatment for exposure by all routes (see pages 108-122).
Fire hazard	Flash point 2 °C (closed cup). Extinguish fires with water spray, carbon dioxide, or dry chemical powder.
Spillage disposal	See general section.

RSC *Chemical Safety Data Sheets* Vol. 1, No. 38, 1988 gives extended coverage.

546. Dioxygen difluoride

Hazardous reactions	Very powerful oxidant reacting with many materials at low temperatures.

547. Dipentene

Colourless liquid with pleasant, lemon-like odour; b.p. 178 ˚C; insoluble in water.

RISKS
Flammable – Irritating to skin (R10, R38)

SAFETY PRECAUTIONS
After contact with skin, wash immediately with plenty of water (S28)

Toxic effects   It is considered to be a moderate skin irritant and sensitizer, but its effects are not well documented. Harmful if inhaled or ingested.

First aid   Standard treatment for exposure by all routes (see pages 108-122).

Fire hazard   Flash point 45 ˚C; Autoignition temperature 237 ˚C. Extinguish fire with foam, dry powder, or carbon dioxide.

Spillage disposal   See general section.

548. Diphenylamine

Colourless or grey crystals with floral smell; m.p. 52-55 ˚C; b.p. 302 ˚C; insoluble in water.

RISKS
Toxic by inhalation, in contact with skin, and if swallowed – Danger of cumulative effects (R23/24/25, R33)

SAFETY PRECAUTIONS
After contact with skin, wash immediately with plenty of water – Wear protective clothing and gloves – If you feel unwell, seek medical advice (show label where possible) (S28, S36/37, S44)

Limit values   OES short-term 20 mg m^{-3}; long-term 10 mg m^{-3}.

Toxic effects   Irritant to mucous membranes, eyes, and skin. Toxic action similar to aniline, but less severe.

First aid   Standard treatment for exposure by all routes (see pages 108-122).

Fire hazard   Flash point 153 ˚C (closed cup); Autoignition temperature 634-635 ˚C. Extinguish fires with water spray, carbon dioxide, dry chemical powder, or alcohol or polymer foam.

Spillage disposal   See general section.

RSC *Chemical Safety Data Sheets* Vol. 4a, No. 64, 1991 gives extended coverage.

549. Diphenyldistibene

Hazardous reactions   Ignites in air and is oxidized explosively by nitric acid.

550. Diphenyl ether

Colourless crystalline solid; m.p. 26-30 ˚C; b.p. 259 ˚C; very slightly soluble in water.

Limit values OES long-term 1 p.p.m. (7 mg m^{-3}) (as vapour).

Toxic effects Irritant to skin and eyes on prolonged exposure.

Fire hazard Flash point 110 ˚C.

551. 1,1-Diphenylethylene

Liquid; m.p. 8 ˚C; b.p. 277 ˚C.

Toxic effects Irritant to skin, eyes, and upper respiratory tract. Harmful on inhalation or ingestion.

Hazardous reactions Forms explosive peroxide with oxygen.

First aid Standard treatment for exposure by all routes (see pages 108-122).

552. Diphenylmagnesium

Solid; m.p. decomposes at 280 ˚C; reacts violently with water.

Hazardous reactions Ignites in moist air; reacts violently with water.

553. Diphenylmercury

White powder; m.p. 128-129 ˚C.

Limit values OES short-term 0.15 mg m^{-3} (as Hg); long-term 0.05 mg m^{-3} (as Hg).

Toxic effects Extremely destructive to tissues of skin, eyes, upper respiratory tract, and mucous membranes. Harmful if inhaled, ingested, or absorbed through the skin. Inhalation may lead to inflammation and oedema of the larynx and bronchi, chemical pneumonitis, and eventually death.

Hazardous reactions Reacts violently with sulfur trioxide or dichlorine monoxide.

First aid Standard treatment for exposure by all routes (see pages 108-122).

Fire hazard Use water spray, carbon dioxide, dry chemical powder, or alcohol or polymer foam to extinguish fires.

Spillage disposal See general section.

554. Diphenyl thiocarbazone

Bluish black, crystalline powder; insoluble in water.

Toxic effects   Irritant to skin, eyes, and upper respiratory tract. Harmful if inhaled, ingested, or absorbed through the skin. Carcinogenic.

First aid   Standard treatment for exposure by all routes (see pages 108-122).

Fire hazard   Extinguish fires with water spray, carbon dioxide, dry chemical powder, or alcohol or polymer foam.

Spillage disposal   See general section.

555. Diphenyltin

Limit values   OES short-term 0.2 mg m^{-3} (as Sn); long-term 0.1 mg m^{-3} (as Sn).

Hazardous reactions   Ignites on contact with fuming nitric acid.

556. Diphenyltriazene

Golden yellow crystals; m.p. 98-99 ˚C with explosive decomposition.

Toxic effects   Toxic by ingestion. Experimental tumourigen.

Hazardous reactions   Decomposes explosively at m.p. 98 ˚C; mixture with acetic anhydride exploded on warming.

557. Diphosphane

Hazardous reactions   Ignites in air; when present 0.2% v/v causes other flammable gases to ignite in air.

558. Dipotassium acetylide

Hazardous reactions   Contact with water may cause ignition and explosion of evolved acetylene.

559. Dipropionyl peroxide

Oily liquid.

Toxic effects   Irritant to skin, eyes, and upper respiratory tract.

Hazardous reactions   Explodes on standing at room temperature.

First aid   Standard treatment for exposure by all routes (see pages 108-122).

560. Dipropylene glycol methyl ether

Colourless liquid with ethereal odour; m.p. -80 °C; b.p. 188.3 °C; miscible with water.

Toxic effects   Irritant to skin and eyes. A mild allergen.

First aid   Standard treatment for exposure by all routes (see pages 108-122).

Fire hazard   Flash point 85 °C. Extinguish fires with water spray, carbon dioxide, dry chemical powder, or foam.

561. Dipyridinesodium

Hazardous reactions   This addition product of sodium and pyridine ignites in air.

562. Disilane

Gas with repulsive odour; m.p. -132.5 °C; b.p. -14.5 °C.

Toxic effects   Toxic by inhalation.

Hazardous reactions   Explodes on contact with some compounds; ignites spontaneously in air.

563. Disodium peroxodisulfate

White, crystalline solid; m.p. decomposes on heating; soluble in water.

Limit values   OES long-term 1 mg m^{-3} (as S_2O_8).

Toxic effects   Irritant to skin, eyes, upper respiratory tract, and mucous membranes. May cause allergic reactions. Certain sensitive individuals may develop asthma or eczema on inhalation.

Hazardous reactions   Some metals may initiate decomposition (not stainless steel).

First aid   Standard treatment for exposure by all routes (see pages 108-122).

Fire hazard   Use water spray to extinguish fires.

Spillage disposal   See general section.

564. Disulfur dichloride

Yellow-brown, fuming liquid with a suffocating odour; b.p. 136 °C; decomposed by water with formation of hydrochloric acid, sulfur dioxide, and sulfur.

RISKS
Reacts violently with water – Causes burns – Irritating to respiratory system (R14, R34, R37)

SAFETY PRECAUTIONS
In case of contact with eyes, rinse immediately with plenty of water and seek medical advice (S26)

Limit values   OES short-term 1 p.p.m. (6 mg m^{-3}).

Toxic effects   The vapour irritates the respiratory system. The vapour and liquid irritate the eyes and skin severely and the liquid may cause burns. Its ingestion would result in severe internal irritation and damage.

Hazardous   Vigorous or violent reactions with a number of inorganic and organic
reactions   substances.

First aid   Standard treatment for exposure by all routes (see pages 108-122).

Spillage   See general section.
disposal

565. Disulfur heptaoxide

Viscous liquid; m.p. 0 °C; decomposed by water.

Hazardous   Explodes in moist air.
reactions

566. Disulfuryl dichloride

Colourless, mobile, fuming liquid; m.p. -37 °C; b.p. 151 °C; reacts violently with water.

Toxic effects   Corrosive to skin, eyes, and mucous membranes. Harmful if absorbed into the body.

Hazardous   Reacts vigorously with red phosphorus; reaction with water can be violent.
reactions

567. Disulfuryl difluoride

Hazardous   Reacts violently with ethanol.
reactions

568. Divanadium pentoxide

Reddish yellow, crystalline solid; m.p. 670 °C; b.p. 1750 °C with decomposition; almost insoluble in water.

RISKS
I larmful by inhalation (R20)

SAFETY PRECAUTIONS
Do not breathe dust (S22)

Limit values OES long-term 0.5 mg m^{-3} (total inhalable dust) (as V), 0.05 mg m^{-3} (fume and respirable dust) (as V).

Toxic effects Irritant to skin, eyes, upper respiratory tract, and mucous membranes. May be fatal following inhalation, ingestion, or skin absorption.

First aid Standard treatment for exposure by all routes (see pages 108-122).

Fire hazard Use extinguishant appropriate to the surroundings.

Spillage disposal See general section.

569. Divinylbenzene

Colourless liquid; b.p. 195 °C.

Limit values OES long-term I0 p.p.m. (50 mg m^{-3}).

Toxic effects Irritant to skin, eyes, upper respiratory tract, and mucous membranes.

First aid Standard treatment for exposure by all routes (see pages 108-122).

Fire hazard Flash point 61 °C; Explosive limits 0.7-6.5 °C; Autoignition temperature 470 °C. Extinguish fires with water spray, carbon dioxide, dry chemical powder, or alcohol or polymer foam.

Spillage disposal See general section.

570. Dodecanoyl peroxide

White powder; m.p. 55 °C; decomposes slowly above 60 °C especially in sunlight; insoluble in water.

Toxic effects It irritates and may burn the skin and eyes; irritant and harmful if taken by mouth.

Hazardous reactions Becomes shock-sensitive on heating.

First aid Standard treatment for exposure by all routes (see pages 108-122).

Spillage disposal See general section.

571. 2,3-Epoxy-1-propanol

Colourless, slightly viscous liquid; m.p. -45 ˚C; b.p. 162 ˚C; completely soluble in water.

RISKS
Harmful in contact with skin and if swallowed – Toxic by inhalation – Irritating to eyes, respiratory system and skin – May cause sensitization by inhalation and skin contact (R21/22, R23, R36/37/38, R42/43)

SAFETY PRECAUTIONS
If you feel unwell, seek medical advice (show label where possible) (S44)

Toxic effects	As above. Experimental animal carcinogen.
Hazardous reactions	May polymerize explosively if stored inappropriately.
First aid	Standard treatment for exposure by all routes (see pages 108-122).
Fire hazard	Flash point 72 ˚C (closed cup); Autoignition temperature 415 ˚C. Extinguish fires with carbon dioxide, dry chemical powder, or alcohol or polymer foam.
Spillage disposal	See general section.

RSC *Chemical Safety Data Sheets* Vol. 4a, No. 65, 1991 gives extended coverage.

572. 2,3-Epoxypropyl acrylate

Colourless liquid; b.p. 115 ˚C at 78 mm Hg; insoluble in water.

RISKS
Toxic by inhalation, in contact with skin, and if swallowed – Causes burns – May cause sensitization by skin contact (R23/24/25, R34, R43)

SAFETY PRECAUTIONS
In case of contact with eyes, rinse immediately with plenty of water and seek medical advice – Wear suitable protective clothing, gloves, and eye/face protection – If you feel unwell, seek medical advice (show label where possible) (S26, S36/37/39, S44)

Toxic effects	Extremely irritating and destructive to eyes, upper respiratory tract, and mucous membranes.
First aid	Standard treatment for exposure by all routes (see pages 108-122), with special attention needed in case of frostbite from skin contact.
Fire hazard	Flash point 61 ˚C (open cup). Extinguish fires with water spray, carbon dioxide, dry chemical powder, or foam extinguishers.
Spillage disposal	See general section.

RSC *Chemical Safety Data Sheets* Vol. 4a, No. 66, 1991 gives extended coverage.

573. Epoxy resins

Toxic effects   Cured resins are relatively harmless. The degree of toxicity of uncured resin mixtures is dependent on the nature and proportion of uncured components. The main effects would be on the skin, *e.g.* dermatitis and eczema.

First aid   Standard treatment for exposure by all routes (see pages 108-122).

Spillage disposal   See general section.

574. Ethane

Colourless gas; b.p. -89 ˚C; insoluble in water.

RISKS
Extremely flammable (R12)

SAFETY PRECAUTIONS
Keep container in a well ventilated place – Keep away from sources of ignition – No Smoking – Take precautionary measures against static discharges (S9, S16, S33)

Toxic effects   Considered to be a simple asphyxiant gas which can act as an anaesthetic at high concentrations.

First aid   For gas inhaled in quantity apply standard treatment.

Fire hazard   Explosive limits 3-12.5%; Autoignition temperature 515 ˚C. As the gas is supplied in a cylinder, turning off the valve will reduce any fire involving it; if possible, cylinders should be removed quickly from an area in which a fire has developed.

Spillage disposal   Surplus gas or leaking cylinder can be vented slowly to air in a safe, open area or gas burnt-off through a suitable burner.

575. Ethane-1,2-dione

Yellow crystals or liquid; m.p. 15 ˚C; b.p. 50.4 ˚C; miscible with water.

RISKS
Irritating to eyes and skin (R36/38)

SAFETY PRECAUTIONS
In case of contact with eyes, rinse immediately with plenty of water and seek medical advice – After contact with skin, wash immediately with plenty of water (S26, S28)

Toxic effects   Irritating to eyes, skin, and mucous membranes. Lung congestion and convulsions may occur following inhalation. Mutagenic.

Hazardous reactions   Oxidation of paraldehyde with nitric acid may become violent.

First aid   Standard treatment for exposure by all routes (see pages 108-122).

Fire hazard   Use carbon dioxide, dry chemical powder, or alcohol or polymer foam to extinguish fires.

Spillage   See general section.
  disposal

RSC *Chemical Safety Data Sheets* Vol. 3, No. 22, 1990 gives extended coverage.

576. Ethanethiol

Colourless liquid with penetrating garlic-like odour; b.p. 35 ˚C; sparingly soluble in water.

RISKS
Highly flammable – Harmful by inhalation (R11, R20)

SAFETY PRECAUTIONS
Keep away from sources of ignition – No Smoking – Avoid contact with eyes (S16, S25)

Limit values   OES short-term 2 p.p.m. (3 mg m^{-3}); long-term 0.5 p.p.m. (1 mg m^{-3}).

Toxic effects   The vapour irritates the respiratory system and may be narcotic in high concentrations. The liquid irritates the eyes and mucous membranes; it is assumed to be poisonous if taken by mouth.

First aid   Standard treatment for exposure by all routes (see pages 108-122).

Fire hazard   Flash point below 21 ˚C; Explosive limits 2.8-18%; Autoignition temperature 299 ˚C. Extinguish fire with water spray, foam, dry powder, or carbon dioxide.

Spillage   See general section.
  disposal

577. Ethanol

Colourless, mobile liquid with characteristic smell. Ethanol is supplied to laboratories in a variety of forms mainly because of Excise control. In the United Kingdom, 'duty paid' material comes as 90% (rectified spirit), 95%, and 99/100% (absolute) grades; denatured ethanol is sold as industrial methylated spirit (supplied against Excise requisitions) and methylated spirit (mineralized). The materials used for denaturing ethanol add substantially to its toxicity and are not taken into account in this monograph which deals with pure (98/100%) material; m.p. -130 ˚C; b.p. 79 ˚C; miscible with water.

RISKS
Harmful if swallowed (R22)

SAFETY PRECAUTIONS
Keep container tightly closed – Keep away from sources of ignition – No Smoking (S7, S16)

Limit values   OES long-term 1000 p.p.m. (1900 mg m^{-3}).

Toxic effects   The intoxicating qualities of ethanol are so well appreciated that a stark summary of them is superfluous.

Hazardous reactions	Reacts with varying degrees of violence with a wide range of oxidants, including silver nitrate with which an explosion was reported.
First aid	Standard treatment for exposure by all routes (see pages 108-122).
Fire hazard	Flash point 12 °C; Explosive limits 3.3-19%; Autoignition temperature 363 °C. Extinguish fire with water spray, dry powder, or carbon dioxide.
Spillage disposal	See general section.

RSC *Chemical Safety Data Sheets* Vol. 1, No. 40, 1988 gives extended coverage.

578. Ethanolamine

Colourless, viscous liquid with mild ammoniacal smell; m.p. 10.5 °C; b.p. 170 °C; miscible with water.

RISKS
Harmful by inhalation – Irritating to eyes, respiratory system, and skin (R20, R36/37/38)

Limit values	OES short-term 6 p.p.m. (15 mg m^{-3}); long-term 3 p.p.m. (8 mg m^{-3}).
Toxic effects	As the vapour pressure is low, it is unlikely to cause irritation of the respiratory system except when the liquid is hot. The liquid irritates the eyes and skin and may irritate the alimentary system if taken by mouth.
First aid	Standard treatment for exposure by all routes (see pages 108-122).
Fire hazard	Flash point 85 °C; Autoignition temperature 410 °C. Extinguish fires by using water, alcohol-resistant foam, carbon dioxide, or dry powder extinguisher.
Spillage disposal	See general section.

RSC *Chemical Safety Data Sheets* Vol. 1, No. 41, 1988 gives extended coverage.

579. Ethoxyacetylene

Liquid; b.p. 61 °C; insoluble in water.

Hazardous reactions	Explodes when heated at around 100 °C in sealed tubes.
Fire hazard	Flash point -7 °C. Extinguish fire with carbon dioxide, dry chemical powder, or foam extinguishers.

580. Ethoxyanilines

Ethoxyanilines are red to brown liquids; m.p. *o-* below -21 ˚C; *p-* 3-4 ˚C; b.p. *o-* 229 ˚C; *p-*254 ˚C; *m-* 248 ˚C; immiscible with water.

RISKS
Toxic by inhalation, in contact with skin, and if swallowed – Danger of cumulative effects (R23/24/25, R33)

SAFETY PRECAUTIONS
After contact with skin, wash immediately with plenty of water – Wear protective clothing and gloves – In case of accident or if you feel unwell, seek medical advice immediately (show label where possible) (S28, S36/37, S45)

Toxic effects Irritant to skin, eyes, and upper respiratory tract. Skin absorption may lead to headache, drowsiness, and cyanosis.

First aid Standard treatment for exposure by all routes (see pages 108-122).

Fire hazard Flash point *p-* 115 ˚C; *m-* over 110 ˚C. Use water spray, carbon dioxide, dry chemical powder, or alcohol or polymer foam to extinguish fires.

Spillage disposal See general section.

RSC *Chemical Safety Data Sheets* Vol. 4b, No. 113, 1991 gives extended coverage of the *p-* isomer.

581. Ethoxypropyne

Liquid with a penetrating odour; b.p. 80 ˚C; moderately soluble in water.

Hazardous reactions Exploded during distillation under vacuum.

582. Ethyl acetate

Colourless, volatile liquid with fragrant odour; m.p. -83.6 ˚C; b.p. 77 ˚C; very slightly soluble water.

RISKS
Highly flammable (R11)

SAFETY PRECAUTIONS
Keep away from sources of ignition – No Smoking – Do not breathe vapour – Do not empty into drains – Take precautionary measures against static discharges (S16, S23, S29, S33)

Limit values OES long-term 400 p.p.m. (1400 mg m^{-3}).

Toxic effects The vapour may irritate the eyes and respiratory system. The liquid irritates the eyes and mucous membranes. Prolonged inhalation may cause kidney and liver damage.

Hazardous reactions May ignite or explode with solid lithium aluminium hydride or potassium t-butoxide.

First aid	Standard treatment for exposure by all routes (see pages 108-122).
Fire hazard	Flash point -4.4 ˚C; Explosive limits 2.2-11.5%; Autoignition temperature 426 ˚C. Extinguish small fires with water spray, foam, dry powder, or carbon dioxide. Large fires are best controlled by alcohol-resistant foam.
Spillage disposal	See general section.

RSC *Chemical Safety Data Sheets* Vol. 1, No. 42, 1988 gives extended coverage.

583. Ethyl acetoacetate

Colourless liquid; m.p. -43 ˚C; b.p. 181 ˚C; soluble in water.

Toxic effects	Irritant to skin, eyes, upper respiratory tract, and mucous membranes. May be absorbed through the skin.
Hazardous reactions	Violent decomposition occurred during the preparation of zinc chelate of tris-2(bromomethyl)ethyl acetoacetate.
First aid	Standard treatment for exposure by all routes (see pages 108-122).
Fire hazard	Flash point 84 ˚C. Extinguish fires with carbon dioxide, dry chemical powder, or alcohol or polymer foam.
Spillage disposal	See general section.

584. Ethyl acrylate

Colourless liquid with acrid, penetrating odour; normally supplied containing stabilizer; m.p. -71.2 ˚C; b.p. 100 ˚C; immiscible with water.

RISKS
Highly flammable – Harmful by inhalation and if swallowed – Irritating to eyes, respiratory system and skin – May cause sensitization by skin contact (R11, R20/22, R36/37/38, R43)

SAFETY PRECAUTIONS
Keep container in a well ventilated place – Keep away from sources of ignition – No Smoking – Take precautionary measures against static discharges (S9, S16, S33)

Limit values	OES short-term 15 p.p.m. (60 mg m^{-3}); long term 5 p.p.m. (20 mg m^{-3}).
Toxic effects	The vapour irritates the eyes and respiratory system. High concentrations may result in lethargy and convulsions. The liquid irritates the skin and eyes. High degree of toxicity if taken by mouth.
Hazardous reactions	Polymerization in a large clear glass bottle burst it.
First aid	Standard treatment for exposure by all routes (see pages 108-122).
Fire hazard	Flash point 15.5 ˚C (open cup); Explosive limits 1.4-14.0%; Autoignition temperature 273 ˚C. Extinguish fire with dry powder or carbon dioxide.
Spillage disposal	See general section.

RSC *Chemical Safety Data Sheets* Vol. 3, No. 23, 1990 gives extended coverage.

585. Ethylamine and solutions

Colourless, volatile liquid with a fishy ammoniacal odour; m.p. -80.6 ˚C; b.p. 17 ˚C; the solutions in water and ethanol are commonly used.

RISKS
Extremely flammable liquefied gas – Irritating to eyes and respiratory system (R13, R36/37)

SAFETY PRECAUTIONS
Keep away from sources of ignition – No Smoking – In case of contact with eyes, rinse immediately with plenty of water and seek medical advice – Do not empty into drains (S16, S26, S29)

Limit values   OES long-term 10 p.p.m. (18 mg m^{-3}).

Toxic effects   The vapour irritates the eyes, mucous membranes, and respiratory system; in high concentrations it may affect the nervous system. The solutions irritate the eyes and skin. Assumed to be irritant and poisonous if taken by mouth.

First aid   Standard treatment for exposure by all routes (see pages 108-122).

Fire hazard   Flash point below -18 ˚C; Explosive limits 3.5-14%; Autoignition temperature 384 ˚C. Extinguish fire with water spray or dry powder.

Spillage   See general section.
disposal

586. *N*-Ethylaniline

Colourless to yellow-brown liquid; m.p. -63 ˚C; b.p. 204 ˚C; insoluble in water.

RISKS
Toxic by inhalation, in contact with skin, and if swallowed – Danger of cumulative effects (R23/24/25, R33)

SAFETY PRECAUTIONS
After contact with skin, wash immediately with plenty of water – Wear suitable gloves – If you feel unwell, seek medical advice (show label where possible) (S28, S37, S44)

Toxic effects   Excessive breathing of the vapour or absorption of the liquid through the skin can cause headache, drowsiness, cyanosis, mental confusion, and, in severe cases, convulsions. The liquid damages the eyes, and the above effects can be expected if it is swallowed. Continued exposure to the vapour, or slight skin exposure to the liquid, over a period may affect the nervous system and the blood, causing fatigue, loss of appetite, headache, and dizziness.

First aid   Standard treatment for exposure by all routes (see pages 108-122).

Fire hazard   Flash point 85 ˚C (open cup); Explosive limits 1.6-9.5. Extinguish fires with water spray, carbon dioxide, dry chemical powder, or alcohol or polymer foam.

Spillage   See general section.
disposal

587. Ethylbenzene

Colourless liquid with aromatic odour; m.p. -95 ˚C; b.p. 136 ˚C; immiscible with water.

RISKS
Highly flammable – Harmful by inhalation (R11, R20)

SAFETY PRECAUTIONS
Keep away from sources of ignition – No Smoking – Avoid contact with skin and eyes – Do not empty into drains (S16, S24/25, S29)

Limit values   OES short-term 125 p.p.m. (545 mg m^{-3}); long-term 100 p.p.m. (435 mg m^{-3}).

Toxic effects   The vapour irritates the eyes and respiratory system and may cause dizziness. The liquid irritates the skin and eyes.

First aid   Standard treatment for exposure by all routes (see pages 108-122).

Fire hazard   Flash point 15 ˚C; Explosive limits 1-6.7%; Autoignition temperature 432 ˚C. Extinguish fire with foam, dry powder, or carbon dioxide.

Spillage disposal   See general section.

RSC *Chemical Safety Data Sheets* Vol. 1, No. 43, 1988 gives extended coverage.

588. Ethyl bromoacetate

Colourless to straw coloured liquid; m.p. -13.8 ˚C; b.p. 168 ˚C; insoluble in and partially decomposed by water.

RISKS
Very toxic by inhalation, in contact with skin, and if swallowed (R26/27/28)

SAFETY PRECAUTIONS
Keep container tightly closed and in a well ventilated place – In case of contact with eyes, rinse immediately with plenty of water and seek medical advice – In case of accident or if you feel unwell, seek medical advice immediately (show label where possible) (S7/9, S26, S45)

Toxic effects   Irritating and destructive to skin, eyes, upper respiratory tract, and mucous membranes. Inhalation can be fatal. Experimental animal carcinogen.

First aid   Standard treatment for exposure by all routes (see pages 108-122). Particular attention should be paid following ingestion as this can lead to convulsions, unconsciousness, cessation of heart, and/or breathing.

Fire hazard   Flash point 47 ˚C. Extinguish fires with carbon dioxide, dry chemical powder, or foam extinguishers.

Spillage disposal   See general section.

RSC *Chemical Safety Data Sheets* Vol. 4a, No. 68, 1991 gives extended coverage.

589. Ethyl butyrate

Colourless liquid with pineapple-like odour; m.p. -100 ˚C; b.p. 121 ˚C; sparingly soluble in water.

Toxic effects   In high concentrations the vapour irritates the eyes and respiratory system, and is narcotic.

First aid   Standard treatment for exposure by all routes (see pages 108-122).

Fire hazard   Flash point 26 ˚C; Autoignition temperature 463 ˚C. Extinguish fire with water spray, foam, dry powder, or carbon dioxide.

Spillage disposal   See general section.

590. Ethyl chloroacetate

Colourless liquid with pungent, fruity odour; m.p. -26 ˚C; b.p. 144 ˚C; immiscible with water.

RISKS
Toxic by inhalation, in contact with skin, and if swallowed (R23/24/25)

SAFETY PRECAUTIONS
Keep container tightly closed and in a well ventilated place – If you feel unwell, seek medical advice (show label where possible) (S7/9, S44)

Toxic effects   The vapour irritates the respiratory system. The vapour and liquid irritate the eyes severely. Repeated or prolonged contact between the liquid and skin causes irritation. Assumed to be irritant if taken by mouth.

First aid   Standard treatment for exposure by all routes (see pages 108-122).

Fire hazard   Flash point 64 ˚C (open cup). Extinguish fire with water spray, foam, dry powder, or carbon dioxide.

Spillage disposal   See general section.

RSC *Chemical Safety Data Sheets* Vol. 4a, No. 69, 1991 gives extended coverage.

591. Ethyl chloroformate

Colourless liquid with irritating odour; m.p. -81 ˚C; b.p. 94 ˚C; immiscible with water.

RISKS
Highly flammable – Toxic by inhalation – Irritating to eyes, respiratory system, and skin (R11, R23, R36/37/38)

SAFETY PRECAUTIONS
Keep container in a well ventilated place – Keep away from sources of ignition – No Smoking – Take precautionary measures against static discharges – If you feel unwell, seek medical advice (show label where possible) (S9, S16, S33, S44)

Limit values   OES long-term 1 p.p.m. (4.4 mg m^{-3}).

| Toxic effects | The vapour irritates the eyes and respiratory system. The liquid is irritant to skin, eyes, and gastro-intestinal system. |

Toxic effects The vapour irritates the eyes and respiratory system. The liquid is irritant to skin, eyes, and gastro-intestinal system.

First aid Standard treatment for exposure by all routes (see pages 108-122).

Fire hazard Flash point 16.1 °C (closed cup); Autoignition temperature 500 °C. Extinguish fire with water spray, foam, dry powder, or carbon dioxide.

Spillage disposal See general section.

RSC *Chemical Safety Data Sheets* Vol. 4a, No. 70, 1991 gives extended coverage.

592. Ethyl diazoacetate

Yellow oily liquid; m.p. -22 °C; b.p. 140-141 °C at 720 mm Hg; slightly soluble in water.

Toxic effects Irritant to skin, eyes, and upper respiratory tract. Absorbed through the skin. Experimental carcinogen.

Hazardous reactions Explosive; distillation may be dangerous.

First aid Standard treatment for exposure by all routes (see pages 108-122).

Fire hazard Flash point 26 °C. Use water spray to extinguish fires.

Spillage disposal See general section.

593. Ethylene

Colourless gas with sweetish smell and taste; m.p. -169 °C; b.p. -103.7 °C; soluble in water.

RISKS
Extremely flammable liquefied gas (R13)

SAFETY PRECAUTIONS
Keep container in a well ventilated place – Keep away from sources of ignition – No Smoking – Take precautionary measures against static discharges (S9, S16, S33)

Toxic effects Considered to be a simple asphyxiant gas which can act as an anaesthetic at high concentrations.

Hazardous reactions Mixtures of ethylene and several oxidants and halogenated substances are liable to explode; highly compressed gas may decompose explosively if sparked or in presence of copper.

First aid Standard treatment for exposure by all routes (see pages 108-122).

Fire hazard Explosive limits 3.1-32%; Autoignition temperature 450 °C. As the gas is supplied in a cylinder, turning off the valve will reduce any fire involving it; if possible, cylinders should be removed quickly from an area in which a fire has developed.

Spillage disposal Surplus gas or leaking cylinder can be vented slowly to air in a safe, open area or gas burnt-off through a suitable burner.

594. Ethylene dimethacrylate

Colourless or pale yellow, mobile liquid with mild smell; b.p. 120 ˚C; insoluble in water.

RISKS
Irritating to eyes and respiratory system (R36/37)

Toxic effects   Irritant to skin, eyes, and upper respiratory tract. May cause skin sensitization.

First aid   Standard treatment for exposure by all routes (see pages 108-122).

Fire hazard   Flash point above 110 ˚C (closed cup). Extinguish fires with water spray, carbon dioxide, dry chemical powder, or foam extinguishers.

Spillage   See general section.
  disposal

RSC *Chemical Safety Data Sheets* Vol. 3, No. 25, 1990 gives extended coverage.

595. Ethylene diperchlorate

Hazardous   Highly sensitive, violently explosive material. Explodes on contact with small
  reactions   amounts of water.

596. Ethylene glycol

Colourless syrupy liquid with sweetish taste. It is the main constituent of 'anti-freeze' mixtures; m.p. -13 ˚C; b.p. 198 ˚C; miscible with water.

RISKS
Harmful if swallowed (R22)

SAFETY PRECAUTIONS
Keep out of reach of children (S2)

Limit values   OES short-term 125 mg m^{-3} (vapour); long-term 10 mg m^{-3} (particulate), 60 mg m^{-3} (vapour).

Toxic effects   Death has followed the drinking of ethylene glycol as a substitute for spirits; 100 cm^3 may prove fatal. Smaller quantities may result in restlessness, unsteady gait, drowsiness, coma, and injury to the kidneys.

Hazardous   Product of reaction with perchloric acid explodes on addition of water;
  reactions   explosion occurred when it was heated wih phosphorus pentasulfide in hexane; may ignite in contact with silvered-copper d.c. conductors.

First aid   Standard treatment for exposure by all routes (see pages 108-122).

Fire hazard   Flash point 111 ˚C (open cup); Explosive limits 3.2-15.5%; Autoignition temperature 413 ˚C. For small fires use carbon dioxide or dry chemical powder extinguishers or extinguish by diluting greatly with water. For larger fires alcohol-resistant foam is preferred.

| Spillage disposal | Mop up with plenty of water and run to waste, diluting greatly with running water. Check procedure with local authority. |

RSC *Chemical Safety Data Sheets* Vol. 1, No. 45, 1988 gives extended coverage.

597. Ethylene glycol monobutyl ether

Colourless liquid; m.p. -74.8 °C; b.p. 168.4 °C; miscible with water.

RISKS
Harmful by inhalation, in contact with skin, and if swallowed – Irritating to respiratory system (R20/21/22, R37)

SAFETY PRECAUTIONS
Avoid contact with skin and eyes (S24/25)

Limit values	MEL long-term 25 p.p.m. (120 mg m^{-3}).
Toxic effects	Potent eye irritant causing intense pain. Inhalation, ingestion, or skin absorption cause changes to the blood system, liver, kidneys, and spleen.
Hazardous reactions	Liable to form explosive peroxides on exposure to air or light which should be decomposed before the ether is distilled to small volume.
First aid	Standard treatment for exposure by all routes (see pages 108-122).
Fire hazard	Flash point 69 °C (open cup); Explosive limits 4-13%; Autoignition temperature 214 °C. Extinguish fires with carbon dioxide, dry chemical powder, or alcohol-resistant foam.
Spillage disposal	See general section.

RSC *Chemical Safety Data Sheets* Vol. 1, No. 46, 1988 gives extended coverage.

598. Ethylene glycol monoethyl ether

Colourless liquid; b.p. 135 °C; miscible with water.

RISKS
Flammable – Irritating to eyes (R10, R36)

SAFETY PRECAUTIONS
Avoid contact with skin (S24)

Limit values	MEL long-term 10 p.p.m. (37 mg m^{-3}).
Toxic effects	The vapour irritates the eyes and respiratory system when in high concentrations. The liquid irritates the eyes. The liquid is poisonous if taken by mouth and may cause blood and kidney damage.
Hazardous reactions	Liable to form explosive peroxides on exposure to light and air which should be decomposed before the ether is distilled to small volume.
First aid	Standard treatment for exposure by all routes (see pages 108-122).

Fire hazard   Flash point 49 °C (open cup); Explosive limits 2.8-18.0%; Autoignition temperature 240 °C. Extinguish fire with water spray, foam, dry powder, or carbon dioxide.

Spillage   See general section.
disposal

RSC *Chemical Safety Data Sheets* Vol. 1, No. 47, 1988 gives extended coverage.

599. Ethylene glycol monoethyl ether acetate

Colourless liquid with pleasant ester-like odour; m.p. -61.7 °C; b.p. 156 °C; sparingly soluble in water.

RISKS
Flammable – Harmful by inhalation and in contact with skin (R10, R20/21)

SAFETY PRECAUTIONS
Avoid contact with skin (S24)

Limit values   MEL long-term 10 p.p.m. (54 mg m^{-3}).

Toxic effects   The vapour irritates the eyes and respiratory system in high concentrations. The liquid irritates the eyes. The liquid is poisonous if taken by mouth and may cause blood and kidney damage.

Hazardous   Liable to form explosive peroxides on exposure to air and light which should
reactions   be decomposed before the ether ester is distilled to small volume.

First aid   Standard treatment for exposure by all routes (see pages 108-122).

Fire hazard   Flash point 56 °C (open cup); Explosive limits 1.7-8.2%; Autoignition temperature 379 °C. Extinguish fire with foam, dry powder, or carbon dioxide.

Spillage   See general section.
disposal

RSC *Chemical Safety Data Sheets* Vol. 1, No. 48, 1988 gives extended coverage.

600. Ethylene oxide

Colourless gas; m.p. -111.3 °C; b.p. 10.7 °C; soluble in water.

RISKS
May cause cancer – May cause heritable genetic damage – Extremely flammable liquefied gas – Toxic by inhalation – Irritating to eyes, respiratory system, and skin (R45, R46, R13, R23, R36/37/38)

SAFETY PRECAUTIONS
Avoid exposure-obtain special instructions before use – Keep container tightly closed, in a cool well ventilated place – Keep away from sources of ignition – No Smoking – Take precautionary measures against static discharges – If you feel unwell, seek medical advice (show label where possible) (S53, S3/7/9, S16, S33, S44)

Limit values	MEL long-term 5 p.p.m. (10 mg m^{-3}).
Toxic effects	If inhaled, the vapour irritates the respiratory tract and may give rise to bronchitis or pulmonary oedema; other possible effects include nausea, vomiting, convulsions, and coma. The vapour irritates the eyes and may cause conjunctivitis and corneal damage. Excessive contact with liquid or solution can cause delayed burns and blistering.
Hazardous reactions	Vapour readily initiated into explosive decomposition; liable to explode in autoclave reactions with some alkanethiols or alcohols; ammonia and other contaminants may cause explosive polymerization. Iron blue pigment reacts to give a pyrophoric solid.
First aid	Standard treatment for exposure by all routes (see pages 108-122).
Fire hazard	Flash point -20 ˚C (closed cup); Explosive limits 3-100%; Autoignition temperature 428.9 ˚C. Extinguish fires with water spray, carbon dioxide, dry chemical powder, or alcohol or polymer foam.
Spillage disposal	The contents of a broken ampoule of ethylene oxide are best dispersed by thorough ventilation of the area concerned; care must be taken to shut off all possible sources of ignition and breathing apparatus must be worn. A leaking cylinder of the substance can be vented slowly to air in a safe, open area.

RSC *Chemical Safety Data Sheets* Vol. 4a, No. 72, 1991 gives extended coverage.

601. Ethylene ozonide

Liquid; b.p. 18 ˚C.

Hazardous reactions	Stable at 0 ˚C. Explodes very violently on heating, friction, or shock. May explode when poured.

602. Ethyl formate

Colourless, volatile liquid with a pleasant aromatic odour; m.p. -79 ˚C; b.p. 54 ˚C; soluble in water.

RISKS
Highly flammable (R11)

SAFETY PRECAUTIONS
Keep container in a well ventilated place – Keep away from sources of ignition – No Smoking – Take precautionary measures against static discharges (S9, S16, S33)

Limit values ·	OES short-term 150 p.p.m. (450 mg m^{-3}); long-term 100 p.p.m. (300 mg m^{-3}).
Toxic effects	The vapour irritates the skin, eyes, and respiratory system. Inhalation of high concentrations can affect the nervous system and cause unconsciousness. The liquid irritates the skin and eyes severely.
First aid	Standard treatment for exposure by all routes (see pages 108-122).
Fire hazard	Flash point 93 ˚C; Explosive limits 2.7-13.5%; Autoignition temperature 456 ˚C. Extinguish small fires with water spray, foam, dry powder, or carbon dioxide. Large fires are best controlled by alcohol-resistant foam.

Spillage   See general section.
disposal

RSC *Chemical Safety Data Sheets* Vol. 1, No. 51, 1988 gives extended coverage.

603. 2-Ethylhexyl acrylate

Colourless liquid with a pleasant smell; m.p. -90 ˚C; b.p. 130 ˚C at 50 mm Hg; insoluble in water.

RISKS
Irritating to respiratory system and skin – May cause sensitization by skin contact (R37/38, R43)

Toxic effects   Irritant to skin, eyes, and upper respiratory tract. Skin contact causes sensitization in humans while skin tumours have been caused in laboratory animals.

First aid   Standard treatment for exposure by all routes (see pages 108-122).

Fire hazard   Flash point 91 ˚C (closed cup); Explosive limits 0.8-6.4%. Extinguish fires with carbon dioxide, dry chemical powder, or alcohol foam.

Spillage   See general section.
disposal

RSC *Chemical Safety Data Sheets* Vol. 3, No. 26, 1990 gives extended coverage.

604. 2-Ethylhexyl chloroformate

Colourless liquid; b.p. 106-107 ˚C at 30 mm Hg; decomposed by water.

Limit values   OES long-term 1 p.p.m. (7.9 mg m^{-3}).

Toxic effects   Extremely destructive to skin, eyes, upper respiratory tract, and mucous membranes. Inhalation may be fatal as a result of spasm, inflammation, chemical pneumonitis, and pulmonary oedema. May also be fatal following ingestion or skin absorption.

First aid   Standard treatment for exposure by all routes (see pages 108-122).

Fire hazard   Flash point 81 ˚C. Extinguish fires with carbon dioxide, dry chemical powder, or alcohol or polymer foam. Do NOT use water.

Spillage   See general section.
disposal

605. Ethyl hypochlorite

Mobile yellow liquid with irritating odour; b.p. 36 ˚C at 752 mm Hg.

Hazardous   Very unstable. Ignition or rapid heating of vapour causes explosion.
reactions

606. Ethylidene norbornene

Colourless liquid.

Toxic effects   Irritant to skin, eyes, upper respiratory tract, and mucous membranes.

First aid   Standard treatment for exposure by all routes (see pages 108-122).

Fire hazard   Flash point 38 ˚C; Explosive limits 0.9-6.4%. Extinguish fires with water spray, carbon dioxide, dry chemical powder, or alcohol or polymer foam.

Spillage disposal   See general section.

607. Ethyl isocyanide

Colourless liquid; m.p. -66 ˚C; b.p. 79 ˚C; soluble in water.

Hazardous reactions   Liable to explode.

608. Ethyl lactate

Colourless liquid; m.p. -25 ˚C; b.p. 154 ˚C; soluble in water with partial hydrolysis.

RISKS
Flammable (R10)

SAFETY PRECAUTIONS
Do not breathe vapour (S23)

Toxic effects   No evidence was found that ethyl lactate has significant toxic properties.

First aid   Standard treatment for exposure by all routes (see pages 108-122).

Fire hazard   Flash point 46 ˚C; Explosive limits 1.5-30%; Autoignition temperature 400 ˚C. Extinguish fire with water spray, foam, dry powder, or carbon dioxide.

Spillage disposal   See general section.

609. Ethyl methacrylate

Clear colourless liquid; b.p. 119 ˚C; insoluble in water.

RISKS
Highly flammable – Irritating to eyes, respiratory system, and skin – May cause sensitization by skin contact (R11, R36/37/38, R43)

SAFETY PRECAUTIONS
Keep container in a well ventilated place – Keep away from sources of ignition – No Smoking – Do not empty into drains – Take precautionary measures against static discharges (S9, S16, S29, S33)

Toxic effects   The vapour irritates the eyes and respiratory system. The liquid irritates the skin and eyes. Less toxic when taken by mouth than ethyl acrylate.

First aid	Standard treatment for exposure by all routes (see pages 108-122).
Fire hazard	Flash point 20 ˚C. Extinguish fire with water spray, foam, dry powder, or carbon dioxide.
Spillage disposal	See general section.

610. Ethyl methyl peroxide

Liquid with an ethereal odour; b.p. 40 ˚C at 740 mm Hg.

Toxic effects	Irritant to skin, eyes, and upper respiratory tract.
Hazardous reactions	Shock-sensitive as liquid or vapour; explodes violently when heated strongly.
First aid	Standard treatment for exposure by all routes (see pages 108-122).

611. Ethylmorpholine

Colourless liquid; b.p. 138 ˚C; miscible with water.

Limit values	OES short-term 20 p.p.m. (95 mg m^{-3}); long-term 5 p.p.m. (23 mg m^{-3}).
Toxic effects	Extremely destructive to tissues of the skin, eyes, upper respiratory tract, and mucous membranes. Toxic by inhalation, ingestion, and skin absorption. Inhalation may be fatal following spasm, chemical pneumonitis, and pulmonary oedema.
Hazardous reactions	Explodes violently when heated strongly.
First aid	Standard treatment for exposure by all routes (see pages 108-122).
Fire hazard	Flash point 32 ˚C. Extinguish fire with water spray, foam, dry powder, or carbon dioxide.
Spillage disposal	See general section.

612. Ethyl nitrate

Colourless liquid with pleasant odour; b.p. 89 ˚C; insoluble in water.

RISKS
Risk of explosion by shock, friction, fire, or other sources of ignition (R2)

SAFETY PRECAUTIONS
Do not breathe vapour – Avoid contact with skin and eyes (S23, S24/25)

| **Toxic effects** | Acute local effects are not documented but it is stated to be moderately toxic systemically by inhalation and ingestion. |

Hazardous reactions	Explodes at 85 ˚C.
First aid	Standard treatment for exposure by all routes (see pages 108-122).
Fire hazard	Flash point 10 ˚C. In view of the probability of explosion shortly after ignition, the area involving an ethyl nitrate fire (or the risk of it occurring) should be evacuated and the fire brigade informed at once of the situation.
Spillage disposal	See general section.

613. Ethyl nitrite

Colourless or yellowish liquid with an ethereal odour; b.p. 17 ˚C; very slightly soluble in water.

RISKS
Risk of explosion by shock, friction, fire, or other sources of ignition – Harmful by inhalation, in contact with skin and if swallowed (R2, R20/21/22)

Toxic effects	Narcotic in high concentrations. Inhalation can result in lowered blood pressure.
Hazardous reactions	Explodes above 90 ˚C.
First aid	Standard treatment for exposure by all routes (see pages 108-122).
Fire hazard	Flash point -1 ˚C; Explosive limits 3.0-50.0%; Autoignition temperature 90 ˚C. Use water spray, carbon dioxide, dry chemical powder, or foam to extinguish fires.
Spillage disposal	See general section.

614. Ethyl perchlorate

Oily liquid; b.p. 74 ˚C; insoluble in water.

Hazardous reactions	Reputedly most explosive substance known; very sensitive to impact, friction and heat.

615. Ethylphosphine

Colourless liquid with unpleasant odour; b.p. 25 ˚C; decomposed by water.

Hazardous reactions	Ignites in air; explodes on contact with chlorine, bromine, or nitric acid.
Fire hazard	To extinguish fires use carbon dioxide, dry chemical powder, or foam extinguishers.

616. *N*-Ethylpiperidine

Colourless liquid; b.p. 131 ˚C; miscible with water.

Toxic effects Irritant to skin, eyes, upper respiratory tract, and mucous membranes. Harmful on inhalation, ingestion, and skin absorption.

First aid Standard treatment for exposure by all routes (see pages 108-122).

Fire hazard Flash point 19 ˚C. Extinguish fire with water spray, foam, dry powder, or carbon dioxide.

Spillage disposal See general section.

617. Ethyl vinyl ether

Colourless liquid; m.p. -115 ˚C; b.p. 35.5 ˚C.

Toxic effects Irritant to skin, eyes, upper respiratory tract, and mucous membranes. Inhalation in high concentrations may lead to unconsciousness and respiratory paralysis. Chronic exposure can lead to liver damage.

Hazardous reactions Methanesulfonic acid catalysed explosive polymerization of the ether.

First aid Standard treatment for exposure by all routes (see pages 108-122).

Fire hazard Flash point -45 ˚C. Extinguish fires with carbon dioxide, dry chemical powder, or alcohol or polymer foam.

Spillage disposal See general section.

618. 2-Ethynylfuran

Liquid; b.p. 105.6 ˚C.

Hazardous reactions Explodes on heating or on contact with concentrated nitric acid.

619. Ethynyl vinyl selenide

Hazardous reactions Explodes on heating.

620. Ferrocene

Burnt orange, crystalline solid; m.p. 174-176 ˚C; b.p. 249 ˚C; insoluble in water.

Limit values OES short-term 20 mg m^{-3}; long-term 10 mg m^{-3}.

Toxic effects	Harmful by inhalation, ingestion, or skin absorption.
First aid	Standard treatment for exposure by all routes (see pages 108-122).
Fire hazard	Extinguish fires with water spray, carbon dioxide, dry chemical powder, or alcohol or polymer foam.
Spillage disposal	See general section.

621. Fluoramine

Hazardous reactions	Impure material is very explosive.

622. Fluorescein

Orange-red, crystalline powder; m.p. 290 °C with decomposition; insoluble in water.

Toxic effects	May be harmful if ingested in quantity, causing nausea and vomiting. Causes temporary discomfort to eyes. Slight allergen.
First aid	Standard treatment for exposure by all routes (see pages 108-122).
Fire hazard	Use water spray to extinguish fires.
Spillage disposal	See general section.

623. Fluorides

Normally colourless crystals or powders; soluble in water.

Limit values	OES long-term 2.5 mg m^{-3} (as F).
Toxic effects	The dust irritates all parts of the respiratory tract, eyes, and skin. Nausea, vomiting, diarrhoea, and abdominal pains follow ingestion. Shortness of breath, cough, elevated temperature, and cyanosis result from prolonged or repeated exposure.
First aid	Standard treatment for exposure by all routes (see pages 108-122), with special reference to fluorides.
Fire hazard	Extinguish fires with water spray, carbon dioxide, dry chemical powder, or foam.
Spillage disposal	See general section.

RSC *Chemical Safety Data Sheets* Vol. 2, Nos. 86,93,103, 1989 and Vol. 3, Nos. 9,55,61, 1990 give extended coverage on individual fluorides.

624. Fluorine

Pale yellow gas with a sharp penetrating odour; m.p. -218 ˚C; b.p. -188 ˚C; reacts vigorously with water.

RISKS
May cause fire – Very toxic by inhalation – Causes severe burns (R7, R26, R35)

SAFETY PRECAUTIONS
Keep container tightly closed and in a well ventilated place – Wear suitable protective clothing – In case of accident or if you feel unwell, seek medical advice immediately (show label where possible) (S7/9, S36, S45)

Limit values	OES short-term 1 p.p.m. (1.5 mg m^{-3}).
Toxic effects	The gas is highly irritant to the eyes and respiratory system; high concentrations may produce thermal-type burns on the skin and chemical-type burns may result from lower concentrations.
Hazardous reactions	Anyone working with fluorine should study the whole section devoted to its dangerous reactions in Bretherick's Handbook of reactive chemical hazards as it is one of the most reactive elements known.
First aid	Standard treatment for exposure by all routes (see pages 108-122).
Fire hazard	Keep containers cool with water spray or remove them from the immediate vicinity.
Spillage disposal	Leaking cylinder should be vented slowly into a water-fed scrubbing tower or column in a fume cupboard, or into a fume cupboard served by such a tower.

625. Fluorine perchlorate

Colourless gas with pungent, acrid odour; m.p. -167.3 ˚C; b.p. -15.9 ˚C.

Toxic effects	Toxic by ingestion and inhalation. Corrosive to the skin, eyes, and mucous membranes.
Hazardous reactions	Liquid explodes on freezing at -167 ˚C; explosion of gas initiated by sparks, flame, or contact with grease, dust, or rubber tube; ignition occurs in excess of hydrogen.

626. Fluoroacetylene

Gas; m.p. -196 ˚C; b.p. -80 ˚C.

Hazardous reactions	Liquid explodes close to its b.p., -80 ˚C; ignition occurs in contact with solution of bromine in carbon tetrachloride; mercury salt decomposes but silver salt explodes on heating.

627. Fluoroboric acid and salts

Fluoroboric acid is a colourless, fuming liquid. The salts are generally colourless, crystalline solids; b.p. the acid decomposes at 130 °C; the acid and salts are soluble in water.

Toxic effects	The acid is extremely destructive to skin and eyes. The acid is also very irritant and destructive to the upper respiratory tract and may be fatal following spasm, inflammation, and oedema. Very harmful if swallowed.
Hazardous reactions	Caution is advised when dehydrating aqueous 48% fluoroboric acid by addition to acetic anhydride as the reaction is rather exothermic.
First aid	Standard treatment for exposure by all routes (see pages 108-122).
Fire hazard	Use dry chemical powder to extinguish fires.
Spillage disposal	See general section.

RSC *Chemical Safety Data Sheets* Vol. 3, No. 27, 1990 gives extended coverage.

628. 1-Fluoro-2,4-dinitrobenzene

Pale yellow crystals; m.p. 26 °C; b.p. 178 °C at 25 mm Hg; insoluble in water.

Toxic effects	The dust, vapour from the molten compound, or sprayed aerosols irritate the respiratory system. These also irritate the skin and may cause blistering, dermatitis, and severe allergic reactions. Absorption into the body by inhalation, ingestion, or through the skin can lead to death.
Hazardous reactions	Explosions with distillation residues.
First aid	Standard treatment for exposure by all routes (see pages 108-122).
Fire hazard	Flash point above 110 °C. Extinguish fires with water spray, carbon dioxide, dry chemical powder, or alcohol or polymer foam.
Spillage disposal	See general section.

629. Fluorosilicic acid and salts

Fluorosilicic acid is a colourless, fuming liquid with sour pungent smell. The salts are generally colourless, crystalline solids. The acid decomposes on heating and is miscible with water. The salts are soluble in water.

RISKS
(over 25%) Causes burns (R34)

SAFETY PRECAUTIONS
(over 25%) In case of contact with eyes, rinse immediately with plenty of water and seek medical advice – Take off immediately all contaminated clothing (S26, S27)

Toxic effects	The vapour and dust irritate all parts of the respiratory system. The acid and salts burn the eyes severely. The acid burns the skin. If taken by mouth there would be severe internal irritation and damage.

First aid   Standard treatment for exposure by all routes (see pages 108-122).

Fire hazard   Extinguish fires with water spray, carbon dioxide, dry chemical powder, or
                     alcohol foam.

Spillage   See general section.
    disposal

RSC *Chemical Safety Data Sheets* Vol. 3, No. 28, 1990 gives extended coverage.

630. Fluorosulfuric acid

Colourless or yellow fuming liquid; m.p. -87.3 °C; b.p. 165.5 °C; soluble in water.

Toxic effects   The vapour irritates all parts of the respiratory system causing inflammation
                     and oedema of the larynx and bronchi, chemical pneumonitis, and pulmonary
                     oedema. The vapour irritates and the liquid burns the eyes severely. The
                     liquid and vapour burn the skin. If taken by mouth there would be severe
                     internal irritation and damage.

First aid   Standard treatment for exposure by all routes (see pages 108-122).

Spillage   See general section.
    disposal

631. Formaldehyde solution

Colourless, sometimes milky, solution with pungent odour generally containing 37-41%
formaldehyde and 11-14% methanol; b.p. 96 °C; miscible with water.

RISKS
(≥ 1%, < 5%) Possible risk of irreversible effects – May cause sensitization by skin contact
(R40, R43)
(≥ 5%, < 25%) Harmful by inhalation, in contact with skin, and if swallowed – Irritating to eyes,
respiratory system and skin – Possible risk of irreversible effects – May cause sensitization by
skin contact – ≥ 25% (R20/21/22, R36/37/38, R40, R43)
(≥25%) Toxic by inhalation, in contact with skin and if swallowed – Causes burns – Possible
risk of irreversible effects – May cause sensitization by skin contact (R23/24/25, R34, R40,
R43)

SAFETY PRECAUTIONS
(≥ 1%, < 5%) Do not breathe vapour – Wear suitable gloves (S23, S37)
(≥ 5%, < 25%) In case of contact with eyes, rinse immediately with plenty of water and seek
medical advice – Wear protective clothing and gloves – Use only in well ventilated areas (S26,
S36/37, S51)
(≥25%) ;S26;S36/37;S44;S51 In case of contact with eyes, rinse immediately with plenty of
water and seek medical advice – Wear protective clothing and gloves – If you feel unwell, seek
medical advice (show label where possible) – Use only in well ventilated areas (S26, S36/37,
S44, S51)

Limit values   MEL short-term 2 p.p.m. (2.5 mg m^{-3}) (formaldehyde); long-term 2 p.p.m. (2.5
                     mg m^{-3}) (formaldehyde).

Toxic effects	The vapour irritates all parts of the respiratory system. High concentrations inhaled for long periods can cause laryngitis, bronchitis, or bronchial pneumonia. The liquid and vapour irritate the eyes severely and prolonged exposure may cause conjunctivitis. The liquid in contact with the skin has a hardening or tanning effect and causes irritation. Contact for long periods will cause cracking of the skin and ulceration, particularly around fingernails. Severe abdominal pains with nausea and vomiting and possibly loss of consciousness follow ingestion. Nasal tumours have been reported.
Hazardous reactions	As an active reducing agent, it has been involved in various reactive incidents with oxidants.
First aid	Standard treatment for exposure by all routes (see pages 108-122).
Fire hazard	Flash point (37%) 50 °C. Extinguish fire with water spray, dry powder, or carbon dioxide.
Spillage disposal	See general section.

RSC *Chemical Safety Data Sheets* Vol. 3, No. 29, 1990 gives extended coverage.

632. Formamide

Colourless liquid; m.p. 2-3 °C; b.p. 210 °C with decomposition; very soluble in water.

Limit values	OES short-term 30 p.p.m. (45 mg m^{-3}); long-term 20 p.p.m. (30 mg m^{-3}).
Toxic effects	Irritant to skin, eyes, upper respiratory tract, and mucous membranes. May cause malformations in the foetus.
Hazardous reactions	Bottles containing a modified Karl Fischer reagent with formamide replacing methanol developed gas pressure over several months and burst.
First aid	Standard treatment for exposure by all routes (see pages 108-122).
Fire hazard	Flash point 154 °C; Autoignition temperature 500 °C. Extinguish fires with water spray, carbon dioxide, dry chemical powder, or alcohol or polymer foam.
Spillage disposal	See general section.

633. Formic acid

Colourless, fuming liquid with pungent odour; m.p. 8.4 °C; b.p. 98/100% acid boils at 100.5 °C; miscible with water.

RISKS
Causes burns (R34)

SAFETY PRECAUTIONS
Keep out of reach of children – Do not breathe vapour – In case of contact with eyes, rinse immediately with plenty of water and seek medical advice (S2, S23, S26)

Limit values	OES long-term 5 p.p.m. (9 mg m^{-3}).

Toxic effects	The vapour irritates all parts of the respiratory system. The vapour irritates the eyes and the liquid causes painful eye burns. The liquid burns the skin. If taken by mouth there is severe internal irritation and damage.
Hazardous reactions	Tightly sealed containers may burst from pressure of carbon monoxide produced during slow decomposition in storage or on freezing at low temperatures; explosive decomposition on nickel.
First aid	Standard treatment for exposure by all routes (see pages 108-122).
Fire hazard	Flash point 90% solution 69 °C (open cup); Explosive limits 90% solution 18-57%; Autoignition temperature 601 °C. Extinguish fires with water spray, carbon dioxide, dry chemical powder, or alcohol foam.
Spillage disposal	See general section.

RSC *Chemical Safety Data Sheets* Vol. 3, No. 30, 1990 gives extended coverage.

634. Fumaric acid

White powder; m.p. 299-300 °C with sublimation; b.p. 165 °C at 1.7 mm Hg; sparingly soluble in water.

RISKS
Irritating to eyes (R36)

SAFETY PRECAUTIONS
In case of contact with eyes, rinse immediately with plenty of water and seek medical advice (S26)

Toxic effects	Irritating to skin, eyes, and mucous membranes.
First aid	Standard treatment for exposure by all routes (see pages 108-122).
Fire hazard	Extinguish fires with water spray, carbon dioxide, dry chemical powder, or alcohol or polymer foam.
Spillage disposal	See general section.

635. Furfural

Colourless to reddish-brown liquid with almond-like smell, which darkens on exposure to light and air; m.p. -36.5 °C; b.p. 161.7 °C; fairly soluble in water.

RISKS
Toxic by inhalation and if swallowed (R23/25)

SAFETY PRECAUTIONS
Avoid contact with skin and eyes – If you feel unwell, seek medical advice (show label where possible) (S24/25, S44)

Limit values	OES short-term 10 p.p.m. (40 mg m^{-3}); long-term 2 p.p.m. (8 mg m^{-3}).
Toxic effects	The vapour irritates the skin, eyes, and respiratory system. Prolonged exposure may result in nervous disturbances and eye inflammation. The liquid irritates the eyes and skin. Stomach irritation follows ingestion.

Hazardous reactions	Reaction with sodium bicarbonate in oil refining operations may give pyrophoric coke.
First aid	Standard treatment for exposure by all routes (see pages 108-122).
Fire hazard	Flash point 60 ˚C (closed cup); Explosive limits 2.1%-19.3%; Autoignition temperature 316 ˚C. Use carbon dioxide, dry chemical powder, or alcohol-resistant foam to extinguish fires.
Spillage disposal	See general section.

RSC *Chemical Safety Data Sheets* Vol. 1, No. 52, 1988 and Vol. 4a, No. 74, 1991 gives extended coverage.

636. Furfuryl alcohol

Colourless to light yellow liquid with an ethereal odour and bitter taste, the colour darkening on exposure to light and air; m.p. -31 ˚C; b.p. 171 ˚C; miscible with water but unstable.

RISKS
Harmful by inhalation, in contact with skin, and if swallowed (R20/21/22)

Limit values	OES short-term 15 p.p.m. (60 mg m^{-3}); long-term 5 p.p.m. (20 mg m^{-3}).
Toxic effects	The vapour is irritant to eyes and respiratory passages. Headaches and nausea may follow inhalation or ingestion. The liquid may be absorbed through the skin.
Hazardous reactions	Reacts violently with formic acid; mixture with cyanoacetic acid exploded on heating; may react violently or explosively on contact with acids or acidic materials.
First aid	Standard treatment for exposure by all routes (see pages 108-122).
Fire hazard	Flash point 75 ˚C (open cup); Explosive limits 1.8-16.3 ˚C; Autoignition temperature 491 ˚C. Use carbon dioxide, dry chemical powder, or alcohol-resistant foam to extinguish fires.
Spillage disposal	See general section.

RSC *Chemical Safety Data Sheets* Vol. 1, No. 53, 1988 gives extended coverage.

637. Furoyl chloride

Colourless liquid; m.p. -2 ˚C; b.p. 170 ˚C; decomposed by water.

Toxic effects	The vapour irritates the eyes.
Hazardous reactions	Freshly distilled material polymerized explosively in storage.

638. Gallium

Lustrous, silver metal or grey solid; m.p. 29.78 °C; b.p. 2403 °C; insoluble in water.

Toxic effects	Although animal experiments have shown certain effects, there is little evidence of toxicity to humans from industrial use.
Hazardous reactions	Reacts violently with chlorine and bromine.
First aid	Standard treatment for exposure by all routes (see pages 108-122).
Fire hazard	Extinguish fires with dry chemical powder.
Spillage disposal	Spillages should be kept in a bag and held for re-use or recycling.

RSC *Chemical Safety Data Sheets* Vol. 2, No. 49, 1989 gives extended coverage.

639. Gallium chloride

Colourless needles; m.p. 77.8 °C; reacts violently with water.

Toxic effects	Irritant to skin and eyes. Toxic by inhalation.
First aid	Standard treatment for exposure by all routes (see pages 108-122).
Fire hazard	Extinguish fires with carbon dioxide, dry chemical powder, alcohol or polymer foam, or by flooding with water.
Spillage disposal	See general section.

RSC *Chemical Safety Data Sheets* Vol. 2, No. 50, 1989 gives extended coverage.

640. Gallium nitrate

White, deliquescent crystals; m.p. 110 °C with decomposition; soluble in water.

Toxic effects	Strongly irritant to skin, eyes, and upper respiratory tract.
First aid	Standard treatment for exposure by all routes (see pages 108-122).
Fire hazard	Extinguish fires with water spray.
Spillage disposal	See general section.

RSC *Chemical Safety Data Sheets* Vol. 2, No. 51, 1989 gives extended coverage.

641. Gallium triperchlorate

Hazardous reactions	Moist product may react violently with any organic material.

642. Germanium

Greyish-white, lustrous, brittle metalloid; m.p. 937 °C; b.p. 2830 °C; insoluble in water.

Toxic effects	Germanium is generally considered to be of low toxicity by inhalation and ingestion. It can irritate the skin and eyes.
Hazardous reactions	Powdered metal ignites in chlorine, lumps ignite on heating in chlorine or bromine; powdered metal reacts violently with nitric acid, mixtures with potassium chlorate and nitrate explode on heating.
First aid	Standard treatment for exposure by all routes (see pages 108-122).
Fire hazard	Use extinguishant appropriate to the surroundings.
Spillage disposal	Spillages should be swept up into a bag and kept for recycling or re-use.

RSC *Chemical Safety Data Sheets* Vol. 2, No. 52, 1989 gives extended coverage.

643. Germanium dioxide

Colourless crystals; m.p. 1115 °C; very slightly soluble in water.

Toxic effects	Irritating to the eyes and highly toxic orally.
First aid	Standard treatment for exposure by all routes (see pages 108-122).
Fire hazard	Use extinguishant appropriate to the surroundings.
Spillage disposal	See general section.

RSC *Chemical Safety Data Sheets* Vol. 2, No. 53, 1989 gives extended coverage.

644. Germanium hydride

Colurless gas with a nauseating smell; m.p. -165 °C; b.p. -88.5 °C; insoluble in water.

Limit values	OES short-term 0.6 p.p.m. (1.8 mg m^{-3}); long-term 0.2 p.p.m. (0.6 mg m^{-3}).
Toxic effects	Inhalation can lead to toxic effects on blood, nervous system, and kidneys. It is mildly irritant to eyes and skin.
Hazardous reactions	Liable to ignite in air; contact with bromine may cause ignition at -112 °C.
First aid	Standard treatment for exposure by all routes (see pages 108-122).
Fire hazard	Extinguish fires with carbon dioxide. Do NOT use water.
Spillage disposal	Disposal of germane should be by licenced contractor.

RSC *Chemical Safety Data Sheets* Vol. 2, No. 54, 1989 gives extended coverage.

645. Germanium monohydride

Colourless liquid.

Hazardous   May decompose explosively in air.
reactions

646. Germanium tetrachloride

Colourless liquid with an acidic smell; m.p. -49.5 °C; b.p. 84 °C; decomposed by water.

Toxic effects   The fumes are irritating to the upper respiratory tract. Prolonged exposure to high concentrations may cause liver, kidney, and other organ damage. The fumes are irritating and highly destructive to the skin and eyes.

Hazardous   Reacts violently with water.
reactions

First aid   Standard treatment for exposure by all routes (see pages 108-122).

Fire hazard   Extinguish fires with carbon dioxide, dry chemical powder, or alcohol or polymer foam. Do NOT use water.

Spillage   See general section.
disposal

RSC *Chemical Safety Data Sheets* Vol. 2, No. 55, 1989 gives extended coverage.

647. Glutaraldehyde

Clear, colourless liquid; m.p. -14 °C; b.p. 187-189 °C; very soluble in water.

Limit values   OES short-term 0.2 p.p.m. (0.7 mg m^{-3}).

Toxic effects Irritant to skin, eyes, and upper respiratory tract. Suspected carcinogen.

First aid   Standard treatment for exposure by all routes (see pages 108-122), but any onset of convulsions and allergic asthma must be dealt with immediately.

Fire hazard   Extinguish fires with water spray, carbon dioxide, dry chemical powder, or alcohol or polymer foam.

Spillage   See general section.
disposal

RSC *Chemical Safety Data Sheets* Vol. 3, No. 31, 1990 gives extended coverage.

648. Glycerol

Colourless, odourless, hygroscopic liquid; m.p. 17.9 °C but solidifies at a much lower temperature; b.p. 290 °C; miscible with water.

Limit values   OES long-term 10 mg m^{-3}.

Toxic effects Generally considered to be of low toxicity, with the mist a nuisance particulate. Ingestion of large doses can cause headaches, nausea, vomiting, and convulsions leading to renal failure. It is irritating to the eyes but not to the skin.

Hazardous reactions Reacts violently or explosively with many oxidizing agents.

First aid Standard treatment for exposure by all routes (see pages 108-122).

Fire hazard Flash point 199 ˚C; Autoignition temperature 370 ˚C. Extinguish fires with carbon dioxide, dry chemical powder, or alcohol foam.

Spillage disposal See general section.

RSC *Chemical Safety Data Sheets* Vol. 1, No. 54, 1988 gives extended coverage.

649. Glycerol trinitrate

Colourless or yellow oily liquid with a sweet taste; m.p. 13 ˚C; b.p. 150 ˚C at 15 mm Hg; explodes at 270 ˚C; sparingly soluble in water.

Limit values OES short-term 0.2 p.p.m. (2 mg m^{-3}); long-term 0.2 p.p.m. (2 mg m^{-3}).

Toxic effects Can be absorbed through the skin, lungs, and mucous membranes. Small amounts can result in headache, nausea, vomiting, and abdominal pain; large amounts in depression, confusion, delirium, methaemoglobinaemia, and cyanosis. Experimental tumourigen and teratogen.

Hazardous reactions Severe explosion hazard when shocked or exposed to ozone, heat, or flame.

First aid Standard treatment for exposure by all routes (see pages 108-122).

Fire hazard Autoignition temperature 270 ˚C.

650. Glycolonitrile

Colourless, odourless, oily liquid with a sweetish taste; b.p. 183 ˚C with slight decomposition; soluble in water.

Toxic effects Irritant to the eyes. Harmful by inhalation, ingestion, and skin absorption.

Hazardous reactions Liable to polymerize explosively on storage, promoted by traces of alkali.

651. Guanidinium nitrate

Hazardous reactions Explosions have occurred in preparation from dicyanodiamide and ammonium nitrate.

652. Hafnium

Shiny, grey metal; m.p. approx. 1500 ˚C; b.p. approx. 2325 ˚C; insoluble in water.

Limit values   OES short-term 1.5 mg m^{-3}; long-term 0.5 mg m^{-3}.

Toxic effects   May cause skin and eye irritation.

Hazardous reactions   Powdered hafnium, if dry or only slightly wetted, is pyrophoric. The dry powder may react explosively with certain other chemicals.

First aid   Standard treatment for exposure by all routes (see pages 108-122).

Fire hazard   Use dry chemical powder to extinguish fires.

Spillage disposal   See general section.

653. Heptane

Colourless liquid; m.p. -90.5 ˚C; b.p. 98 ˚C; insoluble in water.

RISKS
Highly flammable (R11)

SAFETY PRECAUTIONS
Keep container in a well ventilated place – Keep away from sources of ignition – No Smoking – Do not breathe vapour – Do not empty into drains – Take precautionary measures against static discharges (S9, S16, S23, S29, S33)

Limit values   OES short-term 500 p.p.m. (2000 mg m^{-3}); long-term 400 p.p.m. (1600 mg m^{-3}).

Toxic effects   May irritate the respiratory system as vapour which, in high concentrations, is narcotic. Is also irritant to the skin and eyes.

First aid   Standard treatment for exposure by all routes (see pages 108-122).

Fire hazard   Flash point -4 ˚C; Explosive limits 1.0-6.7%; Autoignition temperature 204 ˚C. Extinguish small fires with foam, dry powder, or carbon dioxide. Large spills are best controlled by alcohol-resistant foam.

Spillage disposal   See general section.

RSC *Chemical Safety Data Sheets* Vol. 1, No. 55, 1988 gives extended coverage.

654. 2-Heptyn-1-ol

Liquid; b.p. 113-116 ˚C at 56 mm Hg.

Hazardous reactions   Risk of explosion of distillation residue.

655. Hexacarbonylchromium

Colourless, crystalline solid; m.p. 152-155 °C; insoluble in water.

Limit values	MEL long-term 0.5 mg m^{-3} (as Cr(vi)).
Toxic effects	Slight irritant. May be fatal following inhalation or ingestion.
Hazardous reactions	Explodes at 210 °C. Light sensitive.
First aid	Standard treatment for exposure by all routes (see pages 108-122).
Fire hazard	Extinguish fires with water spray.
Spillage disposal	See general section.

656. Hexacarbonylmolybdenum

Colourless crystalline solid; m.p. 150 °C with decomposition.

Limit values	OES short-term 20 mg m^{-3} (as Mo); long-term 10 mg m^{-3} (as Mo).
Toxic effects	Slight irritant. May be fatal following inhalation or ingestion.
Hazardous reactions	Solutions liable to explode on long storage.
First aid	Standard treatment for exposure by all routes (see pages 108-122).
Fire hazard	Extinguish fires with water spray.
Spillage disposal	See general section.

657. Hexacarbonyltungsten

Colourless, crystalline solid; b.p. above 150 °C with decomposition.

Limit values	OES short-term 10 mg m^{-3} (as W); long-term 5 mg m^{-3} (as W).
Toxic effects	Slight irritant. May be fatal following inhalation or ingestion.
Hazardous reactions	Preparation from tungsten hexachloride, aluminium powder, and carbon monoxide in autoclave can be dangerous.
First aid	Standard treatment for exposure by all routes (see pages 108-122).
Fire hazard	Extinguish fires with water spray.
Spillage disposal	See general section.

658. Hexachlorobutadiene

Colourless to pale yellow liquid; m.p. -19 to -22 ˚C; b.p. 210-220 ˚C.

Toxic effects	Extremely destructive to skin, eyes, upper respiratory tract, and mucous membranes. Inhalation may be fatal following spasm, chemical pneumonitis, and pulmonary oedema. Suspected human carcinogen.
First aid	Standard treatment for exposure by all routes (see pages 108-122).
Fire hazard	Extinguish fires with carbon dioxide, dry chemical powder, or alcohol or polymer foam.
Spillage disposal	See general section.

659. Hexachlorocyclohexane

All stereoisomers are solids.

RISKS
Toxic by inhalation, in contact with skin, and if swallowed – Irritating to eyes and skin (R23/24/25, R36/38)

SAFETY PRECAUTIONS
Keep out of reach of children – Keep away from food, drink, and animal feeding stuffs – If you feel unwell, seek medical advice (show label where possible) (S2, S13, S44)

Limit values	OES short-term 1.5 mg m^{-3}; long-term 0.5 mg m^{-3}.
Toxic effects	Inhalation can cause headache, nausea, vomiting, and fever. Also toxic following ingestion or skin absorption. Experimental carcinogen and mutagen.
Hazardous reactions	Reaction with dimethylformamide in presence of iron may become violent.

660. Hexachlorocyclopentadiene

Yellow liquid; m.p. -10 ˚C; b.p. 239 ˚C at 753 mm Hg.

Toxic effects	Extremely destructive to tissues of the skin, eyes, upper respiratory tract, and mucous membranes. Inhalation may be fatal following spasm, chemical pneumonitis, and pulmonary oedema.
Hazardous reactions	Small portions of sodium and hexachlorocyclopentadiene exploded on shaking the tube for a few minutes after mixing.
First aid	Standard treatment for exposure by all routes (see pages 108-122).
Fire hazard	Extinguish fires with carbon dioxide, dry chemical powder, or alcohol or polymer foam.
Spillage disposal	See general section.

661. Hexachloroethane

White, crystalline powder; m.p. 190-195 °C with decomposition; sparingly soluble in water.

Limit values   OES long-term 5 p.p.m. (50 mg m^{-3}) (vapour), 10 mg m^{-3} (total inhalable dust), 5 mg m^{-3} (respirable dust).

Toxic effects   Irritant to skin, eyes, upper respiratory tract, and mucous membranes. Exposure may cause damage to liver and central nervous system. Carcinogenic.

First aid   Standard treatment for exposure by all routes (see pages 108-122).

662. Hexachloronaphthalene

White solid with aromatic odour; m.p. 137 °C; b.p. 343-387 °C; insoluble in water.

Toxic effects   Harmful by inhalation, ingestion, and skin absorption. Causes severe acne-form eruptions and toxic narcosis of the liver.

First aid   Standard treatment for exposure by all routes (see pages 108-122).

Spillage disposal   See general section.

663. Hexafluoroacetone

Colourless gas; m.p. -129 °C; b.p. -26 °C.

Toxic effects   Extremely destructive to tissues of skin, eyes, upper respiratory tract, and mucous membranes. Inhalation may be fatal following spasm, chemical pneumonitis, and pulmonary oedema. Experimental teratogen.

First aid   Standard treatment for exposure by all routes (see pages 108-122).

Fire hazard   Use water spray to keep cylinders cool or move them out of the danger zone.

664. Hexahydrophthalic anhydride

White powder; m.p. 35-37 °C.

RISKS
Irritating to eyes, respiratory system, and skin (R36/37/38)

SAFETY PRECAUTIONS
Do not breathe fumes – Wear eye/face protection (S23, S39)

Toxic effects   Seems to be a potent industrial sensitizer. Otherwise very little information is available.

First aid   Standard treatment for exposure by all routes (see pages 108-122).

Fire hazard   Extinguish fires with water, carbon dioxide, dry chemical powder, or alcohol or polymer foam.

Spillage   See general section. But note that explosive peroxides may be formed in the
disposal   presence of air.

RSC *Chemical Safety Data Sheets* Vol. 3, No. 32, 1990 gives extended coverage.

665. Hexahydro-1,3,5-trinitro-1,3,5-triazine

White, crystalline powder; m.p. 202 ˚C.

Limit values   OES short-term 3 mg m^{-3}; long-term 1.5 mg m^{-3}.

Toxic effects   Corrosive irritant to skin, eyes, and mucous membranes. Epileptiform convulsions
                    have been reported from exposure. Experimental reproductive effects.

Hazardous   Powerful high explosive.
reactions

First aid   Standard treatment for exposure by all routes (see pages 108-122). There
                may be injuries following fire and/or explosion.

666. Hexalithium disilicide

Hazardous   Reacts violently with water; explodes with nitric acid, ignites on warming in
reactions   fluorine; incandescent reactions with phosphorus, selenium, or tellurium.

667. Hexamethyldiplatinum

Hazardous   Explodes on heating.
reactions

668. Hexamethylene diacrylate

Colourless liquid; sparingly soluble in water.

RISKS
Irritating to eyes and skin – May cause sensitization by skin contact (R36/38, R43)

SAFETY PRECAUTIONS
Wear eye/face protection (S39)

Toxic effects   Irritant to skin and eyes. High concentrations are very destructive to the upper
                    respiratory tract.

First aid   Standard treatment for exposure by all routes (see pages 108-122).

Fire hazard   Flash point above 110 ˚C (closed cup). Extinguish fires with water spray,
                  carbon dioxide, dry chemical powder, or alcohol or polymer foam.

Spillage   See general section.
disposal

RSC *Chemical Safety Data Sheets* Vol. 3, No. 33, 1990 gives extended coverage.

669. Hexamethylphosphoramide

Colourless, mobile liquid with aromatic odour.; m.p. 7 ˚C; b.p. 230 ˚C; miscible with water.

RISKS
May cause cancer – May cause heritable genetic damage (R45, R46)

SAFETY PRECAUTIONS
Avoid exposure – Obtain special instructions before use – If you feel unwell, seek medical advice (show label where possible) (S53, S44)

Toxic effects	No data are available on human toxicity, but animal experiments indicate acute toxic effects on lungs, kidneys, and the nervous and reproductive systems. Animal experiments have led to tumours from inhalation of 0.4 p.p.m. for long periods. Irritant to skin and eyes.
First aid	Standard treatment for exposure by all routes (see pages 108-122).
Fire hazard	Flash point 105 ˚C. Small fires can be extinguished with carbon dioxide, dry chemical powder, alcohol-resistant foam, or plenty of water. Large fires are best controlled by alcohol-resistant foam.
Spillage disposal	See general section.

RSC *Chemical Safety Data Sheets* Vol. 1, No. 56, 1988 gives extended coverage.

670. Hexane

Colourless liquid with a slightly basic odour; m.p. -95.6 ˚C; b.p. 69 ˚C; insoluble in water.

RISKS
Highly flammable – Harmful by inhalation – Danger of serious damage to health by prolonged exposure (R11, R20, R48)

SAFETY PRECAUTIONS
Keep container in a well ventilated place – Keep away from sources of ignition – No Smoking – Avoid contact with skin and eyes – Do not empty into drains – Use only in well ventilated areas (S9, S16, S24/25, S29, S51)

Limit values	OES short-term 1000 p.p.m. (3600 mg m^{-3}) (all isomers except n-); long-term 20 p.p.m. (70 mg m^{-3}) (n-hexane), 500 p.p.m. (1800 mg m^{-3}) (other isomers).
Toxic effects	The vapour may irritate the respiratory system and, in high concentrations, have narcotic action. Chronic exposure can lead to loss of sensation in hands and feet. Also irritant to skin, eyes, and the gastro-intestinal tract.
First aid	Standard treatment for exposure by all routes (see pages 108-122).
Fire hazard	Flash point -22 ˚C; Explosive limits 1.2-7.5%; Autoignition temperature 223 ˚C. Extinguish small fires with foam, dry powder, or carbon dioxide extinguishers. Large fires are best controlled by alcohol-resistant foam.
Spillage disposal	See general section.

RSC *Chemical Safety Data Sheets* Vol. 1, No. 57, 1988 gives extended coverage.

671. Hexanoic acid

Colourless liquid with unpleasant, sweat-like odour.; m.p. -3 ˚C; b.p. 205 ˚C; immiscible with water.

Toxic effects   The liquid burns the skin and eyes. Corrosive if taken by mouth. Inhalation may prove fatal following severe irritation and destruction of the respiratory tissues.

First aid   Standard treatment for exposure by all routes (see pages 108-122).

Fire hazard   Flash point 104 ˚C. Extinguish fires with water spray, carbon dioxide, dry chemical powder, or alcohol or polymer foam.

Spillage disposal   See general section.

672. Hexan-2-one

Colourless liquid with acetone-like odour; m.p. -57 ˚C; b.p. 126 ˚C; slightly soluble in water.

RISKS
Highly flammable – Toxic by inhalation – Danger of serious damage to health by prolonged exposure (R11, R23, R48)

SAFETY PRECAUTIONS
Keep container in a well ventilated place – Keep away from sources of ignition – No Smoking – Do not empty into drains – If you feel unwell, seek medical advice (show label where possible) – Use only in well ventilated areas (S9, S16, S29, S44, S51)

Limit values   OES long-term 5 p.p.m. (20 mg m^{-3}).

Toxic effects   The vapour irritates the eyes and respiratory system. Repeated and prolonged exposure has a depressant action on the nervous system resulting in muscle weakness and loss of sensation in hands and feet. The liquid will irritate the eyes and may cause severe damage. If swallowed may cause gastric irritation. It is rapidly absorbed through the skin.

First aid   Standard treatment for exposure by all routes (see pages 108-122).

Fire hazard   Flash point 35 ˚C (open cup); Explosive limits 1.2-8%; Autoignition temperature 423 ˚C. Extinguish small fires with foam, dry powder, or carbon dioxide. Large fires are best controlled by alcohol-resistant foam.

Spillage disposal   See general section.

RSC *Chemical Safety Data Sheets* Vol. 1, No. 69, 1988 gives extended coverage.

673. Hex-3-enedinitrile

Tan crystals; m.p. 74-76 ˚C.

Toxic effects   Irritant to skin, eyes, upper respiratory tract, and mucous membranes.

Hazardous reactions   Accelerating polymerization-decomposition due to overheating in vacuum evaporator.

First aid   Standard treatment for exposure by all routes (see pages 108-122).

Fire hazard   Extinguish fires with water spray, carbon dioxide, dry chemical powder, or alcohol or polymer foam.

Spillage disposal   See general section.

674. s-Hexyl acetate

Colourless liquid with a pleasant odour; m.p. -80 ˚C; b.p. 168-170 ˚C; slightly soluble in water.

Toxic effects   Irritant to skin and eyes.

First aid   Standard treatment for exposure by all routes (see pages 108-122).

Fire hazard   Flash point 37 ˚C. Extinguish fires with carbon dioxide, dry chemical powder, or alcohol or polymer foam.

Spillage disposal   See general section.

675. Hydrazine

Colourless, fuming, oily liquid with ammonia-like smell; m.p. 2 ˚C; b.p. 113.5 ˚C; miscible with water.

RISKS
Flammable – Very toxic by inhalation, in contact with skin, and if swallowed – Causes burns – Possible risk of irreversible effects (R10, R26/27/28, R34, R40)

SAFETY PRECAUTIONS
Wear suitable protective clothing, gloves, and eye/face protection – In case of accident or if you feel unwell, seek medical advice immediately (show label where possible) (S36/37/39, S45)

Limit values   OES long-term 0.1 p.p.m. (0.1 mg m^{-3}) under review.

Toxic effects   The liquid burns the eyes and skin. If taken by mouth there is severe internal irritation and damage.

Hazardous reactions   Contact with many transition metals, their oxides (rust), or salts causes catalytic decomposition (violent if concentrated solutions) and possibly ignition of evolved hydrogen; reaction with oxidants is violent.

First aid   Standard treatment for exposure by all routes (see pages 108-122).

Fire hazard   Flash point 37.78 ˚C (closed cup); Explosive limits 4.7%-100%; Autoignition temperature 270 ˚C (can be as low as 24 ˚C). Extinguish fires with water spray, carbon dioxide, or dry chemical powder.

Spillage disposal   See general section.

RSC *Chemical Safety Data Sheets* Vol. 4a, No. 75, 1991 gives extended coverage.

676. Hydrazine salts

Colourless crystals; soluble in water.

Toxic effects   The crystals and solutions irritate or burn the eyes and skin. If taken by mouth there would be severe internal irritation and damage.

Hazardous reactions   See the hydrazinium salts below.

First aid   Standard treatment for exposure by all routes (see pages 108-122).

Fire hazard   Use water spray, carbon dioxide, dry chemical powder, or alcohol or polymer foam to extinguish fires.

Spillage disposal   See general section.

677. Hydrazinium chlorate

Hazardous reactions   Explodes violently at mp, 80 ˚C.

678. Hydrazinium chlorite

Hazardous reactions   Spontaneously flammable when dry.

679. Hydrazinium nitrate

Hazardous reactions   Explosive properties have been studied in detail.

680. Hydrazinium perchlorate

Hazardous reactions   Deflagration and thermal decomposition have been studied. Used as a rocket propellant.

681. Hydriodic acid

Yellow to brown, fuming liquid commonly available in concentrations of 55% and 66% HI; b.p. 127 ˚C; miscible with water.

RISKS
Causes burns (R34)

SAFETY PRECAUTIONS
In case of contact with eyes, rinse immediately with plenty of water and seek medical advice (S26)

Toxic effects	The vapour irritates and is severely destructive to the skin, eyes, upper respiratory tract, and mucous membranes. The liquid burns the skin and eyes. If taken by mouth there would be severe internal irritation and damage.
Hazardous reactions	Blockage of condenser in preparation from wet iodine and phosphorus caused explosion.
First aid	Standard treatment for exposure by all routes (see pages 108-122).
Fire hazard	Use extinguishing medium appropriate to the surroundings.
Spillage disposal	See general section.

682. Hydrobromic acid

Colourless, fuming liquid with acrid smell commonly available in concentrations of 47%, 50%, and 60%; miscible with water.

RISKS
Causes burns – Irritating to respiratory system (R34, R37)

SAFETY PRECAUTIONS
Keep container tightly closed and in a well ventilated place – In case of contact with eyes, rinse immediately with plenty of water and seek medical advice (S7/9, S26)

Toxic effects	The vapour irritates and is severely destructive to the skin, eyes, upper respiratory tract, and mucous membranes. The liquid burns the eyes and skin. If taken by mouth there would be severe internal irritation and damage.
First aid	Standard treatment for exposure by all routes (see pages 108-122).
Fire hazard	Extinguish fires with a medium appropriate to the surroundings. Do NOT use dry chemical powder.
Spillage disposal	See general section.

683. Hydrochloric acid

Colourless, fuming liquid with pungent smell, commonly available in concentrations of 32% and 36% HCl; miscible with water.

RISKS
(≥ 10%, ≤ 25%) Irritating to eyes – Irritating to skin (R36, R38)
(≥ 25%) Causes burns – Irritating to respiratory system (R34, R37)

SAFETY PRECAUTIONS
(≥ 10%, ≤ 25%) Keep out of reach of children – After contact with skin, wash immediately with plenty of water (S2, S28)
(≥ 25%) Keep out of reach of children – In case of contact with eyes, rinse immediately with plenty of water and seek medical advice (S2, S26)

Toxic effects	The vapour irritates and is severely destructive to the skin, eyes, upper respiratory tract, and mucous membranes. The liquid burns the eyes and skin. If taken internally there would be severe irritation and damage.

First aid   Standard treatment for exposure by all routes (see pages 108-122).

Fire hazard   Use extinguishant appropriate to the surroundings.

Spillage   See general section.
  disposal

RSC *Chemical Safety Data Sheets* Vol. 3, No. 34, 1990 gives extended coverage.

684. Hydrofluoric acid

Colourless, fuming liquid with pungent smell, commonly available in concentrations of 40%, 42%, and 48% HF; miscible with water.

RISKS
Very toxic by inhalation, in contact with skin, and if swallowed – Causes severe burns (R26/27/28, R35)

SAFETY PRECAUTIONS
Keep container tightly closed and in a well ventilated place – In case of contact with eyes, rinse immediately with plenty of water and seek medical advice – Wear protective clothing and gloves – In case of accident or if you feel unwell, seek medical advice immediately (show label where possible) (S7/9, S26, S36/37, S45)

Toxic effects   The fume irritates severely and is extremely destructive to all parts of the respiratory system. The fume or acid rapidly causes irritation of the eyes and eyelids and burns may develop. Skin burns do not usually cause pain until several hours have elapsed since contact. If taken by mouth there is immediate and severe internal irritation and damage.

First aid   Standard treatment for exposure by all routes (see pages 108-122).

Fire hazard   Extinguish fires using dry chemical powder.

Spillage   See general section.
  disposal

RSC *Chemical Safety Data Sheets* Vol. 3, No. 35, 1990 gives extended coverage.

685. Hydrogen

Colourless, odourless gas; b.p. -253 °C; sparingly soluble in water.

RISKS
Extremely flammable (R12)

SAFETY PRECAUTIONS
Keep container tightly closed and in a well ventilated place (S7/9)

Toxic effects   Classed as a simple asphyxiant gas.

Hazardous   Catalytic platinum and similar metals containing adsorbed oxygen will heat
  reactions   and cause ignition in contact with hydrogen; high-velocity jets may ignite spontaneously particularly if rust is present; reaction hazards with many oxidants.

Fire hazard Explosive limits 4-75%; Autoignition temperature 585 °C. Extinguish small hydrogen fires with dry powder or carbon dioxide. Where the gas is supplied in a cylinder, turning off the valve will reduce fire.

Spillage disposal Surplus gas or leaking cylinder can be vented slowly to air in a safe open space or gas burnt-off through a suitable burner.

686. Hydrogenated terphenyls

A mixture of *o*-, *m*-, and *p*- 40% hydrogenated terphenyls.

Toxic effects Harmful by ingestion.

First aid Standard treatment for exposure by all routes (see pages 108-122).

687. Hydrogen azide

Liquid with intolerant pungent odour; m.p. -80 °C; b.p. 37 °C.

Limit values OES short-term 0.1 p.p.m. (as vapour).

Toxic effects Irritant to the eyes. Harmful by inhalation.

Hazardous reactions Safe in dilute solutions, extremely endothermic and violently explosive in concentrated or pure states; readily forms explosive heavy metal azides.

688. Hydrogen bromide

Colourless gas, fuming in moist air, with pungent acrid smell; m.p. -87 °C; b.p. -67 °C; dissolves readily in water forming hydrobromic acid.

RISKS
Causes severe burns – Irritating to respiratory system (R35, R37)

SAFETY PRECAUTIONS
Keep container tightly closed and in a well ventilated place – In case of contact with eyes, rinse immediately with plenty of water and seek medical advice – If you feel unwell, seek medical advice (show label where possible) (S7/9, S26, S44)

Limit values OES short-term 3 p.p.m. (10 mg m^{-3}).

Toxic effects The gas irritates and is severely destructive to tissues of the respiratory system; it also irritates and may burn the skin and eyes.

Hazardous reactions Ignites on contact with fluorine; explodes with ozone.

First aid Standard treatment for exposure by all routes (see pages 108-122).

Fire hazard Use water spray to keep cylinders cool and move them out of the area if this can be done without risk.

Spillage disposal Surplus gas or leaking cylinder can be vented slowly into a water-fed scrubbing tower or column in a fume cupboard, or into a fume cupboard served by such a tower.

689. Hydrogen chloride

Colourless gas, fuming in moist air, with pungent suffocating smell; m.p. -114 ˚C; b.p. -85 ˚C; dissolves readily in water forming hydrochloric acid.

RISKS
Causes severe burns – Irritating to respiratory system (R35, R37)

SAFETY PRECAUTIONS
Keep container tightly closed and in a well ventilated place – In case of contact with eyes, rinse immediately with plenty of water and seek medical advice – If you feel unwell, seek medical advice (show label where possible) (S7/9, S26, S44)

Limit values OES short-term 5 p.p.m. (7 mg m^{-3}).

Toxic effects The gas irritates and is extremely destructive to the respiratory system; it also irritates the skin and may cause severe burns to both eyes and skin.

Hazardous reactions Violent reaction with aluminium; ignites on contact with fluorine; dangerous reactions with hexalithium disilicide, some acetylides, uranium dicarbide, and tetraselenium tetranitride are recorded.

First aid Standard treatment for exposure by all routes (see pages 108-122).

Fire hazard Use water spray to keep cylinders cool and move them out of the fire area if this can be done without risk.

Spillage disposal Surplus gas or leaking cylinder can be vented slowly into a water-fed scrubbing tower or column in a fume cupboard, or into a fume cupboard served by such a column.

RSC *Chemical Safety Data Sheets* Vol. 3, No.34, 1990 gives extended coverage.

690. Hydrogen cyanide

Colourless liquid or gas with faint odour of bitter almonds; m.p. -13.2 ˚C; b.p. 26 ˚C; very soluble in water, the solution being only weakly acidic.

RISKS
Extremely flammable – Very toxic by inhalation, in contact with skin, and if swallowed (R12, R26/27/28)

SAFETY PRECAUTIONS
Keep container tightly closed and in a well ventilated place – Keep away from food, drink, and animal feeding stuffs – Keep away from sources of ignition – No Smoking – In case of accident or if you feel unwell, seek medical advice immediately (show label where possible) (S7/9, S13, S16, S45)

Limit values MEL short-term 10 p.p.m. (10 mg m^{-3}).

Toxic effects Inhalation of high concentrations leads to shortness of breath, paralysis, unconsciousness, convulsions, and death by respiratory failure. With lethal concentrations, death is extremely rapid although breathing may continue for some time. With low concentrations the effects are likely to be headache, vertigo, nausea, and vomiting. Chronic exposure over long periods may induce fatigue and weakness. The average fatal dose is 55 mg which can also be assimilated by skin contact with the liquid.

Hazardous reactions	In absence of inhibitor, exothermic polymerization occurs at 184 °C and this is explosive.
First aid	Standard treatment for exposure by all routes (see pages 108-122).
Fire hazard	Flash point -18 °C; Explosive limits 6-14%; Autoignition temperature 538 °C. As the liquid is supplied in a cylinder, turning off the valve will reduce any fire involving it; if possible, cylinders should be removed quickly from an area in which a fire has developed. Breathing apparatus must be worn during these operations.
Spillage disposal	Breathing apparatus must be worn. Surplus gas or leaking cylinder can be vented slowly into a water-fed scrubbing tower or column in a fume cupboard.

RSC *Chemical Safety Data Sheets* Vol. 4a, No. 76, 1991 gives extended coverage.

691. Hydrogen fluoride

Colourless, fuming gas or liquid; b.p. 19.5 °C; dissolves in water readily forming hydrofluoric acid.

RISKS
Very toxic by inhalation, in contact with skin, and if swallowed – Causes severe burns (R26/27/28, R35)

SAFETY PRECAUTIONS
Keep container tightly closed and in a well ventilated place – In case of contact with eyes, rinse immediately with plenty of water and seek medical advice – Wear protective clothing and gloves – In case of accident or if you feel unwell, seek medical advice immediately (show label where possible) (S7/9, S26, S36/37, S45)

Limit values	OES short-term 3 p.p.m. (2.5 mg m^{-3}) (as F).
Toxic effects	The gas severely irritates the eyes and respiratory system and may cause burns to the eyes. It irritates the skin and painful burns may develop after an interval. The liquid causes severe, painful burns on contact with all body tissues.
Hazardous reactions	Risk of violent reaction in fluorination of organic substances by passing hydrogen fluoride into stirred suspension of mercury(II) oxide; arsenic trioxide and calcium oxide incandesce in contact with the liquid; violent evolution of silicon tetrafluoride when potassium tetrafluorosilicate comes in contact with the liquid.
First aid	Exposure by all routes requires special treatment as for hydrofluoric acid.
Fire hazard	Extinguish fires with water spray or dry chemical powder.
Spillage disposal	Surplus gas or leaking cylinder can be vented slowly into a water-fed scrubbing tower or column in a fume cupboard, or into a fume cupboard served by such a column.

RSC *Chemical Safety Data Sheets* Vol. 3, No. 35, 1990 gives extended coverage.

692. Hydrogen iodide

Colourless gas; m.p. -50.8 ˚C; b.p. -35.5 ˚C; very soluble in water to form hydroiodic acid.

RISKS
Causes severe burns – Irritating to respiratory system (R35, R37)

SAFETY PRECAUTIONS
Keep container tightly closed and in a well ventilated place – In case of contact with eyes, rinse immediately with plenty of water and seek medical advice – If you feel unwell, seek medical advice (show label where possible) (S7/9, S26, S44)

Toxic effects	The gas irritates the respiratory system; it also irritates and may burn the eyes and skin.
Hazardous reactions	Causes momentary ignition of magnesium metal; mixture of the anhydrous liquid and potassium metal explodes very violently; the gas is ignited by molten potassium chlorate.
First aid	Standard treatment for exposure by all routes (see pages 108-122).
Fire hazard	Extinguish fires with water spray.
Spillage disposal	Surplus gas or leaking cylinder can be vented slowly into a water-fed scrubbing tower or column in a fume cupboard, or into a fume cupboard served by such a column.

693. Hydrogen peroxide

Colourless heavy liquid with a bitter taste; m.p. -0.4 ˚C; b.p. 152 ˚C; miscible with water.

RISKS
($\geq$ 20%, $\leq$ 60%) Causes burns (R34)
(> 60%) Contact with combustible material may cause fire – Causes burns (R8, R34)

SAFETY PRECAUTIONS
($\geq$ 20%, $\leq$ 60%) After contact with skin, wash immediately with plenty of water – Wear eye/face protection (S28, S39)
(> 60%) Keep in a cool place – After contact with skin, wash immediately with plenty of water – Wear suitable protective clothing and eye/face protection (S3, S28, S36/39)

Limit values	OES short-term 2 p.p.m. (3 mg m^{-3}); long-term 1 p.p.m. (1.5 mg m^{-3}).
Toxic effects	Hydrogen peroxide, especially in higher concentrations, is irritant and caustic to the mucous membranes, eyes, and skin. If swallowed, the sudden evolution of oxygen may cause injury by acute distension of the stomach and may cause nausea, vomiting, and internal bleeding.
Hazardous reactions	Hazardous reactions ranging from ignition to explosion are recorded with many substances.
First aid	Standard treatment for exposure by all routes (see pages 108-122).
Fire hazard	Use water spray to extinguish fires.
Spillage disposal	See general section.

694. Hydrogen sulfide

Colourless gas with an offensive odour; m.p. -85.5 °C; b.p. -60.4 °C; soluble in water.

RISKS
Extremely flammable liquefied gas – Very toxic by inhalation (R13, R26)

SAFETY PRECAUTIONS
Keep container tightly closed and in a well ventilated place – Avoid contact with skin – In case of accident or if you feel unwell, seek medical advice immediately (show label where possible) (S7/9, S24, S45)

Limit values	OES short-term 15 p.p.m. (21 mg m^{-3}); long-term 10 p.p.m. (14 mg m^{-3}).
Toxic effects	In high concentrations may cause immediate unconsciousness followed by respiratory paralysis; at lower concentrations causes irritation of all parts of the respiratory system and eyes, headache, dizziness, and also weakness. Irritates the eyes and may cause conjunctivitis.
Hazardous reactions	May ignite on contact with a wide range of metal oxides; finely divided tungsten glows red-hot in stream of the gas; dangerous reactions with a variety of oxidants have been recorded; exothermic reaction with soda lime, which may become explosive in presence of oxygen.
First aid	Standard treatment for exposure by all routes (see pages 108-122).
Fire hazard	Explosive limits 4.3-46%; Autoignition temperature 260 °C. As the gas is supplied in a cylinder, turning off the valve will reduce any fire involving it; if possible, cylinders should be removed quickly from an area in which a fire has developed.
Spillage disposal	Surplus gas or leaking cylinder can be vented slowly into a water-fed scrubbing tower or column in a fume cupboard, or into a fume cupboard served by such a column.

RSC *Chemical Safety Data Sheets* Vol. 4a, No. 77, 1991 gives extended coverage.

695. Hydrogen trisulfide

Yellow liquid; m.p. -52 °C; decomposes at 90 °C.

Hazardous reactions	Some metal oxides cause violent decomposition and ignition; reactions with benzenediazonium chloride, nitrogen trichloride, and pentyl alcohol.

696. Hydroquinone

White crystals; m.p. 170 ˚C; b.p. 285 ˚C at 730 mm Hg; soluble in water.

RISKS
Harmful by inhalation and if swallowed (R20/22)

SAFETY PRECAUTIONS
Keep out of reach of children – Avoid contact with skin and eyes – Wear eye/face protection (S2, S24/25, S39)

Limit values	OES short-term 4 mg m^{-3}; long-term 2 mg m^{-3}.
Toxic effects	Ingestion has led to nausea, shortness of breath, delirium, or collapse. Contact with skin may lead to dermatitis and prolonged exposure of eyes to vapour may lead to corneal damage. Suspect carcinogen.
Hazardous reactions	Decomposition in contact with sodium hydroxide; reaction with pressurized oxygen to form clathrate compound exploded.
First aid	Standard treatment for exposure by all routes (see pages 108-122).
Fire hazard	Flash point 165 ˚C; Autoignition temperature 515 ˚C. Extinguish fires with water spray, carbon dioxide, or dry chemical powder.
Spillage disposal	See general section.

697. 2-Hydroxy-3-butenonitrile

Toxic effects	Severely irritates skin and eyes. Toxic by inhalation, ingestion, and skin contact.
Hazardous reactions	Liable to polymerize explosively when exposed to light and air above 25 ˚C.

698. 2-Hydroxyethyl methacrylate

Pale yellow, mobile liquid, with a distinctive smell; m.p. -12 ˚C; b.p. 87 ˚C at 5 mbar; miscible with water.

RISKS
Irritating to eyes and skin – May cause sensitization by skin contact (R36/38, R43)

SAFETY PRECAUTIONS
In case of contact with eyes, rinse immediately with plenty of water and seek medical advice – After contact with skin, wash immediately with plenty of water (S26, S28)

Toxic effects	Splashes on skin and mucous surfaces can cause severe blistering. Irritating to the upper respiratory tract following inhalation and mildly toxic on ingestion.
First aid	Standard treatment for exposure by all routes (see pages 108-122).

Fire hazard   Flash point 108 ˚C (open cup). Extinguish fires with carbon dioxide or dry chemical powder extinguishers.

Spillage   See general section.
   disposal

RSC *Chemical Safety Data Sheets* Vol. 3, No. 36, 1990 gives extended coverage.

699. 1-Hydroxyethyl peroxyacetate

A low melting point solid.

Hazardous   The solid is explosive.
   reactions

700. Hydroxylamine

Colourless liquid or white crystals; m.p. 34 ˚C; b.p. 110 ˚C; decomposed by hot water.

Toxic effects   Irritant to skin, eyes, and mucous membranes. Harmful by inhalation and ingestion.

Hazardous   Risk of explosion in preparation from hydroxylamine hydrochloride and
   reactions   sodium hydroxide in methanol; ignites on contact with anhydrous copper(II) sulfate; dangerous reactions with some finely divided metals; it reacts violently or explosively with a large number of oxidants.

Fire hazard   Flash point explodes at 130 ˚C.

701. Hydroxylammonium salts

White crystals; m.p. SO_4- 170 ˚C with decomposition; Cl_- 155-157 ˚C with decomposition; soluble in water.

Toxic effects   The salts and solutions irritate and burn the eyes and skin. Continued skin contact can cause dermatitis. If taken by mouth there is severe internal irritation and damage.

Hazardous   Many salts decompose violently above 100 ˚C as solids or solutions.
   reactions

First aid   Standard treatment for exposure by all routes (see pages 108-122).

Fire hazard   Use water spray, carbon dioxide, dry chemical powder, or alcohol or polymer foam to extinguish fires.

Spillage   See general section.
   disposal

702. 4-Hydroxy-4-methylpentan-2-one

Colourless liquid which yellows on ageing with faint, pleasant odour; m.p. -50 ˚C; b.p. 168 ˚C; miscible with water.

RISKS
Irritating to eyes (R36)

SAFETY PRECAUTIONS
Avoid contact with skin and eyes (S24/25)

Limit values   OES short-term 75 p.p.m. (360 mg m^{-3}); long-term 50 p.p.m. (240 mg m^{-3}).

Toxic effects   The vapour irritates the eyes and respiratory system. The liquid irritates the eyes and mucous membranes and is absorbed by the skin, possibly with harmful effects. Taken by mouth, it has a narcotic effect. Kidney and liver injury and anaemia have been produced in experimental animals

First aid   Standard treatment for exposure by all routes (see pages 108-122).

Fire hazard   Flash point 64 ˚C; Explosive limits 1.8-6.9 ˚C; Autoignition temperature 603 ˚C. Extinguish small fires with carbon dioxide, dry chemical powder, or alcohol-resistant foam extinguishers. Large fires are best controlled by alcohol-resistant foam.

Spillage disposal   See general section.

RSC *Chemical Safety Data Sheets* Vol. 1, No. 20, 1988 gives extended coverage.

703. 2-Hydroxy-2-methylpropiononitrile

Colourless liquid; m.p. -19 ˚C; b.p. 82 ˚C at 30 mbar; miscible with water.

Toxic effects   High vapour concentrations, if inhaled, rapidly cause giddiness, headache, unconsciousness, and convulsions; in severe cases, breathing may cease due to hydrogen cyanide poisoning. It must be assumed that skin absorption and taking by mouth will have similar effects.

Hazardous reactions   Sulfuric acid may dehydrate it to methacrylonitrile and catalyse exothermic polymerization.

First aid   Standard treatment for exposure by all routes (see pages 108-122), with particular reference to cyanides.

Fire hazard   Flash point 63 ˚C. Use water spray, carbon dioxide, dry chemical powder, or alcohol or polymer foam to extinguish fires.

Spillage disposal   See general section.

704. 2-Hydroxypropyl acrylate

Liquid; b.p. 191 ˚C; soluble in water.

RISKS
Toxic by inhalation, in contact with skin, and if swallowed – Causes burns – May cause sensitization by skin contact (R23/24/25, R34, R43)

SAFETY PRECAUTIONS
In case of contact with eyes, rinse immediately with plenty of water and seek medical advice – Wear suitable protective clothing and eye/face protection – If you feel unwell, seek medical advice (show label where possible) (S26, S36/39, S44)

Limit values   OES long-term 0.5 p.p.m. (3 mg m^{-3}).

Toxic effects   Irritant to skin, eyes, and upper respiratory tract.

First aid   Standard treatment for exposure by all routes (see pages 108-122).

Fire hazard   Flash point 65 ˚C; Explosive limits lel 1.8%. Alcohol foam has been recommended for extinguishing fires.

705. Hydroxypropyl methacrylate

Colourless liquid with a distinctive smell; m.p. below -70 ˚C; b.p. 96 ˚C at 9.8 mm Hg; sparingly soluble in water.

RISKS
Irritating to eyes and skin (R36/38)

SAFETY PRECAUTIONS
In case of contact with eyes, rinse immediately with plenty of water and seek medical advice – After contact with skin, wash immediately with plenty of water (S26, S28)

Toxic effects   Irritant to skin, eyes, and upper respiratory tract. Prolonged skin contact may cause burns and sensitization.

First aid   Standard treatment for exposure by all routes (see pages 108-122).

Fire hazard   Flash point 121 ˚C (open cup). Extinguish fires with carbon dioxide, dry chemical powder, or foam extinguishants.

Spillage   See general section.
disposal

RSC *Chemical Safety Data Sheets* Vol. 3, No. 37, 1990 gives extended coverage.

706. 8-Hydroxyquinoline

Off-white to light tan, granular powder; m.p. 72-74 ˚C; b.p. 267 ˚C at 752 mm Hg; sparingly soluble in water.

Toxic effects   Irritant to skin, eyes, upper respiratory tract, and mucous membranes. Experimental carcinogen.

First aid	Standard treatment for exposure by all routes (see pages 108-122).
Fire hazard	Extinguish fires with water spray, carbon dioxide, dry chemical powder, or alcohol or polymer foam.
Spillage disposal	See general section.

707. Hypochlorous acid

Greenish-yellow liquid (aqueous solution).

| **Hazardous reactions** | Contact with alcohols forms unstable alkyl hypochlorites; explodes violently on contact with ammonia gas or acetic anhydride; ignites on contact with arsenic. |

708. Hyponitrous acid

Solid.

| **Hazardous reactions** | Extremely explosive solid. |

709. 3,3-Iminobis(propylamine)

Colourless to golden liquid; m.p. -14.5 °C; b.p. 240.6 °C; soluble in water.

RISKS
Harmful in contact with skin and if swallowed – Causes burns – May cause sensitization by skin contact (R21/22, R34, R43)

SAFETY PRECAUTIONS
In case of contact with eyes, rinse immediately with plenty of water and seek medical advice – Wear suitable protective clothing, gloves, and eye/face protection (S26, S36/37/39)

Toxic effects	Irritant to skin, eyes, and mucous membranes. May cause hypersensitivity and cross reactivity to other amines. Localized corrosive injury on ingestion. Inhalation may be fatal following bronchial spasm, inflammation, and oedema.
First aid	Standard treatment for exposure by all routes (see pages 108-122).
Fire hazard	Flash point 118 °C. Extinguish fire with water spray, carbon dioxide, dry chemical powder, or alcohol or polymer foam.
Spillage disposal	See general section.

RSC *Chemical Safety Data Sheets* Vol. 3, No. 38, 1990 gives extended coverage.

710. 2,2'-Iminodiethanol

White solid or viscous colourless liquid; m.p. 27-30 ˚C; b.p. 217 ˚C at 150 mmHg; very soluble in water.

RISKS
Irritating to eyes and skin (R36/38)

SAFETY PRECAUTIONS
In case of contact with eyes, rinse immediately with plenty of water and seek medical advice (S26)

Limit values OES long-term 3 p.p.m. (15 mg m^{-3}).

Toxic effects Extremely destructive to skin, eyes, upper respiratory tract, and mucous membranes. Inhalation may be fatal as a result of spasm, inflammation, chemical pneumonitis, and pulmonary oedema.

First aid Standard treatment for exposure by all routes (see pages 108-122).

Fire hazard Flash point 137 ˚C; Explosive limits lel 1.6%. Extinguish fires with carbon dioxide, dry chemical powder, or alcohol or polymer foam.

Spillage disposal See general section.

711. 2,2'-Iminodi(ethylamine)

Liquid; m.p. -39 ˚C; b.p. 208 ˚C with slight decomposition; very soluble in water.

RISKS
Harmful in contact with skin and if swallowed – Causes burns – May cause sensitization by skin contact (R21/22, R34, R43)

SAFETY PRECAUTIONS
In case of contact with eyes, rinse immediately with plenty of water and seek medical advice – Wear suitable protective clothing, gloves, and eye/face protection (S26, S36/37/39)

Limit values OES long-term 1 p.p.m. (4 mg m^{-3}).

Toxic effects Extremely destructive to skin, eyes, and upper respiratory tract. A skin and possibly pulmonary sensitizer. Inhalation may be fatal following spasm, chemical pneumonitis and pulmonary oedema.

First aid Standard treatment for exposure by all routes (see pages 108-122).

Fire hazard Flash point 98 ˚C; Autoignition temperature 358 ˚C. Extinguish fires with water spray, carbon dioxide, dry chemical powder, or alcohol or polymer foam.

Spillage disposal See general section.

RSC *Chemical Safety Data Sheets* Vol. 1, No. 29, 1988 gives extended coverage.

712. Iminodipropionitrile

Colourless liquid; m.p. -5.5 °C; b.p. 173 °C at 10 mm Hg; soluble in water.

Toxic effects   Little is known regarding toxicity to humans. In animals experiments CNS damage was noted after absorption orally or through the skin. There was also damage to the eyes following contact.

Hazardous   Bottles in store for 18 months exploded.
reactions

713. Indene

Colourless to pale yellow liquid; m.p. -2 °C; b.p. 181.6 °C; insoluble in water.

Limit values   OES short-term 15 p.p.m. (70 mg m^{-3}); long-term 10 p.p.m. (45 mg m^{-3}).

Toxic effects   Irritant to skin and eyes. May be absorbed through the skin.

First aid   Standard treatment for exposure by all routes (see pages 108-122).

Fire hazard   Flash point 58 °C. Extinguish fires with water spray, carbon dioxide, dry chemical powder, or alcohol or polymer foam.

Spillage   See general section.
disposal

714. Indium

Lustrous, silver-white metal, soft, very malleable, and ductile; m.p. 156.6 °C; b.p. 2000 °C; insoluble in water.

Limit values   OES short-term 0.3 mg m^{-3}; long-term 0.1 mg m^{-3}.

Toxic effects   May be irritating to eyes and respiratory tract. There is little information on human toxicity.

First aid   Standard treatment for exposure by all routes (see pages 108-122).

Fire hazard   Extinguish fires with carbon dioxide, dry chemical powder, or foam extinguishants.

Spillage   See general section.
disposal

RSC *Chemical Safety Data Sheets* Vol. 2, No. 56, 1989 gives extended coverage.

715. Indium oxide

White or pale yellow powder.; m.p. 850 °C with decomposition; insoluble in water.

Limit values   OES short-term 0.3 mg m^{-3} (as In); long-term 0.1 mg m^{-3} (as In).

Toxic effects   Irritating to skin, eyes, and upper repsiratory tract.

First aid   Standard treatment for exposure by all routes (see pages 108-122).

Fire hazard   Extinguish fires with carbon dioxide, dry chemical powder, or alcohol or polymer foam.

Spillage disposal   See general section.

RSC *Chemical Safety Data Sheets* Vol. 2, No. 57, 1989 gives extended coverage.

716. Indium sulfate

White, crystalline, deliquescent solid; m.p. 250 °C with decomposition; very soluble in water.

Limit values   OES short-term 0.3 mg m^{-3} (as In); long-term 0.1 mg m^{-3} (as In).

Toxic effects Highly irritant to skin and eyes. Little evidence on human toxicity by inhalation or ingestion.

First aid   Standard treatment for exposure by all routes (see pages 108-122).

Fire hazard   Extinguish fires with carbon dioxide, dry chemical powder, or alcohol or polymer foam.

Spillage disposal   See general section.

RSC *Chemical Safety Data Sheets* Vol. 2, No. 58, 1989 gives extended coverage.

717. Indium trichloride

White, very deliquescent crystals; m.p. 586 °C (sublimes over 400 °C); very soluble in water with some decomposition.

Limit values   OES short-term 0.3 mg m^{-3} (as In); long-term 0.1 mg m^{-3} (as In).

Toxic effects Highly irritating to skin, eyes, and upper respiratory tract. Inhalation may be fatal following spasm, inflammation, and oedema.

First aid   Standard treatment for exposure by all routes (see pages 108-122).

Fire hazard   Use extinguishant appropriate to the surroundings.

Spillage disposal   See general section.

RSC *Chemical Safety Data Sheets* Vol. 2, No. 59, 1989 gives extended coverage.

718. Indole

Brown, crystalline solid with intense faecal odour; m.p. 52-54 °C; b.p. 253-254 °C; soluble in hot water.

Toxic effects Irritant to skin, eyes, upper respiratory tract, and mucous membranes. May be absorbed through the skin. Experimental carcinogen.

First aid	Standard treatment for exposure by all routes (see pages 108-122).
Fire hazard	Extinguish fires with water spray, carbon dioxide, or dry chemical powder.
Spillage disposal	See general section.

719. Iodic acid

Colourless crystals or powder; m.p. 110 °C with decomposition; soluble in water.

Toxic effects	The dust irritates all parts of the respiratory system. The acid and solutions burn the eyes and skin. If taken by mouth is assumed to cause severe internal irritation and damage.
Hazardous reactions	Reacts with boron below 40 °C and incandesces; deflagrates with charcoal, phosphorus, and sulfur on heating.
First aid	Standard treatment for exposure by all routes (see pages 108-122).
Fire hazard	Use extinguishing medium appropriate to surroundings.
Spillage disposal	Site of spillage should be washed thoroughly to remove all oxidant, which is liable to render any organic matter (particularly wood, paper, and textiles) with which it comes into contact, dangerously combustible when dry. Also see general section.

720. Iodine

Bluish-black, crystalline scales with a characteristic odour; m.p. 114 °C; b.p. 154 °C; almost insoluble in water.

RISKS
Harmful by inhalation and in contact with skin (R20/21)

SAFETY PRECAUTIONS
Do not breathe vapour – Avoid contact with eyes (S23, S25)

Limit values	OES short-term 0.1 p.p.m. (1 mg m^{-3}).
Toxic effects	The vapour irritates all parts of the respiratory system. The vapour and solid irritate the eyes. The solid burns the skin. If taken by mouth there is severe internal irritation and damage.
Hazardous reactions	Forms highly explosive addition compounds with ammonia; reaction with ethanol and phosphorus considered dangerous as school experiment; violent or explosive reactions with a variety of substances.
First aid	Standard treatment for exposure by all routes (see pages 108-122).
Fire hazard	Extinguish fires with dry chemical powder.
Spillage disposal	See general section. Iodine stains on flooring can be cleared by mopping with thiosulfate or metabisulfite solution.

721. Iodine bromide

Black, crystalline solid; m.p. 42-50 °C.

Toxic effects	Extremely destructive to skin, eyes, upper respiratory tract, and all tissues. Harmful if inhaled, ingested, or absorbed through the skin. Can be fatal.
Hazardous reactions	Reacts violently or explosively with potassium, sodium, tin, and phosphorus.
First aid	Standard treatment for exposure by all routes (see pages 108-122).
Fire hazard	Extinguish fires with dry chemical powder.
Spillage disposal	See general section.

722. Iodine chloride

Reddish-brown liquid with pungent odour; m.p. 13.9 °C; b.p. 97 °C; soluble in water.

Toxic effects	The vapour irritates all parts of the respiratory system. The vapour and liquid burn the eyes severely. The liquid burns the skin. Assumed to cause severe internal irritation and damage if taken by mouth.
Hazardous reactions	Mixtures with sodium explode on impact, with potassium on contact. Aluminium foil may ignite on long contact.
First aid	Standard treatment for exposure by all routes (see pages 108-122).
Fire hazard	Extinguish fires with dry chemical powder. Do NOT use water.
Spillage disposal	See general section.

723. Iodine heptafluoride

Colourless gas with mouldy, acrid odour; m.p. 6.45 °C but sublimes at 4.77 °C; soluble in water with some decomposition.

Toxic effects	Highly corrosive to all tissues. Harmful on inhalation or ingestion.
Hazardous reactions	Activated carbon ignites in gas; mixtures with methane ignite, those with hydrogen explode on heating or sparking; violent reactions with a variety of substances.
First aid	Standard treatment for exposure by all routes (see pages 108-122), with special reference to fluorides.

724. Iodine isocyanate

Hazardous reactions	Solutions gradually deposit touch-sensitive explosive.

725. Iodine(V) oxide

White, crystalline powder; m.p. 300 °C with decomposition; soluble in water.

Toxic effects	Extremely destructive to tissues of the skin, eyes, upper respiratory tract, and mucous membranes. Inhalation may be fatal following spasm, chemical pneumonitis, and pulmonary oedema.
Hazardous reactions	Reacts violently with bromine pentafluoride; reacts explosively with warm aluminium, carbon, sulfur, resin, sugar, or powdered easily oxidizable elements.
First aid	Standard treatment for exposure by all routes (see pages 108-122).
Fire hazard	Extinguish fires with water spray.
Spillage disposal	See general section.

726. Iodine pentafluoride

Liquid; m.p. 9.4 °C; b.p. 100.5 °C; reacts violently with water.

Toxic effects	Irritant and highly corrosive to eyes, skin, and mucous membranes. Harmful on inhalation or ingestion.
Hazardous reactions	Reacts violently with benzene above 50 °C; reactions with water and dimethyl sulfoxide are violent; incandescence may occur with a variety of substances. Violent reactions with organic compounds or potassium hydroxide.
First aid	Standard treatment for exposure by all routes (see pages 108-122), with special reference to fluorides.

727. Iodine(III) perchlorate

Hazardous reactions	Exploded on laser irradiation at low temperature.

728. Iodine triacetate

Hazardous reactions	Explodes at 140 °C.

729. Iodine trichloride

Fuming, orange-red, crystalline masses with pungent odour; m.p. about 33 °C; soluble in water.

Toxic effects	The vapour irritates all parts of the respiratory system. The vapour and solid irritate and burn the eyes severely. The solid burns the skin. Assumed to cause severe internal irritation and damage if taken by mouth.

Hazardous reactions	Ignites phosphorus.
First aid	Standard treatment for exposure by all routes (see pages 108-122).
Fire hazard	Extinguish fires with water spray.
Spillage disposal	See general section.

730. Iodoacetic acid

Colourless or white crystals; m.p. 82-83 ˚C; soluble in water.

RISKS
Very toxic by inhalation, in contact with skin, and if swallowed – Causes severe burns (R26/27/28, R35)

SAFETY PRECAUTIONS
Do not breathe dust – Wear suitable protective clothing, gloves, and eye/face protection – In case of accident or if you feel unwell, seek medical advice immediately (show label where possible) (S22, S36/37/39, S45)

Toxic effects	Causes severe irritation and severe destruction to all tissues with which it comes in contact. Harmful by inhalation, ingestion, and skin absorption. Symptoms may be delayed. Inhalation can be fatal.
First aid	Standard treatment for exposure by all routes (see pages 108-122).
Fire hazard	Extinguish fires with water spray, carbon dioxide, dry chemical powder, or alcohol or polymer foam.
Spillage disposal	See general section.

RSC *Chemical Safety Data Sheets* Vol. 4a, No. 78, 1991 gives extended coverage.

731. 1-Iodobuta-1,3-diyne

Hazardous reactions	Crude material exploded during vacuum distillation; the pure compound exploded on scratching under illumination. Do not handle above 30 ˚C.

732. Iododimethylarsine

Yellowish oil; b.p. 155-160 ˚C.

Hazardous reactions	Ignites when heated in air.

733. Iodoethane

Colourless to brown liquid, sensitive to air and light; m.p. -110.9 ˚C; b.p. 72 ˚C; slightly soluble in water.

Toxic effects	Harmful on inhalation, ingestion, or skin absorption. Extremely destructive to all tissues. Inhalation can be fatal.
Hazardous reactions	Reaction of ethanol, phosphorus, and iodine considered too dangerous for preparation in school work; explosion with silver chlorite.
First aid	Standard treatment for exposure by all routes (see pages 108-122).
Fire hazard	Use extinguishing medium appropriate to surroundings.
Spillage disposal	See general section.

734. Iodoform

Yellow powder; m.p. 120-123 ˚C; sparingly soluble in water.

Limit values	OES short-term 1 p.p.m. (20 mg m^{-3}); long-term 0.6 p.p.m. (10 mg m^{-3}).
Toxic effects	Irritant to skin, eyes, upper respiratory tract, and mucous membranes. Exposure by all routes can cause headache, nausea, and vomiting.
First aid	Standard treatment for exposure by all routes (see pages 108-122).
Fire hazard	Use extinguishant appropriate to surrounding conditions.
Spillage disposal	See general section.

735. Iodomethane

Colourless liquid with pungent smell, turns brown on exposure to light; m.p. -66.4 ˚C; b.p. 42.5 ˚C; sparingly soluble in water.

RISKS
Harmful in contact with skin – Toxic by inhalation and if swallowed – Irritating to respiratory system and skin – Possible risk of irreversible effects (R21, R23/25, R37/38, R40)

SAFETY PRECAUTIONS
Wear protective clothing and gloves – In case of insufficient ventilation, wear suitable respiratory equipment – If you feel unwell, seek medical advice (show label where possible) (S36/37, S38, S44)

Limit values	OES short-term 10 p.p.m. (56 mg m^{-3}) under review; long-term 5 p.p.m. (28 mg m^{-3}) under review.
Toxic effects	Inhalation of vapour may cause dizziness, drowsiness, mental confusion, muscular twitching, and delirium, leading to death. The vapour and liquid irritate the eyes and distort the vision. The liquid irritates the skin and may cause blistering. The liquid must be assumed to be irritant and poisonous if taken by mouth.

Hazardous reactions	Vigorously reactive with sodium dispersed in toluene.
First aid	Standard treatment for exposure by all routes (see pages 108-122).
Fire hazard	Extinguish fires with water spray, carbon dioxide, dry chemical powder, or alcohol or polymer foam.
Spillage disposal	See general section.

RSC *Chemical Safety Data Sheets* Vol. 4a, No. 79, 1991 gives extended coverage.

736. 3-Iodo-1-phenylpropyne

Liquid; b.p. 137-140 ˚C.

| **Hazardous reactions** | Exploded on distillation at about 180 ˚C. |

737. Iodosylbenzene

Yellow, amorphous powder; m.p. 210 ˚C with decomposition.

| **Hazardous reactions** | Explodes at 210 ˚C. |

738. Iodylbenzene

Crystalline solid; m.p. 230 ˚C.

| **Hazardous reactions** | Explodes at 230 ˚C; extreme care required in heating, compressing, or grinding iodyl compounds. |

739. Iodylbenzene perchlorate

| **Hazardous reactions** | Small sample exploded violently while still damp. |

740. Ion-exchange resins

| **Hazardous reactions** | Expansion on moistening may fracture container; violent oxidation by dichromates or nitric acid with various metal ions. |

741. Iridium and compounds

Toxic effects   The metal is clinically inert and very little toxicity data are available. Most of the compounds are poorly absorbed by the body. The chlorides are eye and skin irritants and are toxic by ingestion.

Hazardous reactions   The powdered metal may ignite spontaneously in air. Violent reaction or ignition follows contact of the powdered metal with interhalogen compounds.

First aid   Standard treatment for exposure by all routes (see pages 108-122).

Fire hazard   Use extinguishant appropriate to the surroundings.

Spillage disposal   See general section.

742. Iron(III) chloride anhydrous

Black-brown crystals and aggregated masses; m.p. 37 ˚C; b.p. 280 ˚C; violently decomposed by water with the formation of hydrogen chloride.

Limit values   OES short-term 2 mg m^{-3} (as Fe); long-term 1 mg m^{-3} (as Fe).

Toxic effects   Inhalation of fine crystals produces irritation or burns of the mucous membranes. Will cause painful eye burns. When moisture is present on skin, heat is produced on contact resulting in thermal and acid burns. If taken by mouth the immediate local reaction would cause severe burns.

First aid   Standard treatment for exposure by all routes (see pages 108-122).

Fire hazard   Use extinguishant appropriate to the surroundings.

Spillage disposal   See general section.

743. Iron(II) maleate

Solid.

Limit values   OES short-term 2 mg m^{-3} (as Fe); long-term 1 mg m^{-3} (as Fe).

Hazardous reactions   The finely divided compound (a by-product of phthalic anhydride manufacture) is rapidly oxidized in air above 150 ˚C and has been involved in plant fires.

744. Iron(III) oxide

Red-brown to black powder; m.p. 1565 ˚C; insoluble in water.

Limit values   OES short-term 10 mg m^{-3}; long-term 5 mg m^{-3} (as Fe) (fume).

Toxic effects   Long term inhalation can cause X-ray changes in the lung. Suspect carcinogen and tumourigen.

Hazardous reactions	Can react violently or explosively with a variety of chemicals.
First aid	Standard treatment for exposure by all routes (see pages 108-122).

745. Iron pentacarbonyl

Orange liquid; m.p. -10 ˚C; b.p. 103 ˚C; insoluble in water.

Toxic effects	Causes severe irritation to all tissues. May be fatal following inhalation, ingestion, or skin absorption.
Hazardous reactions	Brown pyrophoric powder produced if the carbonyl is dissolved in acetic acid containing above 5% water. Heating rapidly to above 50 ˚C with nitric oxide in an autoclave caused explosive reaction.
First aid	Standard treatment for exposure by all routes (see pages 108-122).
Fire hazard	Flash point 15 ˚C. Extinguish fires with water spray, carbon dioxide, dry chemical powder, or alcohol or polymer foam.
Spillage disposal	See general section.

746. Iron(II) perchlorate

White or greyish-white, crystalline solid; b.p. above 100 ˚C with decomposition; very soluble in water.

Limit values	OES short-term 2 mg m^{-3} (as Fe); long-term 1 mg m^{-3} (as Fe).
Hazardous reactions	Violent explosion when mixture of iron(II) sulfate and perchloric acid was being heated.

747. Iron salts

Limit values	OES short-term 2 mg m^{-3} (as Fe); long-term 1 mg m^{-3} (as Fe).
Toxic effects	Ingestion can cause gastro-intestinal irritation. Also irritant to skin and upper respiratory tract.

748. Isobutane

Colourless gas.; b.p. 12 ˚C; somewhat soluble in water.

Toxic effects	The gas has an anaesthetic effect but is not toxic.
First aid	Standard treatment for exposure by all routes (see pages 108-122).
Fire hazard	Explosive limits 1.8-8.4%; Autoignition temperature 462 ˚C. As the gas is supplied in a cylinder, turning off the valve will reduce any fire involving it; if possible cylinders should be removed quickly from an area in which a fire has developed.

749. Isobutanol

Colourless liquid with characteristic sweet odour; m.p. 28 ˚C (open cup); b.p. 107.9 ˚C; soluble in water.

RISKS
Harmful by inhalation (R20)

SAFETY PRECAUTIONS
Keep away from sources of ignition – No Smoking (S16)

Limit values	OES short-term 75 p.p.m. (225 mg m^{-3}); long-term 50 p.p.m. (150 mg m^{-3}).
Toxic effects	The vapour irritates the respiratory system and, in high concentrations, produces symptoms of narcosis and central nervous depression with bone marrow effects and morphological changes to the liver and brain. The liquid irritates the eyes and is harmful if taken internally.
First aid	Standard treatment for exposure by all routes (see pages 108-122).
Fire hazard	Flash point -108 ˚C; Explosive limits 1.6-10.9%; Autoignition temperature 426 ˚C. Extinguish fire with water spray, dry powder, carbon dioxide, or alcohol-resistant foam.
Spillage disposal	See general section.

RSC *Chemical Safety Data Sheets* Vol. 1, No. 59, 1988 gives extended coverage.

750. Isobutyl methacrylate

Clear, colourless liquid; b.p. 155 ˚C; insoluble in water.

RISKS
Flammable – Irritating to eyes, respiratory system, and skin – May cause sensitization by skin contact (R10, R36/37/38, R43)

Toxic effects	Irritant to skin, eyes, and upper respiratory tract. Harmful on inhalation, ingestion, or skin contact.
First aid	Standard treatment for exposure by all routes (see pages 108-122).
Fire hazard	Flash point 41 ˚C (closed cup). Extinguish fires with water spray, carbon dioxide, dry chemical powder, or alcohol or polymer foam.
Spillage disposal	See general section.

RSC *Chemical Safety Data Sheets* Vol. 3, No. 39, 1990 gives extended coverage.

751. Isobutyric acid

Colourless, oily liquid with rancid odour; m.p. -47 °C; b.p. 154 °C; soluble in water.

RISKS
Harmful in contact with skin and if swallowed (R21/22)

Toxic effects The liquid is irritant and extremely destructive to all tissues. Harmful on inhalation, ingestion, or skin absorption. Inhalation may be fatal.

First aid Standard treatment for exposure by all routes (see pages 108-122).

Fire hazard Extinguish fires with water spray, carbon dioxide, dry chemical powder, or alcohol or polymer foam.

Spillage disposal See general section.

752. Isocyanatomethane

Colourless, lachrymatory liquid; m.p. -17 °C; b.p. 39 °C; sparingly soluble in and decomposed by water.

RISKS
Extremely flammable – Toxic by inhalation, in contact with skin, and if swallowed – Irritating to eyes, respiratory system, and skin (R12, R23/24/25, R36/37/38)

SAFETY PRECAUTIONS
Keep container in a well ventilated place – Never add water to this product – In case of fire, use dry powder or carbon dioxide – If you feel unwell, seek medical advice (show label where possible) (S9, S30, S43, S44)

Limit values MEL short-term 0.07 mg m^{-3} (as -NCO); long-term 0.02 mg m^{-3} (as -NCO).

Toxic effects The vapour irritates the respiratory system and eyes severely, causing lachrymation. With high levels of vapour severe and extensive damage is caused, some still evident over a year after the Bhopal disaster. The liquid irritates the eyes and skin severely and is assumed to be highly irritant and damaging if taken internally.

Hazardous reactions The underlying cause of the exothermic reaction in the Bhopal storage tank which led to the vaporization and escape of several thousand kg of the highly toxic material appears to have been presence of a substantial amount of water, possibly accompanied by catalytically active impurities.

First aid Standard treatment for exposure by all routes (see pages 108-122).

Fire hazard Flash point -7 °C; Explosive limits 5.3-26%; Autoignition temperature 534 °C. Extinguish fire with dry powder or carbon dioxide.

Spillage disposal See general section.

RSC *Chemical Safety Data Sheets* Vol. 4b, No. 93, 1991 gives extended coverage.

753. Isooctyl alcohol

Colourless liquid; b.p. approx. 170-180 °C.

Limit values   OES long-term 50 p.p.m. (270 mg m^{-3}) for mixed isomers.

Toxic effects   Irritant to skin, eyes, upper respiratory tract, and mucous membranes. Exposure can cause nausea, dizziness, and headaches.

First aid   Standard treatment for exposure by all routes (see pages 108-122).

Fire hazard   Flash point approx. 70-80 °C. Extinguish fires with water spray, carbon dioxide, dry chemical powder, or alcohol or polymer foam.

**Spillage
disposal**   See general section.

754. Isopentyl acetate

Colourless liquid; m.p. -78 °C; b.p. 142 °C.

Limit values   OES short-term 125 p.p.m. (655 mg m^{-3}); long-term 100 p.p.m. (525 mg m^{-3}).

Toxic effects   Irritant to skin, eyes, upper respiratory tract, and mucous membranes. Harmful on inhalation, ingestion, and skin absorption.

First aid   Standard treatment for exposure by all routes (see pages 108-122).

Fire hazard   Flash point 25 °C; Explosive limits 1.1-7.0%; Autoignition temperature 379 °C. Extinguish fires with carbon dioxide, dry chemical powder, or alcohol or polymer foam.

**Spillage
disposal**   See general section.

755. Isophorone

Colourless liquid; m.p. -8.1 °C; b.p. 215 °C; slightly soluble in water.

RISKS
Irritating to eyes, respiratory system, and skin (R36/37/38)

SAFETY PRECAUTIONS
In case of contact with eyes, rinse immediately with plenty of water and seek medical advice (S26)

Limit values   OES short-term 5 p.p.m. (25 mg m^{-3}).

Toxic effects   The vapour irritates the respiratory system giving rise to complaints of headaches, malaise, dizziness, and feeling of suffocation. The liquid and vapour irritate the eyes and may cause corneal damage. It is assumed to be toxic if taken by mouth. Animal carcinogen.

**Hazardous
reactions**   Hazardous peroxides may be produced on prolonged exposure with air.

First aid   Standard treatment for exposure by all routes (see pages 108-122).

Fire hazard Flash point 96 °C (open cup); Explosive limits 0.8-3.8%; Autoignition temperature 460 °C. Extinguish fires with carbon dioxide, dry powder, or alcohol resistant foam extinguishers. The foam is more suitable for large fires.

Spillage disposal See general section.

RSC *Chemical Safety Data Sheets* Vol. 1, No. 61, 1988 gives extended coverage.

756. Isophorone di-isocyanate

Colourless to slightly yellow liquid; m.p. -60 °C (approx.); b.p. 153 °C at 10 mm Hg; very slightly soluble in water.

RISKS
Toxic by inhalation – Irritating to eyes, respiratory system, and skin – May cause sensitization by inhalation and skin contact (R23, R36/37/38, R42/43)

SAFETY PRECAUTIONS
In case of contact with eyes, rinse immediately with plenty of water and seek medical advice – After contact with skin, wash immediately with plenty of water – In case of insufficient ventilation, wear suitable respiratory equipment – In case of accident or if you feel unwell, seek medical advice immediately (show label where possible) (S26, S28, S38, S45)

Limit values OES short-term 0.07 mg m^{-3} (as -NCO); long-term 0.02 mg m^{-3} (as -NCO).

Toxic effects Inhalation can cause asthma and other long-term effects.

First aid Standard treatment for exposure by all routes (see pages 108-122).

Fire hazard Flash point 162 °C. Extinguish fires with carbon dioxide, dry chemical powder, or alcohol or polymer foam.

Spillage disposal See general section.

757. Isoprene

Colourless, volatile liquid; b.p. 34 °C; insoluble in water.

RISKS
Extremely flammable (R12)

SAFETY PRECAUTIONS
Keep container in a well ventilated place – Keep away from sources of ignition – No Smoking – Do not empty into drains – Take precautionary measures against static discharges (S9, S16, S29, S33)

Toxic effects The vapour is irritating to the respiratory system and is narcotic in high concentrations. The liquid irritates the skin and eyes and may be irritant and harmful on ingestion.

Hazardous reactions Absorbs oxygen from the air forming explosive polymeric peroxides; explosion occurred during ozonization in heptane.

First aid Standard treatment for exposure by all routes (see pages 108-122).

Fire hazard   Flash point -53 ˚C; Autoignition temperature 220 ˚C. Extinguish fire with foam,
                    dry powder, or carbon dioxide.

Spillage   See general section.
  disposal

758. Isopropyl acetate

Colourless liquid; m.p. -73 ˚C; b.p. 93 ˚C; sparingly soluble in water.

RISKS
Highly flammable (R11)

SAFETY PRECAUTIONS
Keep away from sources of ignition – No Smoking – Do not breathe vapour – Do not empty into
drains – Take precautionary measures against static discharges (S16, S23, S29, S33)

Limit values   OES short-term 200 p.p.m. (840 mg m^{-3}).

Toxic effects   The vapour may irritate the eyes and respiratory system and is narcotic in
                    high concentrations. The liquid irritates the eyes and will be irritant and
                    narcotic if taken by mouth.

First aid   Standard treatment for exposure by all routes (see pages 108-122).

Fire hazard   Flash point 4.4 ˚C; Explosive limits 1.7-7.8%; Autoignition temperature
                    460 ˚C. Extinguish fire with dry powder, carbon dioxide, or alcohol-resistant
                    foam. The foam is particularly suitable for large fires.

Spillage   See general section.
  disposal

RSC *Chemical Safety Data Sheets* Vol. 1, No. 62, 1988 gives extended coverage.

759. *N*-Isopropylaniline

Amber liquid; b.p. 112-113 ˚C at 18 mm Hg; insoluble in water.

Toxic effects   Irritant to skin, eyes, upper respiratory tract, and mucous membranes.

First aid   Standard treatment for exposure by all routes (see pages 108-122).

Fire hazard   Flash point 95 ˚C. Extinguish fires with water spray, carbon dioxide, dry
                    chemical powder, or alcohol or polymer foam.

Spillage   See general section.
  disposal

760. Isopropyl chloroformate

Colourless liquid; b.p. 103-105 ˚C at 721 mm Hg; insoluble in and decomposed by water.

Limit values   OES long-term 1 p.p.m. (5 mg m^{-3}).

Toxic effects	Irritant to skin, eyes, upper respiratory tract, and mucous membranes. May cause allergic respiratory and skin reactions. Inhalation may be fatal as a result of spasm, inflammation, chemical pneumonitis, and pulmonary oedema.
First aid	Standard treatment for exposure by all routes (see pages 108-122).
Fire hazard	Flash point -11 ˚C; Explosive limits 4-15%. Extinguish fires with carbon dioxide, dry chemical powder, or alcohol or polymer foam.
Spillage disposal	See general section.

761. Isopropyl glycidyl ether

Colourless liquid; b.p. 127 ˚C; fairly soluble in water.

Limit values	OES short-term 75 p.p.m. (360 mg m^{-3}); long-term 50 p.p.m. (240 mg m^{-3}).
Toxic effects	Irritant to skin, eyes, upper respiratory tract, and mucous membranes.
First aid	Standard treatment for exposure by all routes (see pages 108-122).
Fire hazard	Flash point 33 ˚C. Extinguish fires with water spray, carbon dioxide, dry chemical powder, or alcohol or polymer foam.
Spillage disposal	See general section.

762. Isopropyl hydroperoxide

Liquid; b.p. 107-109 ˚C with explosion.

Hazardous reactions	Explodes just above bp, 107-109 ˚C.

763. Isopropyl hypochlorite

Hazardous reactions	Has extremely low stability – explosions have occurred during preparation.

764. Ketene

Colourless gas with disagreeable taste; m.p. -151 ˚C; b.p. -56 ˚C; decomposed by water.

Limit values	OES short-term 1.5 p.p.m. (3 mg m^{-3}); long-term 0.5 p.p.m. (0.9 mg m^{-3}).
Toxic effects	Irritant to skin, eyes, upper respiratory tract, and mucous membranes. Inhalation can cause shortness of breath, coughing, emphysema, and pulmonary oedema.

| **Hazardous reactions** | The reaction with hydrogen peroxide can rapidly form explosive diacetyl peroxide. |
| **First aid** | Standard treatment for exposure by all routes (see pages 108-122). |

765. Lactic acid

Colourless or slightly yellow, syrupy, hygroscopic liquid; m.p. 16.8 °C; b.p. 122 °C at 14 mm Hg; miscible with water.

SAFETY PRECAUTIONS
Avoid contact with skin and eyes (S24/25)

Toxic effects	Irritates and may burn the eyes, skin, and upper respiratory tract. Harmful if inhaled, ingested, or absorbed through the skin.
First aid	Standard treatment for exposure by all routes (see pages 108-122).
Fire hazard	Flash point above 110 °C (closed cup). Extinguish fires with water spray, carbon dioxide, dry chemical powder, or alcohol or polymer foam.
Spillage disposal	See general section.

766. Lead

Bluish-grey, soft metal; m.p. 327.4 °C; b.p. 1740 °C; insoluble in water.

Limit values	OES long-term 0.15 mg m^{-3}.
Toxic effects	Irritant to mucous membranes and upper respiratory tract.
First aid	Standard treatment for exposure by all routes (see pages 108-122).
Fire hazard	Use extinguishant appropriate to the surroundings.
Spillage disposal	See general section.

767. Lead arsenate

White, crystalline solid; m.p. above 280 °C with decomposition; insoluble in water.

RISKS
Harmful by inhalation and if swallowed – Danger of cumulative effects (R20/22, R33)

SAFETY PRECAUTIONS
Keep away from food, drink, and animal feeding stuffs – When using do not eat, drink, or smoke (S13, S20/21)

| **Limit values** | OES long-term 0.15 mg m^{-3} (as Pb). |
| **Toxic effects** | The acute toxic effects are chiefly due to the arsenic content, while the long term effects are chiefly due to the lead component. |

First aid   Standard treatment for exposure by all routes (see pages 108-122).

Fire hazard   Water may be used to fight fires involving lead arsenate.

768. Lead(II) azide

Colourless, crystalline powder; m.p. explodes at 350 ˚C.

RISKS
Harmful by inhalation and if swallowed – Danger of cumulative effects (R20/22, R33)

SAFETY PRECAUTIONS
Keep away from food, drink, and animal feeding stuffs – When using do not eat, drink, or smoke (S13, S20/21)

Limit values   OES long-term 0.15 mg m^{-3} (as Pb).

Hazardous reactions   A detonator that has been studied in detail; in prolonged contact with copper or zinc may form extremely sensitive azides of these metals.

First aid   Standard treatment for exposure by all routes (see pages 108-122), but treatment is more likely to be required for burns or wounds from an explosion.

769. Lead(IV) azide

Crystalline solid.

RISKS
Harmful by inhalation and if swallowed – Danger of cumulative effects (R20/22, R33)

SAFETY PRECAUTIONS
Keep away from food, drink, and animal feeding stuffs – When using do not eat, drink, or smoke (S13, S20/21)

Limit values   OES long-term 0.15 mg m^{-3} (as Pb).

Hazardous reactions   Liable to spontaneous decomposition which is sometimes explosive.

First aid   Standard treatment for exposure by all routes (see pages 108-122), but treatment is more likely to be required for burns or wounds from an explosion.

770. Lead bromate

Colourless, crystalline powder; m.p. 180 ˚C with decomposition; slightly soluble in cold water, soluble in hot water.

Limit values   OES long-term 0.15 mg m^{-3} (as Pb).

Hazardous reactions   An explosive salt.

771. Lead carbonate

White, heavy powder; m.p. 400 ˚C with decomposition; insoluble in water.

RISKS
Harmful by inhalation and if swallowed – Danger of cumulative effects (R20/22, R33)

SAFETY PRECAUTIONS
Keep away from food, drink, and animal feeding stuffs – When using do not eat, drink, or smoke (S13, S20/21)

Limit values OES long-term 0.15 mg m^{-3} (as Pb).

Toxic effects Early symptoms of lead poisoning include fatigue, sleep disturbances, aching bones and muscles, and reduced appetite. Later, anaemia, lead line on the gums, and colic may occur. Large doses affect the central nervous system and may cause death.

First aid Standard treatment for exposure by all routes (see pages 108-122).

Fire hazard Extinguish fires with water spray, carbon dioxide, or foam extinguishants.

Spillage disposal See general section.

RSC *Chemical Safety Data Sheets* Vol. 2, No. 61, 1989 gives extended coverage.

772. Lead chloride

White crystals; m.p. 501 ˚C; b.p. 950 ˚C; soluble in water.

RISKS
Harmful by inhalation and if swallowed – Danger of cumulative effects (R20/22, R33)

SAFETY PRECAUTIONS
Keep away from food, drink, and animal feeding stuffs – When using do not eat, drink, or smoke (S13, S20/21)

Limit values OES long-term 0.15 mg m^{-3} (as Pb).

Toxic effects Early symptoms of lead poisoning include fatigue, sleep disturbances, aching bones and muscles, and reduced appetite. Later, anaemia, lead line on the gums, and colic may occur. Large doses affect the central nervous system and may cause death.

First aid Standard treatment for exposure by all routes (see pages 108-122).

Fire hazard Use extinguishant appropriate to the surroundings.

Spillage disposal See general section.

RSC *Chemical Safety Data Sheets* Vol. 2, No. 62, 1989 gives extended coverage.

773. Lead chromate

Yellow to orange-yellow powder; m.p. 844 °C; insoluble in water.

RISKS
Danger of cumulative effects – Possible risk of irreversible effects (R33, R40)

SAFETY PRECAUTIONS
Do not breathe dust (S22)

Limit values	OES long-term 0.15 mg m^{-3} (as Pb).
Toxic effects	The dust is irritating to the skin and eyes. There is little evidence of toxicity following inhalation or ingestion, although recent evidence shows that it is carcinogenic.
Hazardous reactions	As an oxidant it has been involved in fires and explosions when mixed with organic pigments.
First aid	Standard treatment for exposure by all routes (see pages 108-122).
Fire hazard	Use carbon dioxide or dry chemical powder extinguishers.
Spillage disposal	See general section.

RSC *Chemical Safety Data Sheets* Vol. 2, No. 63, 1989 gives extended coverage.

774. Lead dichlorite

Yellow, crystalline powder; very slightly soluble in water.

RISKS
Harmful by inhalation and if swallowed – Danger of cumulative effects (R20/22, R33)

SAFETY PRECAUTIONS
Keep away from food, drink, and animal feeding stuffs – When using do not eat, drink, or smoke (S13, S20/21)

Limit values	OES long-term 0.15 mg m^{-3} (as Pb).
Hazardous reactions	Explodes on heating above 100 °C or on rubbing with antimony sulfide or fine sulfur.
First aid	Standard treatment for exposure by all routes (see pages 108-122).
Spillage disposal	See general section.

775. Lead dioxide

Dark brown powder or hexagonal crystals; m.p. 290 ˚C with decomposition; insoluble in water.

RISKS
Harmful by inhalation and in contact with skin – Danger of cumulative effects (R20/21, R33)

SAFETY PRECAUTIONS
Keep away from food, drink, and animal feeding stuffs – When using do not eat, drink, or smoke (S13, S20/21)

Limit values   OES long-term 0.15 mg m^{-3} (as Pb).

Toxic effects   Early symptoms of lead poisoning include fatigue, sleep disturbances, aching bones and muscles and reduced appetite. Later, anaemia, lead line on the gums and colic may occur. Large doses affect the central nervous system and may cause death.

First aid   Standard treatment for exposure by all routes (see pages 108-122).

Fire hazard   Extinguish fires with water spray.

Spillage   See general section.
  disposal

RSC *Chemical Safety Data Sheets* Vol. 2, No. 64, 1989 gives extended coverage.

776. Lead diperchlorate

White, crystalline powder; m.p. 100 ˚C with decomposition; soluble in water.

RISKS
Harmful by inhalation and if swallowed – Danger of cumulative effects (R20/22, R33)

SAFETY PRECAUTIONS
Keep away from food, drink, and animal feeding stuffs – When using do not eat, drink, or smoke (S13, S20/21)

Limit values   OES long-term 0.15 mg m^{-3} (as Pb).

Hazardous   A saturated solution of this anhydrous salt in dry methanol exploded violently.
  reactions

First aid   Standard treatment for exposure by all routes (see pages 108-122).

Spillage   See general section.
  disposal

777. Lead dithiocyanate

Colourless, crystalline powder; m.p. 180 ˚C with decomposition; slightly soluble in water.

RISKS
Harmful by inhalation and if swallowed – Danger of cumulative effects (R20/22, R33)

SAFETY PRECAUTIONS
Keep away from food, drink, and animal feeding stuffs – When using do not eat, drink, or smoke (S13, S20/21)

Limit values	OES long-term 0.15 mg m^{-3} (as Pb).
Hazardous reactions	Explosive.
First aid	Standard treatment for exposure by all routes (see pages 108-122).
Spillage disposal	See general section.

778. Lead imide

Solid; explodes on contact with water.

RISKS
Harmful by inhalation and if swallowed – Danger of cumulative effects (R20/22, R33)

SAFETY PRECAUTIONS
Keep away from food, drink, and animal feeding stuffs – When using do not eat, drink, or smoke (S13, S20/21)

Limit values	OES long-term 0.15 mg m^{-3} (as Pb).
Hazardous reactions	Explodes on heating or in contact with water or dilute acids.
First aid	Standard treatment for exposure by all routes (see pages 108-122).
Spillage disposal	See general section.

779. Lead monoxide

Reddish-yellow, tetragonal crystals, or yellow, orthorhombic crystals; m.p. 888 ˚C; b.p. 1472 ˚C with decomposition; insoluble in water.

RISKS
Harmful by inhalation and if swallowed – Danger of cumulative effects (R20/22, R33)

SAFETY PRECAUTIONS
Keep away from food, drink, and animal feeding stuffs – When using do not eat, drink, or smoke (S13, S20/21)

Limit values	OES long-term 0.15 mg m^{-3} (as Pb).

Toxic effects	Early symptoms of lead poisoning include fatigue, sleep disturbances, aching bones and muscles, and reduced appetite. Later, anaemia, lead line on gums, and colic may occur. Large doses affect the central nervous system and may cause death.
Hazardous reactions	Reacts violently or explosively with a wide variety of substances.
First aid	Standard treatment for exposure by all routes (see pages 108-122).
Fire hazard	Use extinguishant appropriate to the surroundings.
Spillage disposal	See general section.

RSC *Chemical Safety Data Sheets* Vol. 2, No. 65, 1989 gives extended coverage.

780. Lead nitrate

White crystals; m.p. 470 °C with decomposition; soluble in water.

RISKS
Harmful by inhalation and if swallowed – Danger of cumulative effects (R20/22, R33)

SAFETY PRECAUTIONS
Keep away from food, drink, and animal feeding stuffs – When using do not eat, drink, or smoke (S13, S20/21)

Limit values	OES long-term 0.15 mg m^{-3} (as Pb).
Toxic effects	Early symptoms of lead poisoning include fatigue, sleep disturbances, aching bones and muscles and reduced appetite. Later, anaemia, lead line on the gums and colic may occur. Large doses affect the central nervous system and may cause death.
First aid	Standard treatment for exposure by all routes (see pages 108-122).
Fire hazard	Extinguish fires using water spray.
Spillage disposal	See general section.

RSC *Chemical Safety Data Sheets* Vol. 2, No. 66, 1989 gives extended coverage.

781. Lead tetrachloride

Yellow, oily liquid; m.p. -15 °C; b.p. explosive on heating above 100 °C; decomposed by water.

RISKS
Harmful by inhalation and if swallowed – Danger of cumulative effects (R20/22, R33)

SAFETY PRECAUTIONS
Keep away from food, drink, and animal feeding stuffs – When using do not eat, drink, or smoke (S13, S20/21)

Limit values	OES long-term 0.15 mg m^{-3} (as Pb).

Hazardous reactions	Liable to explode with potassium or on heating above 100 °C; explodes on warming with dilute sulfuric acid.
First aid	Standard treatment for exposure by all routes (see pages 108-122).
Spillage disposal	See general section.

782. Linseed oil

Yellowish liquid with a peculiar odour; m.p. -19 °C; b.p. 343 °C; insoluble in water.

Toxic effects	An allergen and skin irritant to humans.
Hazardous reactions	Cloths used to apply this to benches were dropped in waste bin and the laboratory was destroyed by fire some hours later.
First aid	Standard treatment for exposure by all routes (see pages 108-122).
Fire hazard	Flash point 222 °C; Autoignition temperature 343 °C. Extinguish fires with carbon dioxide or dry chemical powder.
Spillage disposal	See general section.

783. Lithium

Silver-white metal, becoming yellowish on exposure to moist air; m.p. 180 °C; b.p. 1336 °C; reacts with water.

RISKS
Reacts violently with water, liberating highly flammable gases – Causes burns (R14/15, R34)

SAFETY PRECAUTIONS
Keep container dry – In case of fire, use dry graphite or appropriate metal fire extinguishing dry powder. Do NOT use water (S8, S43).

Toxic effects	Lithium reacts with moisture forming lithium hydroxide which is irritating and destructive to the skin, eyes, and mucous surfaces and may cause burns.
Hazardous reactions	Finely divided metal may ignite in air, will burn in nitrogen or carbon dioxide, and is difficult to extinguish once alight; reacts violently or explosively with a variety of substances.
First aid	Standard treatment for exposure by all routes (see pages 108-122).
Fire hazard	Extinguish fire with dry graphite or appropriate metal fire extinguishing dry powder. Do NOT use water. Carbon dioxide and dry powder are ineffective.
Spillage disposal	See general section.

RSC *Chemical Safety Data Sheets* Vol. 2, No. 68, 1989 gives extended coverage.

784. Lithium amide

White, crystalline powder; m.p. starts to decompose at 320 °C, melts at 375 °C, and becomes the imide at above 400 °C; violently decomposed by water producing flammable and toxic vapours.

Toxic effects   Strongly irritating to skin, eyes, upper respiratory tract, and all mucous membranes. Inhalation may be fatal due to inflammation and oedema.

First aid   Standard treatment for exposure by all routes (see pages 108-122).

Fire hazard   Extinguish fires with Class D extinguishants only. Do NOT use water.

Spillage disposal   See general section.

RSC *Chemical Safety Data Sheets* Vol. 2, No. 69, 1989 gives extended coverage.

785. Lithium benzenehexoxide

Hazardous reactions   Explodes on contact with water.

786. Lithium borohydride

Off white powder; m.p. 268 °C; b.p. 380 °C with decomposition; soluble in water.

Toxic effects   Harmful if inhaled, ingested, or absorbed through the skin. Extremely destructive to all tissues. Inhalation may be fatal.

Hazardous reactions   Contact with limited amounts of water may cause ignition.

First aid   Standard treatment for exposure by all routes (see pages 108-122).

Fire hazard   Flash point -18 °C (closed cup). Extinguish fires with carbon dioxide, dry chemical powder, or alcohol or polymer foam. Do NOT use water.

Spillage disposal   See general section.

787. Lithium carbonate

White, light, alkaline powder; m.p. 618 °C; sparingly soluble in water.

Toxic effects   Highly irritant and destructive to skin and eyes. Inhalation may result in irritation, coughing, nausea, pneumonitis, and oedema. Animal teratogen.

First aid   Standard treatment for exposure by all routes (see pages 108-122).

Fire hazard   Extinguish fires with carbon dioxide, dry chemical powder, or alcohol or polymer foam.

Spillage disposal   See general section.

RSC *Chemical Safety Data Sheets* Vol. 2, No. 70, 1989 gives extended coverage.

788. Lithium chloride

Cubic, white, deliquescent crystals with sharp saline taste; m.p. 608 °C; b.p. 1360 °C; very soluble in water.

Toxic effects	Irritant to skin, eyes, and upper respiratory tract. Animal teratogen.
First aid	Standard treatment for exposure by all routes (see pages 108-122).
Fire hazard	Use extinguishant appropriate to the surroundings.
Spillage disposal	See general section.

RSC *Chemical Safety Data Sheets* Vol. 2, No. 71, 1989 gives extended coverage.

789. Lithium fluoride

Fine, white powder; m.p. 848 °C; b.p. 1681 °C; almost insoluble in water.

Limit values	OES long-term 2.5 mg m^{-3} (as F).
Toxic effects	Irritant to skin and eyes. Not absorbed through the skin. Prolonged inhalation may result in perforation of the nasal septum.
First aid	Standard treatment for exposure by all routes (see pages 108-122), with special reference to fluorides.
Fire hazard	Extinguish fires with water spray, carbon dioxide, dry chemical powder, or alcohol or polymer foam.
Spillage disposal	See general section.

RSC *Chemical Safety Data Sheets* Vol. 2, No. 72, 1989 gives extended coverage.

790. Lithium hydride

White, translucent crystals which darken rapidly on exposure to light; m.p. 686.4 °C; decomposed by water.

Limit values	OES long-term 0.025 mg m^{-3}.
Toxic effects	Irritant and corrosive to skin, eyes, upper respiratory tract, and all mucous membranes.
Hazardous reactions	Ignites in warm air; mixtures with liquid oxygen are detonable explosives.
First aid	Standard treatment for exposure by all routes (see pages 108-122).
Fire hazard	Extinguish fires by smothering with fire blanket, dry graphite, or ground dolomite. Do NOT use water, carbon dioxide, or dry chemical powder.
Spillage disposal	See general section.

RSC *Chemical Safety Data Sheets* Vol. 2, No. 73, gives extended coverage.

791. Lithium hydroxide

White, granular powder with acid taste; m.p. 471 ˚C; soluble in water.

Limit values	OES short-term 1 mg m⁻³.

Limit values   OES short-term 1 mg m^{-3}.

Toxic effects   The solid and solution are irritant and severely destructive to all tissues.

First aid   Standard treatment for exposure by all routes (see pages 108-122).

Fire hazard   Extinguish fires with carbon dioxide, dry chemical powder, or alcohol or polymer foam.

Spillage disposal   See general section.

RSC *Chemical Safety Data Sheets* Vol. 2, No. 74, 1989, gives extended coverage.

792. Lithium tetrahydroaluminate

White, microcrystalline powder and lumps; b.p. 125 ˚C with decomposition; reacts violently with water with evolution of hydrogen.

RISKS
Contact with water liberates highly flammable gases (R15)

SAFETY PRECAUTIONS
Keep container tightly closed and dry – Avoid contact with skin and eyes – In case of fire, use sand or specially manufactured dry powder extinguisher (S7/8, S24/25, S43)

Toxic effects   Contact with moist tissues forms corrosive lithium hydroxide which is irritant to all tissues.

Hazardous reactions   Vigorous reactions have been reported with various substances.

First aid   Standard treatment for exposure by all routes (see pages 108-122).

Fire hazard   Fires involving this material are best extinguished by smothering with sand, dry powdered limestone, graphite, soda ash, dry sodium chloride or lithium chloride, or specially manufactured dry powder extinguishers. Do NOT use conventional extinguishers.

Spillage disposal   See general section.

RSC *Chemical Safety Data Sheets* Vol. 2. No. 75, 1989 gives extended coverage.

793. Magnesium

Hexagonal, silvery white crystals; m.p. 651 ˚C; b.p. 1100 ˚C; insoluble in water.

RISKS
Highly flammable – Contact with water liberates highly flammable gases (R11, R15)

SAFETY PRECAUTIONS
Keep container tightly closed and dry – In case of fire, use dry graphite or any suitable dry powder. Do NOT use water (S7/8, S43)

Toxic effects	Fine dust is irritant to the eyes and respiratory tract.
Hazardous reactions	Fine powder dispersed in air is a serious explosion hazard; reaction with beryllium fluoride is violent; explosive acetylide may be formed from traces of acetylene in ethylene oxide if magnesium is contained in fittings used in ethylene oxide service; the powdered metal reacts and may explode on contact with chloromethane, chloroform, or carbon tetrachloride and mixtures with carbon tetrachloride or trichloroethylene will flash on heavy impact; mixtures with PTFE used as igniters; ignites, if moist, in fluorine or chlorine; may ignite, if finely divided, on heating in iodine vapour; reacts vigorously with certain cyanides; violently reduces some metal oxides; reaction with metal oxosalts may be explosive; reaction with methanol may become vigorous; heating with numerous oxidants can be hazardous; mixtures of Mg dust with methanol or water are detonable.
First aid	Standard treatment for exposure by all routes (see pages 108-122).
Fire hazard	Explosive limits (lower) 0.04%; Autoignition temperature 560 ˚C. Extinguish fires with dry graphite or any other suitable dry powder. Do NOT use water.
Spillage disposal	See general section.

RSC *Chemical Safety Data Sheets* Vol. 2, No. 76, 1989 gives extended coverage.

794. Magnesium chloride

Thin, white to opaque grey granules and/or flakes; m.p. 708 ˚C; b.p. 1412 ˚C; soluble in water.

Toxic effects	Very destructive to all biological tissues. Highly toxic on inhalation or ingestion.
First aid	Standard treatment for exposure by all routes (see pages 108-122).
Fire hazard	Use extinguishant appropriate to the surroundings.
Spillage disposal	See general section.

RSC *Chemical Safety Data Sheets* Vol. 2, No. 77, 1989 gives extended coverage.

795. Magnesium phosphide

Solid; reacts with water.

RISKS
Contact with water liberates toxic, highly flammable gas – Very toxic if swallowed (R15/29, R28)

SAFETY PRECAUTIONS
Keep locked up and out of reach of children – Do not breathe dust – In case of fire, use carbon dioxide, dry chemical powder, alcohol foam, or polymer foam. Never use water (S1/2, S22, S43, S45).

Toxic effects	Little data available. Moderately toxic on inhalation.

Hazardous reactions	Reacts with water to liberate phosphine which may ignite.
First aid	Standard treatment for exposure by all routes (see pages 108-122).
Spillage disposal	See general section.

RSC *Chemical Safety Data Sheets* Vol. 4b, No. 80, 1991 gives extended coverage.

796. Maleic anhydride

White, crystalline powder or lumps; m.p. 53 ˚C; b.p. 202 ˚C; dissolves in water forming maleic acid.

RISKS
Harmful if swallowed – Irritating to eyes, respiratory system, and skin – May cause sensitization by inhalation (R22, R36/37/38, R42)

SAFETY PRECAUTIONS
Do not breathe dust – After contact with skin, wash immediately with plenty of water – Wear eye/face protection (S22, S28, S39)

Limit values	OES long-term 0.25 p.p.m. (1 mg m^{-3}) under review.
Toxic effects	The dust and vapour irritate the eyes, skin, and respiratory system; prolonged contact with tissues may result in burns. Assumed to be irritant and harmful if taken by mouth.
Hazardous reactions	Decomposes exothermally in presence of alkali or alkaline earth metal or ammonium ions, dimethylamine, triethylamine, pyridine, or quinoline at temperatures above 150 ˚C; sodium ions and pyridine are particularly active.
First aid	Standard treatment for exposure by all routes (see pages 108-122).
Fire hazard	Flash point 110 ˚C (open cup); Explosive limits 1.4-7.1%; Autoignition temperature 447 ˚C. Use carbon dioxide to extinguish fires. Water or foam may cause frothing. Dry chemical powder should NOT be used.
Spillage disposal	See general section.

RSC *Chemical Safety Data Sheets* Vol. 3, No. 40, 1990 gives extended coverage.

797. Maleic anhydride ozonide

| **Hazardous reactions** | Explodes on warming to -40 ˚C. |

798. Malononitrile

Colourless, crystalline solid; m.p. 30.5 ˚C; b.p. 220 ˚C; soluble in water.

RISKS
Toxic by inhalation, in contact with skin, and if swallowed (R23/24/25)

SAFETY PRECAUTIONS
Do not breathe vapour – Take off immediately all contaminated clothing (S23, S27)

Toxic effects	It is said that the toxicity of malononitrile is of the same order as that of hydrogen cyanide though no cases of poisoning of humans have been traced. It is therefore considered wise to treat the compound with the respect given to the alkali cyanides.
Hazardous reactions	May polymerize violently on heating at 130 ˚C or in contact with strong bases at lower temperatures.
First aid	Standard treatment for exposure by all routes (see pages 108-122), with particular reference to cyanides.
Fire hazard	Flash point 130 ˚C (open cup). Extinguish fires with water spray, carbon dioxide, dry chemical powder, or alcohol or polymer foam.
Spillage disposal	See general section.

RSC *Chemical Safety Data Sheets* Vol. 4b, No. 81, 1991 gives extended coverage.

799. Manganese dioxide

Black powder; m.p. 535 ˚C with decomposition; insoluble in water.

RISKS
Harmful by inhalation and if swallowed (R20/22)

SAFETY PRECAUTIONS
Avoid contact with eyes (S25)

Limit values	OES long-term 5 mg m^{-3} (as Mn).
Toxic effects	Inhalation of dust may lead to increased incidence of respiratory infections. Continued inhalation of dust may lead to excessive tiredness and effects on the central nervous system. It is assumed to be harmful on ingestion.
Hazardous reactions	As an oxidant it reacts violently with aluminium powder, calcium hydride, rubidium acetylide, hydrogen sulfide, or potassium azide. It also catalytically decomposes, usually violently or explosively, other oxidants.
First aid	Standard treatment for exposure by all routes (see pages 108-122).
Fire hazard	Use extinguishing medium appropriate to surroundings.
Spillage disposal	See general section.

800. Manganese diperchlorate

Limit values   OES long-term 5 mg m^{-3} (as Mn).

Hazardous   Explodes at 195 °C.
reactions

801. Manganese tetroxide

Brownish-black powder; m.p. 1564 °C; insoluble in water.

Limit values   OES long-term 1 mg m^{-3}.

Toxic effects   Chronic manganism can involve the central nervous system.

First aid   Standard treatment for exposure by all routes (see pages 108-122).

802. Manganese trifluoride

Red, crystalline solid; m.p. decomposes on heating; decomposed by water.

Limit values   OES long-term 5 mg m^{-3} (as Mn).

Toxic effects   Irritant to all tissues.

Hazardous   Glass is attacked violently by heating in contact with it, silicon tetrafluoride
reactions   being evolved.

803. Menthol

White solid with peppermint odour; m.p. 28-30 °C; b.p. 216 °C; sparingly soluble in water.

Toxic effects   Irritant to skin and eyes. May be absorbed through the skin.

First aid   Standard treatment for exposure by all routes (see pages 108-122).

Fire hazard   Flash point 93 °C. Extinguish fires with water spray, carbon dioxide, dry
                     chemical powder, or alcohol or polymer foam.

Spillage   See general section.
disposal

804. Mercaptoacetic acid

Colourless liquid with unpleasant smell; m.p. -16.5 °C; b.p. 123 °C at 39 mbar; miscible with
water.

Limit values   OES long-term 1 p.p.m. (5 mg m^{-3}).

Toxic effects   The liquid irritates the eyes, skin, and upper respiratory tract. Toxic if inhaled,
                     ingested, or absorbed through the skin.

First aid   Standard treatment for exposure by all routes (see pages 108-122).

Fire hazard   Flash point 125 ˚C; Explosive limits lel 5.9%; Autoignition temperature 350 ˚C. Extinguish fires with water spray, carbon dioxide, dry chemical powder, or alcohol or polymer foam.

Spillage disposal   See general section.

HSC *Chemical Safety Data Sheets* Vol. 4b, No. 141, 1991 gives extended coverage.

805. 2-Mercaptoethanol

Colourless liquid; b.p. 157 ˚C with decomposition; miscible with water.

Toxic effects   High concentrations are severely irritant and highly destructive to all body tissues. May be fatal on inhalation, ingestion, or skin absorption.

Fire hazard   Flash point 73 ˚C. Extinguish fires with carbon dioxide, dry chemical powder, or alcohol foam.

806. Mercury

Heavy, silvery, mobile liquid; m.p. -39 ˚C; b.p. 356.5 ˚C; insoluble in water.

RISKS
Toxic by inhalation – Danger of cumulative effects (R23, R33)

SAFETY PRECAUTIONS
Keep container tightly closed – If you feel unwell, seek medical advice (show label where possible) (S7, S44)

Limit values   OES short-term 0.15 mg m^{-3}; long-term 0.05 mg m^{-3}.

Toxic effects   High concentrations of vapour may cause metallic taste, nausea, abdominal pain, vomiting, diarrhoea, and headache. Continued exposure to small concentrations of vapour may result in severe nervous disturbance, including tremor of the hands, insomnia, loss of memory, irritability, and depression; other possible effects are loosening of teeth and excessive salivation. Continued skin contact with mercury may cause dermatitis and the above effects may be caused by absorption through the skin or following ingestion. Kidney damage may ensue.

Hazardous reactions   Prolonged contact between mercury and ammonia may result in formation of explosive solid; ease with which it forms amalgams with laboratory and electrical contact metals can cause severe corrosion problems; reacts violently with dry bromine; chlorine dioxide explodes when shaken with mercury.

First aid   Standard treatment for exposure by all routes (see pages 108-122).

Fire hazard   Use extinguishing medium appropriate to surroundings.

Spillage disposal   Because of the high toxicity of mercury vapour it is important to clean up mercury as thoroughly as possible, especially in confined areas. A small aspirator with a capillary tube and connected to a pump can be used for sucking up droplets. Mercury spilt into floor cracks can be made non-volatile by putting zinc dust down the cracks to form the amalgam. Smooth surfaces may be decontaminated by scattering and sweeping up a mixture of equal weights of zinc dust and sawdust, which should then be buried at an isolated site.

RSC *Chemical Safety Data Sheets* Vol. 4b, No. 82, 1991 gives extended coverage.

807. Mercury alkyls

Some are liquid and some are solid; in general they are stable towards moisture.

RISKS
Very toxic by inhalation, in contact with skin, and if swallowed – Danger of cumulative effects (R26/27/28, R33)

SAFETY PRECAUTIONS
Keep out of reach of children – Keep away from food, drink, and animal feeding stuffs – After contact with skin, wash immediately with plenty of water – Wear suitable protective clothing – In case of accident or if you feel unwell, seek medical advice immediately (show label where possible) (S2, S13, S28, S36, S45)

Limit values	OES short-term 0.03 mg m^{-3} (as Hg); long-term 0.01 mg m^{-3} (as Hg).
Toxic effects	Irritant and corrosive to skin and other tissues, resulting in chemical burns and blistering, sometimes delayed in appearing. Absorption can cause liver and kidney damage, also ataxia, dysarthria, and altered planar reflexes. Intense and widespread degeneration of certain sensory nerve pathways can occur. The aryl compounds appear to be less toxic than the alkyl compounds.
Hazardous reactions	Can react explosively with a variety of substances.
First aid	Standard treatment for exposure by all routes (see pages 108-122).

808. Mercury(II) oxide

Bright red or orange-red, heavy, odourless, crystalline powder or scales; m.p. 300 ˚C with decomposition; practically insoluble in water.

RISKS
Very toxic by inhalation, in contact with skin, and if swallowed – Danger of cumulative effects (R26/27/28, R33)

SAFETY PRECAUTIONS
Keep locked up and out of reach of children – Keep away from food, drink, and animal feeding stuffs – After contact with skin, wash immediately with plenty of water – In case of accident or if you feel unwell, seek medical advice immediately (show label where possible) (S1/2, S13, S28, S45)

Limit values	OES short-term 0.15 mg m^{-3} (as Hg); long-term 0.05 mg m^{-3} (as Hg).

Toxic effects	Strong skin irritant and an allergen. May be irritant to eyes and upper respiratory tract. Acute exposure can cause gastro-intestinal symptoms, anuria, and uraemia. Chronic exposure can cause tremors and neuropsychiatric disturbances.
Hazardous reactions	Can react violently or explosively with a wide variety of substances.
First aid	Standard treatment for exposure by all routes (see pages 108-122).
Fire hazard	Use extinguishant appropriate to the surroundings.
Spillage disposal	See general section.

RSC *Chemical Safety Data Sheets* Vol. 4b, No. 88, 1991 gives extended coverage.

809. Mercury salts

Mercury salts are nearly all crystalline solids. The colours vary from white through yellow, brown, and red, to black. The behaviour on heating and solubility in water vary greatly.

RISKS
Very toxic by inhalation, in contact with skin, and if swallowed – Danger of cumulative effects (R26/27/28, R33)

SAFETY PRECAUTIONS
Keep locked up and out of reach of children – Keep away from food, drink and animal feeding stuffs – After contact with skin, wash immediately with plenty of water – In case of accident or if you feel unwell, seek medical advice immediately (show label where possible) (S1/2, S13, S28, S45)

Limit values	OES short-term 0.15 mg m^{-3} (as Hg); long-term 0.05 mg m^{-3} (as Hg).
Toxic effects	Skin and eye irritants of varying severity. Inhalation and ingestion can give rise to nausea, vomiting, intestinal pain, and diarrhoea. Kidney damage may ensue. Chronic exposure over a long period may cause tremor of the hands, insomnia, loss of memory, irritability, depression, loosening of the teeth, and excessive salivation.
Hazardous reactions	Many mercury salts show explosive instability and can also react violently or explosively with many other substances.
First aid	Standard treatment for exposure by all routes (see pages 108-122), with special reference to cyanide and fluoride where appropriate.
Fire hazard	In most cases extinguishant appropriate to the surroundings is suitable.
Spillage disposal	See general section. Specialist help may be needed to deal with large spills.

RSC *Chemical Safety Data Sheets* Vol. 4b, Nos. 83-87, 1991 give extended coverage.

810. Mesitylene

Colourless liquid; m.p. -45 ˚C; b.p. 162-164 ˚C; practically insoluble in water.

RISKS
Flammable – Irritating to respiratory system (R10, R37)

Limit values   OES short-term 35 p.p.m. (170 mg m^{-3}) under review; long-term 25 p.p.m. (125 mg m^{-3}) under review.

Toxic effects   Primary skin irritant. Inhalation of high concentrations causes central nervous system depression, respiratory irritation, and blood changes.

First aid   Standard treatment for exposure by all routes (see pages 108-122).

Fire hazard   Flash point 50 ˚C; Explosive limits lel 0.88%; Autoignition temperature 550 ˚C. Extinguish fires with water spray, carbon dioxide, or fog foam.

Spillage   See general section.
   disposal

RSC *Chemical Safety Data Sheets* Vol. 1, No. 63, 1988 gives extended coverage.

811. Mesoxalonitrile

Liquid; m.p. -36 ˚C; b.p. 65.5 ˚C.

Hazardous   Reacts explosively with water.
   reactions

812. Metal hydrides

The metal hydrides and alkaline earth metal hydrides, together with a few others, react readily and vigorously with water or acids, evolving explosive hydrogen and, with water, leaving the hydroxides which are often caustic and irritant to skin, eyes, and upper respiratory tract. Some metal hydrides ignite or explode vigorously on heating and can cause difficulties in fire fighting. Particularly hazardous in this respect is lithium hydride (*q.v.*) which is given extended coverage in RSC *Chemical Safety Data Sheets* Vol. 2, No. 73, 1989.

813. Methacrylic acid

Colourless solid or liquid with unpleasant, acrid smell; m.p. 16 ˚C; b.p. 158 ˚C; soluble in warm water.

RISKS
Causes burns (R34)

SAFETY PRECAUTIONS
Keep away from heat – In case of contact with eyes, rinse immediately with plenty of water and seek medical advice (S15, S26)

Limit values   OES short-term 40 p.p.m. (140 mg m^{-3}); long-term 20 p.p.m. (70 mg m^{-3}).

Toxic effects The vapour irritates the eyes and respiratory system. The liquid irritates the eyes and skin and is assumed to be very irritant and harmful if taken internally.

First aid Standard treatment for exposure by all routes (see pages 108-122).

Fire hazard Flash point 76 °C (open cup). Extinguish fires with water spray, carbon dioxide, dry chemical powder, or alcohol foam.

Spillage disposal See general section.

RSC *Chemical Safety Data Sheets* Vol. 3, No. 41, 1990 gives extended coverage.

814. Methacrylonitrile

Clear, colourless liquid with a slight cyanide like odour; m.p. -35.8 °C; b.p. 90.3 °C; slightly soluble in water.

RISKS
Highly flammable – Toxic by inhalation, in contact with skin, and if swallowed – May cause sensitization by skin contact (R11, R23/24/25, R43)

SAFETY PRECAUTIONS
Keep container in a well ventilated place – Keep away from sources of ignition – No Smoking – Handle and open container with care – Do not empty into drains – In case of accident or if you feel unwell, seek medical advice immediately (show label where possible) (S9, S16, S18, S29, S45)

Limit values OES long-term 1 p.p.m. (3 mg m^{-3}).

Toxic effects High concentrations are extremely destructive to all tissues. Highly toxic on inhalation.

First aid Standard treatment for exposure by all routes (see pages 108-122), but with special attention to cyanides.

Fire hazard Flash point 13 °C (open cup); Explosive limits 2%-6.8%. Extinguish fires with water spray, carbon dioxide, dry chemical powder, or alcohol or polymer foam.

Spillage disposal See general section.

RSC *Chemical Safety Data Sheets* Vol. 4b, No. 89, 1991 gives extended coverage.

815. Methane

Colourless gas; m.p. -182.5 °C; b.p. -161.5 °C; sparingly soluble in water.

RISKS
Extremely flammable (R12)

SAFETY PRECAUTIONS
Keep container in a well ventilated place – Keep away from sources of ignition – No Smoking – Take precautionary measures against static discharges (S9, S16, S33)

Toxic effects Methane is non-toxic but can have narcotic effects in high concentrations in the absence of oxygen.

First aid   Standard treatment for exposure by all routes (see pages 108-122).

Fire hazard   Explosive limits 5-15%; Autoignition temperature 537 °C. As the gas is
                    supplied in cylinders, turning off the valve will reduce any fire involving it; if
                    possible, cylinders should be removed quickly from an area in which a fire
                    has developed.

Spillage   Surplus gas or leaking cylinder can be vented slowly to air in a safe area or
disposal   burnt-off in a suitable gas burner.

816. Methanesulfinyl chloride

Liquid; b.p. 59 °C at 42 mm Hg with decomposition.

Hazardous   An unrefrigerated ampoule burst after extended storage.
reactions

817. Methanethiol

Colourless gas with extremely disagreeable smell; b.p. 6 °C; sparingly soluble in water.

Limit values   OES long-term 0.5 p.p.m. (1 mg m^{-3}).

Toxic effects   The gas is nauseous and may be narcotic in high concentrations.

Hazardous   Reaction with mercuric oxide is rather violent.
reactions

First aid   Standard treatment for exposure by all routes (see pages 108-122).

Fire hazard   Flash point below -18 °C; Explosive limits 3.9-21.8%. As the gas is supplied in
                    a cylinder, turning off the valve will reduce any fire in which it is involved; if
                    possible, cylinders should be removed quickly from an area in which a fire
                    has developed.

Spillage   Surplus gas or gas form a leaking cylinder should be burnt through a suitable
disposal   gas burner in a fume cupboard.

818. Methanol

Colourless, volatile liquid; m.p. -97.8 °C; b.p. 65 °C; miscible with water.

RISKS
Highly flammable – Toxic by inhalation and if swallowed (R11, R23/25)

SAFETY PRECAUTIONS
*Keep out of reach of children – Keep container tightly closed – Keep away from sources of
ignition – No Smoking – Avoid contact with skin (S2, S7, S16, S24)*

Limit values   OES short-term 250 p.p.m. (310 mg m^{-3}); long-term 200 p.p.m. (260 mg m^{-3}).

Toxic effects   Inhalation of high concentrations of vapour may cause dizziness, stupor, cramps, and digestive disturbance. Lower concentrations may cause headache, nausea, vomiting, and irritation of the mucous membranes. The vapour and liquid are very dangerous to the eyes, the effects sometimes being delayed for many hours. Ingestion damages the central nervous system, particularly the optic nerve (causing temporary or permanent blindness), and injures the kidneys, liver, heart, and other organs; apart from the effects referred to above, unconsciousness may develop after some hours and this may be followed by death. Continued exposure to low concentrations of vapour may cause many of the above effects, while continued skin contact may cause dermatitis.

Hazardous reactions   Violent explosion occurred when sodium was added to methanol/chloroform mixture; reaction with magnesium can be very vigorous; reaction with bromine can be violent, with sodium hypochlorite explosive; reactions with nitric acid or hydrogen peroxide may become explosive; violent reaction or ignition with several other compounds.

First aid   Standard treatment for exposure by all routes (see pages 108-122).

Fire hazard   Flash point -16 °C; Explosive limits 6-36.5%; Autoignition temperature 385 °C. Extinguish small fires with water spray, dry powder, carbon dioxide, or alcohol resistant foam. The foam is more suitable for tackling large fires.

Spillage disposal   See general section.

RSC *Chemical Safety Data Sheets* Vol. 1, No. 65, 1988 gives extended coverage.

819. *p*-Methoxybenzoyl chloride

Colourless crystalline solid; m.p. 22-23 °C; b.p. 263 °C with slight decomposition; decomposed by water.

Hazardous reactions   Bottles of this exploded on storage at room temperature.

820. 2-Methoxyethanol

Colourless volatile liquid with pleasant smell; m.p. -86.5 °C; b.p. 125 °C; miscible with water.

RISKS
Flammable – Harmful by inhalation, in contact with skin, and if swallowed – Irritating to respiratory system (R10, R20/21/22, R37)

SAFETY PRECAUTIONS
Avoid contact with skin and eyes (S24/25)

Limit values   MEL long-term 5 p.p.m. (16 mg m^{-3}).

Toxic effects   The vapour irritates the respiratory system. The vapour and liquid irritate the eyes. Poisonous if taken by mouth. Anaemia, blood abnormalities, and symptoms of central nervous system damage have resulted from prolonged exposure.

Hazardous reactions	Liable to form explosive peroxides on exposure to light and air which should be decomposed before distilling the ether to small volume.
Fire hazard	Flash point 43 ˚C (open cup); Explosive limits 2.5-14%; Autoignition temperature 285 ˚C. Extinguish fire with dry powder, carbon dioxide, or alcohol-resistant foam.
Spillage disposal	See general section.

RSC *Chemical Safety Data Sheets* Vol. 1, No. 49, 1988 gives extended coverage.

821. 2-Methoxyethyl acetate

Colourless liquid; b.p. 143 ˚C; miscible with water.

RISKS
Flammable – Harmful by inhalation and in contact with skin (R10, R20/21)

SAFETY PRECAUTIONS
Avoid contact with skin (S24)

Limit values	MEL long-term 5 p.p.m. (24 mg m^{-3}).
Toxic effects	The liquid irritates the eyes and may irritate the skin; it may cause headache, dizziness, fatigue, nausea, vomiting, and more serious disorders if taken by mouth or absorbed extensively through the skin.
Hazardous reactions	Liable to form explosive peroxides on exposure to light and air which must be decomposed before the ether is distilled to small volume.
Fire hazard	Flash point 140 ˚C; Explosive limits 1.7-8.2%; Autoignition temperature 392 ˚C. Extinguish fire with dry powder, carbon dioxide, or alcohol-resistant foam.
Spillage disposal	See general section.

RSC *Chemical Safety Data Sheets* Vol. 1, No. 50, 1988 gives extended coverage.

822. 3-Methoxypropyne

Hazardous reactions	Explodes on distillation at 61 ˚C (B428).

823. Methyl acetate

Colourless, volatile liquid with pleasant odour; b.p. 58 ˚C; miscible with water.

RISKS
Highly flammable (R11)

SAFETY PRECAUTIONS
Keep away from sources of ignition – No Smoking – Do not breathe vapour – Do not empty into drains – Take precautionary measures against static discharges (S16, S23, S29, S33)

Limit values	OES short-term 250 p.p.m. (760 mg m^{-3}); long-term 200 p.p.m. (610 mg m^{-3}).

Toxic effects   The vapour irritates the eyes and respiratory system. Inhalation of high concentrations may cause dizziness and palpitations. Assumed to be poisonous if taken by mouth.

First aid   Standard treatment for exposure by all routes (see pages 108-122).

Fire hazard   Flash point -10 °C; Explosive limits 3.1-16%; Autoignition temperature 455 °C. Extinguish small fires with water spray, dry powder, carbon dioxide, or alcohol resistant foam. The foam is more suitable for large fires.

Spillage disposal   See general section.

RSC *Chemical Safety Data Sheets* Vol. 1, No. 66, 1988 gives extended coverage.

824. Methyl acrylate

Colourless liquid with acrid odour; m.p. -76.5 °C; b.p. 70 °C at 608 mm Hg; slightly soluble in water.

RISKS
Highly flammable – Harmful by inhalation and if swallowed – Irritating to eyes, respiratory system and skin (R11, R20/22, R36/37/38)

SAFETY PRECAUTIONS
Keep container in a well ventilated place – Keep away from sources of ignition – No Smoking – Take precautionary measures against static discharges (S9, S16, S33)

Limit values   OES long-term 10 p.p.m. (35 mg m^{-3}).

Toxic effects   The vapour irritates the eyes and respiratory system. Inhalation of high concentrations may cause lethargy, convulsions, dyspnoea, and palpitations. The liquid irritates the skin, eyes, and gastro-intestinal tract.

Hazardous reactions   Peroxidizes and may polymerize violently. Store inhibited but with access of air.

First aid   Standard treatment for exposure by all routes (see pages 108-122).

Fire hazard   Flash point -2.8 °C (open cup); Explosive limits 2.8-25%. Extinguish fire with dry chemical powder or carbon dioxide.

Spillage disposal   See general section.

RSC *Chemical Safety Data Sheets* Vol. 3, No. 42, 1990 gives extended coverage.

825. Methylamine

Colourless gas with pungent, fishy smell; b.p. -6.3 °C; very soluble in water.

RISKS
Extremely flammable liquefied gas – Irritating to eyes and respiratory system (R13, R36/37)

SAFETY PRECAUTIONS
Keep away from sources of ignition – No Smoking – In case of contact with eyes, rinse immediately with plenty of water and seek medical advice – Do not empty into drains (S16, S26, S29)

Limit values   OES long-term 10 p.p.m. (12 mg m^{-3}).

Toxic effects   The gas irritates the skin, eyes, and respiratory system and sustained contact may cause burns.

Hazardous reactions   Addition to nitromethane renders it susceptible to initiation by a detonator.

First aid   Standard treatment for exposure by all routes (see pages 108-122).

Fire hazard   Flash point 0 ˚C; Explosive limits 4.9-20.7%; Autoignition temperature 430 ˚C. As the gas is supplied in a cylinder, turning off the valve will reduce any fire involving it; if possible, cylinders should be removed quickly from an area in which fire has developed.

Spillage disposal   Surplus gas or leaking cylinder can be vented slowly into a water-fed scrubbing tower or column in a fume cupboard, or into a fume cupboard served by such a tower.

826. Methylamine solutions

Methylamine is commonly available in either aqueous or ethanolic solution in a concentration of 25-33%. The solutions are colourless and have a fishy smell; the solutions are miscible with water.

Toxic effects   The vapour irritates the eyes and respiratory system. The solutions irritate the eyes and skin. The solutions will cause irritation and damage if taken internally.

First aid   Standard treatment for exposure by all routes (see pages 108-122).

Fire hazard   Flash point depends upon nature and strength of solution. Extinguish fire with water spray, foam, dry powder, or carbon dioxide.

Spillage disposal   See general section.

827. Methyl n-amyl ketone

Liquid with a penetrating fruity odour; m.p. -35 ˚C; b.p. 151.5 ˚C; slightly soluble in water.

RISKS
Flammable – Harmful if swallowed (R10, R22)

SAFETY PRECAUTIONS
Do not breathe vapour (S23)

Limit values   OES long-term 50 p.p.m. (240 mg m^{-3}).

Toxic effects   Mildly irritant to the skin. Inhalation of high concentrations produces narcosis. Kidney and liver damage may follow ingestion.

First aid   Standard treatment for exposure by all routes (see pages 108-122).

Fire hazard   Flash point 44.4 ˚C (open cup); Explosive limits 1-5.5%; Autoignition temperature 532 ˚C. Small fires can be extinguished with carbon dioxide, dry chemical powder, or alcohol-resistant foam. Large fires are best controlled by alcohol-resistant foam.

Spillage disposal   See general section.

RSC *Chemical Safety Data Sheets* Vol. 1, No. 68, 1988 gives extended coverage.

828. *N*-Methylaniline

Colourless or slightly yellow liquid, becoming brown on standing; m.p. -57 ˚C; b.p. 194-196 ˚C; slightly soluble in water.

RISKS
Toxic by inhalation, in contact with skin, and if swallowed – Danger of cumulative effects (R23/24/25, R33)

SAFETY PRECAUTIONS
After contact with skin, wash immediately with plenty of water – Wear suitable gloves – If you feel unwell, seek medical advice (show label where possible) (S28, S37, S44)

Limit values OES long-term 0.5 p.p.m. (2 mg m^{-3}).

Toxic effects Excessive breathing of the vapour or absorption of the liquid through the skin can cause headache, drowsiness, cyanosis, mental confusion, and, in severe cases, convulsions. The liquid is dangerous to the eyes. The above effects can also be experienced if it is swallowed. Continued exposure to the vapour, or slight skin exposure to the liquid over a period, may affect the nervous system and the blood causing fatigue, loss of appetite, headache, and dizziness.

Fire hazard Flash point 79.4 ˚C (closed cup). Extinguish fires with water spray, carbon dioxide, dry chemical powder, or alcohol or polymer foam.

Spillage disposal See general section.

RSC *Chemical Safety Data Sheets* Vol. 4b, No. 90, 1991 gives extended coverage.

829. 2-Methylaziridine

Colourless, oily liquid, with a strong ammonia-like smell; m.p. -65 ˚C; b.p. 66 ˚C; miscible with water.

RISKS
May cause cancer – Highly flammable – Very toxic by inhalation, in contact with skin, and if swallowed – Risk of serious damage to eyes (R45, R11, R26/27/28, R41)

SAFETY PRECAUTIONS
Avoid exposure – Obtain special instructions before use – In case of contact with eyes, rinse immediately with plenty of water and seek medical advice – In case of accident or if you feel unwell, seek medical advice immediately (show label where possible) (S53, S26, S45)

Toxic effects Extremely destructive to skin, eyes, upper respiratory tract, and mucous membranes. Potent experimental animal carcinogen.

Hazardous reactions May polymerize explosively if exposed to acids or acidic fumes, so it should always be stored over solid alkali.

First aid Standard treatment for exposure by all routes (see pages 108-122).

Fire hazard Flash point -10 ˚C. Extinguish fires with water spray, carbon dioxide, dry chemical powder, or alcohol or polymer foam.

Spillage disposal See general section.

RSC *Chemical Safety Data Sheets* Vol. 4b, No. 91, 1991 gives extended coverage.

830. Methyl benzenediazoate

Hazardous   Explodes on heating or after storage for about an hour in a sealed tube at
reactions   ambient temperature.

831. 2-Methylbutan-2-ol

Colourless, volatile liquid with characteristic odour and burning taste; b.p. 102 ˚C.

SAFETY PRECAUTIONS
Do not breathe vapour – Avoid contact with skin and eyes (S23, S24/25)

Toxic effects   Vapour may irritate the eyes and respiratory system. Liquid irritates the eyes
                     severely and may irritate skin. If swallowed may cause headache, vertigo,
                     nausea, vomiting, excitement, and delirium followed by coma.

Fire hazard   Flash point 19 ˚C; Explosive limits 1.2-9%. Extinguish fires with water spray,
                     dry powder, or carbon dioxide.

Spillage   See general section.
   disposal

832. 1-Methylbutyl acetate

Colourless liquid; b.p. 133.5 ˚C; slightly soluble in water.

Limit values   OES short-term 150 p.p.m. (800 mg m^{-3}).

Toxic effects   Irritant to skin, eyes, upper respiratory tract, and mucous membranes.

First aid   Standard treatment for exposure by all routes (see pages 108-122).

Fire hazard   Flash point 32 ˚C; Explosive limits 1.1-7.5%. Extinguish fires with water spray,
                     carbon dioxide, dry chemical powder, or alcohol or polymer foam.

Spillage   See general section.
   disposal

833. Methyl chloroformate

Colourless liquid; b.p. 71 ˚C; immiscible with water.

RISKS
Highly flammable – Toxic by inhalation – Irritating to eyes, respiratory system, and skin (R11,
R23, R36/37/38)

SAFETY PRECAUTIONS
Keep container in a well ventilated place – Keep away from sources of ignition – No Smoking –
Take precautionary measures against static discharges – If you feel unwell, seek medical
advice (show label where possible) (S9, S16, S33, S44)

Toxic effects   The vapour irritates the eyes and respiratory system. The liquid irritates and
                     burns the skin and eyes. Irritant and poisonous if taken by mouth.

First aid	Standard treatment for exposure by all routes (see pages 108-122).
Fire hazard	Flash point 24.4 °C (open cup); Autoignition temperature 504 °C. Extinguish fire with foam, dry powder, or carbon dioxide. Do NOT use water.
Spillage disposal	See general section.

RSC *Chemical Safety Data Sheets* Vol. 4b, No. 92, 1991 gives extended coverage.

834. Methyl 2-cyanoacrylate

Faint yellow liquid; b.p. 48-49 °C at 2.5 mm Hg.

Limit values	OES short-term 4 p.p.m. (16 mg m^{-3}); long-term 2 p.p.m. (8 mg m^{-3}).
Toxic effects	Irritant to eyes, upper respiratory tract, and mucous membranes. Harmful by inhalation, ingestion, or skin absorption.
First aid	Standard treatment for exposure by all routes (see pages 108-122).
Fire hazard	Extinguish fires with carbon dioxide, dry chemical powder, or alcohol or polymer foam.
Spillage disposal	See general section.

835. Methylcyclohexane

Colourless liquid; m.p. -126.6 °C; b.p. 100.9 °C; almost insoluble in water.

RISKS
Highly flammable (R11)

SAFETY PRECAUTIONS
Keep container dry – Keep away from sources of ignition – No Smoking – Take precautionary measures against static discharges (S8, S16, S33)

Limit values	OES short-term 500 p.p.m. (2000 mg m^{-3}); long-term 400 p.p.m. (1600 mg m^{-3}).
Toxic effects	These are not well documented but animal experiments suggest that it is more toxic than cyclohexane. In high concentrations the vapour causes narcosis and anaesthesia.
First aid	Standard treatment for exposure by all routes (see pages 108-122).
Fire hazard	Flash point -4 °C; Autoignition temperature 285 °C. Extinguish small fires with dry powder, carbon dioxide, or alcohol-resistant foam. The foam is more suitable for large fires.
Spillage disposal	See general section.

RSC *Chemical Safety Data Sheets* Vol. 1, No. 70, 1988 gives extended coverage.

836. Methylcyclohexanol

Colourless, viscous liquid with odour similar to that of menthol; b.p. mixed isomers 155-180 ˚C; slightly soluble in water.

RISKS
Harmful by inhalation (R20)

SAFETY PRECAUTIONS
Avoid contact with skin and eyes (S24/25)

Limit values   OES short-term 75 p.p.m. (350 mg m^{-3}); long-term 50 p.p.m. (235 mg m^{-3}).

Toxic effects   The vapour irritates the eyes and respiratory system. Prolonged exposure may produce headaches and ocular disturbance. The liquid irritates the eyes and may irritate the skin. Assumed to be irritant and harmful if taken by mouth.

First aid   Standard treatment for exposure by all routes (see pages 108-122).

Fire hazard   Extinguish fires with water spray, carbon dioxide, dry chemical powder, or alcohol or polymer foam.

Spillage disposal   See general section.

837. 2-Methylcyclohexanone

Colourless to pale yellow liquid with smell like that of acetone; b.p. 169-170 ˚C.; insoluble in water.

RISKS
Flammable – Harmful by inhalation (R10, R20)

SAFETY PRECAUTIONS
Avoid contact with eyes (S25)

Limit values   OES short-term 75 p.p.m. (345 mg m^{-3}); long-term 50 p.p.m. (230 mg m^{-3}).

Toxic effects   The vapour irritates the eyes and respiratory system. The liquid irritates the eyes and prolonged contact with the skin may result in kidney and liver damage. Assumed to cause irritation and damage if taken internally.

Hazardous reactions   Oxidation of the 4-isomer by addition to nitric acid at about 75 ˚C caused a detonation; reaction with mixtures of hydrogen peroxide and nitric acid caused the 3-isomer to form an oily explosive peroxide.

Fire hazard   Flash point 48 ˚C. Extinguish fires with foam, dry powder, carbon dioxide, or alcohol-resistant foam. The foam is preferred for large fires.

Spillage disposal   See general section.

RSC *Chemical Safety Data Sheets* Vol. 1, No. 71, 1988 gives extended coverage.

838. Methylcyclopentadienylmanganese tricarbonyl

Yellow orange liquid; m.p. -1 ˚C; b.p. 232-233 ˚C.

Toxic effects   May be fatal if inhaled, ingested, or absorbed through the skin. Possible carcinogen.

First aid   Standard treatment for exposure by all routes (see pages 108-122).

Fire hazard   Flash point 96 ˚C. Extinguish fires with water spray, carbon dioxide, dry chemical powder, or alcohol or polymer foam.

Spillage disposal   See general section.

839. 3-Methyldiazirine

Hazardous reactions   The gas explodes on heating.

840. Methyl diazoacetate

Hazardous reactions   Explodes with violence when heated.

841. Methylenebis (4-cyclohexyl isocyanate)

White powder; m.p. 19-23 ˚C; b.p. decomposes; insoluble in water.

RISKS
Toxic by inhalation – Irritating to eyes, respiratory system, and skin – May cause sensitization by inhalation and skin contact (R23, R36/37/38, R42/43)

SAFETY PRECAUTIONS
In case of contact with eyes, rinse immediately with plenty of water and seek medical advice – After contact with skin, wash immediately with plenty of water – In case of insufficient ventilation, wear suitable respiratory equipment – In case of accident or if you feel unwell, seek medical advice immediately (show label where possible) (S26, S28, S38, S45)

Limit values   MEL short-term 0.07 mg m^{-3} (as -NCO); long-term 0.02 mg m^{-3} (as -NCO).

Toxic effects   Irritant to skin, eyes, and upper respiratory tract. Inhalation can cause dry throat, cough, bronchitis, and respiratory sensitization.

First aid   Standard treatment for exposure by all routes (see pages 108-122).

Fire hazard   Flash point above 202 ˚C.

Spillage disposal   See general section.

842. Methyl fluorosulfate

Liquid, ethereal odour.; m.p. -95 ˚C; b.p. 92 ˚C; decomposed by water.

Toxic effects   Vapour irritates eyes severely and may lead to temporary corneal damage. Inhalation of vapour in small amounts has led to a temporary cough, succeeded after six hours by fatal lung inflammation. Potent experimental mutagen.

First aid   Standard treatment for exposure by all routes (see pages 108-122).

Spillage disposal   See general section.

843. Methyl formate

Colourless liquid with pleasant odour; m.p. -99.8 ˚C; b.p. 32 ˚C; soluble in water.

Limit values   OES short-term 150 p.p.m. (375 mg m^{-3}); long-term 100 p.p.m. (250 mg m^{-3}).

Toxic effects   The vapour irritates the upper respiratory system and inhalation of high concentrations may result in retching, narcosis, and pulmonary irritation leading to death. It is also irritant to eyes and skin and is assumed to cause severe irritation and damage if taken internally.

First aid   Standard treatment for exposure by all routes (see pages 108-122).

Fire hazard   Flash point -19 ˚C; Explosive limits 5.9-20%; Autoignition temperature 449 ˚C. Extinguish small fires with dry powder, carbon dioxide, or alcohol resistant foam. The foam is more suitable for large fires.

Spillage disposal   See general section.

RSC *Chemical Safety Data Sheets* Vol. 1, No. 73, 1988 gives extended coverage.

844. 5-Methyl-3-heptanone

Colourless liquid with penetrating fruity odour; b.p. 157-162 ˚C; insoluble in water.

RISKS
Flammable – Irritating to eyes and respiratory system (R10, R36/37)

SAFETY PRECAUTIONS
Do not breathe vapour (S23)

Limit values   OES long-term 25 p.p.m. (130 mg m^{-3}).

Toxic effects   Also irritant to the skin and may be absorbed through it. Inhalation of high concentrations can cause narcosis and death.

First aid   Standard treatment for exposure by all routes (see pages 108-122).

Fire hazard   Flash point 43 ˚C. Extinguish fires with water spray, carbon dioxide, dry chemical powder, or alcohol or polymer foam.

Spillage disposal   See general section.

845. 5-Methylhexan-2-one

Colourless liquid; m.p. -74 °C; b.p. 145 °C; very slightly soluble in water.

RISKS
Flammable (R10)

SAFETY PRECAUTIONS
Do not breathe vapour (S23)

Limit values	OES short-term 75 p.p.m. (360 mg m^{-3}); long-term 50 p.p.m. (240 mg m^{-3}).
Toxic effects	Irritant to skin and eyes. Harmful on inhalation, ingestion, and skin absorption.
First aid	Standard treatment for exposure by all routes (see pages 108-122).
Fire hazard	Flash point 41 °C. Extinguish fires with water spray, carbon dioxide, dry chemical powder, or alcohol or polymer foam.
Spillage disposal	See general section.

846. Methylhydrazine

Colourless, hygroscopic liquid; m.p. below -80 °C; b.p. 87 °C at 745 mm Hg; slightly soluble in water.

Toxic effects	Material is extremely destructive to all tissues. May be fatal if inhaled, ingested, or absorbed through the skin. At lower concentrations may cause allergic reactions or damage to the liver and kidneys. Experimental carcinogen.
Hazardous reactions	May ignite in air when extended (on fibre or as film). A powerful reducing agent and fuel, hypergolic with many oxidants.
First aid	Standard treatment for exposure by all routes (see pages 108-122).
Fire hazard	Flash point 21 °C (closed cup); Explosive limits 2.5-97%. Extinguish fires with water spray, carbon dioxide, dry chemical powder, or alcohol or polymer foam.
Spillage disposal	See general section.

847. Methyl hydroperoxide

Liquid; b.p. 38-40 °C; soluble in water.

Toxic effects	Irritant to skin, eyes, and mucous membranes.
Hazardous reactions	Violently explosive, shock-sensitive, especially on warming.

848. Methyl hypochlorite

Gas or liquid; b.p. 12 ˚C at 726 mm Hg.

Hazardous   Superheated vapour explodes as does liquid on ignition.
reactions

849. Methyl isocyanide

Colourless liquid; m.p. -45 ˚C; b.p. 59.6 ˚C; soluble in water.

Toxic effects   Little information is available on human toxicity but it is generally regarded as
                     hazardous.

Hazardous   Exploded when heated in a sealed ampoule. Has exploded on distillation.
reactions

850. Methyllithium

Solid; m.p. 250 ˚C with decomposition; reacts violently with water.

Toxic effects   Harmful if inhaled, ingested, or absorbed through the skin. Extremely
                     destructive to all tissues. Inhalation may be fatal.

Hazardous   Ignites and burns in air.
reactions

First aid   Standard treatment for exposure by all routes (see pages 108-122).

Fire hazard   Extinguish fires with dry chemical powder. Do NOT use water.

Spillage   See general section.
disposal

851. Methyl methacrylate

Colourless liquid; m.p. -50 ˚C; b.p. 101 ˚C; slightly soluble in water.

RISKS
*Highly flammable – Irritating to eyes, respiratory system, and skin – May cause sensitization by
skin contact (R11, R36/37/38, R43)*

SAFETY PRECAUTIONS
*Keep container in a well ventilated place – Keep away from sources of ignition – No Smoking –
Do not empty into drains – Take precautionary measures against static discharges (S9, S16,
S29, S33)*

Limit values   OES short-term 125 p.p.m. (510 mg m^{-3}); long-term 100 p.p.m. (410 mg m^{-3}).

Toxic effects   The vapour may irritate the eyes and respiratory system. Inhalation may
                     cause CNS depression, pulmonary oedema, and narcosis leading to death.
                     The liquid will irritate the skin, eyes, and alimentary system.

Hazardous reactions	Exposure to air at room temperature for two months generated an explosive ester/oxygen interpolymer; ignition occurred when dibenzoyl peroxide was added to a small amount of the ester. Store inhibited monomer with slight access of air.
First aid	Standard treatment for exposure by all routes (see pages 108-122).
Fire hazard	Flash point 10 ˚C (open cup); Explosive limits 2.1-12.5%; Autoignition temperature 421 ˚C. Extinguish fire with foam, dry powder, or carbon dioxide.
Spillage disposal	See general section.

RSC *Chemical Safety Data Sheets* Vol. 3, No. 43, 1990 gives extended coverage.

852. Methyl nitrate

Colourless liquid; m.p. -83 ˚C; b.p. decomposes explosively at 65 ˚C; slightly soluble in water.

Toxic effects	Very little information is available on human toxicity. Harmful on inhalation or ingestion.
Hazardous reactions	Explodes at 65 ˚C and has high shock-sensitivity.

853. Methyl-2-nitrobenzenediazoate

Hazardous reactions	Explodes on heating or on disturbing after 24 hours in a sealed tube at ambient temperature.

854. 1-Methyl-3-nitro-1-nitrosoguanidine

Pale yellow, orange, or pink crystals; m.p. 118-123.5 ˚C with decomposition; slightly soluble in water.

RISKS
May cause cancer – Harmful by inhalation – Irritating to eyes and skin (R45, R20, R36/38)

SAFETY PRECAUTIONS
Avoid exposure – Obtain special instructions before use – If you feel unwell, seek medical advice (show label where possible) (S53, S44)

Toxic effects	Severe irritant to all tissues. Experimental carcinogen and mutagen.
Hazardous reactions	A former diazomethane precursor, this compound detonates on high impact; sample exploded when heated in sealed capillary tube.
First aid	Standard treatment for exposure by all routes (see pages 108-122).
Fire hazard	Extinguish fires with carbon dioxide, dry chemical powder, or alcohol or polymer foam.
Spillage disposal	See general section.

RSC *Chemical Safety Data Sheets* Vol. 4b, No. 94, 1991 gives extended coverage.

855. 2-Methyl-2-nitropropane

Crystalline solid; m.p. 24 ˚C; insoluble in water.

Hazardous   Sample exploded during distillation.
reactions

856. *N*-Methyl-*N*-nitrosotoluene-4-sulfonamide

Yellow crystals; insoluble in water.

SAFETY PRECAUTIONS
Avoid contact with skin and eyes (S24/25)

Toxic effects   No evidence has been found that this reagent is irritant or otherwise toxic, but its close association with the methylating technique involving highly toxic diazomethane which is generated when it is treated with alkalies suggests that protection against skin and eye contact should be used when it is being handled.

Hazardous   May explode when heated to above 45 ˚C.
reactions

First aid   Standard treatment for exposure by all routes (see pages 108-122).

Fire hazard   Use water spray to extinguish fires.

Spillage   See general section.
disposal

857. *N*-Methyl-*N*-nitrosourea

Pale yellow, crystalline solid; m.p. 124 ˚C with decomposition; slightly soluble in water.

Toxic effects   Experimental carcinogen. Toxic effects occur after skin contact and ingestion.

Hazardous   Material stored at 20 ˚C exploded after 6 months. Alkaline hydrolysis
reactions   produces the explosive gas diazomethane.

First aid   Standard treatment for exposure by all routes (see pages 108-122).

Fire hazard   Extinguish fires with carbon dioxide, dry chemical powder, or foam extinguishers.

Spillage   See general section.
disposal

858. 2-Methylpentane-2,4-diol

Colourless liquid with a mild odour; m.p. -40 ˚C; b.p. 198 ˚C; miscible with water.

RISKS
Irritating to eyes and skin (R36/38)

Limit values   OES short-term 25 p.p.m. (125 mg m^{-3}); long-term 25 p.p.m. (125 mg m^{-3}).

Toxic effects	Irritant to skin, eyes, upper respiratory tract, and mucous membranes. Harmful by inhalation, ingestion, or skin absorption.
First aid	Standard treatment for exposure by all routes (see pages 108-122).
Fire hazard	Flash point 93 °C; Explosive limits 1.3-7.4 °C. Extinguish fires with water spray, carbon dioxide, dry chemical powder, or alcohol or polymer foam.
Spillage disposal	See general section.

859. 2-Methyl-1-pentan-1-ol

Colourless liquid; b.p. 148 °C; slightly soluble in water.

Toxic effects	Irritant to skin, eyes, upper respiratory tract, and mucous membranes.
First aid	Standard treatment for exposure by all routes (see pages 108-122).
Fire hazard	Flash point 50 °C; Explosive limits 1.1-9.6%. Extinguish fires with carbon dioxide, dry chemical powder, or alcohol or polymer foam.
Spillage disposal	See general section.

860. 4-Methyl-2-pentanol

Colourless liquid; m.p. -90 °C; b.p. 132 °C; sparingly soluble in water.

RISKS
Flammable – Irritating to respiratory system (R10, R37)

SAFETY PRECAUTIONS
Avoid contact with skin and eyes (S24/25)

Toxic effects	Irritant to eyes and upper respiratory tract. Damage to the respiratory and gastro-intestinal tracts have been noted in animal experiments.
First aid	Standard treatment for exposure by all routes (see pages 108-122).
Fire hazard	Flash point 46 °C (open cup); Explosive limits 1-5.5%. Extinguish small fires with carbon dioxide, dry chemical powder, or alcohol-resistant foam extinguishers. Large fires are best controlled by foam.
Spillage disposal	See general section.

RSC *Chemical Safety Data Sheets* Vol. 1, No. 75, 1988 gives extended coverage.

861. 4-Methylpentan-2-one

Colourless liquid with faint camphor-like smell; b.p. 126 °C; slightly soluble in water.

SAFETY PRECAUTIONS
Do not breathe vapour – Avoid contact with skin and eyes (S23, S24/25)

Toxic effects	The vapour is somewhat irritating to the eyes and respiratory system and

narcotic in high concentrations. The liquid irritates the eyes and will cause irritation and damage if taken internally.

Hazardous   Unusual peroxide explosion.
reactions

Fire hazard   Flash point 17 ˚C; Explosive limits 1.2-8%; Autoignition temperature 460 ˚C. Extinguish fire with water spray, dry powder, or carbon dioxide.

Spillage   See general section.
disposal

862. 4-Methylpent-3-en-2-one

Colourless, oily liquid with smell somewhat like honey. Darkens on standing; m.p. -59 ˚C; b.p. 130 ˚C; sparingly soluble in water.

RISKS
Flammable – Harmful by inhalation, in contact with skin, and if swallowed (R10, R20/21/22)

SAFETY PRECAUTIONS
Avoid contact with eyes (S25)

Limit values   OES short-term 25 p.p.m. (100 mg m^{-3}); long-term 15 p.p.m. (60 mg m^{-3}).

Toxic effects   The vapour irritates the eyes and respiratory system. The liquid is highly irritating to the eyes and skin and will cause internal irritation and damage if taken by mouth.

Fire hazard   Flash point 31 ˚C; Explosive limits 1.4-7.2%; Autoignition temperature 344 ˚C. Extinguish fire with foam, dry powder, or carbon dioxide. Foam is recommended for large fires.

Spillage   See general section.
disposal

RSC *Chemical Safety Data Sheets* Vol. 1, No. 64, 1988 gives extended coverage.

863. Methyl perchlorate

Hazardous   Explosive.
reactions

864. Methylphosphine

Colourless gas; b.p. -14 ˚C at 759 mm Hg; insoluble in water.

Toxic effects   Harmful on inhalation or ingestion.

Hazardous   Readily ignites in air. Can react vigorously with oxidizing materials.
reactions

865. Methylpotassium

Crystalline solid; decomposed by moisture.

Toxic effects   Irritant to all mucous membranes because of its decomposition by water.

Hazardous   Dry material is highly pyrophoric.
reactions

866. 2-Methylpropan-2-ol

Colourless, crystalline solid or liquid with camphor-like odour; m.p. 25 °C; b.p. 83 °C; miscible with water.

SAFETY PRECAUTIONS
Do not breathe vapour – Avoid contact with skin and eyes (S23, S24/25)

Toxic effects   Vapour may irritate eyes and respiratory system. The liquid irritates the eyes and may irritate the skin causing dermatitis. If taken by mouth may cause headache, dizziness, drowsiness, and narcosis.

Fire hazard   Explosive limits 2.4-8%; Autoignition temperature 478 °C. Extinguish fire with dry powder or carbon dioxide.

Spillage   See general section.
disposal

867. 2-Methylpropene

Colourless gas with smell like that of coal gas; b.p. -7 °C; practically insoluble in water.

SAFETY PRECAUTIONS
Do not breathe gas (S23)

Toxic effects   The gas has an anaesthetic effect but is not toxic.

Fire hazard   Flash point <-7 °C; Explosive limits 1.8-8.8%; Autoignition temperature 465 °C. Since the gas is supplied in a cylinder, turning off the valve will reduce any fire involving it; if possible, cylinders should be removed quickly from an area in which a fire has developed.

Spillage   Surplus gas or leaking cylinder can be vented slowly to air in a safe, open
disposal   area, or gas burnt-off through a suitable burner.

868. Methylpyridines

The 2-, 3-, and 4-isomers are colourless liquids, the 2- and 4-isomers have unpleasant smells; b.p. 129 ˚C, 144 ˚C, and 145 ˚C, respectively; all are very soluble in water.

RISKS
Flammable – Harmful by inhalation, in contact with skin, and if swallowed – Irritating to eyes and respiratory system (R10, R20/21/22, R36/37)

SAFETY PRECAUTIONS
In case of contact with eyes, rinse immediately with plenty of water and seek medical advice – Wear suitable protective clothing (S26, S36)

Toxic effects	The vapours irritate the respiratory tract. The liquids irritate the skin and eyes and are assumed to cause irritation and damage if ingested.
First aid	Standard treatment for exposure by all routes (see pages 108-122).
Fire hazard	Flash point 28 ˚C, 40 ˚C, and 57 ˚C, respectively; Autoignition temperature 2-isomer 538 ˚C; 4-isomer 500 ˚C. Extinguish fire with alcohol-resistant foam.
Spillage disposal	See general section.

RSC *Chemical Safety Data Sheets* Vol. 4b, No. 121, 1991 gives extended coverage of the 4-isomer.

869. *N*-Methylpyrrolidone

Hygroscopic, colourless liquid with a slight odour of amines; m.p. -24 ˚C; b.p. 202 ˚C; miscible with water.

RISKS
Irritating to eyes and skin (R36/38)

SAFETY PRECAUTIONS
In case of fire and/or explosion do not breathe fumes (S41)

Limit values	OES long-term 100 p.p.m. (400 mg m^{-3}).
Toxic effects	Irritant to skin and eyes. Low toxicity on inhalation or ingestion. Skin absorption may cause systemic injury.
First aid	Standard treatment for exposure by all routes (see pages 108-122).
Fire hazard	Flash point 96 ˚C (open cup); Explosive limits 1.3-9.5%; Autoignition temperature 346 ˚C. Small fires can be extinguished with carbon dioxide, dry chemical powder, or alcohol-resistant foam. Large spills are best controlled with alcohol-resistant foam.
Spillage disposal	See general section.

RSC *Chemical Safety Data Sheets* Vol. 1, No. 76, 1988 gives extended coverage.

870. Methylsodium

Powdery solid; m.p. decomposes without melting.

Toxic effects   Irritant and corrosive to all tissues.

Hazardous   Ignites immediately in air.
 reactions

871. α-Methylstyrene

Colourless liquid; b.p. 167 ˚C; insoluble in water

RISKS
Flammable – Irritating to eyes and respiratory system (R10, R36/37)

Limit values   OES short-term 100 p.p.m. (480 mg m^{-3}).

Toxic effects   The vapour irritates the eyes and respiratory system. The liquid irritates the eyes and skin, causing conjunctivitis and dermatitis respectively. It is assumed to be irritant and harmful if taken by mouth.

First aid   Standard treatment for exposure by all routes (see pages 108-122).

Fire hazard   Flash point 54 ˚C; Explosive limits 1.9-6.1%; Autoignition temperature 494 ˚C. Extinguish fire with foam, dry powder, or carbon dioxide.

Spillage   See general section.
 disposal

872. 2-, 3-, and 4-Methylstyrenes

All colourless or yellow liquids; m.p. (2-) -69 ˚C; (3-) -82.5 ˚C; b.p. (2-) 169-170 ˚C; (3-) 170-171 ˚C; (4-) 170-175 ˚C.

RISKS
Harmful by inhalation (R20)

SAFETY PRECAUTIONS
Avoid contact with skin (S24)

Limit values   OES short-term 150 p.p.m. (720 mg m^{-3}); long-term 100 p.p.m. (480 mg m^{-3}).

Toxic effects   All irritant to skin, eyes, upper respiratory tract, and mucous membranes. Harmful by inhalation, ingestion, and skin absorption.

First aid   Standard treatment for exposure by all routes (see pages 108-122).

Fire hazard   Flash point (2-) 51 ˚C; (3-) 51 ˚C; (4-) 45 ˚C. Extinguish fires with water spray, carbon dioxide, dry chemical powder, or alcohol or polymer foam.

Spillage   See general section.
 disposal

873. *N*-Methyl-*N*,2,4,6-tetranitroaniline

Yellow, crystalline solid; m.p. 130 ˚C; b.p. explodes at 187 ˚C; insoluble in water.

Limit values   OES short-term 3 mg m⁻³; long-term 1.5 mg m⁻³.

Toxic effects   Irritant to skin, eyes, upper respiratory tract, and mucous membranes. Exposure can cause sensitization.

First aid   Standard treatment for exposure by all routes (see pages 108-122).

874. Methyltrichlorosilane

Colourless to pale yellow, volatile liquid with pungent smell, which fumes strongly in moist air; b.p. 65.5 ˚C; it reacts vigorously with water forming hydrochloric acid and polymeric gels.

RISKS
Highly flammable – Reacts violently with water – Irritating to eyes, respiratory system, and skin (R11, R14, R36/37/38)

SAFETY PRECAUTIONS
In case of contact with eyes, rinse immediately with plenty of water and seek medical advice – Wear eye/face protection (S26, S39)

Toxic effects   The vapour irritates the eyes and is destructive to the respiratory system. Inhalation may be fatal. The liquid burns the skin and eyes and will cause severe internal damage if taken by mouth.

First aid   Standard treatment for exposure by all routes (see pages 108-122).

Fire hazard   Flash point -15 ˚C (closed cup); Explosive limits 5.1-20%; Autoignition temperature 404 ˚C. Extinguish fire with dry powder or carbon dioxide.

Spillage   See general section.
disposal

RSC *Chemical Safety Data Sheets* Vol. 3, No. 71, 1990 gives extended coverage.

875. 1-Methyltrimethylene diacrylate

Toxic effects   Irritant to skin and eyes. Moderately toxic on ingestion.

First aid   Standard treatment for exposure by all routes (see pages 108-122).

Fire hazard   Extinguish fires with water spray, carbon dioxide, dry chemical powder, or alcohol or polymer foam.

Spillage   See general section.
disposal

RSC *Chemical Safety Data Sheets* Vol. 3, No. 44, 1990 gives extended coverage.

876. Methyl vinyl ether

Colourless liquid or gas with sweetish odour; b.p. 8 °C; slightly soluble in water.

RISKS
Extremely flammable liquefied gas (R13)

SAFETY PRECAUTIONS
Keep container in a well ventilated place – Keep away from sources of ignition – No Smoking – Take precautionary measures against static discharges (S9, S16, S33)

Toxic effects These have not been fully investigated. It has narcotic properties.

Hazardous reactions Forms peroxides; acids hydrolyse the latter to acetaldehyde and cause rapid polymerization.

First aid Standard treatment for exposure by all routes (see pages 108-122).

Fire hazard Flash point -51 °C. As the liquid is supplied in a cylinder, turning off the valve will reduce any fire involving it; if possible, cylinders should be removed from an area in which a fire has developed.

Spillage disposal Surplus gas or leaking cylinder can be vented slowly to air in a safe, open area or gas burnt-off through a suitable burner.

877. Molecular sieves

Hazardous reactions Catalysed ignition of ethylene; fire when heated to regenerate, explosions with benzyl bromide and nitromethane.

878. Molybdenum and compounds

The metal is a shiny grey solid; m.p. (metal) 2620 °C; b.p. (metal) 4800 °C; insoluble in water. The salts are mainly coloured, crystalline solids, some soluble and some insoluble in water.

Limit values OES short-term 10 mg m^{-3} (soluble compounds), 20 mg m^{-3} (insoluble compounds) (as Mo); long-term 5mg m^{-3} (soluble compounds), 10 mg m^{-3} (insoluble compounds) (as Mo).

Toxic effects The metal is an experimental teratogen. Some of the compounds are highly irritant and all, or nearly all, are harmful by inhalation or ingestion. Symptoms of acute poisoning include gastro-intestinal irritation with diarrhoea and can lead to death.

Hazardous reactions Powdered molybdenum metal is explosive and reacts violently with certain oxidants.

First aid Standard treatment for exposure by all routes (see pages 108-122).

Fire hazard With the exception of powdered molybdenum metal all fires can be treated with carbon dioxide, dry chemical powder, or alcohol or polymer foam.

Spillage disposal See general section.

879. Morpholine

Colourless, mobile liquid with amine-like odour; m.p. -4.9 ˚C; b.p. 128 ˚C; miscible with water.

RISKS
Flammable – Harmful by inhalation, in contact with skin, and if swallowed – Causes burns
(R10, R20/21/22, R34)

SAFETY PRECAUTIONS
Do not breathe vapour – In case of contact with eyes, rinse immediately with plenty of water
and seek medical advice (S23, S26)

Limit values   OES short-term 30 p.p.m. (105 mg m^{-3}); long-term 20 p.p.m. (70 mg m^{-3}).

Toxic effects   The vapour irritates the eyes and respiratory system. The liquid irritates the
eyes and skin; it is also irritant when taken internally and may cause kidney
and liver injury.

Hazardous   Its addition to nitromethane makes it susceptible to initiation by a detonator.
    reactions

First aid   Standard treatment for exposure by all routes (see pages 108-122).

Fire hazard   Flash point 38 ˚C (open cup); Explosive limits 1.4 – 11.2%; Autoignition
temperature 290-310 ˚C. Extinguish fire with dry powder, carbon dioxide, or
alcohol foam.

Spillage   See general section.
    disposal

RSC *Chemical Safety Data Sheets* Vol. 1, No. 77, 1988 gives extended coverage.

880. Naphtha

Colourless or yellowish liquids; m.p. depends on formulation; b.p. (Stoddard solvent)
200-300 ˚C; (ligroin) 40-80 ˚C; insoluble in water.

Limit values   OES short-term white spirit 125 p.p.m. (720 mg m^{-3}); long-term white spirit
100 p.p.m. (575 mg m^{-3}).

Toxic effects   Irritant to skin and eyes. Can permeate the skin following defatting and then
cause systemic effects. Low toxicity on ingestion but serious risk of aspiration
into the lungs if vomiting occurs.

First aid   Standard treatment for exposure by all routes (see pages 108-122).

Fire hazard   Flash point (Stoddard solvent) 100-120 ˚C;(ligroin) -20 to -10 ˚C; Explosive
limits depend on formulation; Autoignition temperature 200-260 ˚C.
Extinguish fires with carbon dioxide, dry chemical powder, or alcohol foam.

Spillage   See general section.
    disposal

RSC *Chemical Safety Data Sheets* Vol. 1, No. 78, 1988 gives extended coverage.

881. Naphthalene

Colourless solid; m.p. 80.2 °C; b.p. 218 °C; slightly soluble in water.

Limit values	OES short-term 15 p.p.m. (75 mg m^{-3}); long-term 10 p.p.m. (50 mg m^{-3}).
Toxic effects	Irritant to the skin but there is little information about skin absorption. Inhalation of the vapour may cause headache, confusion, nausea, vomiting, and, in prolonged exposure, optic neuritis and haematuria.
Hazardous reactions	Reacts explosively with dinitrogen pentaoxide.
First aid	Standard treatment for exposure by all routes (see pages 108-122).
Fire hazard	Flash point 80 °C; Explosive limits 0.88-5.9%; Autoignition temperature 529 °C. Extinguish fires with carbon dioxide or dry chemical powder.
Spillage disposal	See general section.

882. Naphthalene-1- and -2-diazonium salts

The chlorides are yellow, crystalline solids; m.p. the 1-chloride melts at 96 °C with decomposition; the 2-chloride explodes on heating; the chlorides are soluble in water.

Hazardous reactions	They react with ammonium sulfide or hydrogen sulfide to form explosive compounds.

883. 1,5-Naphthalene diisocyanate

White to light yellow crystals.

RISKS
Harmful by inhalation – Irritating to eyes, respiratory system, and skin – May cause sensitization by inhalation (R20, R36/37/38, R42)

SAFETY PRECAUTIONS
In case of contact with eyes, rinse immediately with plenty of water and seek medical advice – After contact with skin, wash immediately with plenty of water – In case of insufficient ventilation, wear suitable respiratory equipment – In case of accident or if you feel unwell, seek medical advice immediately (show label where possible) (S26, S28, S38, S45)

Limit values	MEL short-term 0.07 mg m^{-3} (as NCO-); long-term 0.02 mg m^{-3} (as NCO-).
Toxic effects	Irritant to body tissues. Powerful allergen and possible carcinogen.
First aid	Standard treatment for exposure by all routes (see pages 108-122).

884. 1-Naphthol

Brown, crystalline solid with phenolic odour; m.p. 95-96 °C; b.p. 278-280 °C; slightly soluble in water.

Toxic effects   High concentrations are extremely destructive to all body tissues. Experimental teratogen.

First aid   Standard treatment for exposure by all routes (see pages 108-122).

Fire hazard   Explosive limits 0.8-5.0%; Autoignition temperature 540 °C. Extinguish fires with water spray, carbon dioxide, dry chemical powder, or alcohol or polymer foam.

Spillage disposal   See general section.

885. 2-Naphthol

White to yellowish white, crystalline, solid with slight phenolic odour; m.p. 122-123 °C; b.p. 285-286 °C; slightly soluble in water.

RISKS
Harmful by inhalation and if swallowed (R20/22)

SAFETY PRECAUTIONS
Avoid contact with skin and eyes (S24/25)

Toxic effects   Irritant to all tissues, the extent of irritation and destruction depending on the intensity and duration of exposure.

First aid   Standard treatment for exposure by all routes (see pages 108-122).

Fire hazard   Extinguish fires with water spray, carbon dioxide, dry chemical powder, or alcohol or polymer foam.

Spillage disposal   See general section.

886. 1-Naphthylamine and salts

Colourless crystals when pure, darkening on exposure to light and air; the base has an unpleasant smell; m.p. 50 °C (the base); b.p. 301 °C (the base); the base is insoluble in water, the hydrochloride is soluble.

RISKS
Very toxic by inhalation, in contact with skin, and if swallowed – Danger of very serious irreversible effects (R26/27/28, R39)

SAFETY PRECAUTIONS
Do not breathe dust – Take off immediately all contaminated clothing – Wear suitable protective clothing – In case of accident or if you feel unwell, seek medical advice immediately (show label where possible) (S22, S27, S36, S45)

Toxic effects   The salts or their solutions irritate the eyes. Exposure to the dust or absorption through the skin may cause bladder tumours. The use of 1-naphthylamine and its salts is controlled in the United Kingdom by the Carcinogenic Substances Regulations 1967.

First aid	Standard treatment for exposure by all routes (see pages 108-122).
Fire hazard	Flash point (base) 157 ˚C. Extinguish fires with water spray, carbon dioxide, or dry chemical powder.
Spillage disposal	See general section.

887. 2-Naphthylamine

Inhalation or absorption through the skin of the dust has been recognized as a cause of bladder cancer. The use of these compounds in the United Kingdom is therefore prohibited under The Carcinogenic Substances Regulations 1967. It is not therefore considered appropriate to deal with their hazards more fully in this book.

888. Neopentyl glycol diacrylate

RISKS
Toxic if swallowed – Irritating to eyes and skin – May cause sensitization by skin contact (R25, R36/38, R43)

SAFETY PRECAUTIONS
After contact with skin, wash immediately with plenty of water – Wear eye/face protection – If you feel unwell, seek medical advice (show label where possible) (S28, S39, S44)

Toxic effects	Animal carcinogen. Irritant to skin and eyes.
First aid	Standard treatment for exposure by all routes (see pages 108-122).
Fire hazard	Extinguish fires with water spray, carbon dioxide, dry chemical powder, or alcohol or polymer foam.
Spillage disposal	See general section.

RSC *Chemical Safety Data Sheets* Vol. 4b, No. 96, 1991 gives extended coverage.

889. Nickel metal

Grey solid; m.p. 1455 ˚C; b.p. 2835 ˚C; insoluble in water.

Limit values	MEL long-term 0.5 mg m^{-3}.
Toxic effects	Irritant to skin, eyes, upper respiratory tract, and mucous membranes. Harmful on inhalation, ingestion, and skin absorption. Carcinogenic.
Hazardous reactions	Finely divided nickel powder is pyrophoric. Reactions involving Raney nickel catalysts can be explosive.
First aid	Standard treatment for exposure by all routes (see pages 108-122).
Fire hazard	Fires involving nickel powder should be tackled with dry chemical powder, those involving solid metal can be treated with any extinguishant.
Spillage disposal	See general section.

890. Nickel, organic compounds

RISKS
(for nickel tetracarbonyl) Highly flammable – Very toxic by inhalation – Possible risk of irreversible effects (R11, R26, R40)

SAFETY PRECAUTIONS
(for nickel tetracarbonyl) Keep container in a well ventilated place – Do not breathe dust – In case of accident or if you feel unwell, seek medical advice immediately (show label where possible) (S9, S23, S45)

Limit values   OES short-term 3 mg m^{-3} (as Ni); long-term 1 mg m^{-3} (as Ni).

891. Nickel salts

Green crystals or powder; mostly soluble in water.

Limit values   MEL long-term 0.1 mg m^{-3} (as soluble compounds), 0.5 mg m^{-3} (as insoluble compounds).

Toxic effects   The salts and their solutions will irritate the eyes. Assumed to be poisonous if taken by mouth. Continued skin contact can cause dermatitis.

First aid   Standard treatment for exposure by all routes (see pages 108-122).

Fire hazard   For most nickel salts use extinguishant appropriate to the surroundings.

Spillage   See general section.
   disposal

892. Nicotine

Thick, colourless or pale yellow oil with a slight fishy smell when warm; m.p. <80 ˚C; b.p. 247.3 ˚C; soluble in water.

RISKS
Very toxic by inhalation, in contact with skin, and if swallowed (R26/27/28)

SAFETY PRECAUTIONS
Keep locked up – Keep away from food, drink, and animal feeding stuffs – After contact with skin, wash immediately with plenty of water – In case of accident or if you feel unwell, seek medical advice immediately (show label where possible) (S1, S13, S28, S45)

Limit values   OES short-term 1.5 mg m^{-3}; long-term 0.5 mg m^{-3}.

Toxic effects   Eye irritant. Can be absorbed from the gut, lungs, and skin. It acts very rapidly and can be fatal. The major effects are stimulation followed by depression or paralysis of the central nervous system.

First aid   Standard treatment for exposure by all routes (see pages 108-122). Convulsions and shock may need special consideration.

Fire hazard   Flash point 95 ˚C (closed cup); Explosive limits 0.75%-4%; Autoignition temperature 243.9 ˚C. Extinguish fires with water, carbon dioxide, or dry chemical powder.

Spillage   See general section.
disposal

RSC *Chemical Safety Data Sheets* Vol. 4b, No. 98, 1991 gives extended coverage.

893. Nitric acid

Colourless or pale yellow fuming liquid; m.p. -42 ˚C; b.p. 86 ˚C; miscible with water.

RISKS
(≥ 20%, ≤ 70%) Causes severe burns (R35)
(> 70%) Contact with combustible material may cause fire – Causes severe burns (R8, R35)

SAFETY PRECAUTIONS
(≥ 20%, ≤ 70%) Keep out of reach of children – Do not breathe vapour – In case of contact with eyes, rinse immediately with plenty of water and seek medical advice – Take off immediately all contaminated clothing (S2, S23, S26, S27)
(> 70%) Do not breathe vapour – In case of contact with eyes, rinse immediately with plenty of water and seek medical advice – Wear suitable protective clothing (S23, S26, S36)

Limit values   OES short-term 4 p.p.m. (10 mg m^{-3}); long-term 2 p.p.m. (5 mg m^{-3}).

Toxic effects   The vapour and liquid are severely irritant and corrosive to all tissues. Inhalation can be fatal. Experimental teratogen.

Hazardous   The range of vigorous, violent, and explosive reactions, in which the stronger
reactions   forms of nitric acid participate, is very wide. This powerful oxidant is the compound most frequently involved in hazardous reactions. Fuming nitric acid will attack unprotected plastic screw caps of storage bottles, and causes combustible materials to ignite.

First aid   Standard treatment for exposure by all routes (see pages 108-122).

Fire hazard   Use an extinguishing agent appropriate to the surroundings.

Spillage   See general section.
disposal

RSC *Chemical Safety Data Sheets* Vol. 3, No. 46, 1990 gives extended coverage.

894. Nitric amide

Shiny, white crystals; m.p. 72-75 ˚C with decomposition; soluble in water.

Hazardous   Various preparations have been violent or explosive; drop of concentrated
reactions   alkali added to solid causes a flame and explosive decomposition; it explodes on contact with concentrated sulfuric acid.

895. Nitric oxide

Colourless gas which, on release to atmosphere, is rapidly oxidized to nitrogen dioxide, a red gas with a pungent odour; m.p. -164 °C; b.p. -152 °C; slightly soluble in water.

RISKS
Very toxic by inhalation – Very toxic in contact with skin (R26, R27)

SAFETY PRECAUTIONS
Keep container tightly closed and in a well ventilated place – In case of contact with eyes, rinse immediately with plenty of water and seek medical advice – In case of accident or if you feel unwell, seek medical advice immediately (show label where possible) (S7/9, S26, S45)

Limit values   OES short-term 35 p.p.m. (45 mg m^{-3}); long-term 25 p.p.m. (30 mg m^{-3}).

Toxic effects   The toxic effects must be assumed to be those of nitrogen dioxide which is rapidly formed when nitric oxide mixes with air; nitrogen dioxide is a particularly dangerous gas because of its insidious mode of attack – several hours may elapse before the person exposed to it develops lung irritation and great discomfort. If gassing has been extensive, pulmonary oedema (flooding of lungs) may develop, and this can result, several days later, in death.

Hazardous reactions   Highly endothermic and an active oxidant (53% O). The liquid is sensitive to detonation in the absence of fuel; when mixed with carbon disulfide an explosion occurred; it ignited hydrogen/oxygen mixtures; pyrophoric chromium incandesces in the gas; acts as initiator to explosion of NCl_3; reacts with boron at ambient temperature with brilliant flashes while charcoal and phosphorus burn more brilliantly than in air; carbon black ignites at 100 °C if potassium hydrogen tartrate is present.

First aid   It is important to treat any case of considerable exposure as serious and to obtain medical attention even if symptoms of respiratory irritation have not shown themselves. If exposed to other than very low concentrations of the gas, the casualty should be made to rest and be kept warm until medical attention is received.

Fire hazard   Do NOT use water to extinguish fires.

Spillage disposal   Surplus gas or leaking cylinder can be vented slowly into a water-fed scrubbing tower or a column in a fume cupboard served by such a tower.

896. Nitrides

As a class the nitrides constitute a moderate explosive hazard. Fine powders of the nitrides of the transition metals might be pyrophoric. Nitrides of group I B and II B metals should be handled with extreme care as they are highly unstable and react readily with moisture evolving toxic and flammable ammonia and forming caustic hydroxides. The toxicity of the group has not been well documented but the nitrides of the actinide metals are said to be carcinogenic.

897. 5-Nitroacenaphthene

Yellow powder; m.p. 103-104 ˚C; insoluble in water.

RISKS
May cause cancer (R45)

SAFETY PRECAUTIONS
Avoid exposure – Obtain special instructions before use – If you feel unwell, seek medical advice (show label where possible) (S53, S44)

Toxic effects   Mutagenic and carcinogenic in animal experiments.

First aid   Standard treatment for exposure by all routes (see pages 108-122).

Fire hazard   Extinguish fires with water spray, carbon dioxide, dry chemical powder, or alcohol or polymer foam.

Spillage disposal   Special treatment will be required because of the carcinogenic nature of this substance.

RSC *Chemical Safety Data Sheets* Vol. 4b, No. 99, 1991 gives extended coverage.

898. Nitroanilines

The nitroanilines are yellow to orange-red crystals or powders; m.p. (*o-*) 69-71 ˚C; (*m-*) 114 ˚C; (*p-*)148.5 ˚C; b.p. (*o-*) 284 ˚C; (*m-*) 306 ˚C with decomposition; (*p-*) 332 ˚C; slightly soluble in water.

RISKS
Toxic by inhalation, in contact with skin, and if swallowed – Danger of cumulative effects (R23/24/25, R33)

SAFETY PRECAUTIONS
After contact with skin, wash immediately with plenty of water – Wear protective clothing and gloves – If you feel unwell, seek medical advice (show label where possible) (S28, S36/37, S44)

Limit values   OES long-term 6 mg m^{-3} (4-isomer).

Toxic effects   Inhalation of dusts or excessive skin absorption of the solids may result in headache, flushing of the face, difficulty in breathing, nausea, and vomiting; weakness, drowsiness, irritability, and cyanosis may follow. Dermatitis may follow skin contact. The dusts will damage the eyes and effects similar to the above may be expected if the substances are taken by mouth.

Hazardous reactions   Thermal stability of the isomers is reduced by various impurities.

First aid   Standard treatment for exposure by all routes (see pages 108-122).

Fire hazard   Flash point (*o-*) 168 ˚C; (*m-*) 199 ˚C; (*p-*) 199 ˚C. Extinguish fires with water spray, carbon dioxide, dry chemical powder, or alcohol or polymer foam.

Spillage disposal   See general section.

RSC *Chemical Safety Data Sheets* Vol. 4b, No. 100, 101, and 102, 1991 give extended coverage of the isomers.

899. 2-Nitro-*p*-anisidine

Orange powder; m.p. 123-126 ˚C.

RISKS
Very toxic by inhalation, in contact with skin, and if swallowed – Danger of cumulative effects
(R26/27/28, R33)

SAFETY PRECAUTIONS
After contact with skin, wash immediately with plenty of water – Wear protective clothing and
gloves – In case of accident or if you feel unwell, seek medical advice immediately (show label
where possible) (S28, S36/37, S45)

Toxic effects   Irritant to skin, eyes, and upper respiratory tract.

First aid   Standard treatment for exposure by all routes (see pages 108-122), with
particular attention to the possible occurrence of convulsions.

Fire hazard   Extinguish fires with water spray, carbon dioxide, dry chemical powder, or
alcohol or polymer foam.

Spillage   See general section.
disposal

RSC *Chemical Safety Data Sheets* Vol. 4b, No. 103, 1991 gives extended coverage.

900. Nitrobenzene

Greenish-yellow crystals or yellow liquid with odour of bitter almonds; m.p. 6 ˚C; b.p. 210 ˚C;
slightly soluble in water.

RISKS
Very toxic by inhalation, in contact with skin, and if swallowed – Danger of cumulative effects
(R26/27/28, R33)

SAFETY PRECAUTIONS
After contact with skin, wash immediately with plenty of water – Wear protective clothing and
gloves – In case of accident or if you feel unwell, seek medical advice immediately (show label
where possible) (S28, S36/37, S45)

Limit values   OES short-term 2 p.p.m. (10 mg m^{-3}); long-term 1 p.p.m. (5 mg m^{-3}).

Toxic effects   Inhalation of the vapour or skin absorption may cause a burning sensation in the
chest, difficulty in breathing, cyanosis, and, in severe cases, unconsciousness.
Drowsiness, vomiting, cyanosis, and unconsciousness may follow ingestion with
a delay of several hours. The vapour and liquid are irritant to the eyes.

Hazardous   Nitrobenzene reacts with oxidizing agents and several other substances to
reactions   give detonable or explosive mixtures.

First aid   Standard treatment for exposure by all routes (see pages 108-122).

Fire hazard   Flash point 88 ˚C; Explosive limits lel 1.8%; Autoignition temperature 482 ˚C.
Extinguish fires with carbon dioxide, dry chemical powder, or alcohol-resistant foam.

Spillage   See general section.
disposal

RSC *Chemical Safety Data Sheets* Vol. 1, No. 79, 1988 and Vol. 4b, No. 104, 1991 give
extended coverage.

901. *m*-Nitrobenzenediazonium perchlorate

Hazardous reactions   Explosive, very sensitive to heat and shock.

902. 2-Nitrobenzonitrile

Off-white to tan powder; m.p. 107-109 °C.

Toxic effects   Irritant to skin, eyes, upper respiratory tract, and mucous membranes. May be harmful by inhalation, ingestion, or skin absorption.

Hazardous reactions   Explosion occurred.

First aid   Standard treatment for exposure by all routes (see pages 108-122).

Fire hazard   Extinguish fires with water spray, carbon dioxide, dry chemical powder, or alcohol or polymer foam.

Spillage disposal   See general section.

903. 2-Nitrobenzoyl chloride

Orange liquid; m.p. 25 °C; b.p. 148-149 °C at 9 mm Hg; decomposed by water.

Toxic effects   Extremely irritant and destructive to skin, eyes, upper respiratory tract, and mucous membranes. Inhalation may result in coughing, nausea, laryngitis, and could be fatal following spasm, chemical pneumonitis, and pulmonary oedema.

Hazardous reactions   Explosion of distillation residue.

First aid   Standard treatment for exposure by all routes (see pages 108-122).

Fire hazard   Flash point 110 °C (closed cup). Extinguish fires with carbon dioxide or dry chemical powder. Do NOT use water.

Spillage disposal   See general section.

904. 4-Nitrobenzoyl chloride

Yellow crystals with pungent odour.; m.p. 75 °C; b.p. 202-205 °C at 105 mm Hg; reacts with water forming benzoic and hydrochloric acids.

Toxic effects   These result mainly from its reaction with moisture on the tissues to form hydrochloric acid, which is the primary irritant. Thus irritation or burns may be caused at the point of contact.

First aid   Standard treatment for exposure by all routes (see pages 108-122).

Fire hazard   Extinguish fires with carbon dioxide or dry chemical powder. Do NOT use water.

Spillage   See general section.
disposal

905. 3-Nitrobenzoyl nitrate

Hazardous   Explodes if heated rapidly.
reactions

906. 4-Nitrobiphenyl

Toxic effects   Inhalation or absorption through the skin has been suggested as a cause of bladder tumours. Its use therefore is prohibited under the Carcinogenic Substances Regulations 1967. It is not considered appropriate to deal with the hazards more fully in this book.

907. Nitroethane

Colourless, oily liquid with pleasant odour; m.p. -90 ˚C; b.p. 114 ˚C; slightly soluble in water.

RISKS
Flammable – Harmful by inhalation – Harmful if swallowed (R10, R20, R22)

SAFETY PRECAUTIONS
Keep container in a well ventilated place – Avoid contact with eyes – Keep away from living quarters (S9, S25, S4)

Limit values   OES long-term 100 p.p.m. (310 mg m^{-3}).

Toxic effects The vapour irritates the respiratory system, but it is rapidly metabolized, and is a hazard to health only from prolonged exposure. The liquid and vapour irritate the eyes. Ingestion has given rise to liver and kidney damage in experimental animals.

Hazardous   May form explosive compounds with certain amines and alkalies.
reactions

First aid   Standard treatment for exposure by all routes (see pages 108-122).

Fire hazard   Flash point 28 ˚C; Explosive limits (lower) 3.4%; Autoignition temperature 414 ˚C. Extinguish fire with dry powder, carbon dioxide, or alcohol-resistant foam.

Spillage   See general section.
disposal

RSC *Chemical Safety Data Sheets* Vol. 1, No. 80, 1988 gives extended coverage.

908. 2-Nitroethanol

Colourless or light yellow liquid; m.p. below -80 °C; b.p. 194 °C; very soluble in water.

Toxic effects   Irritant to skin, eyes, upper respiratory tract, and mucous membranes. May be harmful by inhalation, Ingestion, or skin absorption.

Hazardous reactions   Explosion towards end of vacuum distillation.

First aid   Standard treatment for exposure by all routes (see pages 108-122).

Fire hazard   Flash point above 110 °C. Extinguish fires with water spray, carbon dioxide, dry chemical powder, or alcohol or polymer foam.

Spillage disposal   See general section.

909. Nitrogen trichloride

Yellow oil; m.p. explodes at 93 °C; insoluble in water.

Hazardous reactions   Wide variety of solids, liquids, and gases will initiate the violent and often explosive decomposition of nitrogen trichloride.

910. Nitrogen trifluoride

Colourless gas with pungent, 'mouldy' smell; m.p. -206.6 °C; b.p. -129 °C; insoluble in water.

Limit values   OES short-term 15 p.p.m. (45 mg m^{-3}); long-term 10 p.p.m. (30 mg m^{-3}).

Toxic effects   The gas irritates eyes, skin, and respiratory system severely. Prolonged exposure to low concentrations may cause mottling of the teeth and skeletal change.

Hazardous reactions   Explosion when adsorbed on activated charcoal at -100 °C. Violent explosion when mixtures with various gases are sparked. Shock exposure of the gas to heat, flame, or electric spark, or active contact with organic material, may cause fire and possibly explosion.

First aid   Standard treatment for exposure by all routes (see pages 108-122).

Spillage disposal   If the cylinder develops a leak it should be vented slowly in a well-ventilated fume cupboard until discharged.

911. Nitroguanidine

Crystalline solid; m.p. 246 °C with decomposition.

Toxic effects   Irritant to skin, eyes, upper respiratory tract, and mucous membranes. May be harmful by inhalation, ingestion, or skin absorption.

Hazardous reactions   Explosive though difficult to detonate.

First aid	Standard treatment for exposure by all routes (see pages 108-122).
Fire hazard	Extinguish fires with water spray or carbon dioxide.
Spillage disposal	See general section.

912. Nitroindane

(4-) Solid; (5-) yellow solid; m.p. (4-) 44-44.5 ˚C; (5-) 40-40.5 ˚C; b.p. (4-) 139 ˚C at 10 mm Hg; (5-)152 ˚C at 14 mm Hg.

| **Hazardous reactions** | Crude mixture of 4- and 5-isomers obtained by nitration of indane is explosive in final stages of preparation. |

913. Nitromethane

Colourless, oily liquid; m.p. -28.5 ˚C; b.p. 101 ˚C; slightly soluble in water.

RISKS
Heating may cause an explosion – Flammable – Harmful if swallowed (R5, R10, R22)

SAFETY PRECAUTIONS
In case of fire and/or explosion do not breathe fumes (S41)

Limit values	OES short-term 150 p.p.m. (375 mg m^{-3}); long-term 100 p.p.m. (250 mg m^{-3}).
Toxic effects	The vapour irritates the respiratory system, while there will be irritation and damage internally if the liquid is taken by mouth. It is mildly irritant to the skin and eyes.
Hazardous reactions	May explode by detonation, heat, or shock; addition of bases or acids renders it susceptible to initiation by detonator; risk of explosion in preparation of 2-nitroethanol after reaction with formaldehyde; risk of explosion on heating with hydrocarbons.
First aid	Standard treatment for exposure by all routes (see pages 108-122).
Fire hazard	Flash point 35 ˚C; Explosive limits (lower) 7.3%; Autoignition temperature 418 ˚C. Extinguish fire with dry powder, carbon dioxide, or alcohol-resistant foam. Shock or heat may cause nitromethane to explode.
Spillage disposal	See general section.

RSC *Chemical Safety Data Sheets* Vol. 1, No. 81, 1988 gives extended coverage.

914. *N*-Nitromethylamine

Crystalline solid; m.p. 38 ˚C; b.p. 158 ˚c; very soluble in water.

| **Hazardous reactions** | Decomposed explosively by concentrated sulfuric acid. |

915. 2-Nitronaphthalene

Yellow crystalline solid; m.p. 79 ˚C; b.p. 165 ˚C at 15 mm Hg; insoluble in water.

RISKS
May cause cancer (R45)

SAFETY PRECAUTIONS
Avoid exposure – Obtain special instructions before use – If you feel unwell, seek medical advice (show label where possible) (S53, S44)

Toxic effects   Irritant to skin and lungs. Metabolized to 2-naphthylamine, a carcinogen.

First aid   Standard treatment for exposure by all routes (see pages 108-122), with particular attention to the possible occurrence of convulsions or shock.

Fire hazard   Extinguish fires with water spray, carbon dioxide, dry chemical powder, or alcohol or polymer foam.

Spillage   See general section.
  disposal

RSC *Chemical Safety Data Sheets* Vol. 4b, No. 105, 1991 gives extended coverage.

916. 3-Nitroperchlorylbenzene

Hazardous   Explosive; shock-sensitive.
  reactions

917. Nitrophenols

Pale yellow or yellow crystals or powders; m.p. (*o-*) 45 ˚C; (*m-*) 96 ˚C; (*p-*) 114 ˚C; b.p. (*o-*)214.5 ˚C; (*m-*) 174 ˚C at 70 mm Hg; (*p-*) 279 ˚C with decomposition; sparingly soluble in water.

RISKS
Harmful by inhalation, in contact with skin, and if swallowed – Danger of cumulative effects (R20/21/22, R33)

SAFETY PRECAUTIONS
After contact with skin, wash immediately with plenty of water (S28)

Toxic effects   Excessive intake by inhalation of the dust, skin absorption, or ingestion may cause irritation, headache, drowsiness, and cyanosis. Effects may be cumulative. Assumed to irritate and injure the eyes.

Hazardous   Violent reaction with potassium hydroxide, and reaction product from
  reactions   *o*-nitrophenol and chlorosulfuric acid decomposed violently; *p*-nitrophenol reacts vigorously with diethyl phosphate.

First aid   Standard treatment for exposure by all routes (see pages 108-122).

Fire hazard   Extinguish fires with water spray, carbon dioxide, dry chemical powder, or alcohol or polymer foam.

Spillage   See general section.
  disposal

918. 2-Nitrophenylacetyl chloride

Hazardous   Distillation residue liable to explode.
reactions

919. *p*-Nitrophenylhydrazine

Orange-yellow powder; m.p. 156-158 ˚C; sparingly soluble in water.

Toxic effects   Irritant to skin, eyes, upper respiratory tract, and mucous membranes.

First aid   Standard treatment for exposure by all routes (see pages 108-122).

Fire hazard   Extinguish fires with water spray, carbon dioxide, dry chemical powder, or alcohol or polymer foam.

Spillage   See general section.
disposal

920. 3-Nitrophthalic acid

Pale yellow crystalline solid; m.p. 218 ˚C.

Hazardous   Eruptive decomposition in nitration of phthalic anhydride.
reactions

921. 1-Nitropropane

Colourless, oily liquid; m.p. -108 ˚C; b.p. 131.6 ˚C; slightly soluble in water.

RISKS
Flammable – Harmful by inhalation, in contact with skin, and if swallowed (R10, R20/21/22)

SAFETY PRECAUTIONS
Keep container in a well ventilated place (S9)

Limit values   OES long-term 25 p.p.m. (90 mg m^{-3}).

Toxic effects   Irritant to the eyes, skin, and mucous membranes. Inhalation can cause headache, dizziness, nausea, vomiting, diarrhoea, and muscular incoordination.

Hazardous   Reacts violently with certain substances and may explode on heating.
reactions

First aid   Standard treatment for exposure by all routes (see pages 108-122).

Fire hazard   Flash point 49 ˚C; Autoignition temperature 420 ˚C. Extinguish fires with water spray, carbon dioxide, dry chemical powder, or alcohol foam.

Spillage   See general section.
disposal

922. 2-Nitropropane

Colourless liquid with a mild odour; m.p. -93 ˚C; b.p. 120 ˚C; slightly soluble in water.

RISKS
May cause cancer – Flammable – Harmful by inhalation and if swallowed (R45, R10, R20/22)

SAFETY PRECAUTIONS
Avoid exposure – Obtain special instructions before use – Keep container in a well ventilated place – If you feel unwell, seek medical advice (show label where possible) (S53, S9, S44)

Limit values	OES short-term 20 p.p.m. (72 mg m^{-3}) under review; long-term 10 p.p.m. (36 mg m^{-3}) under review.
Toxic effects	2-Nitropropane is a suspected human carcinogen and exposure by all routes should be kept to the minimum. It is irritant to the eyes but not to the skin.
Hazardous reactions	2-Nitropropane may explode on heating. Dried salts from nitropropane and certain alkalies are explosive.
First aid	Standard treatment for exposure by all routes (see pages 108-122).
Fire hazard	Flash point 28 ˚C (open cup); Explosive limits 2.6-11.0%; Autoignition temperature 428 ˚C. Extinguish fires with carbon dioxide, dry chemical powder, or alcohol-resistant foam.
Spillage disposal	See general section.

RSC *Chemical Safety Data Sheets* Vol. 1, No. 82, 1988 and Vol. 4b, No. 106, 1991 gives extended coverage.

923. 1-Nitroso-2-naphthol

Yellow-brown powder; m.p. 106-108 ˚C with decomposition; insoluble in water.

Toxic effects	The dust irritates the respiratory system, eyes, and skin. In the latter case it may cause dermatitis. Assumed to be irritant and poisonous if taken by mouth.
Hazardous reactions	Becomes unstable on heating and may ignite spontaneously.
First aid	Standard treatment for exposure by all routes (see pages 108-122).
Fire hazard	Extinguish fires with water spray, carbon dioxide, dry chemical powder, or alcohol or polymer foam.
Spillage disposal	See general section.

924. Nitrosophenols

Hazardous reactions	The *o*-isomer explodes on heating or in contact with concentrated acids. Barrels of the *p*-isomer heated spontaneously and caused a fire: this tendency increases after compaction. These materials are now supplied wet with 10% water.

925. Nitrosyl chloride

Reddish-brown gas with irritant, penetrating smell; m.p. - 64.5 °C; b.p. -5.8 °C; decomposes on contact with water or moisture.

Toxic effects	The gas is intensely irritating to skin, eyes, and mucous membranes but provides good warning of its dangers because of its penetrating odour.
Hazardous reactions	Cold sealed tube containing nitrosyl chloride, platinum wire, and traces of acetone exploded on warming up.
First aid	Standard treatment for exposure by all routes (see pages 108-122).
Spillage disposal	Surplus gas or leaking cylinder may be vented slowly into a water-fed scrubbing tower or column in a fume cupboard, or into a fume cupboard served by such a tower.

926. Nitrosyl fluoride

Colourless gas; m.p. -134 °C; b.p. -56 °C; decomposed by water.

Toxic effects	Highly irritating to skin, eyes, and mucous membranes.
Hazardous reactions	Reaction of mixture with unspecified haloalkene in pressure vessel at -78 °C caused it to rupture when moved; incandescent reactions with various substances; explodes on mixing with oxygen difluoride.

927. Nitrosyl perchlorate

Hazardous reactions	Below 100 °C it slowly decomposes, but at about 115-120 °C it speeds up and explodes. Explodes on contact with pinene; it ignites acetone and ethanol and then explodes; with ether there is gassing followed by explosion; small amounts of primary aromatic amines are ignited on contact while explosions result with larger amounts; urea ignites on stirring with it; reaction mixture with phenyl isocyanate and pentaammineazidocobalt(III) perchlorate exploded when stirring stopped.

928. Nitrosylsulfuric acid

White to pale yellow moist crystals; m.p. 73.5 °C with decomposition; decomposed by water.

Toxic effects	Extremely destructive to tissues of the skin, eyes, upper respiratory tract, and gastro-intestinal tract. Can be absorbed through the skin. Inhalation may be fatal following spasm, inflammation, and pulmonary oedema.
Hazardous reactions	In preparation from sulfur dioxide and nitric acid, absence of dinitrogen tetroxide may result in explosion; explosion occurred during plant-scale diazotization of a dinitroaniline hydrochloride; explosion during plant-scale diazotization of 6-chloro-2,4-dinitroaniline.
First aid	Standard treatment for exposure by all routes (see pages 108-122).

Fire hazard   Use extinguishing medium, but NOT water, appropriate for surrounding conditions.

Spillage   See general section.
  disposal

929. Nitrosyl tribromide

Toxic effects   Severely irritant to skin, eyes, and mucous membranes.

Hazardous   Powdered sodium antimonide ignites when dropped into the vapour.
  reactions

930. 5-Nitrotetrazole

Hazardous   An evaporated solution of the sodium salt exploded after 2 weeks. The silver
  reactions   and mercury salts are explosive.

931. Nitrotoluenes

o-Nitrotoluene is a yellow liquid; *m*-nitrotoluene a yellow or brown-yellow, liquid or crystalline solid and *p*-nitrotoluene is a pale yellow or yellow, crystalline solid; m.p. (*m*-) 15 ˚C; (*p*-) 52 ˚C; b.p. (*o*-) 221.7 ˚C; (*m*-) 233 ˚C; (*p*-) 238.5 ˚C; all isomers are slightly soluble in water.

RISKS
Toxic by inhalation, in contact with skin, and if swallowed – Danger of cumulative effects (R23/24/25, R33)

SAFETY PRECAUTIONS
After contact with skin, wash immediately with plenty of water – Wear suitable gloves – If you feel unwell, seek medical advice (show label where possible) (S28, S37, S44)

Limit values   OES short-term 10 p.p.m. (60 mg m^{-3}); long-term 5 p.p.m. (30 mg m^{-3}).

Toxic effects   Inhalation of vapour or dust, ingestion, or skin absorption may cause difficulty in breathing, cyanosis, and, in severe cases, unconsciousness. Assumed to injure the eyes.

Hazardous   Distillation residues exploded; mixtures of *p*-isomer with oleum decomposed
  reactions   at 52 ˚C.

First aid   Standard treatment for exposure by all routes (see pages 108-122).

Fire hazard   Flash point 106 ˚C (closed cup); Explosive limits (*o*-) lel 2.2%; (*p*-) lel 1.6%; Autoignition temperature (*o*-) 305 ˚C; (*p*-) 390 ˚C. Extinguish fires with water spray, carbon dioxide, dry chemical powder, or alcohol or polymer foam.

Spillage   See general section.
  disposal

RSC *Chemical Safety Data Sheets* Vol. 4b, Nos. 107 and 108 give extended coverage on the *o*- and *p*-isomers.

932. 5-Nitro-*o*-toluidine

Bright yellow powder; m.p. 104-107 ˚C; almost insoluble in water.

RISKS
Toxic by inhalation, in contact with skin, and if swallowed – Danger of cumulative effects (R23/24/25, R33)

SAFETY PRECAUTIONS
After contact with skin, wash immediately with plenty of water – Wear protective clothing and gloves – If you feel unwell, seek medical advice (show label where possible) (S28, S36/37, S44)

Toxic effects   Mutagenic and carcinogenic in animal studies.

First aid   Standard treatment for exposure by all routes (see pages 108-122), but special attention may have to be paid to convulsions and shock.

Fire hazard   Extinguish fires with water spray, carbon dioxide, dry chemical powder, or alcohol or polymer foam.

Spillage   See general section.
  disposal

RSC *Chemical Safety Data Sheets* Vol. 4b, No. 109, 1991 gives extended coverage.

933. Nitrourea

Crystalline solid; m.p. 158-159 ˚C with decomposition; soluble in hot water.

Hazardous   Unstable explosive. Can react vigorously with oxidizing materials.
  reactions

First aid   Standard treatment for exposure by all routes (see pages 108-122).

934. Nitrous acid

Exists only as a solution, pale blue in colour.

Hazardous   Reacts violently or explosively with several substances.
  reactions

935. Nitryl chloride

Colourless gas; m.p. -145 ˚C; b.p. -14.3 ˚C; decomposed by water.

Toxic effects   Irritant to skin, eyes, and mucous membranes. Toxic by inhalation.

Hazardous   Reacts violently with ammonia or sulfur trioxide even at -75 ˚C; it attacks
  reactions   organic matter rapidly, sometimes explosively.

936. Nitryl fluoride

Colourless gas with pungent odour; m.p. -166 ˚C; b.p. -72.4 ˚C; rapidly hydrolysed in water to form nitric and hydrofluoric acids.

Toxic effects Severe irritant to skin, eyes, and mucous membranes. Toxic by inhalation.

Hazardous reactions Reaction with various metals causes incandescence or glowing to occur, and hydrogen explodes at 200 ˚C.

937. Nitryl hypochlorite

Hazardous reactions Reacts explosively with alcohols, ethers, and most organic materials. Explosive reactions with metal halides if above -40 ˚C.

938. Nitryl hypofluorite

Toxic effects Irritant to skin, eyes, and mucous membranes.

Hazardous reactions Dangerously explosive as solid, liquid, or gas. Ignition in gas phase with ammonia, nitrous oxide, or hydrogen sulfide; it may explode on contact with alcohol, ether, aniline, or grease.

939. Nonacarbonyldiiron

Gold flakes.

Toxic effects The toxicity has not been thoroughly investigated, but it is harmful by inhalation, ingestion, or skin absorption.

First aid Standard treatment for exposure by all routes (see pages 108-122).

Fire hazard Flash point 35 ˚C (commercial grade); Autoignition temperature 93 ˚C (commercial grade). Extinguish fires with water spray, carbon dioxide, dry chemical powder, or alcohol or polymer foam.

Spillage disposal See general section.

940. Nonane

Colourless liquid; m.p. -53 ˚C; b.p. 151 ˚C; insoluble in water.

Toxic effects Irritant to skin, eyes, upper respiratory tract, and mucous membranes. Narcotic in high concentrations.

First aid Standard treatment for exposure by all routes (see pages 108-122).

Fire hazard Flash point 31 ˚C; Explosive limits 0.8-2.9%; Autoignition temperature 205 ˚C. Extinguish fires with carbon dioxide, dry chemical powder, or alcohol or polymer foam.

Spillage disposal See general section.

941. Octachloronaphthalene

Pale yellow solid with aromatic odour; m.p. 197-198 ˚C; b.p. 440 ˚C; insoluble in water.

Limit values   OES short-term 0.3 mg m^{-3}; long-term 0.1 mg m^{-3}.

Toxic effects   Produces acne-like skin rash. Absorption into the body results in liver damage.

First aid   Standard treatment for exposure by all routes (see pages 108-122).

Spillage disposal   See general section.

942. Octane

Colourless liquid; m.p. -56.8 ˚C; b.p. 125.8 ˚C; insoluble in water.

RISKS
Highly flammable (R11)

SAFETY PRECAUTIONS
Keep container in a well ventilated place – Keep away from sources of ignition – No Smoking – Do not empty into drains – Take precautionary measures against static discharges (S9, S16, S29, S33)

Limit values   OES short-term 375 p.p.m. (1800 mg m^{-3}); long-term 300 p.p.m. (1450 mg m^{-3}).

Toxic effects   Liquid and high concentrations of vapour are irritating to the eyes. The liquid will defat and irritate the skin. The vapour irritates the upper respiratory tract and in high concentrations can cause narcosis and death.

First aid   Standard treatment for exposure by all routes (see pages 108-122).

Fire hazard   Flash point 13 ˚C; Explosive limits 1-6.5%; Autoignition temperature 206 ˚C. Extinguish small fires with carbon dioxide, dry chemical powder, or alcohol-resistant foam. Use the foam for large fires.

Spillage disposal   See general section.

RSC *Chemical Safety Data Sheets* Vol. 1, No. 83, 1988 gives extended coverage.

943. 1-Octanol

Viscous colourless liquid; m.p. -15 ˚C; b.p. 196 ˚C; very slightly soluble in water.

Toxic effects   Irritant to skin, eyes, upper respiratory tract, and mucous membranes. May cause allergic reactions.

First aid   Standard treatment for exposure by all routes (see pages 108-122).

Fire hazard   Flash point 81 ˚C; Autoignition temperature 273 ˚C. Extinguish fires with carbon dioxide, dry chemical powder, or alcohol or polymer foam.

Spillage disposal   See general section.

944. Oleic acid

Clear colourless liquid; m.p. 16.3 ˚C; b.p. 286 ˚C at 100 mm Hg; insoluble in water.

Toxic effects   Irritant to skin, eyes, and all mucous membranes.

First aid   Standard treatment for exposure by all routes (see pages 108-122).

Fire hazard   Extinguish fires with carbon dioxide or dry chemical powder.

Spillage   See general section.
disposal

945. Oleum

Colourless to yellow, viscous, fuming liquid; m.p. -11 ˚C to 35 ˚C according to concentration; reacts violently with water.

RISKS
Reacts violently with water – Causes severe burns – Irritating to respiratory system (R14, R35, R37)

SAFETY PRECAUTIONS
In case of contact with eyes, rinse immediately with plenty of water and seek medical advice – Never add water to this product (S26, S30)

Toxic effects   The vapour irritates all parts of the respiratory system. The vapour irritates and the liquid burns the eyes severely. The vapour and liquid burn the skin. If swallowed there would be most severe internal irritation and damage.

Hazardous   These are covered under sulfur trioxide.
reactions

First aid   Standard treatment for exposure by all routes (see pages 108-122).

Fire hazard   Use dry chemical powder to extinguish fires. Do NOT use water.

Spillage   See general section.
disposal

946. Osmic acid

Colourless or pale yellow crystals with pungent odour; m.p. 40 ˚C; b.p. 130 ˚C (sublimes); soluble in water.

RISKS
Very toxic by inhalation, in contact with skin, and if swallowed – Causes burns (R26/27/28, R34)

SAFETY PRECAUTIONS
Keep container tightly closed and in a well ventilated place – In case of contact with eyes, rinse immediately with plenty of water and seek medical advice – In case of accident or if you feel unwell, seek medical advice immediately (show label where possible) (S7/9, S26, S45)

Limit values   OES short-term 0.0006 p.p.m. (0.006 mg m^{-3}) (as Os); long-term 0.0002 p.p.m. (0.002 mg m^{-3}) (as Os).

Toxic effects	The vapour irritates all parts of the respiratory system with continued exposure causing disturbances of vision. The vapour, solid, and solution irritate and burn the eyes severely. The acid and its solution burn the skin. Continued skin contact results in dermatitis and ulceration. If taken by mouth there would be severe internal irritation and damage.
Hazardous reactions	Amorphous form, prepared by low temperature dehydration, is pyrophoric; explosion with 1-methylimidazole.
First aid	Standard treatment for exposure by all routes (see pages 108-122).
Fire hazard	Use extinguishing medium appropriate to surroundings.
Spillage disposal	See general section.

RSC *Chemical Safety Data Sheets* Vol. 4b, No. 110, 1991 gives extended coverage.

947. Osmium

Blue-grey powder; m.p. approx. 2700 ˚C; b.p. 5027 ˚C; insoluble in water.

Limit values	OES short-term 0.0006 p.p.m. (0.006 mg m^{-3}) (as Os); long-term 0.0002 p.p.m. (0.002 mg m^{-3}) (as Os).
Toxic effects	Irritant to skin, eyes, upper respiratory tract, and mucous membranes. Skin contact may lead to dermatitis.
Hazardous reactions	Incandesces in phosphorus vapour.
First aid	Standard treatment for exposure by all routes (see pages 108-122).
Fire hazard	Use dry chemical powder to extinguish fires.
Spillage disposal	See general section.

948. Oxalates

Most oxalates are colourless solids; ammonium, sodium, and potassium oxalates are soluble in water.

RISKS
Harmful in contact with skin and if swallowed (R21/22)

SAFETY PRECAUTIONS
Keep out of reach of children – Avoid contact with skin and eyes (S2, S24/25)

Toxic effects	If swallowed there would be severe internal pain followed by collapse.
First aid	Standard treatment for exposure by all routes (see pages 108-122).
Fire hazard	For most oxalates use water spray, carbon dioxide, dry chemical powder, or alcohol or polymer foam.
Spillage disposal	See general section.

949. Oxalic acid

Colourless crystals; m.p. 186-187 ˚C with decomposition; soluble in water.

RISKS
Harmful in contact with skin and if swallowed (R21/22)

SAFETY PRECAUTIONS
Keep out of reach of children – Avoid contact with skin and eyes (S2, S24/25)

Limit values   OES short-term 2 mg m^{-3}; long-term 1 mg m^{-3}.

Toxic effects   The dust irritates the respiratory system. The dust and solutions irritate the eyes. If swallowed there would be severe internal pain followed by collapse.

Hazardous reactions   Dry mixture of oxalic acid and sodium chlorite exploded on addition of water.

First aid   Standard treatment for exposure by all routes (see pages 108-122).

Fire hazard   Extinguish fires using carbon dioxide, dry chemical powder, or alcohol or polymer foam.

Spillage disposal   See general section.

950. Oxalyl chloride

Colourless, fuming liquid with acrid odour; m.p. -10 to -8 ˚C; b.p. 64 ˚C; reacts vigorously with water forming hydrochloric and oxalic acids.

RISKS
Harmful in contact with skin and if swallowed (R21/22)

SAFETY PRECAUTIONS
Keep out of reach of children – Avoid contact with skin and eyes (S2, S24/25)

Toxic effects   The vapour irritates the respiratory system severely. The vapour and liquid irritate the eyes. The liquid irritates the skin and may cause burns; must be assumed to be very irritant and poisonous if taken by mouth.

Hazardous reactions   Mixture with potassium or with dimethyl sulfoxide is explosive.

First aid   Standard treatment for exposure by all routes (see pages 108-122).

Fire hazard   Extinguish fires with carbon dioxide or dry chemical powder. Do NOT use water.

Spillage disposal   See general section.

951. Oxodisilane

Hazardous reactions   Ignites spontaneously in air.

952. Oxosilane

Hazardous reactions   Ignites spontaneously in air or chlorine.

953. Oxygen

Colourless gas; m.p. -219 °C; b.p. -183 °C.

Toxic effects   Inhalation of high concentrations (greater than 75 molar per cent) can lead to cramps, nausea, dizziness, respiratory difficulties, convulsions, and even death.

Hazardous reactions   There have been many hazardous incidents with liquid and gaseous oxygen. Leakage of the latter into unventilated spaces to cause oxygen enrichment by only 3-4% above the normal atmospheric level of 21% dramatically increases fire risk.

First aid   If inhaled at high concentrations remove to fresh air. If not breathing give them artificial respiration.

Fire hazard   Use water spray to keep cylinder cool or move it away from the area.

Spillage disposal   See general section.

954. Oxygen difluoride

Colourless gas; m.p. -223.8 °C; b.p. -144.8 °C; reacts slowly with water.

Toxic effects   Irritant to skin, eyes, upper respiratory tract, and mucous membranes. On inhalation produces pulmonary oedema or congestion. Appearance of symptoms may be delayed.

Hazardous reactions   An explosive hazard when reacted with several inorganic compounds.

First aid   Standard treatment for exposure by all routes (see pages 108-122).

955. Ozone

Colourless gas or dark blue liquid with characteristic odour; m.p. -193 °C; b.p. -111 °C; almost insoluble in water.

RISKS
Very toxic by inhalation (R26)

Limit values   OES short-term 0.3 p.p.m. (0.6 mg m^{-3}); long-term 0.1 p.p.m. (0.2 mg m^{-3}).

Toxic effects	The gas irritates the upper respiratory system strongly and may cause headache. High concentrations have caused death by lung congestion in animals.
Hazardous reactions	Solid and liquid ozone are highly explosive; reacts explosively with many compounds or produces compounds which themselves can be explosive.
First aid	Standard treatment for exposure by all routes (see pages 108-122).
Fire hazard	Use water spray to keep cylinders cool.

956. Palladium compounds

Toxic effects	Limited information available on toxicity.
Fire hazard	Choice of extinguishant depends on particular compound involved. In most cases extinguishant appropriate to the surroundings would be suitable.
Spillage disposal	See general section.

957. Paracetamol

Odourless, white, crystalline powder; m.p. about 170 °C; slightly soluble in water.

Toxic effects	Although used as an analgesic and antipyretic, it can cause systemic effects if absorbed into the body. Experimental carcinogen and tumourigen.

958. Paraffin wax

Colourless or white, translucent, odourless mass; m.p. approx 50-57 °C; insoluble in water.

Limit values	OES short-term 6 mg m^{-3}; long-term 2 mg m^{-3}.
Toxic effects	Experimental tumourigen on implantation. Repeated skin contact may cause irritation.
First aid	Standard treatment for exposure by all routes (see pages 108-122).
Fire hazard	Flash point 199 °C; Autoignition temperature 245 °C. Extinguish fires with carbon dioxide or dry chemical powder.
Spillage disposal	See general section.

959. Paraformaldehyde

White, crystalline powder; m.p. 163-165 °C with decomposition to form formaldehyde; slowly soluble in cold water, more readily in hot.

Toxic effects	The dust irritates the eyes, skin, and respiratory system and is irritant and damaging if taken internally. May cause allergic respiratory and skin

reactions. Inhalation can be fatal following spasm, chemical pneumonitis, and pulmonary oedema.

First aid   Standard treatment for exposure by all routes (see pages 108-122).

Fire hazard   Flash point 70 °C; Explosive limits 7-73%; Autoignition temperature 300 °C. Extinguish fire with water spray, dry chemical powder, carbon dioxide, or foam.

Spillage   See general section.
disposal

RSC *Chemical Safety Data Sheets* Vol. 3, No. 29, 1990 gives extended coverage of formaldehyde.

960. Paraldehyde

Colourless liquid with characteristic, aromatic odour; b.p. 128 °C; soluble in water.

RISKS
Highly flammable (R11)

SAFETY PRECAUTIONS
Keep container in a well ventilated place – Keep away from sources of ignition – No Smoking – Do not empty into drains – Take precautionary measures against static discharges (S9, S16, S29, S33)

Toxic effects   The vapour has narcotic effects, and large doses taken internally cause prolonged unconsciousness, respiratory difficulty, and pulmonary oedema.

Hazardous   Reaction with nitric acid to form glyoxal is liable to become violent.
reactions

First aid   Standard treatment for exposure by all routes (see pages 108-122).

Fire hazard   Flash point 36 °C; Autoignition temperature 238 °C. Extinguish fire with water spray, dry powder, or carbon dioxide.

Spillage   See general section.
disposal

961. Pentaboranes

(9) Colourless gas or liquid with a bad odour; (11) liquid; m.p. (9) -46.6 °C; (11) -123 °C; b.p. (9) 60 °C; (11) 63 °C; (9) hydrolyses in water after long heating; (11) hydrolyses in water.

Toxic effects   (9) Inhalation of low concentrations causes behavioural changes and loss of judgement. At higher concentrations it can affect the central nervous system.

Hazardous B_5H_9 ignites spontaneously if impure; forms shock-sensitive solution in
reactions   solvents containing carbonyl, ether, or ester functional groups and/or halogen substituents. B_5H_{11} ignites in air.

First aid   Standard treatment for exposure by all routes (see pages 108-122).

Fire hazard   Flash point (9) 30 °C (closed cup); Explosive limits (lower) (9) 0.42%. (9) Extinguish fires with water spray, carbon dioxide, or dry chemical powder.

Spillage   See general section.
disposal

962. Pentachloroethane

Colourless, heavy liquid with a chloroform-like odour; m.p. -29 ˚C; b.p. 162 ˚C; immiscible with water.

RISKS
Very toxic by inhalation and in contact with skin (R26/27)

SAFETY PRECAUTIONS
Keep locked up – In case of insufficient ventilation, wear suitable respiratory equipment – In case of accident or if you feel unwell, seek medical advice immediately (show label where possible) (S1, S38, S45)

Toxic effects	The vapour irritates the eyes, nose, and lungs and may cause drowsiness, giddiness, headache, and in high concentrations, unconsciousness. Continuous breathing of low concentrations of vapour over a long period may cause jaundice by action on the liver. It may also affect the nervous system and the blood. Assumed to be poisonous if taken by mouth.
Hazardous reactions	Mixture with potassium may explode after short delay.
First aid	Standard treatment for exposure by all routes (see pages 108-122).
Fire hazard	Flash point 75 ˚C. Extinguish fires with water spray, carbon dioxide, dry chemical powder, or alcohol or polymer foam.
Spillage disposal	See general section.

963. Pentachlorophenol

Colourless to yellow, crystalline solid with a phenolic (carbolic) odour. Technical grade is dark grey to brown; m.p. 191 ˚C; b.p. 309 ˚C with decomposition; insoluble in water. The sodium salt is a buff powder or flaked solid, soluble in water.

RISKS
Toxic by inhalation, in contact with skin, and if swallowed (R23/24/25)

SAFETY PRECAUTIONS
After contact with skin, wash immediately with plenty of water – Wear suitable protective clothing and eye/face protection – If you feel unwell, seek medical advice (show label where possible) (S28, S36/39, S44)

Limit values	OES short-term 1.5 mg m^{-3}; long-term 0.5 mg m^{-3}.
Toxic effects	The dusts irritate the nose and eyes. Absorption through the skin or ingestion may cause accelerated breathing, feverishness, and muscular weakness; in severe cases, convulsions and unconsciousness may follow. Skin contact may cause dermatitis.
First aid	Standard treatment for exposure by all routes (see pages 108-122).
Fire hazard	Extinguish fires with water spray, carbon dioxide, dry chemical powder, or alcohol or polymer foam.
Spillage disposal	See general section.

964. Penta-1,3-diyne

Liquid; b.p. 54-56 °C.

Hazardous reactions   Explodes on distillation.

965. Pentaerythritol

White, crystalline solid; m.p. 262 °C; b.p. 276 °C at 30 mm Hg; slightly soluble in water.

Limit values   OES short-term 20 mg m^{-3} (total inhalable dust); long-term 10 mg m^{-3} (total inhalable dust), 5 mg m^{-3} (respirable dust).

Toxic effects   Irritant to skin and eyes. May be harmful if absorbed into the body.

First aid   Standard treatment for exposure by all routes (see pages 108-122).

Fire hazard   Extinguish fires with water spray, carbon dioxide, dry chemical powder, or alcohol or polymer foam.

Spillage disposal   See general section.

966. Pentaerythritol tetraacrylate

Crystalline solid; m.p. 83-84 °C.

RISKS
Irritating to eyes and skin – May cause sensitization by skin contact (R36/38, R43)

SAFETY PRECAUTIONS
In case of contact with eyes, rinse immediately with plenty of water and seek medical advice – Wear eye/face protection (S26, S39)

Toxic effects   Irritant to the eyes. Very little other information available.

First aid   Standard treatment for exposure by all routes (see pages 108-122).

Fire hazard   Extinguish fires with water spray, carbon dioxide, dry chemical powder, or alcohol or polymer foams.

Spillage disposal   See general section.

RSC *Chemical Safety Data Sheets* Vol. 3, No. 47, 1990 gives extended coverage.

967. Pentaerythritol triacrylate

White semi-solid.

RISKS
Irritating to eyes and skin – May cause sensitization by skin contact (R36/38, R43)

SAFETY PRECAUTIONS
Wear eye/face protection (S39)

Toxic effects   Irritant to skin and eyes. Extremely corrosive to upper respiratory tract causing burning feeling, coughing, dyspnoea, nausea, and vomiting. Harmful by ingestion and skin absorption.

First aid   Standard treatment for exposure by all routes (see pages 108-122).

Fire hazard   Flash point above 110 ˚C (closed cup). Extinguish fires with water spray, carbon dioxide, dry chemical powder, or alcohol or polymer foams.

Spillage disposal   See general section.

RSC *Chemical Safety Data Sheets* Vol. 3, No. 48, 1990 gives extended coverage.

968. Pentaethylenehexamine

Amber liquid; m.p. -30 ˚C; b.p. 180-280 ˚C at 5.04 mm Hg; soluble in water.

RISKS
Causes burns – May cause sensitization by skin contact (R34, R43)

SAFETY PRECAUTIONS
In case of contact with eyes, rinse immediately with plenty of water and seek medical advice – Wear suitable protective clothing, gloves, and eye/face protection (S26, S36/37/39)

Toxic effects   Irritant to skin, eyes, respiratory tract, and all mucous membranes. May produce hypersensitivity and can be absorbed through the skin. Chronic exposure may result in gastric disturbances, anaemia, or weight loss.

First aid   Standard treatment for exposure by all routes (see pages 108-122).

Fire hazard   Flash point 166 ˚C. Extinguish fires with water spray, carbon dioxide, dry chemical powder, or alcohol from extinguishants.

Spillage disposal   See general section.

RSC *Chemical Safety Data Sheets* Vol. 3, No. 49, 1990 gives extended coverage.

969. Pentafluoroguanidine

Colourless liquid or gas; b.p. -1.1 ˚C.

Hazardous reactions   Extremely explosive.

970. Pentane

Colourless liquid.; b.p. 30 ˚C; almost insoluble in water.

RISKS
Highly flammable (R11)

SAFETY PRECAUTIONS
Keep container in a well ventilated place – Keep away from sources of ignition – No Smoking – Do not empty into drains – Take precautionary measures against static discharges (S9, S16, S29, S33)

Limit values   OES short-term 750 p.p.m. (2250 mg m^{-3}); long-term 600 p.p.m. (1800 mg m^{-3}).

Toxic effects   The vapour is narcotic in high concentrations.

First aid   Standard treatment for exposure by all routes (see pages 108-122).

Fire hazard   Flash point -49 ˚C; Explosive limits 1.4-8%; Autoignition temperature 309 ˚C. Extinguish fire with foam, dry powder, or carbon dioxide.

Spillage   See general section.
disposal

971. Pentane-2,4-dione

Colourless or slightly yellow liquid with pleasant smell; b.p. 140 ˚C; slightly soluble in water.

RISKS
Flammable – Harmful if swallowed (R10, R22)

SAFETY PRECAUTIONS
When using do not smoke – Do not breathe vapour – Avoid contact with skin and eyes (S21, S23, S24/25)

Toxic effects   The vapour may irritate the respiratory system. The liquid irritates the skin and eyes. If swallowed, may cause internal irritation and more severe damage.

First aid   Standard treatment for exposure by all routes (see pages 108-122).

Fire hazard   Flash point 34 ˚C. Extinguish fire with water spray, dry powder, or carbon dioxide.

Spillage   See general section.
disposal

972. Pentan-2-one

Colourless liquid with acetone/ether-like odour; b.p. 102 ˚C; slightly soluble in water.

Limit values   OES short-term 250 p.p.m. (875 mg m^{-3}); long-term 200 p.p.m. (700 mg m^{-3}).

Toxic effects   The vapour may irritate the eyes and respiratory system; it is narcotic in high concentrations. The liquid will irritate the eyes and is assumed to be irritant and narcotic if taken internally.

| **Fire hazard** | Flash point 7 ˚C; Explosive limits 1.5-8%; Autoignition temperature 505 ˚C. Extinguish fire with water spray, dry powder, or carbon dioxide. |
| **Spillage disposal** | See general section. |

973. Pentan-3-one

Colourless liquid; m.p. -39.9 ˚C; b.p. 102 ˚C; slightly soluble in water.

RISKS
Highly flammable (R11)

SAFETY PRECAUTIONS
Keep container in a well ventilated place – Keep away from sources of ignition – No Smoking – Take precautionary measures against static discharges (S9, S16, S33)

Limit values	OES short-term 250 p.p.m. (875 mg m^{-3}); long-term 200 p.p.m. (700 mg m^{-3}).
Toxic effects	May be harmful by inhalation, ingestion, or skin absorption. Prolonged exposure can cause nausea, dizziness, headache, narcotic effect, and dermatitis.
First aid	Standard treatment for exposure by all routes (see pages 108-122).
Fire hazard	Flash point 12 ˚C; Explosive limits (lower) 1.6%; Autoignition temperature 451 ˚C. Extinguish fires with carbon dioxide, dry chemical powder, or alcohol or polymer foam.
Spillage disposal	See general section.

974. Pent-2-en-4-yn-3-ol

| **Hazardous reactions** | Distillation residue exploded at over 90 ˚C. |

975. Pentyl acetate

Colourless liquids with pear-like odour. The commercial grade consists mainly of 3-methylbutyl acetate; b.p. (n-) 148 ˚C; (commercial grade) 128-132 ˚C; slightly soluble in water.

RISKS
Flammable (R10)

SAFETY PRECAUTIONS
Do not breathe vapour (S23)

Limit values	OES short-term 150 p.p.m. (800 mg m^{-3}); long-term 100 p.p.m. (530 mg m^{-3}).
Toxic effects	High concentrations of the vapour irritate the eyes and may cause headache and fatigue. Must be considered harmful if taken by mouth.
First aid	Standard treatment for exposure by all routes (see pages 108-122).

Fire hazard   Flash point (n-) 25 ˚C; (commercial grade) 23 ˚C; Explosive limits 1-7.5% for both; Autoignition temperature (n-) 379 ˚C; (commercial grade) 380 ˚C. Extinguish fire with dry powder or carbon dioxide.

Spillage   See general section.
disposal

976. Peracetic acid

The 40% solution is a colourless liquid with an acrid odour; m.p. 0-1 ˚C; b.p. 105 ˚C; soluble in water.

RISKS
(> 10%) Heating may cause an explosion – Harmful if swallowed – Causes burns (R5, R22, R34)

SAFETY PRECAUTIONS
(> 10%) Keep in a cool place – Take off immediately all contaminated clothing – Wear suitable protective clothing (S3, S27, S36)

Toxic effects   Strongly irritant to skin, eyes, respiratory tract, and all mucous membranes. Systemic effects can follow inhalation, ingestion, or skin contact. Possible animal carcinogen.

Hazardous   Reacts violently or explosively with a wide variety of substances.
reactions

First aid   Standard treatment for exposure by all routes (see pages 108-122).

Fire hazard   Flash point 40.5 ˚C; Autoignition temperature 200 ˚C. Severe fire hazard. Fires should be fought from explosion-resistant locations. Water should be used to fight the fire and cool containers.

Spillage   See general section.
disposal

RSC *Chemical Safety Data Sheets* Vol. 3, No. 50, 1990 gives extended coverage.

977. Perchlorates

The perchlorates of ammonium, magnesium, sodium, and potassium are colourless, crystalline solids; all except that of potassium being readily soluble in water.

Toxic effects   The dust and strong solutions will irritate the skin, eyes, and respiratory system. They are also irritant and harmful if taken internally.

Hazardous   Ammonium perchlorate is explosive when dry. Mixtures of inorganic
reactions   perchlorates with combustible materials are readily ignited. Organic perchlorates are self-contained explosives.

First aid   Standard treatment for exposure by all routes (see pages 108-122).

Spillage   See general section. Site of spillage should be washed thoroughly to remove
disposal   all oxidant, which is liable to render any organic matter (particularly paper, textiles, and wood) with which it comes into contact dangerously combustible when dry. Clothing wetted with the solution should be washed thoroughly.

RSC *Chemical Safety Data Sheets* Vol. 2, No. 90, 1989 gives extended coverage of the potassium salt.

978. Perchloric acid

Colourless liquid. Commonly sold commercially as 60-62% or 73% constant boiling acid; miscible with water.

RISKS
(> 50%) Heating may cause an explosion – Contact with combustible material may cause fire – Causes severe burns (R5, R8, R35)
(≥ 10%, ≤ 50%) Causes burns (R34)

SAFETY PRECAUTIONS
(> 50%) Do not breathe vapour – In case of contact with eyes, rinse immediately with plenty of water and seek medical advice – Wear suitable protective clothing (S23, S26, S36)
(≥ 10%, ≤ 50%) Do not breathe vapour – After contact with skin, wash immediately with plenty of water – Wear suitable protective clothing (S23, S28, S36)

Toxic effects	The acid burns eyes and skin severely. Assumed to cause severe internal irritation and damage if taken by mouth.
Hazardous reactions	There is a very long history of accidents with perchloric acid with many organic and inorganic substances. The distillation of 70-72% commercial perchloric acid under vacuum, concentrates the acid to 75% and over, and these stronger acids are liable to explode on heating. Wooden benches which have been subject to occasional perchloric acid spills have been known to explode on percussion.
First aid	Standard treatment for exposure by all routes (see pages 108-122).
Fire hazard	Use dry chemical powder to extinguish fires.
Spillage disposal	See general section.

979. Perchlorylbenzene

Hazardous reactions	After an interval, a mixture with aluminium chloride suddenly exploded.

980. Perchloryl fluoride

Colourless gas with characteristic sweet odour; m.p. -146 °C; b.p. -46.8 °C; very soluble in water.

Limit values	OES short-term 6 p.p.m. (28 mg m^{-3}); long-term 3 p.p.m. (14 mg m^{-3}).
Toxic effects	Moderately toxic by inhalation leading to anaemia, anorexia, and cyanosis due to formation of methaemoglobin. Can be absorbed through the skin. Irritant to all mucous membranes.
Hazardous reactions	Violent or explosive reactions with a wide variety of substances.
First aid	Standard treatment for exposure by all routes (see pages 108-122).

981. 1-Perchlorylpiperidine

Hazardous   Has exploded on storage, on heating, or in contact with piperidine.
reactions

982. Periodic acid

White to off-white powder; m.p. 122 °C; miscible with water.

Toxic effects   Irritating and corrosive to skin, eyes, upper respiratory tract, and mucous
membranes. Inhalation may be fatal following spasm, chemical pneumonitis,
and pulmonary oedema.

Hazardous   1.5 M solutions in dimethyl sulfoxide explode after short interval; evaporation
reactions   with triethylammonium hydroxide led to explosion.

First aid   Standard treatment for exposure by all routes (see pages 108-122).

Fire hazard   Use water spray to extinguish fires.

Spillage   See general section.
disposal

983. Permanganic acid

Exists only in solution; soluble in water.

Hazardous   Likely to explode at room temperature; solution of acid produced by action of
reactions   sulfuric acid on permanganates will explode on contact with some other
organic substances.

984. Peroxodisulfuric acid

Hazardous   An extreme oxidant, which may cause aniline, benzene, ethanol, ether,
reactions   nitrobenzene, or phenol to explode on contact.

985. Peroxomonophosphoric acid

Toxic effects   Very irritating and corrosive to skin, eyes, and mucous membranes.

Hazardous   80% solution may ignite organic matter.
reactions

986. Peroxomonosulfuric acid

Toxic effects   Strong irritant.

Hazardous   Explosions have occurred alone, or with a variety of organic compounds; it
reactions   rapidly carbonizes wool or cellulose, while cotton is ignited.

987. Peroxonitric acid

Toxic effects   Very irritating and corrosive to tissues.

Hazardous reactions   Decomposes explosively at -30 ˚C.

988. Peroxybenzoic acid

Crystalline solid with an acrid odour; m.p. 42 ˚C; b.p. explodes at 80-100 ˚C; insoluble in water.

Toxic effects   Moderately irritating to skin, eyes, and mucous membranes. Experimental tumourigen.

Hazardous reactions   Explodes weakly on heating.

First aid   Standard treatment for exposure by all routes (see pages 108-122).

989. Peroxyhexanoic acid

Liquid; m.p. 15 ˚C; b.p. 41-43 ˚C.

Hazardous reactions   Explodes and ignites on rapid heating.

990. Petroleum spirit

Petroleum spirits are supplied in a variety of fractions boiling between 30 ˚C and 160 ˚C. All are colourless liquids whose smell varies with the volatilities of the fractions; insoluble in water.

Toxic effects   Inhalation of high concentrations of the vapour, particularly of the lower boiling fractions, can cause intoxication, headache, nausea, and coma. The liquids irritate the eyes and skin contact results in defatting of the area of contact, increasing the risk of dermatitis from other agents. If taken by mouth they may cause burning sensation, vomiting, diarrhoea, and drowsiness.

First aid   Standard treatment for exposure by all routes (see pages 108-122).

Fire hazard   Flash point (lower fractions) below -17 ˚C; Explosive limits approx. 1-6%; Autoignition temperature range from about 290 ˚C. Extinguish fire with foam, dry powder, or carbon dioxide.

Spillage disposal   See general section.

RSC *Chemical Safety Data Sheets* Vol. 1, No. 78, 1988 gives extended coverage on naphtha.

991. Phenacyl bromide

Colourless to greenish crystals; m.p. 50 ˚C; b.p. 135 ˚C at 18 mm Hg; insoluble in water.

Toxic effects   Extremely destructive to skin, eyes, upper respiratory tract, and mucous membranes. Harmful if inhaled, ingested, or absorbed through the skin. Inhalation may be fatal as a result of spasm, inflammation, and pulmonary oedema.

First aid   Standard treatment for exposure by all routes (see pages 108-122).

Fire hazard   Flash point above 110 ˚C. Extinguish fires with carbon dioxide, dry chemical powder, or alcohol or polymer foam.

Spillage disposal   See general section.

992. Phenacyl chloride

Colourless to yellow crystals or powder; m.p. 54 ˚C; b.p. 244-245 ˚C; insoluble in water.

Toxic effects   The vapour and dust irritate the respiratory system, eyes, and skin. Assumed to be very irritant if taken by mouth. Harmful following inhalation, ingestion, or skin absorption.

First aid   Standard treatment for exposure by all routes (see pages 108-122).

Fire hazard   Extinguish fires with carbon dioxide or dry chemical powder.

Spillage disposal   See general section.

993. Phenanthroline

White powder; m.p. 114-117 ˚C; b.p. above 300 ˚C; slightly soluble in water.

First aid   Standard treatment for exposure by all routes (see pages 108-122).

Fire hazard   Extinguish fires with water spray, carbon dioxide, dry chemical powder, or alcohol or polymer foam.

Spillage disposal   See general section.

994. Phenol

Colourless to pink, crystalline substance with distinctive odour; m.p. 40.6 °C; b.p. 181.9 °C; somewhat soluble in water.

RISKS
Toxic in contact with skin and if swallowed – Causes burns (R24/25, R34)

SAFETY PRECAUTIONS
Keep out of reach of children – After contact with skin, wash immediately with plenty of water – If you feel unwell, seek medical advice (show label where possible) (S2, S28, S44)

Limit values	OES short-term 10 p.p.m. (38 mg m^{-3}); long-term 5 p.p.m. (19 mg m^{-3}).
Toxic effects	The vapour irritates the respiratory system and eyes. The inhalation of vapour over a long period may cause digestive disturbances, nervous disorders, skin eruptions, and damage to the liver and kidneys. Skin contact causes softening and whitening followed by the development of painful burns; its rapid absorption through the skin may cause headache, dizziness, rapid, and difficult breathing, weakness, and collapse. Dermatitis may result from prolonged contact with weak solutions. If taken by mouth, it causes severe burns, abdominal pain, nausea, vomiting, and internal damage.
Hazardous reactions	Violent or explosive reactions with a variety of substances.
First aid	Standard treatment for exposure by all routes (see pages 108-122).
Fire hazard	Flash point 80 °C (closed cup); Explosive limits 1.7-8.6%; Autoignition temperature 715 °C. Extinguish fires with water, carbon dioxide, dry chemical powder, or alcohol foam.
Spillage disposal	See general section.

RSC *Chemical Safety Data Sheets* Vol. 3, No. 51, 1990 gives extended coverage.

995. Phenothiazine

Yellow crystalline powder; m.p. 180-185 °C; b.p. 371 °C; almost insoluble in water.

Toxic effects	Irritant to skin, eyes, upper respiratory tract, and mucous membranes. Harmful by inhalation, ingestion, and skin absorption. May cause convulsions.
First aid	Standard treatment for exposure by all routes (see pages 108-122).
Fire hazard	Extinguish fires with water spray, carbon dioxide, dry chemical powder, or alcohol or polymer foam.
Spillage disposal	See general section.

996. Phenoxyacetylene

Liquid; b.p. 64-65 °C at 20 mm Hg.

Hazardous   Samples rapidly heated in sealed tubes to about 100 °C exploded.
reactions

997. Phenylacetic acid

White chips with honey-like odour; m.p. 77-78 °C; b.p. 265 °C; slightly soluble in cold water, freely soluble in hot water.

Toxic effects   Experimental teratogen. May cause irritation. May be harmful if absorbed into the body.

First aid   Standard treatment for exposure by all routes (see pages 108-122).

Fire hazard   Extinguish fires with water spray, carbon dioxide, dry chemical powder, or alcohol or polymer foam.

Spillage   See general section.
disposal

998. Phenylacetonitrile

Colourless to yellow liquid; m.p. -24 °C; b.p. 234 °C; insoluble in water.

Toxic effects   Inhalation of the vapour may cause pallor, faintness, headache, and possibly vomiting. The liquid irritates the eyes and may irritate the skin. If taken by mouth, internal irritation and poisoning must be assumed.

Hazardous   When sodium hypochlorite solution was used to destroy acidified residues, a
reactions   violent explosion occurred.

First aid   Standard treatment for exposure by all routes (see pages 108-122).

Fire hazard   Flash point 101 °C. Extinguish fires with water spray, carbon dioxide, dry chemical powder, or alcohol or polymer foam.

Spillage   See general section.
disposal

999. Phenylchlorodiazirine

Pale yellow oil.

Hazardous   Is about three times as shock-sensitive as glyceryl nitrate.
reactions

1000. Phenylenediamines

p-Phenylenediamine is a pale mauve to mauve, crystalline powder. *m*-Phenylenediamine is a colourless to brown or black, crystalline solid; m.p. (*m*-) 62.8 °C; (*p*-) 139.7 °C; (*o*-) 100-102 °C; b.p. (*m*-) 287 °C; (*p*-) 267 °C; (*o*-) 256-258 °C; the *m*-isomer is moderately soluble in water, the *p*-isomer only sparingly so.

RISKS
Toxic by inhalation, in contact with skin, and if swallowed – May cause sensitization by skin contact (R23/24/25, R43)

SAFETY PRECAUTIONS
After contact with skin, wash immediately with plenty of water – If you feel unwell, seek medical advice (show label where possible) (S28, S44)

Limit values   OES long-term (*p*-isomer) 0.1 mg m^{-3}.

Toxic effects   Eye irritation and injury follow contact. May irritate the skin causing dermatitis; the skin becomes blackened. Assumed to be poisonous if taken by mouth. Inhalation of dust may cause bronchial asthma.

First aid   Standard treatment for exposure by all routes (see pages 108-122).

Fire hazard   Extinguish fires with water spray, carbon dioxide powder, or alcohol or polymer foam.

Spillage   See general section.
   disposal

1001. Phenyl ethanol

Colourless liquid with floral odour; m.p. -25.8 °C; b.p. 219.5-220 °C; slightly soluble in water.

Toxic effects   Irritant to skin and eyes. Harmful on inhalation or ingestion.

First aid   Standard treatment for exposure by all routes (see pages 108-122).

Fire hazard   Flash point 102 °C. Extinguish fires with water spray, carbon dioxide, or dry chemical powder.

Spillage   See general section.
   disposal

1002. Phenyl glycidyl ether

Colourless liquid; m.p. 3.5 °C; b.p. 245 °C; sparingly soluble in water.

RISKS
Harmful in contact with skin – May cause sensitization by skin contact (R21, R43)

SAFETY PRECAUTIONS
Avoid contact with skin and eyes (S24/25)

Toxic effects   Irritant to skin, eyes, and upper respiratory tract.

First aid   Standard treatment for exposure by all routes (see pages 108-122).

Fire hazard   Flash point >112 ˚C.

Spillage   See general section.
  disposal

1003. Phenylhydrazine and hydrochloride

Monoclinic prisms or oil; m.p. (base) 19.6 ˚C;(salt) 249 ˚C with decomposition; b.p. (base) 243.5 ˚C; the base is insoluble and the salt is soluble in water.

RISKS
Toxic by inhalation, in contact with skin, and if swallowed – Irritating to eyes (R23/24/25, R36)

SAFETY PRECAUTIONS
After contact with skin, wash immediately with plenty of water – If you feel unwell, seek medical advice (show label where possible) (S28, S44)

Limit values   OES short-term 10 p.p.m. (45 mg m^{-3}) under review; long-term 5 p.p.m. (20 mg m^{-3}) under review.

Toxic effects Absorption resulting from inhalation, skin contact, or ingestion may result in blood and liver damage, giving rise to nausea, vomiting, and jaundice. Dermatitis or hypersensitization may follow skin contact. Considered injurious to the eyes.

Hazardous   Reacts violently with several compounds.
  reactions

First aid   Standard treatment for exposure by all routes (see pages 108-122).

Fire hazard   Flash point 88.9 ˚C (closed cup). Extinguish fires with carbon dioxide, dry chemical powder, or alcohol or polymer foam.

Spillage   See general section.
  disposal

1004. *N*-Phenylhydroxylamine

Colourless, crystalline solid; m.p. 81.2 ˚C; slightly soluble in water.

Toxic effects Irritant to skin and eyes. Toxic by inhalation, ingestion, and skin absorption.

Hazardous   Reduction of nitrobenzene with zinc gives pyrophoric residues, and the
  reactions   hydrochloride salt may decompose in storage.

First aid   Standard treatment for exposure by all routes (see pages 108-122).

1005. Phenyl isocyanate

Colourless to light yellow liquid; m.p. -30 ˚C; b.p. 162-163 ˚C; decomposed by water.

Toxic effects Vapour or mist is irritating to skin, eyes, upper respiratory tract, and mucous membranes. May cause asthma or allergic reaction. Prolonged contact can be fatal.

First aid	Standard treatment for exposure by all routes (see pages 108-122).
Fire hazard	Flash point 55 ˚C. Extinguish fires with carbon dioxide or dry chemical powder.
Spillage disposal	See general section.

1006. *N*-Phenyl-2-naphthylamine

Off-white, crystalline powder; m.p. 107-100 ˚C; b.p. 395.5 ˚C; insoluble in water.

Toxic effects	Irritant to skin, eyes, upper respiratory tract, and mucous membranes. Harmful if inhaled, ingested, or absorbed through the skin. Suspected human carcinogen.
First aid	Standard treatment for exposure by all routes (see pages 108-122).
Fire hazard	Extinguish fires with water spray, carbon dioxide, dry chemical powder, or alcohol or polymer foam.
Spillage disposal	See general section.

1007. Phenyl oxirane

Colourless liquid; m.p. -36.6 ˚C; b.p. 194.2 ˚C; almost insoluble in water.

RISKS
May cause cancer – Harmful in contact with skin – Irritating to eyes (R45, R21, R36)

SAFETY PRECAUTIONS
Avoid exposure – Obtain special instructions before use – If you feel unwell, seek medical advice (show label where possible) (S53, S44)

Toxic effects	Irritant to skin, eyes, and upper respiratory tract. Absorption into the body may cause central nervous system depression and hepatic lesions. Mutagenic, teratogenic, and carcinogenic in experimental studies.
Hazardous reactions	Will polymerize exothermically or react vigorously with compounds with a labile hydrogen atom in the presence of catalysts.
First aid	Standard treatment for exposure by all routes (see pages 108-122).
Fire hazard	Flash point 74 ˚C (open cup); Autoignition temperature 498 ˚C. Extinguish fires with carbon dioxide, dry chemical powder, or alcohol or polymer foam.
Spillage disposal	See general section.

RSC *Chemical Safety Data Sheets* Vol. 4b, No. 118, 1991 gives extended coverage.

1008. Phenyl phosphine

Clear, colourless liquid; b.p. 160-161 °C; sparingly soluble in water.

Toxic effects	Inhalation may cause irritation, dyspnoea, nausea, vomiting, back pain, chills, and pulmonary oedema.
Hazardous reactions	Pyrophoric in air.
First aid	Standard treatment for exposure by all routes (see pages 108-122).
Fire hazard	Flash point 73 °C.
Spillage disposal	See general section.

1009. Phenylsilver

Solid; m.p. 74 °C with decomposition.

Hazardous reactions	Explodes on warming to room temperature or on light friction.

1010. 5-Phenyltetrazole

Crystalline solid; m.p. 217-218 °C with decomposition.

Hazardous reactions	Explodes on attempted distillation.

1011. Phosgene

A pale yellow liquid or a colourless gas with a musty smell. Solution of the gas in toluene is available commercially; m.p. -118 °C; b.p. 8.3 °C; slightly soluble in and decomposed by water.

RISKS
Very toxic by inhalation (R26)

SAFETY PRECAUTIONS
Keep container tightly closed and in a well ventilated place – Avoid contact with skin and eyes – In case of accident or if you feel unwell, seek medical advice immediately (show label where possible) (S7/9, S24/25, S45)

Limit values	OES long-term 0.1 p.p.m. (0.4 mg m^{-3}).
Toxic effects	The gas produces delayed secretion of fluid into the lung (pulmonary oedema) when inhaled and there may be delay of several hours before effects develop. These include breathlessness, cyanosis, and the coughing up of frothy fluid.

Hazardous reactions	Mixture with potassium is shock-sensitive.
First aid	Standard treatment for exposure by all routes (see pages 108-122).
Fire hazard	Use water to keep cylinders cool and remove from immediate area if possible.
Spillage disposal	Surplus gas or leaking cylinder can be vented slowly into a water-fed scrubbing tower or a column in a fume cupboard, or into a fume cupboard served by such a tower. For the solution in toluene: see general section.

1012. Phosphine

Colourless gas with smell somewhat like that of rotting fish; m.p. -134 °C; b.p. -88 °C; slightly soluble in water.

Limit values	OES short-term 0.3 p.p.m. (0.4 mg m^{-3}).
Toxic effects	The effects of inhalation have been stated to be pain in chest, sensation of coldness, weakness, vertigo, shortness of breath, lung damage, convulsions, coma, and death.
Hazardous reactions	The impure gas ignites spontaneously in air, the liquified gas can be detonated; ignition occurs on contact with chlorine or bromine or their aqueous solutions; oxidized explosively by fuming nitric acid.
First aid	Standard treatment for exposure by all routes (see pages 108-122).
Fire hazard	Impure phosphine must be regarded as spontaneously flammable. As the gas is supplied in a cylinder, turning off the valve will reduce any fire involving it. If possible, cylinders should be removed quickly from an area in which a fire has developed.
Spillage disposal	Surplus gas or leaking cylinder can be vented slowly to air in a safe, open area, or gas burnt-off through a suitable burner in a fume cupboard.

1013. Phosphinic acid

Colourless liquid (50% solution in water).

Toxic effects	Extremely destructive to skin, eyes, upper respiratory tract, and all mucous membranes. Harmful if inhaled, ingested, or absorbed through the skin. Inhalation may be fatal following spasm, chemical pneumonitis, and pulmonary oedema.
Hazardous reactions	Redox reaction with mercury(II) oxide is explosive.
First aid	Standard treatment for exposure by all routes (see pages 108-122).
Fire hazard	Extinguish fires with dry chemical powder.
Spillage disposal	See general section.

1014. Phosphonium perchlorate

Toxic effects   Strongly irritant to tissues. Toxic by inhalation and ingestion.

Hazardous   An explosive, sensitive to moist air, friction, or heat.
reactions

1015. Phosphoric acid

Colourless viscous liquid (88-93%) or moist white crystals (100%); m.p. 42.3 °C; miscible with water.

RISKS
(≥ 10%, ≤ 25%) Irritating to eyes (R36)
(>25%) Causes burns (R34)

SAFETY PRECAUTIONS
(≥ 10%, ≤ 25%) Avoid contact with eyes (S25)
(> 25%) In case of contact with eyes, rinse immediately with plenty of water and seek medical advice (S26)

Toxic effects   The liquid burns the eyes and skin severely. If taken by mouth there would be severe internal irritation and damage. It is irritating to the upper respiratory tract with a small risk of causing pulmonary oedema.

First aid   Standard treatment for exposure by all routes (see pages 108-122).

Fire hazard   Extinguish fires with carbon dioxide, dry chemical powder, or alcohol or polymer foam.

Spillage   See general section.
disposal

RSC *Chemical Safety Data Sheets* Vol. 3, No. 52, 1990 gives extended coverage.

1016. Phosphorus, red

Solid; m.p. sublimes at 416 °C; insoluble in water.

RISKS
Highly flammable – Explosive when mixed with oxidizing substances (R11, R16)

SAFETY PRECAUTIONS
Keep container tightly closed – In case of fire, use water (S7, S43)

Toxic effects   Red phosphorus is relatively harmless physiologically unless it contains the white allotrope. The vapour from ignited phosphorus irritates the nose, throat, lungs, and eyes.

Hazardous   Reacts violently or explosively with a wide variety of inorganic substances.
reactions

First aid   Standard treatment for exposure by all routes (see pages 108-122).

Fire hazard   Spontaneously flammable in air. Water is the best medium for fighting a phosphorus fire caused by its spontaneous ignition.

Spillage   See general section.
disposal

1017. Phosphorus, white (yellow)

Pale yellow, waxy, translucent solid; m.p. 44 ˚C; b.p. 280 ˚C; insoluble in water. Yellow phosphorus usually stored under water.

RISKS
Spontaneously flammable in air – Very toxic by inhalation and if swallowed – Causes severe burns (R17, R26/28, R35)

SAFETY PRECAUTIONS
Keep contents under water – In case of contact with eyes, rinse immediately with plenty of water and seek medical advice – After contact with skin, wash immediately with plenty of water – In case of accident or if you feel unwell, seek medical advice immediately (show label where possible) (S5, S26, S28, S45)

Limit values OES short-term 0.3 mg m^{-3} (yellow); long-term 0.1 mg m^{-3} (yellow).

Toxic effects The vapour from ignited phosphorus irritates the nose, throat, lungs, and eyes. Solid white phosphorus burns the skin and eyes and causes severe internal damage if taken by mouth. Continued absorption of small amounts can cause anaemia, intestinal weakness, pallor, bone, and liver damage.

Hazardous reactions Reacts violently or explosively with a wide variety of inorganic substances.

First aid Standard treatment for exposure by all routes (see pages 108-122).

Fire hazard Spontaneously flammable in air. Water is the best medium for fighting a phosphorus fire caused by its spontaneous ignition.

Spillage disposal See general section.

1018. Phosphorus(III) oxide

White, deliquescent, crystalline solid; m.p. 22.5 ˚C; b.p. 173 ˚C; reacts with water forming phosphoric acid.

Hazardous reactions Reacts violently with a variety of organic and inorganic substances.

1019. Phosphorus pentachloride

White to pale yellow, fuming, crystalline masses with pungent, unpleasant odour; sublimes at 160 ˚C; violently decomposed by water with formation of hydrochloric acid and phosphoric acid.

RISKS
Causes burns – Irritating to respiratory system (R34, R37)

SAFETY PRECAUTIONS
Keep container tightly closed and dry – In case of contact with eyes, rinse immediately with plenty of water and seek medical advice (S7/8, S26)

Limit values OES long-term 0.1 p.p.m. (1 mg m^{-3}).

Toxic effects	Vapour and dust severely irritate the mucous membranes and all parts of the respiratory system. Continued exposure to low concentrations of vapour may cause damage to lungs. The vapour severely irritates and the solid burns the eyes. The vapour and solid burn the skin. If taken by mouth there would be severe internal irritation and damage.
Hazardous reactions	Reacts violently or explosively with a variety of substances.
First aid	Standard treatment for exposure by all routes (see pages 108-122).
Fire hazard	Use extinguishing medium, but NOT water, appropriate to the surroundings.
Spillage disposal	See general section.

1020. Phosphorus pentafluoride

Colourless gas, fuming strongly in moist air, with a pungent smell; b.p. -85 °C; hydrolysed by water with formation of hydrogen fluoride and phosphoric acid.

Toxic effects	Because of its ready hydrolysis by moisture or water to hydrogen fluoride its toxic effects are similar to the latter. It irritates severely the eyes and respiratory system and may cause burns to the eyes. It irritates the skin and, if exposure has been considerable, painful burns may develop after an interval.
First aid	Exposure by all routes requires special treatment for hydrofluoric acid.
Fire hazard	Use extinguishing medium, but NOT water, appropriate to the surroundings.
Spillage disposal	Surplus gas or leaking cylinder can be vented slowly into a water-fed scrubbing tower or column in a fume cupboard, or into a fume cupboard served by such a tower.

1021. Phosphorus pentaoxide

White, crystalline, deliquescent powder; sublimes at 250 °C; reacts violently with water.

RISKS
Causes severe burns (R35)

SAFETY PRECAUTIONS
Do not breathe dust – In case of contact with eyes, rinse immediately with plenty of water and seek medical advice (S22, S26)

Toxic effects	The dust irritates all parts of the respiratory system, burns the eyes severely, and burns the skin. If taken by mouth, there would be severe internal irritation and damage.
Hazardous reactions	Reacts violently or explosively with a wide variety of substances.
First aid	Standard treatment for exposure by all routes (see pages 108-122).
Fire hazard	Use extinguishing medium, but NOT water, appropriate to the surroundings.
Spillage disposal	See general section.

1022. Phosphorus pentasulfide

Grey to yellow-green, crystalline solid; m.p. 286-290 ˚C; b.p. 514 ˚C; reacts with water.

RISKS
Highly flammable – Harmful by inhalation and if swallowed – Contact with water liberates toxic gas (R11, R20/22, R29)

Limit values   OES short-term 3 mg m^{-3}; long-term 1 mg m^{-3}.

Toxic effects   Irritant to skin, eyes, and upper resiratory tract. Harmful if inhaled, ingested, or absorbed through the skin giving rise to stomach pains, vomiting, and diarrheoa.

Hazardous reactions   Reacts with water, steam, or acids producing toxic and flammable gases. Reacts vigorously with oxidizing materials.

First aid   Standard treatment for exposure by all routes (see pages 108-122).

Fire hazard   Explosive limits lel 0.5%; Autoignition temperature 286 ˚C. Extinguish fires with carbon dioxide, dry chemical powder, or sand.

Spillage disposal   See general section.

1023. Phosphorus sesquisulfide

Yellow, crystalline solid; m.p. 173-175 ˚C; b.p. 407-408 ˚C; insoluble in water.

RISKS
Highly flammable – Harmful if swallowed (R11, R22)

SAFETY PRECAUTIONS
Keep container tightly closed – Keep away from sources of ignition – No Smoking – Avoid contact with skin and eyes (S7, S16, S24/25)

Toxic effects   Irritant to skin, eyes, and upper resiratory tract. Toxic by ingestion.

First aid   Standard treatment for exposure by all routes (see pages 108-122).

Fire hazard   Autoignition temperature 100 ˚C. Extinguish fires with carbon dioxide, dry chemical powder, or sand.

Spillage disposal   See general section.

1024. Phosphorus(III) and (V) sulfides

Yellow, crystalline powders with a peculiar odour; m.p. (v) 286-290 ˚C; b.p. (v) 513-515 ˚C; both react with water forming phosphorous acid and hydrogen sulfide.

Toxic effects   The dust irritates the mucous membranes and all parts of the respiratory system; it irritates the eyes and burns the skin. Assumed to cause severe internal irritation and damage, if taken by mouth.

Hazardous reactions	The pentasulfide ignites by friction, sparks, or flames; it heats and may ignite with limited amounts of water.
First aid	Standard treatment for exposure by all routes (see pages 108-122).
Fire hazard	Autoignition temperature (III) 282 ˚C; (V) 142 ˚C. Extinguish fire with dry powder, carbon dioxide, or sand.
Spillage disposal	See general section.

1025. Phosphorus tribromide

Colourless, fuming liquid; m.p. -40 ˚C; b.p. 175 ˚C; reacts violently with water, forming hydrobromic acid and phosphorous acid.

RISKS
Causes burns – Irritating to respiratory system (R34, R37)

SAFETY PRECAUTIONS
Keep container tightly closed and dry – In case of contact with eyes, rinse immediately with plenty of water and seek medical advice (S7/8, S26)

Toxic effects	Vapour severely irritates the mucous membranes and all parts of the respiratory system. The vapour irritates and the liquid burns the eyes. The vapour and liquid burn the skin. Assumed to cause severe internal irritation and damage if taken by mouth.
Hazardous reactions	Sodium floats on the liquid without reaction but the addition of a little water causes violent explosion; potassium ignites on contact with the liquid or vapour.
First aid	Standard treatment for exposure by all routes (see pages 108-122).
Fire hazard	Extinguish fires with dry chemical powder. Do NOT use water.
Spillage disposal	See general section.

1026. Phosphorus trichloride

Colourless, fuming liquid; m.p. -112 ˚C; b.p. 75 ˚C; violently decomposed by water, forming hydrochloric acid and phosphorous acid.

RISKS
Causes burns – Irritating to respiratory system (R34, R37)

SAFETY PRECAUTIONS
Keep container tightly closed and dry – In case of contact with eyes, rinse immediately with plenty of water and seek medical advice (S7/8, S26)

Limit values	OES short-term 0.5 p.p.m. (3 mg m^{-3}); long-term 0.2 p.p.m. (1.5 mg m^{-3}).
Toxic effects	Vapour severely irritates the mucous membranes and all parts of the respiratory system. The vapour and liquid burn the eyes and skin. Assumed to cause severe internal irritation and damage if taken by mouth.

Hazardous reactions	Reacts violently or explosively with a variety of substances: reacts violently with water with liberation of some diphosphane which ignites; aluminium powder incandesces.
First aid	Standard treatment for exposure by all routes (see pages 108-122).
Fire hazard	Use extinguishing medium, but NOT water, appropriate to the surroundings.
Spillage disposal	See general section.

1027. Phosphorus tricyanide

Hazardous reactions	Explosions occurred during vacuum sublimation.

1028. Phosphorus trifluoride

Colourless gas; m.p. -151 ˚C; b.p. -101 ˚C; slowly hydrolysed by water.

Toxic effects	Highly corrosive to all tissues. Can be absorbed through the skin.
Hazardous reactions	Ignites on contact with fluorine; explosion has occurred at low temperatures with dioxygen difluoride.
First aid	Standard treatment for exposure by all routes (see pages 108-122), with special reference to fluorides.

1029. Phosphoryl chloride

Colourless, fuming liquid; m.p. 1.25 ˚C; b.p. 107 ˚C; violently decomposed by water, forming hydrochloric acid and phosphoric acid.

RISKS
Causes burns – Irritating to respiratory system (R34, R37)

SAFETY PRECAUTIONS
Keep container tightly closed and dry – In case of contact with eyes, rinse immediately with plenty of water and seek medical advice (S7/8, S26)

Limit values	OES short-term 0.6 p.p.m. (3.6 mg m^{-3}); long-term 0.2 p.p.m. (1.2 mg m^{-3}).
Toxic effects	The vapour severely irritates the mucous membranes and all parts of the respiratory system; there may be sudden or delayed pulmonary oedema. The vapour severely irritates the eyes and skin. The liquid burns the eyes and skin. Assumed to cause severe internal irritation and damage if taken by mouth.
Hazardous reactions	There is a considerable delay in its reaction with water which ultimately becomes violent – closed or only slightly vented vessels therefore represent a hazard; may react explosively with several other substances.
First aid	Standard treatment for exposure by all routes (see pages 108-122).
Fire hazard	Use extinguishing medium, but NOT water, appropriate to the surroundings.
Spillage disposal	See general section.

1030. Phthalic anhydride

White, crystalline needles; m.p. 131 °C; dissolves in hot water forming phthalic acid.

RISKS
Irritating to eyes, respiratory system, and skin (R36/37/38)

Limit values OES short-term 4 p.p.m. (24 mg m^{-3}) under review; long-term 1 p.p.m. (6 mg m^{-3}) under review.

Toxic effects The dust irritates the eyes, skin, and respiratory system. Numerous symptoms of respiratory tract injury, including possible pulmonary sensitization, have followed inhalation. It will cause internal irritation if taken by mouth.

First aid Standard treatment for exposure by all routes (see pages 108-122).

Fire hazard Extinguish fires with water, carbon dioxide, dry chemical powder, or foam.

Spillage disposal See general section.

RSC *Chemical Safety Data Sheets* Vol. 3, No. 53, 1990 gives extended coverage.

1031. Phthalimide

White to light tan powder; m.p. 232-235 °C; sublimes on further heating; slightly soluble in water.

Toxic effects Irritant to skin, eyes, upper respiratory tract, and mucous membranes. Experimental teratogen.

First aid Standard treatment for exposure by all routes (see pages 108-122).

Fire hazard Extinguish fires with water spray, carbon dioxide, dry chemical powder, or alcohol or polymer foam.

Spillage disposal See general section.

1032. *m*-Phthalodinitrile

Crystalline solid; m.p. 141 °C; sublimes on further heating; slightly soluble in hot water.

Toxic effects Irritant to skin, eyes, and upper respiratory tract. Harmful if inhaled, ingested, or absorbed through the skin.

First aid Standard treatment for exposure by all routes (see pages 108-122).

Fire hazard Extinguish fires with water spray, carbon dioxide, dry chemical powder, or alcohol or polymer foam.

Spillage disposal See general section.

1033. Phthaloyl peroxide

Solid; m.p. 133.5 ˚C with decomposition.

Toxic effects	Irritant to skin, eyes, and mucous membranes.
Hazardous reactions	Detonatable by impact or by melting at 123 ˚C.

1034. Picric acid

Picric acid (yellow crystals) should be kept moist with not less than half its own weight of water. It is commonly used as an alcoholic solution in the laboratory; m.p. 120-122 ˚C.

Limit values	OES short-term 0.3 mg m^{-3}; long-term 0.1 mg m^{-3}.
Toxic effects	Skin contact may result in dermatitis. Poisonous if taken by mouth. Absorption through the skin or inhalation of dust over a long period may result in skin eruptions, headache, nausea, vomiting, or diarrhoea; the skin may become yellow.
Hazardous reactions	Explosive which is usually stored as water-wet paste; forms salts with many metals some of which are sensitive to heat, friction, or impact; contact of acid with concrete floors may form friction-sensitive calcium salt.
First aid	Standard treatment for exposure by all routes (see pages 108-122).
Fire hazard	Extinguish fires with water spray, carbon dioxide, dry chemical powder, or alcohol or polymer foam.
Spillage disposal	See general section.

1035. Piperazine and piperazine hydrate

Both take the form of colourless crystals; m.p. 106 ˚C and (hydrate) 44 ˚C; b.p. 146 ˚C; both are very soluble in water, the solutions being strongly alkaline.

RISKS
Causes burns (R34)

SAFETY PRECAUTIONS
In case of contact with eyes, rinse immediately with plenty of water and seek medical advice –
Wear suitable protective clothing (S26, S36)

Toxic effects	The vapours irritate the eyes and respiratory system. Inhalation of large quantities may be fatal as a result of pulmonary oedema. Repeated exposures may give rise to allergic reactions. The solutions irritate and may burn the eyes and skin. The solids and aqueous solutions will cause internal burning and damage if taken by mouth.
First aid	Standard treatment for exposure by all routes (see pages 108-122).
Fire hazard	Flash point 109 ˚C. Extinguish fires with water spray, dry chemical powder, or alcohol foam.
Spillage disposal	See general section.

RSC *Chemical Safety Data Sheets* Vol. 3, No. 54, 1990 gives extended coverage.

1036. Piperazine dihydrochloride

White powder.

Limit values	OES long-term 5 mg m^{-3}.
Toxic effects	Irritant to skin, eyes, upper respiratory tract, and mucous membranes. May cause allergic respiratory and skin reactions.
First aid	Standard treatment for exposure by all routes (see pages 108-122).
Fire hazard	Extinguish fires with water spray, carbon dioxide, dry chemical powder, or alcohol or polymer foam.
Spillage disposal	See general section.

RSC *Chemical Safety Data Sheets* Vol. 3, No. 54, 1990 gives extended coverage of piperazine.

1037. Piperidine

Colourless liquid with amine-like odour; b.p. 106 ˚C; miscible with water.

RISKS
Highly flammable – Toxic by inhalation and in contact with skin – Causes burns (R11, R23/24, R34)

SAFETY PRECAUTIONS
Keep away from sources of ignition – No Smoking – In case of contact with eyes, rinse immediately with plenty of water and seek medical advice – Take off immediately all contaminated clothing – If you feel unwell, seek medical advice (show label where possible) (S16, S26, S27, S44)

Limit values	OES long-term 1 p.p.m. (3.5 mg m^{-3}).
Toxic effects	The vapour irritates the eyes and respiratory system. The liquid irritates the eyes and skin and is assumed to be irritant and harmful if taken internally.
First aid	Standard treatment for exposure by all routes (see pages 108-122).
Fire hazard	Flash point 16 ˚C. Extinguish fire with dry powder or carbon dioxide.
Spillage disposal	See general section.

1038. Platinum

Silvery metal; m.p. 1769 ˚C; b.p. 3825 ˚C; insoluble in water.

Limit values	OES long-term 5 mg m^{-3}.
Toxic effects	May be harmful following inhalation or ingestion. May cause irritation.
Hazardous reactions	Used Pt catalysts present explosion risk; finely divided Pt and Pt sponge can react violently with some compounds.

First aid   Standard treatment for exposure by all routes (see pages 108-122).

Fire hazard   Use extinguishing medium appropriate to surroundings.

Spillage   See general section.
  disposal

1039. Polyethylene glycol

White crystals or clear, colourless, viscous liquid; soluble in water.

Toxic effects   May be irritant to skin and eyes. May be harmful if absorbed into the body.

First aid   Standard treatment for exposure by all routes (see pages 108-122).

Fire hazard   Autoignition temperature 305 °C. Extinguish fires with water spray, carbon dioxide, dry chemical powder, or alcohol or polymer foam.

Spillage   See general section.
  disposal

1040. Potassium

Soft, silvery white masses normally coated with a grey oxide or hydroxide skin; m.p. 63.2 °C; b.p. 765.5 °C; reacts violently with water with the formation of potassium hydroxide and hydrogen gas, which will ignite.

Toxic effects   The metal in contact with moisture on the skin, eyes, and other tissues can cause thermal and caustic burns. In the same way the potassium hydroxide resulting from the reaction of the metal with water can cause severe burning of the skin, eyes, and other tissues.

Hazardous   Potassium does not react with air or oxygen in complete absence of moisture,
  reactions   but in its presence oxidation becomes fast and melting and ignition takes place; prolonged but restricted access to air results in the formation of coatings of yellow superoxide on top of the monoxide – percussion or dry cutting of the metal brings traces of residual oil into contact with the superoxide and a very violent explosion occurs; the metal reacts vigorously with various organic and inorganic substances.

First aid   Standard treatment for exposure by all routes (see pages 108-122).

Fire hazard   A fire resulting from the ignition of potassium metal is best extinguished by smothering it with dry soda ash, dry graphite, powdered sodium chloride, or appropriate dry powder. Do NOT use water or carbon dioxide.

Spillage   See general section.
  disposal

RSC *Chemical Safety Data Sheets* Vol. 2, No. 79, 1989 gives extended coverage.

1041. Potassium amide

Yellow-green solid; m.p. 338 ˚C; sublimes at 400 ˚C; decomposed by water.

Hazardous   More violently reactive than sodium amide; reaction with water is violent and
reactions   ignition may occur; explodes when heated with potassium nitrite under vacuum.

1042. Potassium benzenehexoxide

Hazardous   Reacts violently with oxygen; explodes on heating in air or on contact with
reactions   water.

1043. Potassium bromate

White crystals or powder; m.p. 350 ˚C; soluble in water.

RISKS
Explosive when mixed with combustible material (R9)

SAFETY PRECAUTIONS
*Avoid contact with skin and eyes – Take off immediately all contaminated clothing (S24/25,
S27)*

Toxic effects   Inhalation may cause dyspnoea. Absorption from the lungs and also ingestion
may result in nausea, vomiting, abdominal pain, headache, unconsciousness,
and convulsions. Liver and kidney damage may also occur. It is irritating to
the skin and eyes.

Hazardous   Sulfur, selenium, and sulfur bromide all ignite with the salt; a heated mixture
reactions   with aluminium and dinitrotoluene evolves huge volumes of gas rapidly.

First aid   Standard treatment for exposure by all routes (see pages 108-122).

Fire hazard   Extinguish fires with flooding amounts of water.

Spillage   See general section.
disposal

RSC *Chemical Safety Data Sheets* Vol. 2, No. 81, 1989 gives extended coverage.

1044. Potassium chlorate

Colourless crystals or white powder; m.p. 368 ˚C; decomposes at 400 ˚C; soluble in water.

RISKS
*Explosive when mixed with combustible material – Harmful by inhalation and if swallowed (R9,
R20/22)*

SAFETY PRECAUTIONS
*Keep out of reach of children – Keep away from food, drink, and animal feeding stuffs – Keep
away from sources of ignition – No Smoking – Take off immediately all contaminated clothing
(S2, S13, S16, S27)*

Toxic effects   When taken internally may irritate the intestinal tract and kidneys. Otherwise it
is not a serious health hazard.

Hazardous reactions	Fabric gloves (wrongly used) became impregnated with the chlorate during handling operations and were subsequently ignited by cigarette ash; reacts vigorously or explosively with a wide variety of substances.
First aid	Standard treatment for exposure by all routes (see pages 108-122).
Fire hazard	Flood with water to extinguish fires.
Spillage disposal	See general section.

RSC *Chemical Safety Data Sheets* Vol. 2, No. 82, 1080 gives extended coverage.

1045. Potassium dioxide

Hazardous reactions	Stable when pure, impact-sensitivity of old peroxidized potassium now known to be due to presence of oil or organic residues; reacts violently with diselenium dichloride; explosions may occur with several organic and inorganic substances.

1046. Potassium ferricyanide

Orange-red, crystalline solid; decomposes on heating; very soluble in water.

Toxic effects	Irritant to skin and eyes. Harmful if inhaled or ingested, although the oral toxicity is low as the cyanide ion is bound.
Hazardous reactions	Reacts violently or explosively with a variety of inorganic substances.
First aid	Standard treatment for exposure by all routes (see pages 108-122).
Fire hazard	Use extinguishant appropriate to the surroundings.
Spillage disposal	See general section.

1047. Potassium ferrocyanide

Yellow, crystalline solid; m.p. 70 °C, but looses water at 60 °C; soluble in water.

Toxic effects	Irritant to skin and eyes. Harmful if inhaled or ingested, although the oral toxicity is low as the cyanide ion is bound.
Hazardous reactions	Reacts violently or explosively with a variety of inorganic substances.
First aid	Standard treatment for exposure by all routes (see pages 108-122).
Fire hazard	Use extinguishant appropriate to the surroundings.
Spillage disposal	See general section.

1048. Potassium fluoride

White, crystalline, deliquescent powder with sharp saline taste; m.p. 857 °C; b.p. 1505 °C; soluble in water.

RISKS
Toxic by inhalation, in contact with skin, and if swallowed (R23/24/25)

SAFETY PRECAUTIONS
Keep locked up and out of reach of children – In case of contact with eyes, rinse immediately with plenty of water and seek medical advice – If you feel unwell, seek medical advice (show label where possible) (S1/2, S26, S44)

Limit values   OES short-term 2.5 mg m^{-3} (as F).

Toxic effects   Toxic following inhalation, ingestion, or skin absorption. Highly irritating and corrosive to skin, eyes, upper respiratory tract, and all mucous membranes.

First aid   Standard treatment for exposure by all routes (see pages 108-122), with particular reference to fluorides.

Fire hazard   Extinguish fires with water fog.

Spillage   See general section.
 disposal

RSC *Chemical Safety Data Sheets* Vol. 2, No. 86, 1989 gives extended coverage.

1049. Potassium hexacyanoferrate

3-Ruby red crystals or orange red crystalline powder;4-lemon yellow crystals; both soluble in water.

Toxic effects   Irritating to eyes and skin. May be harmful if ingested in large quantities.

Hazardous   Contact of the hexacyanoferrate(4 -) with ammonia may be explosive; mixtures
 reactions   of the (4 -) salt with CrO_3 explode on heating above 196 °C or with $Cu(NO_3)_2$ at 220 °C; both of the salts mixed with sodium nitrite explode on heating.

First aid   Standard treatment for exposure by all routes (see pages 108-122).

Fire hazard   Use extinguishant appropriate to surroundings.

Spillage   See general section.
 disposal

1050. Potassium hydride

Crystalline solid; m.p. decomposed on heating; reacts readily with water liberating hydrogen gas.

Toxic effects   Extremely destructive to skin, eyes, upper respiratory tract, and all mucous membranes. Harmful on inhalation, ingestion, or skin absorption. Inhalation may be fatal following spasm, chemical pneumonitis, and pulmonary oedema.

Hazardous reactions	Ignites on contact with fluorine; reacts slowly with oxygen in dry air, more rapidly in moist, explosion with fluoroalkene at 12 ˚C.
First aid	Standard treatment for exposure by all routes (see pages 108-122).
Fire hazard	Use class D fire extinguishants only. Do NOT use water.
Spillage disposal	See general section.

1051. Potassium hydrogen sulfate

Colourless crystals or fused masses; m.p. 210 ˚C; decomposes on further heating; soluble in water.

Toxic effects	The solid and its solutions severely irritate and burn the eyes and skin. If taken by mouth there is severe internal irritation and damage.
First aid	Standard treatment for exposure by all routes (see pages 108-122).
Fire hazard	Extinguish fires with dry chemical powder.
Spillage disposal	See general section.

1052. Potassium hydroxide

White or slightly yellow sticks, flakes, powder, or pellets; m.p. 360 ˚C; b.p. 1320 ˚C; soluble in water.

RISKS
Causes severe burns (R35)

SAFETY PRECAUTIONS
Keep out of reach of children – In case of contact with eyes, rinse immediately with plenty of water and seek medical advice – Wear suitable gloves and eye/face protection (S2, S26, S37/39)

Limit values	OES short-term 2 mg m^{-3}.
Toxic effects	The solid and its solutions are strongly irritant and burn the eyes, skin, and other tissues. If taken by mouth there would be severe internal irritation and damage, possibly leading to collapse or coma.
Hazardous reactions	Reacts vigorously or explosively with a variety of substances.
First aid	Standard treatment for exposure by all routes (see pages 108-122).
Fire hazard	Use flooding amounts of water to fight fires.
Spillage disposal	See general section.

RSC *Chemical Safety Data Sheets* Vol. 2, No. 87, 1989 and Vol. 3, No. 58, 1990 give extended coverage.

1053. Potassium methylamide

Hazardous   Extremely hygroscopic and pyrophoric; may explode on contact with air.
reactions

1054. Potassium nitrate

Transparent, colourless crystals or white, crystalline powder; m.p. 334 °C; decomposes at 400 °C; soluble in water.

Toxic effects   Irritating to the skin, eyes, and upper respiratory tract. Ingestion of large amounts may have serious effects and may be fatal.

Hazardous   Mixtures with a wide variety of substances are explosive; in general, mixtures
reactions   with combustible materials or organic impurities are readily ignited, if finely divided they can be explosive.

First aid   Standard treatment for exposure by all routes (see pages 108-122).

Fire hazard   Mixtures of potassium nitrate and combustible materials are readily ignited; mixtures with finely divided combustible materials can react explosively. Use flooding amounts of water in the early stages of a fire.

Spillage   See general section.
disposal

RSC *Chemical Safety Data Sheets* Vol. 2, No. 88, 1989 gives extended coverage.

1055. Potassium nitrite

White or slightly yellow, deliquescent, crystalline solid; m.p. 387 °C; b.p. decomposes; soluble in water.

RISKS
Contact with combustible material may cause fire – Toxic if swallowed (R8, R25)

SAFETY PRECAUTIONS
If you feel unwell, seek medical advice (show label where possible) (S44)

Toxic effects   Irritating to the skin, eyes, and upper respiratory tract. Ingestion of large amounts may cause nausea, vomiting, cyanosis, collapse, and coma. May be carcinogenic on chronic oral exposure.

First aid   Standard treatment for exposure by all routes (see pages 108-122).

Fire hazard   Use water to extinguish fires.

Spillage   See general section. Also clothing wetted by a solution of the nitrite should be
disposal   washed thoroughly.

RSC *Chemical Safety Data Sheets* Vol. 2, No. 89, 1989 gives extended coverage.

1056. Potassium permanganate

Purple, crystalline solid; m.p. below 240 °C with decomposition; soluble in water.

RISKS
Contact with combustible material may cause fire – Harmful if swallowed (R8, R22)

SAFETY PRECAUTIONS
Keep out of reach of children (S2)

Toxic effects   Extremely destructive to skin, eyes, upper respiratory tract, and mucous membranes. Harmful if inhaled, ingested, or absorbed through the skin. Inhalation may be fatal following spasm, inflammation, and pulmonary oedema.

Hazardous reactions   Mixtures with a wide variety of organic and inorganic substances can react violently or explosively.

First aid   Standard treatment for exposure by all routes (see pages 108-122).

Fire hazard   Use water spray to extinguish fires.

Spillage disposal   See general section.

1057. Potassium peroxodisulfate

White, crystalline powder; m.p. 100 °C with decomposition; decomposes if heated above 50 °C with water.

Toxic effects   Irritant to skin and eyes. Harmful by ingestion and if inhaled as dust.

Hazardous reactions   Dry salt gives off oxygen rapidly at 100 °C, but at only 50 °C when wet.

First aid   Standard treatment for exposure by all routes (see pages 108-122).

Fire hazard   Use water spray to extinguish fires.

Spillage disposal   See general section.

1058. Propadiene

Colourless gas; m.p. -146 °C; b.p. -32 °C.

Hazardous reactions   May decompose explosively under a pressure of 2 bar (2×10^5 Pa).

First aid   Standard treatment for exposure by all routes (see pages 108-122).

Fire hazard   Explosive limits 2.1-13.0%. Use water spray to keep cylinder cool, or move it out of the immediate area.

1059. Propane

Colourless gas which burns with a smoky flame; b.p. -42 ˚C; sparingly soluble in water.

RISKS
Extremely flammable liquefied gas (R13)

SAFETY PRECAUTIONS
Keep container in a well ventilated place – Keep away from sources of ignition – No Smoking –
Take precautionary measures against static discharges (S9, S16, S33)

Toxic effects	The gas is anaesthetic in high concentrations but is not toxic.
First aid	Standard treatment for exposure by all routes (see pages 108-122).
Fire hazard	Flash point -104 ˚C; Explosive limits 2.2-9.5%; Autoignition temperature 468 ˚C. As the gas is supplied in a cylinder, turning off the valve will reduce any fire involving it; if possible, cylinders should be removed quickly from an area in which a fire has developed.
Spillage disposal	Surplus gas or leaking cylinder can be vented slowly to air in a safe, open area, or gas burnt-off through a suitable burner.

1060. 1,3-Propanesultone

Off-white solid or colourless liquid; m.p. 30-33 ˚C; b.p. 112 ˚C; moderately soluble in water.

RISKS
May cause cancer – Harmful in contact with skin and if swallowed (R45, R21/22)

SAFETY PRECAUTIONS
Avoid exposure – Obtain special instructions before use – If you feel unwell, seek medical
advice (show label where possible) (S53, S44)

Toxic effects	Irritant to skin, eyes, and upper respiratory tract. Animal carcinogen and teratogen.
First aid	Standard treatment for exposure by all routes (see pages 108-122).
Fire hazard	Flash point above 110 ˚C. Extinguish fires with water spray, carbon dioxide, dry chemical powder, or alcohol or polymer foam.
Spillage disposal	See general section. Disposal or destruction of the collected material must be done with great care.

RSC *Chemical Safety Data Sheets* Vol. 4b, No. 123, 1991 gives extended coverage.

1061. Propan-1-ol

Colourless liquid with alcoholic odour; m.p. -127 ˚C; b.p. 97 ˚C; miscible with water.

RISKS
Highly flammable (R11)

SAFETY PRECAUTIONS
Keep container tightly closed – Keep away from sources of ignition – No Smoking (S7, S16)

Limit values	OES short-term 250 p.p.m. (625 mg m^{-3}); long-term 200 p.p.m. (500 mg m^{-3}).

Toxic effects The vapour may irritate the eyes and respiratory system and may be narcotic in high concentrations. The liquid irritates the eyes and is narcotic if taken internally.

First aid Standard treatment for exposure by all routes (see pages 108-122).

Fire hazard Flash point 15 ˚C (open cup); Explosive limits 2.1-13.5%; Autoignition temperature 370 ˚C. Extinguish fire with water spray, dry powder, carbon dioxide, or alcohol-resistant foam.

Spillage disposal See general section.

RSC *Chemical Safety Data Sheets* Vol. 1, No. 84, 1988 gives extended coverage.

1062. Propan-2-ol

Colourless liquid with somewhat alcoholic odour; b.p. 82 ˚C; miscible with water.

RISKS
Highly flammable (R11)

SAFETY PRECAUTIONS
Keep container tightly closed – Keep away from sources of ignition – No Smoking (S7, S16)

Limit values OES short-term 500 p.p.m. (1225 mg m^{-3}); long-term 400 p.p.m. (980 mg m^{-3}).

Toxic effects Inhalation of the vapour in high concentrations and ingestion of the liquid may result in headache, dizziness, mental depression, nausea, vomiting, narcosis, anaesthesia, and coma; the fatal dose is about 100 cm^3. The liquid may damage the eyes severely.

Hazardous reactions In certain circumstances the liquid will form explosive peroxides.

First aid Standard treatment for exposure by all routes (see pages 108-122).

Fire hazard Flash point 12 ˚C; Explosive limits 2.3-12.7%; Autoignition temperature 399 ˚C. Extinguish fire with water spray, dry powder, or carbon dioxide.

Spillage disposal See general section.

1063. Propene

Colourless gas which burns with a sooty flame; b.p. -48 ˚C; slightly soluble in water.

RISKS
Extremely flammable liquefied gas (R13)

SAFETY PRECAUTIONS
Keep container in a well ventilated place – Keep away from sources of ignition – No Smoking – Take precautionary measures against static discharges (S9, S16, S33)

Toxic effects The gas is a simple asphyxiant which is anaesthetic if inhaled in high concentrations; it has no significant toxic properties.

First aid   Standard treatment for exposure by all routes (see pages 108-122).

Fire hazard   Flash point -108 ˚C.; Explosive limits 2-11.1%; Autoignition temperature 460 ˚C. As the gas is supplied in a cylinder, turning off the valve will reduce any fire involving it; if possible, cylinders should be removed quickly from an area in which a fire has developed.

Spillage   Surplus gas or leaking cylinder can be vented slowly to air in a safe, open
disposal   area, or gas burnt-off through a suitable burner.

1064. Propene ozonide

Hazardous   Liable to explode at ambient temperature.
reactions

1065. Propiolactone

Colourless liquid; m.p. -33 ˚C; b.p. 155 ˚C with decomposition; slowly hydrolysed in water to hydracrylic acid.

RISKS
May cause cancer – Very toxic by inhalation – Irritating to eyes and skin (R45, R26, R36/38)

SAFETY PRECAUTIONS
Avoid exposure – Obtain special instructions before use – In case of accident or if you feel unwell, seek medical advice immediately (show label where possible) (S53, S45)

Toxic effects   It is highly irritant to the skin and has produced skin cancer in experimental animals. It is assumed to be similarly irritant and dangerous if taken by mouth. It must be regarded as a potential human carcinogen.

First aid   Standard treatment for exposure by all routes (see pages 108-122).

Fire hazard   Flash point 70 ˚C. Extinguish fires with carbon dioxide, dry chemical powder, or alcohol or polymer foam.

Spillage   See general section.
disposal

1066. Propiolaldehyde

Oily liquid; b.p. 59-61 ˚C; very soluble in water.

Hazardous   Undergoes almost explosive polymerization in the presence of alkalies or
reactions   pyridine.

1067. Propionaldehyde

Colourless liquid with suffocating odour; b.p. 49 °C; fairly soluble in water.

RISKS
Highly flammable – Irritating to eyes, respiratory system, and skin (R11, R36/37/00)

SAFETY PRECAUTIONS
Keep container in a well ventilated place – Keep away from sources of ignition – No Smoking –
Do not empty into drains (S9, S16, S29)

Toxic effects	The vapour irritates the respiratory system. The liquid irritates the eyes and is assumed to be irritant and damaging if taken internally.
Hazardous reactions	It may form explosive peroxides.
First aid	Standard treatment for exposure by all routes (see pages 108-122).
Fire hazard	Flash point -9 °C; Explosive limits 2.9-17%; Autoignition temperature 207 °C. Extinguish fire with water spray, dry powder, or carbon dioxide.
Spillage disposal	See general section.

1068. Propionic acid

Colourless, oily liquid with rancid odour; m.p. -20.8 °C; b.p. 141 °C; miscible with water.

RISKS
((≥ 10%, ≤ 25%) Irritating to eyes, respiratory system, and skin (R36/37/38)
(> 25%) Causes burns (R34)

SAFETY PRECAUTIONS
(≥ 10%, ≤ 25%) Keep out of reach of children (S2)
(≥ 25%) Keep out of reach of children – Do not breathe vapour – In case of contact with eyes,
rinse immediately with plenty of water and seek medical advice (S2, S23, S26)

Limit values	OES short-term 15 p.p.m. (45 mg m^{-3}); long-term 10 p.p.m. (30 mg m^{-3}).
Toxic effects	It is irritating and corrosive to the skin, eyes, and upper respiratory tract. It is moderately toxic if ingested. No cumulative effects are known.
First aid	Standard treatment for exposure by all routes (see pages 108-122).
Fire hazard	Flash point 54 °C (closed cup); Explosive limits 2.1-12.0%; Autoignition temperature 485 °C. Extinguish fire with water spray, dry powder, carbon dioxide, or alcohol foam.
Spillage disposal	See general section.

RSC Chemical Safety Data Sheets Vol. 3, No. 59, 1990 gives extended coverage.

1069. Propiononitrile

Colourless liquid, ethereal odour; b.p. 97 ˚C; miscible with water.

Toxic effects	Inhalation of vapour may cause headache, dizziness, rapid breathing, nausea, and, in severe cases, unconsciousness, convulsions, and death. Evidence is lacking on the effects of skin absorption and ingestion, but these may be similar to those from inhalation.
First aid	Special treatment as for hydrogen cyanide.
Fire hazard	Flash point 16 ˚C. Extinguish fire with water spray, foam, dry powder, or carbon dioxide.
Spillage disposal	See general section.

1070. Propionyl chloride

Colourless liquid with acrid odour; b.p. 80 ˚C; reacts with water forming propionic and hydrochloric acids.

RISKS
Highly flammable – Reacts violently with water – Causes burns (R11, R14, R34)

SAFETY PRECAUTIONS
Keep container in a well ventilated place – Keep away from sources of ignition – No Smoking – In case of contact with eyes, rinse immediately with plenty of water and seek medical advice (S9, S16, S26)

Toxic effects	The vapour irritates the eyes and respiratory system. The liquid burns the skin and eyes. Assumed to cause severe internal irritation and damage if taken by mouth.
Hazardous reactions	A mixture with di-isopropyl ether evolved gas and burst the container.
First aid	Standard treatment for exposure by all routes (see pages 108-122).
Fire hazard	Flash point 12 ˚C. Extinguish fire with dry powder or carbon dioxide.
Spillage disposal	See general section.

1071. Propyl acetate

Colourless liquid with pleasant odour; m.p. -92.5 ˚C; b.p. 101.6 ˚C; slightly soluble in water.

RISKS
Highly flammable (R11)

SAFETY PRECAUTIONS
Keep away from sources of ignition – No Smoking – Do not breathe vapour – Do not empty into drains – Take precautionary measures against static discharges (S16, S23, S29, S33)

Limit values	OES short-term 250 p.p.m. (1050 mg m^{-3}); long-term 200 p.p.m. (840 mg m^{-3}).

Toxic effects   The vapour is irritant to eyes and upper respiratory tract. Inhalation can cause nausea, vomiting, tightness of the chest, and narcosis. Prolonged inhalation of non-narcotic concentrations has caused liver damage.

First aid   Standard treatment for exposure by all routes (see pages 108-122).

Fire hazard   Flash point 13 ˚C; Explosive limits 2-8%; Autoignition temperature 450 ˚C. Extinguish small fires with carbon dioxide, dry chemical powder, or alcohol-resistant foam. Use the foam for large fires.

Spillage disposal   See general section.

RSC *Chemical Safety Data Sheets* Vol. 1, No. 85, 1988 gives extended coverage.

1072. Propylamines

The primary propylamines (n- and iso-) are colourless, volatile liquids with an ammoniacal odour; b.p. (n-) 49 ˚C;(iso-) 32 ˚C; miscible with water.

Toxic effects   The vapour irritates the respiratory system. The vapour and liquid irritate and may burn the eyes. The liquid may cause skin burns. Assumed to be very irritant and poisonous if taken by mouth.

First aid   Standard treatment for exposure by all routes (see pages 108-122).

Fire hazard   Flash point (n- and iso-) -37 ˚C; Explosive limits (n- and iso-) about 2-10%; Autoignition temperature (n-) 318 ˚C; (iso-) 420 ˚C. Extinguish fire with water spray, dry powder, or carbon dioxide.

Spillage disposal   See general section.

1073. Propyl chloroformate

Colourless liquid; b.p. 114-115 ˚C at 768mm Hg; insoluble in and slightly decomposed by water.

RISKS
Flammable – Toxic by inhalation – Causes burns (R10, R23, R34)

SAFETY PRECAUTIONS
In case of contact with eyes, rinse immediately with plenty of water and seek medical advice – Wear suitable protective clothing – If you feel unwell, seek medical advice (show label where possible) (S26, S36, S44)

Toxic effects   Irritant to skin, eyes, and upper respiratory tract. Can be absorbed through the skin.

First aid   Standard treatment for exposure by all routes (see pages 108-122), but attention should be paid to the possibility of lung congestion occurring.

Fire hazard   Flash point 28 ˚C. Extinguish fires with carbon dioxide, dry chemical powder, or alcohol or polymer foam. Do NOT use water.

Spillage disposal   See general section.

RSC *Chemical Safety Data Sheets* Vol. 4b, No. 126, 1991 gives extended coverage.

1074. 3-Propyldiazirine

Liquid; b.p. 60-78 ˚C.

Hazardous reactions   Exploded on attempted distillation at about 75 ˚C.

1075. Propylene glycol

Colourless, hygroscopic, viscous liquid with slightly acrid taste; m.p. -60 ˚C; b.p. 187.2 ˚C; soluble in water.

Limit values   OES long-term 150 p.p.m. (470 mg m^{-3}) (total vapour and particulates), 10 mg m^{-3} (particulates).

Toxic effects   Generally considered to be of low toxicity.

First aid   Standard treatment for exposure by all routes (see pages 108-122).

Fire hazard   Flash point 99 ˚C (open cup); Explosive limits 2.6-12.6%; Autoignition temperature 371 ˚C. Extinguish small fires with carbon dioxide or dry chemical powder. Large fires are best controlled with alcohol-resistant foam.

Spillage disposal   See general section.

RSC *Chemical Safety Data Sheets* Vol. 1, No. 86, 1988 gives extended coverage.

1076. Propylene glycol monomethyl ether

Colourless liquid with a mild ethereal odour; b.p. 120.1 ˚C; completely miscible with water.

Limit values   OES short-term 300 p.p.m. (1080 mg m^{-3}); long-term 100 p.p.m. (360 mg m^{-3}).

Toxic effects   Irritant to skin, eyes, upper respiratory tract, and mucous membranes. Harmful by inhalation, ingestion, and skin absorption.

Hazardous reactions   Forms explosive mixtures with air.

First aid   Standard treatment for exposure by all routes (see pages 108-122).

Fire hazard   Flash point 38 ˚C. Extinguish fires with carbon dioxide, dry chemical powder, or alcohol or polymer foam.

Spillage disposal   See general section.

1077. Propylene oxide

Colourless, volatile liquid with ethereal odour; b.p. 35 ˚C; somewhat soluble in water.

RISKS
May cause cancer – Extremely flammable – Harmful by inhalation, in contact with skin, and if swallowed – Irritating to eyes, respiratory system, and skin (R45, R12, R20/21/22, R36/37/38)

Limit values	OES short-term 100 p.p.m. (240 mg m^{-3}); long-term 20 p.p.m. (50 mg m^{-3}).
Toxic effects	The vapour irritates the eyes and respiratory system. The liquid irritates the eyes and has a delayed blistering action upon the skin in which it is rapidly absorbed. Assumed to be irritant and poisonous when taken by mouth.
Hazardous reactions	Use as a biological sterilant and is hazardous because of ready formation of explosive mixtures with air.
First aid	Standard treatment for exposure by all routes (see pages 108-122).
Fire hazard	Flash point -37 ˚C; Explosive limits 2.1-28.5%. Extinguish fire with dry powder or carbon dioxide.
Spillage disposal	See general section.

1078. Propyl nitrate

Pale yellow liquid with sickly odour; m.p. -100 ˚C; b.p. 110.5 ˚C; slightly soluble in water.

Toxic effects	Inhalation can give rise to hypotension and methaemoglobinaemia.
Hazardous reactions	Shock-sensitive liquid.
First aid	Standard treatment for exposure by all routes (see pages 108-122).
Fire hazard	Flash point 20 ˚C; Explosive limits 2-100 ˚C; Autoignition temperature 175 ˚C. Extinguish fires with carbon dioxide, dry chemical powder, or foam extinguishants.
Spillage disposal	See general section.

1079. Propyne

Colourless gas; b.p. -23 ˚C.; slightly soluble in water.

Toxic effects	The gas is anaesthetic in high concentrations but has low toxicity.
First aid	Standard treatment for exposure by all routes (see pages 108-122).
Fire hazard	Explosive limits about 2.4-11%. As the gas is supplied in a cylinder, turning off the valve will reduce any fire involving it; if possible, cylinders should be removed quickly from an area in which a fire has developed.
Spillage disposal	Surplus gas or leaking cylinder can be vented slowly to air in a safe, open area, or burnt-off through a suitable burner.

1080. Prop-2-yn-1-ol

Colourless to yellow liquid; m.p. -17 ˚C; b.p. 114-115 ˚C; soluble in water.

RISKS
Flammable – Toxic by inhalation, in contact with skin, and if swallowed – Causes burns (R10, R23/24/25, R34)

SAFETY PRECAUTIONS
In case of contact with eyes, rinse immediately with plenty of water and seek medical advice – After contact with skin, wash immediately with plenty of water – Wear suitable protective clothing – If you feel unwell, seek medical advice (show label where possible) (S26, S28, S36, S44)

Limit values   OES short-term 3 p.p.m. (6 mg m^{-3}); long-term 1 p.p.m. (2 mg m^{-3}).

Toxic effects   Causes severe irritation and is extremely destructive to skin, eyes, upper respiratory tract, and mucous membranes. May be fatal if inhaled, ingested, or absorbed through the skin.

Hazardous reactions   If dried with alkali or phosphorus pentoxide before distillation, the residue is liable to explode.

First aid   Standard treatment for exposure by all routes (see pages 108-122).

Fire hazard   Flash point 38 ˚C. Extinguish fires with carbon dioxide, dry chemical powder, or alcohol or polymer foam.

Spillage disposal   See general section.

1081. Prop-2-yn-1-thiol

Hazardous reactions   Polymerized explosively when distilled at atmospheric pressure.

1082. Prop-2-ynyl vinyl sulfide

Hazardous reactions   Decomposed explosively above 85 ˚C.

1083. Pseudocumene

Colourless liquid with an aromatic odour; m.p. -44 ˚C; b.p. 169-171 ˚C; practically insoluble in water.

Limit values   OES short-term 35 p.p.m. (170 mg m^{-3}) under review; long-term 25 p.p.m. (125 mg m^{-3}) under review.

Toxic effects   Irritant to skin and eyes. Inhalation of high concentrations causes central nervous system depression, respiratory irritation, and blood changes.

First aid   Standard treatment for exposure by all routes (see pages 108-122).

Fire hazard   Flash point 45.5 ˚C; Explosive limits (lower) 0.88%; Autoignition temperature 487 ˚C. Extinguish fires with water spray, fog foam, or carbon dioxide. Decomposition produces acrid smoke and toxic fumes.

Spillage disposal   See general section.

RSC *Chemical Safety Data Sheets* Vol. 1, No. 87, 1988 gives extended coverage.

1084. Pyridine

Colourless or slightly yellow liquid with a sharp penetrating odour; m.p. -42 ˚C; b.p. 115 ˚C; miscible with water.

RISKS
Highly flammable – Harmful by inhalation, in contact with skin, and if swallowed (R11, R20/21/22)

SAFETY PRECAUTIONS
In case of contact with eyes, rinse immediately with plenty of water and seek medical advice – After contact with skin, wash immediately with plenty of water (S26, S28)

Limit values   OES short-term 10 p.p.m. (30 mg m^{-3}); long-term 5 p.p.m. (15 mg m^{-3}).

Toxic effects   The vapour irritates the respiratory system and may cause headache, nausea, giddiness, and vomiting. The vapour and liquid irritate the eyes and may cause conjunctivitis. The liquid may irritate the skin causing dermatitis and can be absorbed through the skin. Affects the central nervous system if taken by mouth and large doses act as a heart poison.

Hazardous reactions   Reacts violently with a variety of substances.

First aid   Standard treatment for exposure by all routes (see pages 108-122).

Fire hazard   Flash point 20 ˚C; Explosive limits 1.8-12.4%; Autoignition temperature 482 ˚C. Extinguish small fires with dry powder or carbon dioxide extinguishers. For large fires alcohol-resistant foam is best.

Spillage disposal   See general section.

RSC *Chemical Safety Data Sheets* Vol. 1, No. 88, 1988 gives extended coverage.

1085. Pyridinium dichromate

Orange powder; m.p. 152-153 ˚C.

Toxic effects   Material is irritant to mucous membranes and upper respiratory tract.

Hazardous reactions   Preparation from pyridine and chromium trioxide in water is explosion-prone.

First aid   Standard treatment for exposure by all routes (see pages 108-122).

Fire hazard   Extinguish fires with water spray, carbon dioxide, dry chemical powder, or alcohol or polymer foam.

Spillage   See general section.
  disposal

1086. Pyridinium nitrate

Hazardous   Explodes on heating.
  reactions

1087. Pyridinium perchlorate

Hazardous   Can be detonated by impact and has occasionally exploded when disturbed.
  reactions

1088. Pyrogallol

Off-white, crystalline solid; m.p. 133-134 °C; b.p. 309 °C; soluble in water.

Toxic effects   Severely irritant and highly destructive to all mucous membranes. Harmful if inhaled, ingested, or absorbed through the skin.

First aid   Standard treatment for exposure by all routes (see pages 108-122).

Fire hazard   Extinguish fires with water spray, carbon dioxide, dry chemical powder, or alcohol or polymer foam.

Spillage   See general section.
  disposal

1089. Pyromellitic dianhydride

White powder; m.p. 286 °C; slightly soluble in water.

RISKS
Irritating to eyes, respiratory system, and skin (R36/37/38)

SAFETY PRECAUTIONS
Avoid contact with eyes (S25)

Toxic effects   Irritant to skin, eyes, upper respiratory tract, and mucous membranes.

First aid   Standard treatment for exposure by all routes (see pages 108-122).

Fire hazard   Extinguish fires with water spray, carbon dioxide, dry chemical powder, or alcohol or polymer foam.

Spillage   See general section.
  disposal

RSC *Chemical Safety Data Sheets* Vol. 3, No. 60, 1990 gives extended coverage.

1090. Pyrrolidine

Colourless mobile liquid; b.p. 89 ˚C; miscible in water.

Toxic effects	The vapour irritates the respiratory system; the liquid irritates the skin and mucous membranes.
First aid	Standard treatment for exposure by all routes (see pages 108-122).
Fire hazard	Flash point 3 ˚C. Extinguish fire with water spray, foam, or dry powder.
Spillage disposal	See general section.

1091. Quinoline

Colourless to yellow, hygroscopic liquid, with a penetrating odour; m.p. -16 ˚C; b.p. 238 ˚C.; sparingly soluble in cold water.

Toxic effects	Irritates skin, eyes, and mucous membranes. Moderately toxic by inhalation, ingestion, or skin absorption.
First aid	Standard treatment for exposure by all routes (see pages 108-122).
Fire hazard	Flash point 101 ˚C. Extinguish fires with water spray, carbon dioxide, dry chemical powder, or alcohol or polymer foam.
Spillage disposal	See general section.

1092. Resorcinol

Colourless crystals which turn pink on exposure to light and air; m.p. 110 ˚C.; b.p. 276 ˚C; soluble in water.

RISKS
Harmful if swallowed – Irritating to eyes and skin (R22, R36/38)

SAFETY PRECAUTIONS
In case of contact with eyes, rinse immediately with plenty of water and seek medical advice (S26)

Limit values	OES short-term 20 p.p.m. (90 mg m^{-3}); long-term 10 p.p.m. (45 mg m^{-3}).
Toxic effects	Absorption by the skin may result in itching and dermatitis; in severe cases, restlessness, cyanosis, convulsions, and death may follow. The solid irritates the eyes. It may cause dizziness, drowsiness, and tremors if taken by mouth.
First aid	Standard treatment for exposure by all routes (see pages 108-122).
Fire hazard	Flash point 127 ˚C (closed cup). Extinguish fires with water spray, carbon dioxide, or dry chemical powder.
Spillage disposal	See general section.

1093. Rhodium (and salts)

The metal is a greyish-white solid; m.p. 1960 °C; b.p. 3725 °C; the metal is insoluble in water.

Limit values	OES short-term 0.3 mg m^{-3} (metal fume or dust) (as Rh), 0.003 mg m^{-3} (soluble salts) (as Rh); long-term 0.1 mg m^{-3} (metal fume or dust) (as Rh), 0.001 mg m^{-3} (soluble salts) (as Rh).
Toxic effects	Some of the salts are mutagens and/or suspect carcinogens.
Hazardous reactions	When the metal is prepared by heating its compounds in hydrogen it must be allowed to cool in an inert atmosphere to prevent catalytic ignition of the sorbed hydrogen on exposure to air.
First aid	Standard treatment for exposure by all routes (see pages 108-122).
Fire hazard	For most of the salts extinguish fires with water spray, carbon dioxide, dry chemical powder, or alcohol or polymer foam.
Spillage disposal	See general section.

1094. Rubidium

Lustrous, soft, silvery white metal which tarnishes rapidly on exposure to air; m.p. 38.89 °C; b.p. 688 °C; decomposed by hot or cold water.

Toxic effects	Rubidium is very destructive to tissues of the skin, eyes, and upper respiratory tract. Inhalation may cause coughing, dyspnoea, and headaches.
Hazardous reactions	Ignites on exposure to air or dry oxygen; ignites on contact with fluorine or chlorine or the vapours of bromine or iodine; reaction with mercury may be violent; hydrogen is evolved vigorously on contact with cold water and ignites.
First aid	Standard treatment for exposure by all routes (see pages 108-122).
Fire hazard	Use only class-D fire extinguishing agents. Do NOT use water.
Spillage disposal	See general section.

RSC *Chemical Safety Data Sheets* Vol. 2, No. 92, 1989 gives extended coverage.

1095. Rubidium fluoride

Colourless crystals; m.p. 775 °C; b.p. 1410 °C; soluble in water.

Limit values	OES long-term 2.5 mg m^{-3} (as F).
Toxic effects	Severely irritating to skin and eyes. May be toxic by inhalation and ingestion.
First aid	Standard treatment for exposure by all routes (see pages 108-122), with special reference to fluorides.
Fire hazard	Use extinguishant appropriate to the surroundings.
Spillage disposal	See general section.

RSC *Chemical Safety Data Sheets* Vol. 2, No. 93, 1989 gives extended coverage.

1096. Rubidium hydride

Crystalline solid; m.p. 300 °C with decomposition; reacts violently with water.

Hazardous reactions   Its reaction with acetylene is vigorous at -60 °C.

1097. Rubidium hydroxide

Greyish-white deliquescent mass; m.p. 301 °C; soluble in water.

Toxic effects   Irritant to abraded skin, eyes, and upper respiratory tract.

First aid   Standard treatment for exposure by all routes (see pages 108-122).

Fire hazard   Extinguish fires with dry chemical powder.

Spillage disposal   See general section.

RSC *Chemical Safety Data Sheets* Vol. 2, No. 94, 1989 gives extended coverage.

1098. Ruthenium compounds

Toxic effects   There is limited information on the toxicity of ruthenium compounds, but they should all be treated with caution.

1099. Salicylic acid

White, crystalline powder; m.p. 158-160 °C; b.p. 211 °C; slightly soluble in water.

Toxic effects   Irritant to skin, eyes, upper respiratory tract, and mucous membranes.

First aid   Standard treatment for exposure by all routes (see pages 108-122).

Fire hazard   Flash point 157 °C; Autoignition temperature 545 °C. Extinguish fires with water spray, carbon dioxide, dry chemical powder, or alcohol or polymer foam.

Spillage disposal   See general section.

1100. Sebacoyl dichloride

Colourless, fuming liquid; m.p. -2.5 °C; b.p. 168 °C at 12 mm Hg; slowly decomposes in water.

Toxic effects   Irritant and destructive to skin, eyes, upper respiratory tract, and mucous membranes. Harmful on inhalation, ingestion, or skin absorption. May be fatal due to spasm, inflammation, and oedema.

Hazardous reactions	At the end of vacuum distillation, the residue frequently decomposes spontaneously, producing voluminous black foam.
First aid	Standard treatment for exposure by all routes (see pages 108-122).
Fire hazard	Flash point above 110 ˚C. Extinguish fires with carbon dioxide or dry chemical powder. Do NOT use water.
Spillage disposal	See general section.

1101. Seleninyl bromide

Red-yellow solid; m.p. 41.5-41.7 ˚C; b.p. 217 ˚C with decomposition; decomposed by water.

RISKS
Toxic by inhalation and if swallowed – Danger of cumulative effects (R23/25, R33)

SAFETY PRECAUTIONS
When using do not eat, drink, or smoke – After contact with skin, wash immediately with plenty of water – If you feel unwell, seek medical advice (show label where possible) (S20/21, S28, S44)

Limit values	OES long-term 0.1 mg m^{-3} (as Se).
Hazardous reactions	The liquid bromide reacts explosively with sodium and potassium, and ignites zinc dust; red phosphorus ignites and the white allotrope explodes in contact with the liquid bromide.

1102. Seleninyl chloride

Nearly colourless or yellowish liquid; m.p. about 5 ˚C; b.p. 180 ˚C; decomposed by water.

Limit values	OES long-term 0.1 mg m^{-3} (as Se).
Toxic effects	Corrosive to skin, eyes, and all mucous membranes. May cause death or permanent injury after small exposures.
Hazardous reactions	Potassium˙ and phosphorus (white) explode on contact with the liquid, powdered antimony ignites.
First aid	Standard treatment for exposure by all routes (see pages 108-122).
Fire hazard	Extinguish fires with water spray, carbon dioxide, dry chemical powder, or foam extinguishants.
Spillage disposal	See general section.

1103. Selenium and compounds

Selenium is a steel grey or purplish powder, also fabricated into pellets, sticks, or plates; insoluble in water. Selenium dioxide, selenous acid, and the alkali metal selenites and selenates are colourless powders or crystals, soluble in water. Selenium chloride, selenyl chloride, and selenic acid are liquids. Selenium tetrachloride is a cream coloured, crystalline solid.

RISKS
Toxic by inhalation and if swallowed – Danger of cumulative effects (R23/25, R33)

SAFETY PRECAUTIONS
When using do not eat, drink, or smoke – After contact with skin, wash immediately with plenty of water – If you feel unwell, seek medical advice (show label where possible) (S20/21, S28, S44)

Limit values OES long-term 0.1 mg m^{-3} (as Se).

Toxic effects The chloride and solutions of the acids and salts may burn the skin – severe pain may be experienced under the finger-nails by skin absorption at the finger-tips. Selenium dioxide dust is particularly penetrating and irritates the respiratory system, eyes, and skin. Assumed to be irritant and poisonous if taken by mouth. Inhalation of selenium dust over a prolonged period may cause fatigue, loss of appetite, digestive disturbance, and bronchitis; dermatitis may result from prolonged exposure of skin to small amounts of selenium and its compounds. Some salts have shown signs of carcinogenicity and mutagenicity.

Hazardous reactions Selenium itself may react violently or explosively with a wide variety of substances.

First aid Standard treatment for exposure by all routes (see pages 108-122).

Fire hazard Use extinguishant appropriate to the surroundings.

Spillage disposal See general section.

RSC *Chemical Safety Data Sheets* Vol. 4b, Nos. 128-135, 1991 give extended coverage of selenium and several of its compounds.

1104. Selenium hexafluoride

Colourless gas; m.p. -39 °C (sublimes at -40.6 °C); b.p. -34.5 °C; insoluble in water.

RISKS
Toxic by inhalation and if swallowed – Danger of cumulative effects (R23/25, R33)

SAFETY PRECAUTIONS
When using do not eat, drink, or smoke – After contact with skin, wash immediately with plenty of water – If you feel unwell, seek medical advice (show label where possible) (S20/21, S28, S44)

Limit values OES long-term 0.1 mg m^{-3} (as Se).

Toxic effects The gas has a high acute toxicity.

First aid   Standard treatment for exposure by all routes (see pages 108-122), with special reference to fluorides.

Fire hazard   Use extinguishant appropriate to the surroundings.

Spillage   Eliminate all sources of ignition. Wear breathing apparatus and appropriate
disposal   protective clothing. Subsequently ventilate the area.

RSC *Chemical Safety Data Sheets* Vol. 4b, No. 131, 1991 gives extended coverage.

1105. Selenium hydride

Colourless, flammable gas with a disagreeable smell; m.p. -64 ˚C; b.p. -41.4 ˚C.

RISKS
Toxic by inhalation and if swallowed – Danger of cumulative effects (R23/25, R33)

SAFETY PRECAUTIONS
When using do not eat, drink, or smoke – After contact with skin, wash immediately with plenty of water – If you feel unwell, seek medical advice (show label where possible) (S20/21, S28, S44)

Limit values   OES long-term 0.05 p.p.m. (0.2 mg m^{-3}) (as Se).

Toxic effects   Severe irritant to skin, eyes, and upper respiratory tract. Inhalation can cause nausea, dizziness, violent sneezing, coughing, oedema, bronchitis, and dyspnoea.

Hazardous   Forms explosive mixtures with air.
reactions

First aid   Standard treatment for exposure by all routes (see pages 108-122).

Fire hazard   Use extinguishant appropriate to the surroundings.

Spillage   See general section.
disposal

RSC *Chemical Safety Data Sheets* Vol. 4b, No. 132, 1991 gives extended coverage.

1106. Silane

Colourless gas; m.p. -185 ˚C; b.p. -112 ˚C; insoluble in water.

RISKS
Toxic by inhalation and if swallowed – Danger of cumulative effects (R23/25, R33)

SAFETY PRECAUTIONS
When using do not eat, drink, or smoke – After contact with skin, wash immediately with plenty of water – If you feel unwell, seek medical advice (show label where possible) (S20/21, S28, S44)

Limit values   OES short-term 1 p.p.m. (1.5 mg m^{-3}); long-term 0.5 p.p.m. (0.7 mg m^{-3}).

Toxic effects   Inhalation can lead to nausea, headache, and dizziness.

Hazardous   The very pure material ignites in air; it burns in contact with bromine, chlorine,
reactions   or some covalent chlorides

First aid   Standard treatment for exposure by all routes (see pages 108-122).

Fire hazard   Do NOT extinguish burning gas if flow cannot be shut off immediately. Use water spray to keep cylinders cool or move them from the threatened area.

1107. Silica

Crystalline forms are white or colourless crystals with high melting points; insoluble in water.

Limit values   OES long-term (*amorphous*) 6 mg m^{-3} (total inhalable dust), 3 mg m^{-3} (respirable dust); (*crystalline*) MEL 0.4 mg m^{-3} (respirable dust); (*fused*) OES 0.1 mg m^{-3} (respirable dust).

Toxic effects The pure amorphous hydrated form has very low toxicity. The crystalline forms are acute irritants, being the chief cause of silicosis, a pulmonary dust disease. This can be associated with cough, dyspnoea, fatigue, cyanosis, liver effects, and total incapacity to work. Experimental carcinogen.

1108. Silicon

Grey, metallic chips or purple grey powder; m.p. 1420 °C; b.p. 2600 °C; decomposed by hot water.

Limit values   OES long-term 10 mg m^{-3} (total inhalable dust), 5 mg m^{-3} (respirable dust).

Toxic effects Irritant to skin, eyes, upper respiratory tract, and mucous membranes. Harmful by inhalation, ingestion, or skin absorption.

Hazardous reactions Reacts with water to liberate toxic and flammable or explosive gas.

First aid   Standard treatment for exposure by all routes (see pages 108-122).

Fire hazard   Extinguish fires with dry chemical powder. Do NOT use water.

Spillage disposal See general section.

1109. Silicon carbide

Blue-black, crystalline solid; b.p. greater than 2700 °C; insoluble in water.

Limit values   OES long-term 10 mg m^{-3} (total inhalable dust); 5 mg m^{-3} (respirable dust).

Toxic effects Irritant to skin, eyes, upper respiratory tract, and mucous membranes. Harmful by inhalation, ingestion, or skin absorption.

First aid   Standard treatment for exposure by all routes (see pages 108-122).

Fire hazard   Extinguish fires with water spray, carbon dioxide, dry chemical powder, or alcohol or polymer foam.

Spillage disposal See general section.

1110. Silicon tetrachloride

Colourless, fuming liquid; m.p. -70 ˚C; b.p. 59 ˚C; reacts violently with water, forming hydrochloric acid and silica.

RISKS
Reacts violently with water – Irritating to eyes, respiratory system, and skin (R14, R36/37/38)

SAFETY PRECAUTIONS
Keep container tightly closed and dry – In case of contact with eyes, rinse immediately with plenty of water and seek medical advice (S7/8, S26)

Toxic effects	The vapour severely irritates the eyes and all parts of the respiratory system. The liquid burns the eyes and skin. If taken by mouth there would be severe internal irritation and damage.
First aid	Standard treatment for exposure by all routes (see pages 108-122).
Fire hazard	Extinguish fires with carbon dioxide, dry chemical powder, or alcohol or polymer foam.
Spillage disposal	See general section.

1111. Silver

Silvery metal; m.p. 961 ˚C; b.p. 2193 ˚C; insoluble in water.

Limit values	OES long-term 0.1 mg m^{-3} under review.
Toxic effects	May cause irritation. May be harmful if swallowed.
Hazardous reactions	Silver powder reacts vigorously with several halides and halogen compounds, catalytically decomposes several oxidants, and forms unstable N-Ag compounds.
First aid	Standard treatment for exposure by all routes (see pages 108-122).
Fire hazard	Use extinguishant appropriate to the surroundings.
Spillage disposal	Sweep up and place in a container for re-use or recycling.

1112. Silver amide

Limit values	OES long-term 0.01 mg m^{-3} (as Ag).
Hazardous reactions	Very explosive when dry.

1113. Silver chlorate

White, crystalline solid; m.p. 230 ˚C; b.p. 270 ˚C with decomposition; slightly soluble in water.

Limit values   OES long-term 0.01 mg m^{-3} (as Ag).

Hazardous   Explosively unstable and a powerful oxidant.
reactions

1114. Silver chlorite

Yellow, crystalline powder; m.p. explodes at 105 ˚C; sparingly soluble in water.

Limit values   OES long-term 0.01 mg m^{-3} (as Ag).

Hazardous   Reacted explosively with iodomethane with or without solvent dilution; the salt
reactions   itself is impact-sensitive, cannot be ground. It explodes in contact with
hydrochloric acid or on rubbing with sulfur.

1115. Silver fluoride

Yellow, deliquescent crystals; m.p. 435 ˚C; soluble in water.

Limit values   OES long-term 0.01 mg m^{-3} (as Ag).

Toxic effects   Irritant and corrosive to skin, eyes, upper respiratory tract, and mucous
membranes. Harmful if inhaled, ingested, or absorbed through the skin.
Inhalation may be fatal due to spasm, inflammation, and oedema.

Hazardous   Reactions with calcium hydride (on grinding mixture) and titanium (at 320 ˚C)
reactions   are incandescent; silicon reacts violently, boron explosively when ground with
silver fluoride.

First aid   Standard treatment for exposure by all routes (see pages 108-122), with
particular reference to fluorides.

Fire hazard   Extinguish fires with dry chemical powder.

Spillage   See general section. Material can be recovered for recycling.
disposal

1116. Silver fulminate

Crystalline solid; m.p. explodes on heating; almost insoluble in water.

Limit values   OES long-term 0.01 mg m^{-3} (as Ag).

Hazardous   A powerful detonator, it explodes violently in contact with hydrogen sulfide.
reactions

1117. Silver nitrate

White crystals; m.p. 212 ˚C; b.p. 444 ˚C with decomposition; soluble in water.

RISKS
Causes burns (R34)

SAFETY PRECAUTIONS
Keep out of reach of children – In case of contact with eyes, rinse immediately with plenty of water and seek medical advice (S2, S26)

Limit values	OES long-term 0.01 mg m^{-3} (as Ag).
Toxic effects	The solid and its solutions severely irritate the eyes and can cause skin burns. If taken by mouth silver nitrate can cause internal damage due to absorption in the blood followed by deposition of silver in various tissues of the body.
Hazardous reactions	Reacts violently or explosively with a wide variety of materials.
First aid	Standard treatment for exposure by all routes (see pages 108-122).
Fire hazard	Extinguish fires with water spray.
Spillage disposal	See general section.

1118. Silver oxalate

White, crystalline solid; m.p. explodes at 140 ˚C; insoluble in water.

Limit values	OES long-term 0.01 mg m^{-3} (as Ag).

1119. Silver(I) oxide

Brown, crystalline, odourless solid; m.p. 300 ˚C with decomposition; insoluble in water.

Limit values	OES long-term 0.01 mg m^{-3} (as Ag).
Toxic effects	Moderately toxic by ingestion.
Hazardous reactions	Slowly forms explosive silver nitride with ammonia or hydrazine; oxidizes carbon monoxide exothermically and may ignite hydrogen sulfide.
First aid	Standard treatment for exposure by all routes (see pages 108-122).
Spillage disposal	See general section.

1120. Silver perchlorate

White, deliquescent crystals; m.p. 486 ˚C with decomposition; freely soluble in water.

Limit values	OES long-term 0.01 mg m^{-3} (as Ag).

Toxic effects   It irritates the skin, eyes, and all mucous surfaces. Harmful if inhaled, ingested, or absorbed through the skin. Inhalation may be fatal following spasm, inflammation, and oedema.

Hazardous reactions   Forms explosive solvates with several organic solvents; that with ether exploded violently on crushing in a mortar; explodes with ethylene diamine.

First aid   Standard treatment for exposure by all routes (see pages 108-122).

Fire hazard   Extinguish fires with water spray.

Spillage disposal   See general section.

1121. Soda asbestos

Grey or brown granules; the fibre content is not asbestos; largely soluble in water.

Toxic effects   The dust severely irritates the nose and mouth. The solid and dust burn the eyes and skin. If swallowed there would be severe internal irritation and damage.

First aid   Standard treatment for exposure by all routes (see pages 108-122).

Spillage disposal   See general section.

1122. Soda lime

Mixed calcium/sodium hydroxide granules.

Hazardous reactions   Fires have been caused in waste bins into which soda lime that has absorbed hydrogen sulfide has been thrown – considerable heat develops when this spent material is exposed to moisture and air.

1123. Sodium amide

White or greyish-white, crystalline powder, smelling of ammonia; m.p. 210 °C; b.p. 400 °C; reacts violently with water with the formation of sodium hydroxide and ammonia.

Toxic effects   The dust severely irritates the mouth and nose. Death could result from inflammation, oedema, and spasms of the larynx and bronchi, pulmonary oedema, and chemical pneumonitis. The solid and dust severely irritate or burn the eyes. The solid in contact with moisture on the skin can cause thermal and caustic burns and is absorbed through the skin. If taken by mouth there would be severe internal irritation and damage.

Hazardous reactions   May ignite or explode on heating or grinding in air; explodes with potassium chlorate or sodium nitrite; fresh material behaves like sodium on contact with water but old material may give delayed explosion.

First aid   Standard treatment for exposure by all routes (see pages 108-122).

Fire hazard   Extinguish fires with carbon dioxide or dry chemical powder. Do NOT use water.

Spillage disposal   See general section.

RSC *Chemical Safety Data Sheets* Vol. 2, No. 96, 1989 gives extended coverage.

1124. Sodium and sodium amalgam

Sodium metal – soft, ductile, malleable, silvery white sticks, pellets, wire, or granules, normally coated with a grey oxide or hydroxide skin. Sodium amalgam – silvery or grey spongy mass; m.p. (metal) 97.81 ˚C; b.p. (metal) 881.4 ˚C; both react vigorously with water with formation of sodium hydroxide and hydrogen gas which may ignite.

RISKS
Reacts violently with water, liberating highly flammable gases – Causes burns (R14/15, R34)

SAFETY PRECAUTIONS
Keep contents under paraffin or light oil – Keep container dry – In case of fire, use dry graphite, soda ash, powdered sodium chloride, or dry powder. Do NOT use water (S5, S8, S43).

Toxic effects	The metal or amalgam in contact with moisture on the skin can cause thermal and caustic burns. In the same way the sodium hydroxide resulting from the reaction of the metal with water can cause burning of the skin and eyes. Fumes from burning sodium are highly irritant to all tissues.
Hazardous reactions	Reacts violently or explosively with a wide variety of substances.
First aid	Standard treatment for exposure by all routes (see pages 108-122).
Fire hazard	Autoignition temperature over 115 ˚C. Extinguish fires with dry graphite, soda ash, powdered sodium chloride, or appropriate dry powder. Do NOT use water or carbon dioxide.
Spillage disposal	See general section. With suitable treatment of the amalgam the mercury can be separated and recovered.

RSC *Chemical Safety Data Sheets* Vol. 2, No. 95, 1989 gives extended coverage of metallic sodium.

1125. Sodium azide

Colourless, crystalline powder; m.p. 300 ˚C with decomposition; soluble in water.

RISKS
Very toxic if swallowed – Causes burns (R28, R34)

SAFETY PRECAUTIONS
After contact with skin, wash immediately with plenty of water (S28)

Limit values	OES short-term 0.3 mg m^{-3} (as NaN_3).
Toxic effects	The dust and solution irritate the skin and eyes. Very irritant and poisonous if taken by mouth. If inhaled, ingested, or absorbed through the skin it can cause death by its effects on the central nervous system.
Hazardous reactions	Decomposes somewhat explosively above its melting point, particularly if heated rapidly; liable to explode with bromine, carbon disulfide, or chromyl chloride; when water is added to the strongly heated azide there is a violent reaction; heavy metals form explosive deposits.
First aid	Standard treatment for exposure by all routes (see pages 108-122).

Fire hazard   Extinguish fires with dry chemical powder. Do NOT use water.

Spillage   See general section.
  disposal

RSC *Chemical Safety Data Sheets* Vol. 2, No. 99, 1989 gives extended coverage.

1126. Sodium benzenehexoxide

Hazardous   Explodes on heating in air or on contact with water.
  reactions

1127. Sodium borate

Hard, odourless crystals, powder, or granules; m.p. 75 °C; slowly soluble in water.

Limit values   OES long-term 5 mg m^{-3} (decahydrate), 1 mg m^{-3} (anhydrous, pentahydrate).

Toxic effects   Irritant to skin, eyes, and upper respiratory tract. Toxic by inhalation, ingestion, or skin absorption. Ingestion of large doses has caused deaths.

First aid   Standard treatment for exposure by all routes (see pages 108-122).

Fire hazard   Extinguish fires with carbon dioxide, dry chemical powder, or alcohol or polymer foam

Spillage   See general section.
  disposal

RSC *Chemical Safety Data Sheets* Vol. 2, No. 100, 1989 gives extended coverage.

1128. Sodium borohydride

White to pale grey, microcrystalline powder or lumps; m.p. above 300 °C; decomposed by water with evolution of hydrogen.

Toxic effects   Irritating to skin, respiratory system, and all mucous membranes. Harmful if inhaled, ingested, or absorbed through the skin.

Hazardous   A large volume of the alkaline solution spontaneously heated and
  reactions   decomposed, liberating hydrogen; anhydrous acids form diborane violently; ruthenium salts give explosive precipitates; hot solutions in dimethyl formamide decompose violently.

First aid   Standard treatment for exposure by all routes (see pages 108-122).

Fire hazard   Use only class D fire extinguishing materials. Do NOT use water.

Spillage   See general section.
  disposal

1129. Sodium chlorate

Colourless crystals; m.p. 248 °C; soluble in water.

Toxic effects	The dust or strong solutions may irritate the eyes and skin. The solid or solutions are damaging if taken internally, symptoms of poisoning being nausea, vomiting, and abdominal pain; kidney damage may follow.
Hazardous reactions	Mixtures with a variety of substances can react violently or explosively.
First aid	Standard treatment for exposure by all routes (see pages 108-122).
Fire hazard	Mixtures of sodium chlorate and combustible materials are readily ignited; mixtures with finely divided combustible materials can react explosively. Extinguish fire with water spray.
Spillage disposal	See general section. Site of spillage should be washed thoroughly to remove all oxidant, which is liable to render any organic matter (particularly wood, paper, and textiles) with which it comes into contact, dangerously combustible when dry. Clothing wetted with the solution should be washed thoroughly.

1130. Sodium chlorite

White flakes; m.p. 180-200 °C with decomposition.

Toxic effects	Irritant to skin, eyes, upper respiratory tract, and mucous membranes. Harmful if inhaled, ingested, or absorbed through the skin.
Hazardous reactions	Explodes on impact; intimate mixtures with finely divided or fibrous organic matter may be very sensitive to heat, impact, or friction.
First aid	Standard treatment for exposure by all routes (see pages 108-122).
Fire hazard	Extinguish fires with water spray.
Spillage disposal	See general section.

1131. Sodium *N*-chloro-*p*-toluenesulfonamide

White or pale cream crystals; m.p. decomposes on heating; soluble in water.

Toxic effects	Inhalation of dust may lead to asthma. Ingestion would be assumed to cause internal irritation and damage.
First aid	Standard treatment for exposure by all routes (see pages 108-122).
Spillage disposal	See general section.

1132. Sodium dihydrogen phosphide

m.p. 27 °C; b.p. 162 °C.

Toxic effects	Arise because of the allyl alcohol formed on hydrolysis.
Hazardous reactions	Ignites spontaneously in air.
Spillage disposal	See general section.

1133. Sodium disulfite

White or off-white, crystalline powder; m.p. decomposes on heating; very soluble in water.

Toxic effects	Irritant to skin and eyes. May be harmful by inhalation, ingestion, or skin absorption. Persons with allergies and/or asthma may exhibit hypersensitivity to the substance.
Hazardous reactions	Large scale addition of solid sodium disulfite to an unstirred and too-concentrated solution of sodium nitrite caused a vigorous exothermic reaction.
First aid	Standard treatment for exposure by all routes (see pages 108-122).
Fire hazard	Use extinguishant appropriate to the surroundings.
Spillage disposal	See general section.

1134. Sodium dithionite

White or off-white, crystalline powder; m.p. above 300 °C with decomposition; very soluble in water, but addition of small amount of water to the salt causes hazardous reaction.

RISKS
May cause fire – Harmful if swallowed – Contact with acids liberates toxic gas (R7, R22, R31)

SAFETY PRECAUTIONS
Keep container tightly closed – Keep container dry – In case of contact with eyes, rinse immediately with plenty of water and seek medical advice – After contact with skin, wash immediately with plenty of water – In case of fire, use carbon dioxide or dry chemical powder extinguishers (S7, S8, S26, S28, S43)

Toxic effects	Irritant to skin, eyes, upper respiratory tract, and mucous membranes. Harmful if inhaled, ingested, or absorbed through the skin.
Hazardous reactions	Addition of 10% of water caused heating and spontaneous ignition.
First aid	Standard treatment for exposure by all routes (see pages 108-122).
Fire hazard	Extinguish fires with carbon dioxide or dry chemical powder.
Spillage disposal	See general section.

1135. Sodium ethoxide and sodium methoxide

White solids. Although these compounds are sometimes used industrially as solids, they are more frequently prepared in the laboratory by reacting sodium metal with ethanol or methanol; m.p. (both) above 300 ˚C; both are decomposed by water.

Toxic effects The solutions are extremely irritant to the skin and eyes, and may cause burns. They will cause severe internal irritation and damage if taken by mouth.

Hazardous reactions Mixture of solid sodium methoxide, methyl alcohol, and chloroform boiled violently, then exploded.

First aid Standard treatment for exposure by all routes (see pages 108-122).

Fire hazard Explosive limits (methoxide) 7.3-36.0 %. Extinguish fires with dry chemical powder. Do NOT use water.

Spillage disposal See general section.

1136. Sodium fluoride

Clear, lustrous crystals or white powder or balls. Insecticide grade is sometimes coloured blue; m.p. 993 ˚C; b.p. 1700 ˚C; soluble in water.

RISKS
Toxic by inhalation, in contact with skin, and if swallowed (R23/24/25)

SAFETY PRECAUTIONS
Keep locked up and out of reach of children – In case of contact with eyes, rinse immediately with plenty of water and seek medical advice – If you feel unwell, seek medical advice (show label where possible) (S1/2, S26, S44)

Limit values OES long-term 2.5 mg m^{-3} (as F).

Toxic effects Highly toxic and irritant to all biological tissues.

First aid Standard treatment for exposure by all routes (see pages 108-122), with particular reference to fluorides.

Fire hazard Extinguish fires with water.

Spillage disposal See general section.

RSC *Chemical Safety Data Sheets* Vol. 2, No. 103, 1989 gives extended coverage.

1137. Sodium fluoroacetate

White powder; m.p. 200 ˚C; soluble in water.

Limit values OES short-term 0.15 mg m^{-3}; long-term 0.05 mg m^{-3}.

Toxic effects Highly toxic by ingestion being rapidly absorbed from the gastro-intestinal tract. Very slowly absorbed through the unbroken skin.

First aid Standard treatment for exposure by all routes (see pages 108-122), with particular reference to fluorides.

Fire hazard   Extinguish fires with water spray, carbon dioxide, dry chemical powder, or foam.

Spillage disposal   See general section.

1138. Sodium hydride

Silvery, crystalline solid; m.p. 800 °C with decomposition; decomposed by water.

Toxic effects   Irritant and destructive to skin, eyes, upper respiratory tract, and mucous membranes. Harmful if inhaled, ingested, or absorbed through the skin.

Hazardous reactions   Addition to small amount of water causes explosion; reacts vigorously with acetylene in the presence of moisture even at -60 °C; the finely divided dry powder ignites in dry air; may react explosively with a variety of substances.

First aid   Standard treatment for exposure by all routes (see pages 108-122).

Fire hazard   Extinguish fires with dry chemical powder. Do NOT use water or carbon dioxide.

Spillage disposal   See general section.

1139. Sodium hydrogen sulfate

Colourless crystals or fused masses; m.p. 58.5 °C; soluble in water.

Toxic effects   The solid and its strong solutions in water cause severe burns of the eyes and skin. Causes severe internal irritation and damage if taken by mouth.

First aid   Standard treatment for exposure by all routes (see pages 108-122).

Fire hazard   Use extinguishant appropriate to the surroundings.

Spillage disposal   See general section.

1140. Sodium hydrogen sulfite

White, crystalline powder; m.p. decomposes on heating; very soluble in water.

Limit values   OES long-term 5 mg m^{-3}.

Toxic effects   Extremely destructive to skin, eyes, upper respiratory tract, and mucous membranes. Certain individuals with pre-existing respiratory conditions may experience hypersensitivity. Inhalation may be fatal following spasm, inflammation, chemical pneumonitis, and pulmonary oedema.

First aid   Standard treatment for exposure by all routes (see pages 108-122).

Fire hazard   Use dry chemical powder to extinguish fires.

Spillage disposal   See general section.

1141. Sodium hydroxide

Colourless sticks, flakes, powder, or pellets; m.p. 318.4 °C; b.p. 1390 °C; soluble in water.

RISKS
(Anhydrous) Causes severe burns (R35)
(Solution, ≥5%) Causes severe burns (R35)
(Solution, ≥1%, ≤5%) Irritating to eyes and skin (R36/38)

SAFETY PRECAUTIONS
(Anhydrous) Keep out of reach of children – In case of contact with eyes, rinse immediately with plenty of water and seek medical advice – Wear suitable gloves and eye/face protection (S2, S26, S37/39)
(Solution, ≥5%) – Keep out of reach of children – In case of contact with eyes, rinse immediately with plenty of water and seek medical advice – Take off immediately all contaminated clothing – Wear suitable gloves and eye/face protection (S2, S26, S27, S37/39)
(Solution, ≥1%, ≤5%) Keep out of reach of children – In case of contact with eyes, rinse immediately with plenty of water and seek medical advice (S2;S26)

Limit values   OES short-term 2 mg m^{-3}.

Toxic effects   The solid and its strong solutions are irritant and corrosive to all tissues and can cause severe burns of the eyes and skin. Solutions as weak as 2.5 M can damage eyes severely. If taken by mouth there would be severe internal irritation and damage. Inhalation of the dust or mist may cause irritation and damage to the respiratory tract.

Hazardous reactions   Very exothermic reaction with limited amounts of water; reacts vigorously with chloroform/methanol; explosion results when it is heated with zirconium; accidental contamination of metal scoop with flake sodium hydroxide, prior to its use with zinc dust, caused the latter to ignite.

First aid   Standard treatment for exposure by all routes (see pages 108-122).

Fire hazard   Water can be used to extinguish fires in an area where sodium hydroxide is stored, provided that the water does not come into contact with the material itself.

Spillage disposal   See general section.

RSC *Chemical Safety Data Sheets* Vol. 2, No. 104, 1989 and Vol. 3, No. 63, 1990 give extended coverage.

1142. Sodium hypochlorite solution

Colourless solution smelling of chlorine.

RISKS
Solution (≥ 10% active chlorine) Contact with acids liberates toxic gas – Causes burns (R31, R34)
Solution (≥ 5%, ≤ 10% active chlorine) Contact with acids liberates toxic gas – Irritating to eyes and skin (R31, R36/38)

SAFETY PRECAUTIONS
Solution (≥ 10% active chlorine) Keep out of reach of children – After contact with skin, wash immediately with plenty of water (S2, S28)
Solution (≥ 5%, ≤ 10% active chlorine) Keep out of reach of children – Avoid contact with eyes (S2, S25)

Toxic effects	Extremely irritating to the eyes and gives rise to burns. Bleaches and may burn the skin. Will cause internal irritation and damage if ingested.
Hazardous reactions	Reacts with nitrogen compounds to form unstable or explosive *N*-chloro compounds; explosions with methanol and with benzyl cyanide have been reported; violent reaction with hot formic acid.
First aid	Standard treatment for exposure by all routes (see pages 108-122).
Spillage disposal	See general section.

RSC *Chemical Safety Data Sheets* Vol. 2, No. 105, 1989 gives extended coverage of the solid salt.

1143. Sodium iodate

White crystals; m.p. decomposes on heating; slightly soluble in water.

Toxic effects	Irritant to skin, eyes, upper respiratory tract, and mucous membranes. May be harmful following ingestion, inhalation, or absorption through the skin.
Hazardous reactions	Mixture of sodium iodate and combustible materials are readily ignited; mixtures with finely divided combustible materials can react explosively.
First aid	Standard treatment for exposure by all routes (see pages 108-122).
Fire hazard	Extinguish fire with water spray.
Spillage disposal	See general section. Site of spillage should be washed thoroughly to remove all oxidant, which is liable to render any organic matter (particularly wood, paper, and textiles) with which it comes into contact, dangerously combustible when dry. Clothing wetted with the solution should be washed thoroughly.

1144. Sodium nitrate

Colourless, deliquescent crystals; m.p. 308 ˚C; b.p. 380 ˚C with decomposition; soluble in water.

Toxic effects	Ingestion may cause gastro-enteritis, abdominal pains, vomiting, muscular weakness, irregular pulse, convulsions, and collapse; 15-30 g in one dose may be fatal.
Hazardous reactions	Reacts violently or explosively with a variety of substances. Mixture of sodium nitrate and combustible materials are readily ignited; mixtures with finely divided, combustible materials can react explosively.
Fire hazard	Extinguish fire with water spray.
Spillage disposal	See general section. Site of spillage should be washed thoroughly to remove all oxidant, which is liable to render any organic matter (particularly wood, paper, and textiles) with which it comes into contact, dangerously combustible when dry. Clothing wetted with the solution should be washed thoroughly.

1145. Sodium nitrite

White or slightly yellow granules, rods, or powder; m.p. 271 ˚C; b.p. 320 ˚C with decomposition; hygroscopic and soluble in water.

RISKS
Contact with combustible material may cause fire – Toxic if swallowed (R8, R25)

SAFETY PRECAUTIONS
If you feel unwell, seek medical advice (show label where possible) (S44)

Toxic effects   Irritant to skin, eyes, and respiratory tract. Inhalation of large concentrations of dust or liquid mist can be fatal. It is also toxic by ingestion.

Hazardous reactions   Explosions are likely to occur on heating mixture of the nitrite with a variety of substances; wood impregnated with solutions of nitrite over a long period may be accidentally ignited and burn fiercely.

First aid   Standard treatment for exposure by all routes (see pages 108-122).

Fire hazard   Extinguish fire with water spray.

Spillage disposal   See general section. Site of spillage should be washed thoroughly to remove all oxidant, which is liable to render any organic matter (particularly wood, paper, and textiles) with which it comes into contact, dangerously combustible when dry. Clothing wetted with the solution should be washed thoroughly.

RSC *Chemical Safety Data Sheets* Vol. 2, No. 106, 1989 gives extended coverage.

1146. Sodium periodate

White crystals; m.p. decomposes at 300 ˚C; soluble in cold water.

Toxic effects   Irritant to skin, eyes, upper respiratory tract, and mucous membranes. Harmful if inhaled or ingested.

Hazardous reactions   Mixtures with combustible materials are readily ignited; mixtures with finely divided combustible materials can react explosively.

First aid   Standard treatment for exposure by all routes (see pages 108-122).

Fire hazard   Extinguish fire with water spray.

Spillage disposal   See general section. Site of spillage should be washed thoroughly to remove all oxidant, which is liable to render any organic matter (particularly wood, paper, and textiles) with which it comes into contact, dangerously combustible when dry. Clothing wetted with the solution should be washed thoroughly.

1147. Sodium peroxide

White powder, becoming yellow on heating; m.p. 460 ˚C with decomposition; decomposed by water.

RISKS
Contact with combustible material may cause fire – Causes severe burns (R8, R35)

SAFETY PRECAUTIONS
Keep container dry – Take off immediately all contaminated clothing – Wear eye/face protection (S8, S27, S39)

Toxic effects   The dust is highly irritating and corrosive to skin, eyes, and all biological tissues. Because of the vigour of its reaction with water it may cause both thermal and caustic burns on moist skin. Would cause severe internal irritation and damage if taken by mouth.

Hazardous reactions   Reacts violently or explosively with a variety of substances; reacts vigorously or explosively with water depending on relative quantities.

First aid   Standard treatment for exposure by all routes (see pages 108-122).

Fire hazard   Small fires may be smothered with a suitable dry chemical. Water may be used on combustible material in the vicinity of the fire.

Spillage disposal   See general section.

RSC *Chemical Safety Data Sheets* Vol. 2, No. 107, 1989 gives extended coverage.

1148. Sodium peroxyacetate

Toxic effects   Experimental carcinogen.

Hazardous reactions   Dry salt exploded at room temperature.

1149. Sodium phosphate

Colourless crystals or white powder; soluble in water.

Toxic effects   Irritant to skin, eyes, upper respiratory tract, and mucous membranes. Inhalation can cause vomiting, lethargy, diarrhoea, carpal spasm, acidois, and coma.

First aid   Standard treatment for exposure by all routes (see pages 108-122).

Fire hazard   Use extinguishant appropriate to the surroundings.

Spillage disposal   See general section.

RSC *Chemical Safety Data Sheets* Vol. 2, No. 108, 1989 gives extended coverage.

1150. Sodium phosphide

Red crystals; m.p. decomposes on heating; decomposed by water or moist air, evolving phosphine which often ignites.

Toxic effects   Moderately toxic by inhalation and skin contact.

First aid   Standard treatment for exposure by all routes (see pages 108-122).

Spillage   See general section.
  disposal

1151. Sodium phosphinate

Colourless, pearly crystals; m.p. 200 ˚C; soluble in water.

Hazardous   Evaporation of aqueous solution by heating may cause explosion, phosphine
  reactions   being evolved; reaction of this powerful reducing agent with oxidants is often
    violent or explosive.

1152. Sodium pyrosulfate

A colourless, deliquescent crystalline salt of variable composition; m.p. about 400 ˚C; b.p. 460 ˚C with decomposition; soluble in water.

Toxic effects   The solid and its strong solutions in water cause burns to the skin and eyes.
    Causes severe internal irritation and damage if taken by mouth.

First aid   Standard treatment for exposure by all routes (see pages 108-122).

Spillage   See general section.
  disposal

1153. Sodium silicide

Solid; reacts explosively with water.

Hazardous   Ignites in air.
  reactions

1154. Sodium sulfide

The hydrated salt consists of colourless crystalline masses. The fused salt forms brownish lumps or powder; m.p. 950 ˚C; soluble in water.

RISKS
Contact with acids liberates toxic gas – Causes burns (R31, R34)

SAFETY PRECAUTIONS
In case of contact with eyes, rinse immediately with plenty of water and seek medical advice (S26)

Toxic effects	The solid or solution are irritant and corrosive to skin and eyes. Irritant and poisonous if taken by mouth.
Hazardous reactions	After exposure to moisture and air, small lumps of fused sodium sulfide are liable to spontaneous heating; mixtures with finely divided carbon react exothermally.
First aid	Standard treatment for exposure by all routes (see pages 108-122).
Fire hazard	Use extinguishant appropriate to the surroundings.
Spillage disposal	See general section.

1155. Sodium tetrahydroaluminate

White or grey, crystalline powder; m.p. begins to melt at 183 °C and decomposes completely at 230-240 °C; decomposed by water with evolution of hydrogen.

Toxic effects	Contact with moist tissues forms corrosive sodium hydroxide which is irritant and causes burns to all tissues.
First aid	Standard treatment for exposure by all routes (see pages 108-122).
Fire hazard	Fires involving this material are best extinguished by smothering with dry sand, dry powdered limestone, graphite, soda ash, dry sodium chloride, or lithium chloride or specially manufactured dry powder extinguishers. Do NOT use conventional extinguishants.
Spillage disposal	See general section.

1156. Stearic acid

White, amorphous powder; m.p. 67-69 °C; b.p. 361 °C; slightly soluble in water.

Toxic effects	Irritant to skin, eyes, upper respiratory tract, and mucous membranes.
First aid	Standard treatment for exposure by all routes (see pages 108-122).
Fire hazard	Autoignition temperature 395 °C. Extinguish fires with water spray, carbon dioxide, dry chemical powder, or alcohol or polymer foam.
Spillage disposal	See general section.

1157. Strontium

Silvery-white metal, yellows rapidly on exposure to air; m.p. 769 °C; b.p. 1384 °C; reacts violently with water.

Toxic effects	Irritant to skin, eyes, and upper respiratory tract. Following inhalation it is retained primarily in the bones but it can affect the heart. Ingestion causes gastro-intestinal upsets.

First aid   Standard treatment for exposure by all routes (see pages 108-122).

Fire hazard   Use only class D fire extinguishants. Do NOT use water.

Spillage   See general section.
  disposal

RSC *Chemical Safety Data Sheets* Vol. 2, No. 109, 1989 gives extended coverage.

1158. Strontium chromate

Monoclinic, yellow crystals; very slightly soluble in water.

RISKS
May cause cancer – Harmful if swallowed (R45, R22)

SAFETY PRECAUTIONS
Avoid exposure – Obtain special instructions before use – If you feel unwell, seek medical advice (show label where possible) (S53, S44)

Toxic effects   Irritant to skin, eyes, and upper respiratory system. May cause skin and nasal ulcers. Probable human carcinogen.

First aid   Standard treatment for exposure by all routes (see pages 108-122).

Fire hazard   Extinguish fires with water spray, carbon dioxide, dry chemical powder, or alcohol or polymer foam.

Spillage   See general section.
  disposal

RSC *Chemical Safety Data Sheets* Vol. 4b, No. 136, 1991 gives extended coverage.

1159. Strontium nitrate

White granules or crystalline powder; m.p. 570 °C; soluble in water.

Toxic effects   Increased frequency of nervous and respiratory diseases have been reported. It is a skin and eye irritant in experimental animals.

Hazardous   Mixtures of strontium nitrate and combustible materials are readily ignited;
  reactions   mixtures with finely divided combustible materials can react explosively.

First aid   Standard treatment for exposure by all routes (see pages 108-122).

Fire hazard   Extinguish fire with water spray.

Spillage   See general section. Site of spillage should be washed thoroughly to remove all
  disposal   oxidant, which is liable to render any organic matter (particularly wood, paper, and textiles) with which it comes into contact, dangerously combustible when dry. Clothing wetted with the solution should be washed thoroughly.

RSC *Chemical Safety Data Sheets* Vol. 2, No. 110, 1989 gives extended coverage.

1160. Strychnine

Hard, white, crystalline alkaloid with a very bitter taste; m.p. 268 ˚C; b.p. 270 ˚C.

RISKS
Very toxic by inhalation and it swallowed (H26/28)

SAFETY PRECAUTIONS
Keep locked up – Keep away from food, drink and animal feeding stuffs – In case of accident or if you feel unwell, seek medical advice immediately (show label where possible) (S1, S13, S45)

Limit values   OES short-term 0.45 mg m^{-3}; long-term 0.15 mg m^{-3}.

Toxic effects   Irritant to skin and eyes. When ingested the time of action depends on the quality and quantity of food in the stomach.

First aid   Standard treatment for exposure by all routes (see pages 108-122).

Fire hazard   Extinguish fires with water spray, carbon dioxide, dry chemical powder, or alcohol or polymer foam.

Spillage disposal   See general section.

1161. Styrene

Colourless to yellow oily liquid with penetrating disagreeable odour; m.p. -31 ˚C; b.p. 145 ˚C; very sparingly soluble in water.

RISKS
Flammable – Harmful by inhalation – Irritating to eyes and skin (R10, R20, R36/38)

SAFETY PRECAUTIONS
Do not breathe vapour (S23)

Limit values   MEL short-term 250 p.p.m. (1050 mg m^{-3}); long-term 100 p.p.m. (420 mg m^{-3}).

Toxic effects   The vapour irritates the eyes and respiratory system. The liquid irritates the eyes and is reported to cause severe eye injuries. Has low oral toxicity. Prolonged skin contact can cause irritation, swelling, and blistering.

Hazardous reactions   Autocatalytic exothermic polymerization becomes self-sustaining above 65 ˚C; on exposure to oxygen at 40-60 ˚C an interpolymeric peroxide was formed which exploded on gentle heating.

First aid   Standard treatment for exposure by all routes (see pages 108-122).

Fire hazard   Flash point 31 ˚C; Explosive limits 1.1-6.1%; Autoignition temperature 485 ˚C. Extinguish fire with alcohol-resistant foam, dry powder, or carbon dioxide.

Spillage disposal   See general section.

RSC *Chemical Hazard Data Sheets* Vol. 1, No. 89, 1988 gives extended coverage.

1162. Succinic anhydride

Colourless or white crystals; m.p. 119.6 °C; b.p. 261 °C; very slightly soluble in water.

RISKS
Irritating to eyes and respiratory system (R36/37)

SAFETY PRECAUTIONS
Avoid contact with eyes (S25)

Toxic effects   Severely irritating to skin, eyes, and respiratory tract. Can be absorbed through the skin.

First aid   Standard treatment for exposure by all routes (see pages 108-122).

Fire hazard   Extinguish fires with carbon dioxide, dry chemical powder, or alcohol or polymer foam.

Spillage disposal   See general section.

RSC *Chemical Safety Data Sheets* Vol. 3, No. 64, 1990 gives extended coverage.

1163. Sulfanilamide

Colourless, crystalline solid; m.p. 165 °C; slightly soluble in water.

Toxic effects   Irritant to skin, eyes, and mucous membranes. Harmful by inhalation, ingestion, or skin absorption. Experimental carcinogen and teratogen.

First aid   Standard treatment for exposure by all routes (see pages 108-122).

Spillage disposal   See general section.

1164. Sulfolane

Colourless, highly polar liquid; m.p. 28.5 °C; b.p. 287.3 °C; miscible with water.

RISKS
Harmful if swallowed (R22)

SAFETY PRECAUTIONS
Avoid contact with eyes (S25)

Toxic effects   Low toxicity by most routes of exposure. Ingestion has caused hyperactivity, clonic-tonic convulsions, and reduction in metabolic rate and body temperature.

First aid   Standard treatment for exposure by all routes (see pages 108-122).

Fire hazard   Flash point 177 °C (open cup). Small fires can be extinguished by carbon dioxide, dry chemical powder, alcohol-resistant foam, or by flooding with water. Large fires are best controlled by alcohol-resistant foam.

Spillage disposal   See general section.

RSC *Chemical Safety Data Sheets* Vol. 1, No. 90, 1988 gives extended coverage.

1165. Sulfonic acids

The simpler sulfonic acids, such as benzenesulfonic, benzenedisulfonic, phenolsulfonic, phenoldisulfonic, and cresolsulfonic acids are generally supplied as solutions in water or sulfuric acid.

Toxic effects The solutions irritate the skin and eyes and may cause burns. Will cause internal irritation and damage if taken by mouth.

First aid Standard treatment for exposure by all routes (see pages 108-122).

Fire hazard Extinguish fires with carbon dioxide, dry chemical powder, or alcohol or polymer foam.

Spillage disposal See general section.

RSC *Chemical Safety Data Sheets* Vol. 3, No. 69, 1990 gives extended coverage on *p*-toluenesulfonic acid.

1166. Sulfur

Yellow, crystalline solid; m.p. 120 ˚C; b.p. 444.6 ˚C; insoluble in water.

Toxic effects Sulfur dust is a mild irritant to skin and upper respiratory tract. May be harmful following inhalation or ingestion.

Hazardous reactions Evaporation of an ethereal extract of sulfur exploded violently; reacts with varying degrees of vigour with a variety of substances.

First aid Standard treatment for exposure by all routes (see pages 108-122).

Fire hazard Explosive limits (lel) 35 mg/litre in air; Autoignition temperature 190 ˚C. Extinguish fires with water spray, carbon dioxide, dry chemical powder, or alcohol or polymer foam.

Spillage disposal See general section.

1167. Sulfur dichloride

Red-brown, fuming liquid; m.p. -78 ˚C; b.p. 59 ˚C; decomposed by water with the liberation of sulfur dioxide and hydrogen chloride.

RISKS
Reacts violently with water – Causes burns – Irritating to respiratory system (R14, R34, R37)

SAFETY PRECAUTIONS
In case of contact with eyes, rinse immediately with plenty of water and seek medical advice (S26)

Toxic effects The vapour irritates the eyes and respiratory system. The liquid irritates the skin and eyes severely and may cause burns. Ingestion would result in severe internal irritation and damage.

Hazardous reactions	Reacts violently or explosively with a variety of substances.
First aid	Standard treatment for exposure by all routes (see pages 108-122).
Fire hazard	Extinguish fires with dry chemical powder. Do NOT use water.
Spillage disposal	See general section.

1168. Sulfur dioxide

Colourless gas with a distinctive odour; supplied in liquefied form in canisters or cylinders; m.p. - 72.7 °C; b.p. -10 °C; somewhat soluble in water.

RISKS
Toxic by inhalation – Irritating to eyes and respiratory system (R23, R36/37)

SAFETY PRECAUTIONS
Keep container tightly closed and in a well ventilated place – If you feel unwell, seek medical advice (show label where possible) (S7/9, S44)

Limit values	OES short-term 5 p.p.m. (13 mg m^{-3}); long-term 2 p.p.m. (5 mg m^{-3}).
Toxic effects	The gas irritates the respiratory system and may cause bronchitis and asphyxia. High concentrations of the gas irritate the eyes and may cause conjunctivitis.
Hazardous reactions	Sulfur dioxide reacts violently or explosively with a variety of substances.
First aid	Standard treatment for exposure by all routes (see pages 108-122).
Fire hazard	Use water spray to keep cylinders cool or remove them from the immediate area.
Spillage disposal	Surplus gas or leaking cylinder can be vented slowly into a water-fed scrubbing tower or column in a fume cupboard, or into a fume cupboard served by such a tower.

RSC *Chemical Safety Data Sheets* Vol. 4b, No. 137, 1991 gives extended coverage.

1169. Sulfur hexafluoride

Colourless gas; m.p. -50.8 °C under pressure; b.p. -63.5 °C with sublimation; very slightly soluble in water.

Limit values	OES short-term 1250 p.p.m. (7500 mg m^{-3}); long-term 1000 p.p.m. (6000 mg m^{-3}).
Toxic effects	Exposure can cause nausea, dizziness, headache, and central nervous system depression. At high concentrations it can act as an asphyxiant by exclusion of oxygen.
Hazardous reactions	Explodes on contact with disilane.
First aid	Standard treatment for exposure by all routes (see pages 108-122).
Fire hazard	Use water spray to keep cylinders cool or move them out of the threatened area.

1170. Sulfuric acid

Concentrated sulfuric acid is a colourless, odourless, viscous liquid; m.p. 10.49 ˚C; b.p. 290 ˚C; miscible with water with great evolution of heat. Always add acid to water to prevent local boiling.

RISKS
(≥ 5%, ≤ 15%) – Irritating to eyes and skin (R36/38)
(≥ 15%) Causes severe burns (R35)

SAFETY PRECAUTIONS
(≥ 5%, ≤ 15%) – Keep out of reach of children – In case of contact with eyes, rinse immediately with plenty of water and seek medical advice (S2, S26)
(≥ 15%) – Keep out of reach of children – In case of contact with eyes, rinse immediately with plenty of water and seek medical advice – Never add water to this product (S2, S26, S30)

Limit values OES long-term 1 mg m⁻³.

Limit values	OES long-term 1 mg m^{-3}.
Toxic effects	The concentrated acid burns the eyes and skin severely. The dilute acid irritates the eyes and may cause burns; it will irritate the skin and may give rise to dermatitis. The concentrated acid, if taken by mouth, will cause severe internal irritation and damage. Inhalation of high concentrations of vapour or mist is highly toxic.
Hazardous reactions	Many substances and classes of substance react with concentrated sulfuric acid (a powerful oxidizing desiccant) with varying degrees of violence.
First aid	Standard treatment for exposure by all routes (see pages 108-122).
Fire hazard	Extinguish fires with dry chemical powder. Do NOT use water.
Spillage disposal	See general section.

RSC *Chemical Safety Data Sheets* Vol. 3, No. 65, 1990 gives extended coverage.

1171. Sulfur pentafluoride

Colourless liquid; m.p. -92 ˚C; b.p. 29 ˚C; insoluble in water.

Limit values	OES short-term 0.075 p.p.m. (0.75 mg m^{-3}); long-term 0.025 p.p.m. (0.25 mg m^{-3}).
Toxic effects	Severe irritant. Inhalation can cause cough, shortness of breath, and pulmonary oedema.
First aid	Standard treatment for exposure by all routes (see pages 108-122), with special reference to fluorides.

1172. Sulfur tetrafluoride

Colourless gas with smell resembling that of sulfur dioxide; b.p. -40 ˚C; reacts violently with water forming hydrogen fluoride and sulfur dioxide.

Limit values	OES short-term 0.3 p.p.m. (1 mg m^{-3}); long-term 0.1 p.p.m. (0.4 mg m^{-3}).

Toxic effects	The gas reacts with moisture on the body tissues forming highly toxic and corrosive hydrogen fluoride. It is thus extremely irritant to the eyes, skin, and respiratory system, and will cause severe burns if exposure is considerable.
Hazardous reactions	Attacks glass.
First aid	Standard treatment for exposure by all routes (see pages 108-122), with special reference to fluorides.
Fire hazard	Use water spray to keep cylinders cool or move them out of the immediate area.
Spillage disposal	Surplus gas or leaking cylinder can be vented slowly into a water-fed scrubbing tower or column in a fume cupboard, or into a fume cupboard served by such a tower.

1173. Sulfur trioxide

Colourless crystals or liquid; m.p. 16.8 °C; b.p. 44.6 °C; reacts vigorously, sometimes explosively, with water.

Toxic effects	Extremely destructive to skin, eyes, respiratory tract, and mucous membranes. May be fatal if inhaled, ingested, or absorbed through the skin.
Hazardous reactions	Reacts violently or explosively with a variety of substances.
First aid	Standard treatment for exposure by all routes (see pages 108-122).
Fire hazard	Extinguish fires with carbon dioxide or dry chemical powder. Do NOT use water.
Spillage disposal	See general section.

1174. Sulfuryl chloride

Colourless or yellow, fuming liquid with a pungent odour; m.p. -54.1 °C; b.p. 69 °C; decomposed by water with formation of hydrochloric and sulfuric acids.

RISKS
Reacts violently with water – Causes burns – Irritating to respiratory system (R14, R34, R37)

SAFETY PRECAUTIONS
In case of contact with eyes, rinse immediately with plenty of water and seek medical advice (S26)

Toxic effects	The vapour irritates the eyes and respiratory system. The liquid irritates the skin and eyes causing burns. Ingestion would result in severe internal irritation and damage.
Hazardous reactions	Reactions with alkalies may be violently explosive; solution of sulfuryl chloride in ether decomposed vigorously; reacts violently or explosively with some materials.
First aid	Standard treatment for exposure by all routes (see pages 108-122).

Fire hazard   Extinguish fires with dry chemical powder. Do NOT use water.

Spillage   See general section.
disposal

1175. Sulfuryl fluoride

Colourless gas; m.p. -137 ˚C; b.p. -55 ˚C; reacts violently with water.

Limit values   OES short-term 10 p.p.m. (40 mg m^{-3}); long-term 5 p.p.m. (20 mg m^{-3}).

Toxic effects   Harmful by inhalation or ingestion causing nausea, vomiting, and cramps. May be narcotic in high concentrations.

1176. Tantalum

Black-grey solid; m.p. 2977 ˚C; b.p. 5425 ˚C; insoluble in water.

Limit values   OES short-term 10 mg m^{-3}; long-term 5 mg m^{-3}.

Toxic effects   Irritant to skin, eyes, upper respiratory tract, and mucous membranes. May be harmful by inhalation, ingestion, or skin absorption.

Hazardous   Finely divided metal may be pyrophoric in air.
reactions

First aid   Standard treatment for exposure by all routes (see pages 108-122).

Fire hazard   For fires involving metal powder use only dry chemical powder, but for solid metal use any extinguishant appropriate to the surroundings.

Spillage   See general section. The material may be collected and saved for re-use or
disposal   recycling.

1177. Tellurium and compounds

Silvery, metallic ingots which break with a white, lustrous fracture, or a greyish-black powder. Telluric acid and the alkali-metal tellurites and tellurates are colourless powders or crystals; the oxide is a colourless solid and the tetrachloride a cream coloured solid; the alkali-metal tellurites and tellurates are soluble in water.

Limit values   OES long-term 0.1 mg m^{-3} (as Te) (all compounds except hydrogen telluride).

Toxic effects   Inhalation of dust or tellurium fume gives rise to a dry mouth, metallic taste, drowsiness, loss of appetite, excessive salivation, nausea, vomiting, and a foul garlic-like odour of the breath. Skin absorption and ingestion also result in foul breath and it can be assumed that effects similar to those of inhalation will be experienced. Ingestion of soluble tellurium salts may, in addition, lead to cyanosis and unconsciousness. Prolonged absorption of very small amounts of tellurium compounds can cause symptoms similar to the above.

Hazardous   The tetrachloride interacts with liquid ammonia at -15 ˚C to form an explosive
reactions   nitride.

First aid   Standard treatment for exposure by all routes (see pages 108-122).

Fire hazard   Extinguish fires with carbon dioxide, dry chemical powder, or alcohol or
                       polymer foam.

Spillage   See general section.
disposal

1178. Terphenyls

All are white, crystalline powders; m.p. (*o*-) 58- 59 ˚C; (*m*-) 86-87 ˚C; (*p*-) 212- 213 ˚C; b.p. (*o*-)
337 ˚C; (*m*-) 379 ˚C; (*p*-) 383 ˚C; all isomers insoluble in water.

Limit values   OES short-term 0.5 p.p.m. (5 mg m^{-3}) (all isomers).

Toxic effects   Irritant to skin, eyes, upper respiratory tract, and mucous membranes. May
                       be harmful by inhalation, ingestion, or skin absorption.

First aid   Standard treatment for exposure by all routes (see pages 108-122).

Fire hazard   Flash point (*o*-) 110 ˚C. Extinguish fires with water spray, carbon dioxide, dry
                       chemical powder, or alcohol or polymer foam.

Spillage   See general section.
disposal

1179. Tetraborane

Gas with a disagreeable odour; m.p. -120 ˚C; b.p. 18 ˚C; hydrolysed by water.

Hazardous   Ignites in air or oxygen and explodes with concentrated nitric acid.
reactions

1180. 1,1,2,2-Tetrabromoethane

Very dense, colourless or yellowish liquid with chloroform- like odour; m.p. 0 ˚C; b.p.
239-242 ˚C with decomposition; immiscible with water.

RISKS
Very toxic by inhalation – Irritating to eyes (R26, R36)

SAFETY PRECAUTIONS
*Keep locked up – Avoid contact with skin – Take off immediately all contaminated clothing – In
case of accident or if you feel unwell, seek medical advice immediately (show label where
possible) (S1, S24, S27, S45)*

Limit values   OES long-term 0.5 p.p.m. (7 mg m^{-3}).

Toxic effects   Irritant to the upper respiratory tract producing changes in the liver, kidneys,
                       and central nervous system. In very large doses it may cause narcosis,
                       coma, and death.

First aid   Standard treatment for exposure by all routes (see pages 108-122).

Fire hazard   Autoignition temperature 335 °C. Use extinguishant appropriate to the surroundings.

Spillage disposal   See general section.

RSC *Chemical Safety Data Sheets* Vol. 4b, No. 138, 1991 gives extended coverage.

1181. Tetracarbonylnickel

Colourless, mobile liquid normally supplied in cylinders; m.p. -25 °C; b.p. 43 °C; practically insoluble in water.

Limit values   OES short-term 0.1 p.p.m. (0.24 mg m^{-3}) (as Ni).

Toxic effects   The vapour is exceedingly poisonous when inhaled. With low concentrations the initial symptoms are giddiness and slight headache. Heavier exposure causes nausea, tightness of the chest, weakness of limbs, perspiring, coughing, vomiting, cold and clammy skin, and shortness of breath. The toxicity is believed to be derived from both the nickel and carbon monoxide liberated in the lungs. Cases of lung cancer have occurred as a result of prolonged exposure to vapour.

Hazardous reactions   Reacts explosively with liquid bromine, but smoothly in the gaseous state; a mixture with mercury will explode on vigorous shaking; on exposure to air, the carbonyl produces a deposit which becomes peroxidized and may ignite.

First aid   Standard treatment for exposure by all routes (see pages 108-122).

Fire hazard   The liquid is highly flammable and is liable to explode if heated to 60 °C or above. Extinguish fire with dry powder or carbon dioxide.

Spillage disposal   See general section.

1182. 1,2,4,5-Tetrachlorobenzene

White, crystalline solid; m.p. 138-140 °C; b.p. 240-246 °C; insoluble in water.

Toxic effects   Irritant to skin, eyes, upper respiratory tract, and mucous membranes. May be harmful by inhalation, ingestion, or skin absorption.

Hazardous reactions   Explosions have occurred in the commercial production of 2,4,5-trichlorophenol by alkaline hydrolysis of tetrachlorobenzene, leading to formation of the extremely toxic tetrachlorodibenzodioxin (TCDD) as in the Seveso incident.

First aid   Standard treatment for exposure by all routes (see pages 108-122).

Fire hazard   Extinguish fires with water spray, carbon dioxide, dry chemical powder, or alcohol or polymer foam.

Spillage disposal   See general section.

1183. Tetrachlorodifluoroethanes

(1,1,1,2-) Colourless crystalline solid; m.p. 41 ˚C; b.p. 91 ˚C.

Limit values　　OES short-term 100 p.p.m. (834 mg m⁻³) (both isomers); long-term 100 p.p.m. (834 mg m⁻³) (both isomers).

Toxic effects　　Irritant to skin and eyes. Harmful by inhalation, ingestion, or skin absorption. Prolonged exposure can cause narcotic effect.

First aid　　Standard treatment for exposure by all routes (see pages 108-122).

Fire hazard　　Use extinguishant appropriate to the surroundings.

Spillage disposal　　See general section.

1184. 1,1,2,2-Tetrachloroethane

Colourless, dense liquid with suffocating, chloroform-like odour; m.p. -36 ˚C; b.p. 146.2 ˚C; slightly soluble in water.

RISKS
Very toxic by inhalation and in contact with skin (R26/27)

SAFETY PRECAUTIONS
Keep out of reach of children – In case of insufficient ventilation, wear suitable respiratory equipment – In case of accident or if you feel unwell, seek medical advice immediately (show label where possible) (S2, S38, S45)

Toxic effects　　The vapour irritates the eyes, nose, and lungs. Inhalation may cause drowiness, giddiness, headache, and, in high concentrations, unconsciousness. Continuous breathing of low concentrations of vapour over a period may cause damage to the liver, kidneys and central nervous system. Toxic on ingestion producing gastric disturbances and eventually damage to other internal organs.

Hazardous reactions　　When heated with solid sodium hydroxide, chloro- or dichloro-acetylene, which ignite in air, are formed; forms impact-sensitive mixtures with potassium and sodium.

First aid　　Standard treatment for exposure by all routes (see pages 108-122).

Fire hazard　　Extinguish fires with water spray, carbon dioxide, dry chemical powder, or alcohol or polymer foam.

Spillage disposal　　See general section.

RSC *Chemical Safety Data Sheets* Vol. 1, No. 91, 1988 and Vol. 4b, No. 139, 1991 give extended coverage.

1185. Tetrachloroethylene

Colourless liquid with faint ethereal odour; m.p. -23.3 °C; b.p. 121.2 °C; immiscible with water.

RISKS
Possible risk of irreversible effects (R40)

SAFETY PRECAUTIONS
Do not breathe vapour – Wear protective clothing and gloves (S23, S36/37)

Limit values   OES short-term 150 p.p.m. (1000 mg m^{-3}); long-term 50 p.p.m. (335 mg m^{-3}).

Toxic effects   Inhalation of the vapour may cause dizziness, nausea, vomiting, and, in high concentrations, stupor. The liquid irritates the eyes and has an irritant and degreasing action on the skin. Ingestion of large quantities will cause symptoms similar to those of inhalation.

Hazardous reactions   Impure material containing trichloroethylene, when treated with solid NaOH and subsequently distilled, may have a volatile, explosive fore-run.

First aid   Standard treatment for exposure by all routes (see pages 108-122).

Fire hazard   Extinguish fires with water spray. Will produce toxic fumes on decomposition.

Spillage disposal   See general section.

RSC *Chemical Safety Data Sheets* Vol. 1, No. 92, 1988 gives extended coverage.

1186. Tetrachloronaphthalene

Colourless to pale yellow, crystalline solid; m.p. 182 °C; b.p. 311-360 °C; insoluble in water.

Limit values   OES short-term 4 mg m^{-3} (all isomers); long-term 2 mg m^{-3} (all isomers).

Toxic effects   Irritant to skin and eyes. Prolonged skin contact can cause chloracne.

First aid   Standard treatment for exposure by all routes (see pages 108-122).

Fire hazard   Flash point 210 °C.

1187. 2,3,4,6-Tetrachlorophenol

Crystalline solid; m.p. 70 °C; b.p. 150 °C at 15 mm Hg; almost insoluble in water.

RISKS
Toxic if swallowed – Irritating to eyes and skin (R25, R36/38)

SAFETY PRECAUTIONS
In case of contact with eyes, rinse immediately with plenty of water and seek medical advice – After contact with skin, wash immediately with plenty of water – Wear suitable gloves – If you feel unwell, seek medical advice (show label where possible) (S26, S28, S37, S44)

Toxic effects   Irritant to upper respiratory tract. Repeated exposure of the skin to the solid or 10% aqueous dispersion may cause acneform dermatitis, but exposure to solutions in organic solvents can lead to systemic toxicity.

First aid   Standard treatment for exposure by all routes (see pages 108-122), but particular attention should be paid to the possibility of convulsions, shock, and the occurrence of lung congestion.

Fire hazard   Extinguish fires with carbon dioxide, dry chemical powder, or alcohol or polymer foam.

Spillage   See general section.
disposal

RSC *Chemical Safety Data Sheets* Vol. 4b, No. 140, 1991 gives extended coverage.

1188. Tetraethylenepentamine

Pale yellow-brown, viscous, hygroscopic liquid; m.p. -40 °C; b.p. 340 °C; miscible with water.

RISKS
Harmful in contact with skin and if swallowed – Causes burns – May cause sensitization by skin contact (R21/22, R34, R43)

SAFETY PRECAUTIONS
In case of contact with eyes, rinse immediately with plenty of water and seek medical advice – Wear suitable protective clothing, gloves, and eye/face protection (S26, S36/37/39)

Toxic effects   The liquid irritates the eyes and skin and may cause burns. It is irritating and corrosive if taken internally. Inhalation of the vapour is irritant to the mucous membranes and may be fatal following bronchial spasm, oedema, and inflammation.

First aid   Standard treatment for exposure by all routes (see pages 108-122).

Fire hazard   Flash point 185 °C; Autoignition temperature 300 °C. Extinguish fires with carbon dioxide, dry chemical powder, or alcohol or polymer foam.

Spillage   See general section.
disposal

RSC *Chemical Safety Data Sheets* Vol. 3, No. 66, 1990 gives extended coverage.

1189. Tetraethyl lead

Colourless oily liquid with pleasant characteristic odour; m.p. -130 °C; b.p. approx. 200 °C with decomposition; insoluble in water.

RISKS
Very toxic by inhalation, in contact with skin, and if swallowed – Danger of cumulative effects (R26/27/28, R33)

SAFETY PRECAUTIONS
Keep away from food, drink, and animal feeding stuffs – In case of contact with eyes, rinse immediately with plenty of water and seek medical advice – Wear protective clothing and gloves – In case of accident or if you feel unwell, seek medical advice immediately (show label where possible) (S13, S26, S36/37, S45)

Limit values   OES long-term 0.1 mg m^{-3}.

Toxic effects   Experimental carcinogen and teratogen. Inhalation can cause headache, anxiety, nervous excitation, and irritation.

Hazardous reactions   Failure to cover the residue with water, after emptying a tank of the compound, caused explosive decomposition after several days.

First aid   Standard treatment for exposure by all routes (see pages 108-122).

Fire hazard   Flash point 94 °C. Extinguish fires with water spray, carbon dioxide, dry chemical powder, or foam.

Spillage disposal   See general section.

1190. Tetraethyl silicate

Colourless liquid; b.p. 166 °C; practically insoluble in water, by which it is slowly hydrolysed.

RISKS
Flammable – Harmful by inhalation – Irritating to eyes and respiratory system (R10, R20, R36/37)

Limit values   OES short-term 30 p.p.m. (255 mg m^{-3}); long-term 10 p.p.m. (85 mg m^{-3}).

Toxic effects   The vapour irritates the eyes and respiratory system and is narcotic in high concentrations. The liquid irritates the eyes and may irritate the skin; it is irritant and damaging if taken by mouth.

First aid   Standard treatment for exposure by all routes (see pages 108-122).

Fire hazard   Flash point 52 °C. Extinguish fire with foam, dry powder, or carbon dioxide.

Spillage disposal   See general section.

1191. Tetrafluoroethylene

Colourless gas; m.p. -142 °C; b.p. -78.4 °C; insoluble in water.

Toxic effects   Irritant to skin, eyes, upper respiratory tract, and mucous membranes.

Hazardous reactions   The monomer explodes spontaneously at pressures above 27 bar (2.7 x 10^5 Pa); the inhibited monomer will explode if ignited; the liquid monomer, exposed to air, will form an explosive peroxidic polymer; explosive reactions with several other substances.

First aid   Standard treatment for exposure by all routes (see pages 108-122).

Fire hazard   Explosive limits 11-60%. Use water spray to keep cylinders cool or move them out of the immediate area.

1192. Tetrafluorohydrazine

Colourless gas; m.p. -163 °C; b.p. -73 °C.

Toxic effects   Irritant to skin, eyes, and upper respiratory tract. Harmful on inhalation.

Hazardous reactions	Explodes on contact with air or combustible vapours or on irradiation; mixtures with hydrocarbons are potentially highly explosive.
First aid	Standard treatment for exposure by all routes (see pages 108-122).
Fire hazard	Keep cylinders cool with water spray or remove them from the immediate area.

1193. Tetrahydrofuran

Colourless, volatile liquid with ethereal odour; m.p. -108 ˚C; b.p. 66 ˚C; miscible with water.

RISKS
Highly flammable – May form explosive peroxides – Irritating to eyes and respiratory system (R11, R19, R36/37)

SAFETY PRECAUTIONS
Keep away from sources of ignition – No Smoking – Do not empty into drains – Take precautionary measures against static discharges (S16, S29, S33)

Limit values	OES short-term 250 p.p.m. (735 mg m^{-3}); long-term 200 p.p.m. (590 mg m^{-3}).
Toxic effects	The vapour irritates the eyes and respiratory system; high concentrations have a narcotic effect. Skin exposure can cause dermatitis.
Hazardous reactions	Readily forms peroxides by autoxidation. Peroxidized material should not be dried with sodium or potassium hydroxides as explosions may occur.
First aid	Standard treatment for exposure by all routes (see pages 108-122).
Fire hazard	Flash point -17 ˚C; Explosive limits 1.8- 11.8%; Autoignition temperature 321 ˚C. Extinguish fire with foam, dry powder, or carbon dioxide.
Spillage disposal	See general section.

RSC *Chemical Safety Data Sheets* Vol. 1, No. 93, 1988 gives extended coverage.

1194. Tetrahydrophthalic anhydride

White, crystalline powder; m.p. 99-101 ˚C; b.p. 195 ˚C at 50 mm Hg; reacts exothermically with water.

RISKS
Irritating to eyes and respiratory system (R36/37)

SAFETY PRECAUTIONS
Avoid contact with eyes (S25)

Toxic effects	Irritant and corrosive to skin, eyes, and respiratory tract. Mildly toxic orally. Inhalation may be fatal following inflammation, chemical pneumonitis, spasm, and oedema.
First aid	Standard treatment for exposure by all routes (see pages 108-122).
Fire hazard	Flash point 157 ˚C (open cup). Extinguish fires with carbon dioxide, dry chemical powder, or alcohol or polymer foam.
Spillage disposal	See general section.

RSC *Chemical Safety Data Sheets* Vol. 3, No. 68, 1990 gives extended coverage.

1195. Tetrahydrothiophene

Colourless liquid with smell like that of coal gas, for which it is an established odorant; b.p. 121 ˚C; immiscible with water.

Toxic effects The evidence available is derived from animal experiments which showed that it was not irritant to the eyes and skin of rabbits and caused no permanent damage to the eyes. It is thought wisest to handle with caution and avoid breathing the vapour, which advertises its presence unmistakeably.

First aid Standard treatment for exposure by all routes (see pages 108-122).

Fire hazard Flash point 18 ˚C. Extinguish fire with foam, dry powder, or carbon dioxide.

Spillage disposal See general section.

1196. Tetralin

Colourless liquid with menthol-like odour; m.p. -35 ˚C; b.p. 207 ˚C; insoluble in water.

Toxic effects Mildly irritant to skin, eyes, and respiratory tract. Narcotic at high concentrations. Ingestion has caused gastro-intestinal disturbances and transient damage to liver and kidneys.

Hazardous reactions Produces peroxides on prolonged exposure to air.

First aid Standard treatment for exposure by all routes (see pages 108-122).

Fire hazard Flash point 77 ˚C; Explosive limits 0.8-5%; Autoignition temperature 425 ˚C. Extinguish fires with water, carbon dioxide, dry chemical powder, or foam extinguishants.

Spillage disposal See general section.

RSC *Chemical Safety Data Sheets* Vol. 1, No. 94, 1988 gives extended coverage.

1197. Tetramethylammonium chlorite

Solid.

Hazardous reactions The dry solid explodes on impact.

1198. Tetramethylene diacrylate

RISKS
Harmful in contact with skin – Causes burns – May cause sensitization by skin contact (R21, R34, R43)

SAFETY PRECAUTIONS
In case of contact with eyes, rinse immediately with plenty of water and seek medical advice – Wear suitable protective clothing, gloves, and eye/face protection (S26, S36/37/39)

Toxic effects   Irritant to skin and eyes. Low acute oral toxicity.

First aid   Standard treatment for exposure by all routes (see pages 108-122).

Fire hazard   Extinguish fires with water spray, carbon dioxide, dry chemical powder, or alcohol or polymer foam.

Spillage disposal   See general section.

RSC *Chemical Safety Data Sheets* Vol. 3, No. 68, 1990 gives extended coverage.

1199. Tetramethyllead

Colourless liquid; m.p. -27.5 °C; b.p. 110 °C; insoluble in water.

RISKS
Very toxic by inhalation, in contact with skin, and if swallowed – Danger of cumulative effects (R26/27/28, R33)

SAFETY PRECAUTIONS
Keep away from food, drink, and animal feeding stuffs – In case of contact with eyes, rinse immediately with plenty of water and seek medical advice – Wear protective clothing and gloves – In case of accident or if you feel unwell, seek medical advice immediately (show label where possible) (S13, S26, S36/37, S45)

Toxic effects   Toxic on inhalation, ingestion, or skin absorption, produing nausea, vomiting, mental confusion, and generalized weakness. Severe restlessness and aggressive behaviour may develop. Experimental teratogen. Onset of symptoms may be delayed for several hours.

Hazardous reactions   Liable to explode above 90 °C.

First aid   Standard treatment for exposure by all routes (see pages 108-122).

Fire hazard   Flash point 37 °C. Extinguish fires with water spray, carbon dioxide, dry chemical powder, or foam extinguishants.

Spillage disposal   See general section.

1200. Tetramethyl orthosilicate

Clear liquid; m.p. 14 °C; b.p. 121-122 °C; hydrolysed by water.

Limit values   OES short-term 5 p.p.m. (30 mg m^{-3}); long-term 1 p.p.m. (6 mg m^{-3}).

Toxic effects   A severe eye irritant. Harmful by inhalation, ingestion, and skin absorption causing systemic effects.

Hazardous reactions   During the preparation of the hexamethoxides of rhenium, molybdenum, and tungsten by co-condensation with excess tetramethoxy silane on a cold surface, simultaneous co-condensation is necessary to avoid the danger of explosion present when sequential condensation of the reactants is employed.

First aid   Standard treatment for exposure by all routes (see pages 108-122).

Fire hazard   Flash point 45 °C. Extinguish fires with carbon dioxide, dry chemical powder, or alcohol or polymer foam.

Spillage disposal   See general section.

1201. Tetramethyl succinonitrile

White, crystalline solid; m.p. 169 °C with sublimation; almost insoluble in water.

Limit values   OES short-term 2 p.p.m. (9 mg m^{-3}); long-term 0.5 p.p.m. (3 mg m^{3}).

Toxic effects   Irritant to skin and eyes. Inhalation can lead to headache, nausea, vomiting, convulsions, coma, and death. Can be absorbed through the skin. Experimental teratogen.

First aid   Standard treatment for exposure by all routes (see pages 108-122).

Spillage disposal   See general section.

1202. Tetramethylurea

Colourless liquid, slight mint odour; m.p. -1 °C; b.p. 176 °C; miscible with water.

Toxic effects   Contact of liquid with eyes may lead to conjunctivitis or temporary opacity of the cornea. Prolonged inhalation or skin contact may lead to systemic injury.

First aid   Standard treatment for exposure by all routes (see pages 108-122).

Fire hazard   Flash point 65 °C. Extinguish fires with water spray, carbon dioxide, dry chemical powder, or alcohol or polymer foam.

Spillage disposal   See general section.

1203. Tetrasilane

Liquid; m.p. approx -90 ˚C; b.p. 109 ˚C; decomposes in water.

Hazardous　　Ignites and explodes in air or oxygen; reacts vigorously with carbon
reactions　　tetrachloride.

1204. Thallium and salts

Thallium metal is a bluish-white, soft, malleable, heavy metal often stored under water; the common salts are colourless, white, or off-white, crystalline solids; the oxide is black; the common salts are soluble in water; the metal and the oxide are insoluble in water.

RISKS
Very toxic by inhalation and if swallowed – Danger of cumulative effects (R26/28, R33)

SAFETY PRECAUTIONS
Keep out of reach of children – Keep away from food, drink, and animal feeding stuffs – After contact with skin, wash immediately with plenty of water – In case of accident or if you feel unwell, seek medical advice immediately (show label where possible) (S2, S13, S28, S45)

Limit values　　OES long-term 0.1 mg m^{-3} (soluble compounds) (as Tl).

Toxic effects　　The dusts irritate the nose and eyes and may cause nausea and abdominal pain by absorption. The metal on contact with moist skin produces a white film of the hydroxide. Skin absorption of the soluble salts and ingestion may cause nausea, vomiting, abdominal pains, weakness of the legs, mental confusion, coma, convulsions, and death. Exposure over a long period to small amounts of the dust or solutions may result in loss of appetite, falling out of hair, pain or weakness of limbs, insomnia, and mental disturbance.

Hazardous　　The nitrate may react violently or explosively with a variety of inorganic
reactions　　compounds.

First aid　　Standard treatment for exposure by all routes (see pages 108-122).

Fire hazard　　Use extinguishant appropriate to the immediate surroundings.

Spillage　　See general section.
disposal

RSC *Chemical Safety Data Sheets* Vol. 2, Nos. 111-117, 1989 give extended coverage for thallium metal and several of its salts.

1205. Thiocyanogen

Liquid or yellow solid; m.p. -2 to -3 ˚C; polymerizes explosively above m.p.; decomposed by water.

1206. Thionyl chloride

Pale yellow or yellow, fuming liquid with an odour like sulfur dioxide; m.p. -105 °C; b.p. 79 °C; reacts with water to form hydrochloric acid and sulfur dioxide.

RISKS
Reacts violently with water – Causes burns – Irritating to respiratory system (R14, R34, R37)

SAFETY PRECAUTIONS
In case of contact with eyes, rinse immediately with plenty of water and seek medical advice (S26)

Limit values	OES short-term 1 p.p.m. (5 mg m^{-3}).
Toxic effects	Extremely destructive to skin, eyes, upper respiratory tract, and mucous membranes. Harmful if inhaled, ingested, or absorbed through the skin. Inhalation may be fatal as a result of spasm, inflammation, and oedema.
Hazardous reactions	Addition of concentrated ammonia may cause an explosion; reacts violently or explosively with several substances; reaction with water generates 990 volumes of mixed corrosive gases.
First aid	Standard treatment for exposure by all routes (see pages 108-122).
Fire hazard	Extinguish fires with dry chemical powder. Do NOT use water.
Spillage disposal	See general section.

1207. Thiophene

Clear, colourless liquid; m.p. -38 °C; b.p. 84 °C; insoluble in water.

Toxic effects	Causes severe eye irritation. May be harmful on inhalation, ingestion, or skin absorption.
Hazardous reactions	Reacts very violently with fuming nitric acid.
First aid	Standard treatment for exposure by all routes (see pages 108-122).
Fire hazard	Flash point -6 °C; Explosive limits (lower) 2.3%. Extinguish fire with dry chemical powder, carbon dioxide, or alcohol or polymer foam.
Spillage disposal	See general section.

1208. Thiophosphoryl fluoride

Liquid or gas; m.p. 3.8 °C at 7.6 atm. pressure; decomposes on further heating.

Toxic effects	Irritant to skin, eyes, and mucous membranes.
Hazardous reactions	Ignites or explodes on contact with air; heated sodium ignites in the gas.

1209. Thiourea

White, crystalline powder; m.p. 175-178 ˚C; decomposes on further heating; soluble in water.

RISKS
Harmful if swallowed – Possible risk of irreversible effects (R22, R40)

SAFETY PRECAUTIONS
Do not breathe dust – Avoid contact with skin (S22, S24)

Toxic effects	May cause eye and skin irritation and allergic reactions. Carcinogenic.
Hazardous reactions	Exposure to acrylaldehyde causes exothermic and violent polymerization. The solid peroxide produced by the action of hydrogen peroxide and nitric acid on thiourea decomposed violently on drying in air.
First aid	Standard treatment for exposure by all routes (see pages 108-122).
Fire hazard	Extinguish fires with water spray, carbon dioxide, dry chemical powder, or alcohol or polymer foam.
Spillage disposal	See general section.

1210. Thorium

White, soft, ductile metal; m.p. 1695 ˚C; b.p. approx. 4225 ˚C; insoluble in water.

Toxic effects	The acute and chronic toxicity of thorium by all routes of exposure is relatively low.
Hazardous reactions	Finely divided metal is pyrophoric in air; incandesces in chlorine, bromine, and iodine vapour; reactions with phosphorus and sulfur are also incandescent.
First aid	Standard treatment for exposure by all routes (see pages 108-122).
Fire hazard	Use extinguishant appropriate to the surroundings.
Spillage disposal	See general section.

1211. Tin

Silvery-white, lustrous, soft, ductile metal. May have a yellowish oxide film when cast; m.p. 232 ˚C; b.p. 2507 ˚C; insoluble in water.

Limit values	OES short-term 4 mg m^{-3}; long-term 2 mg m^{-3}.
Toxic effects	Moderately irritating to skin, eyes, and upper respiratory tract. Poorly absorbed from the gastro-intestinal tract.
Hazardous reactions	Reacts dangerously or explosively with a variety of inorganic compounds.
First aid	Standard treatment for exposure by all routes (see pages 108-122).

Fire hazard	Use extinguishant appropriate to the surroundings.
Spillage disposal	See general section.

RSC *Chemical Safety Data Sheets* Vol. 2, No. 118, 1989 gives extended coverage.

1212. Tin, organic compounds of

Certain alkyltin compounds – notably di-n-butyltin diacetate, dilaurate, maleate, and oxide – are used quite extensively as stabilizers for PVC resins, as catalysts, and biocides.

Limit values	OES short-term 0.2 mg m^{-3} (as Sn); long-term 0.1 mg m^{-3} (as Sn).
Toxic effects	Harmful by skin absorption. Some, as dust, irritate the respiratory system.
First aid	Standard treatment for exposure by all routes (see pages 108-122).
Spillage disposal	See general section.

1213. Tin(II) chloride

Colourless, crystalline solid; m.p. 246 °C; b.p. 652 °C; soluble in water.

Limit values	OES short-term 4 mg m^{-3} (as Sn); long-term 2 mg m^{-3} (as Sn).
Toxic effects	Irritant to skin, eyes, and mucous membranes. Poorly absorbed from the gut. Inhalation may cause dyspnoea, coughing, slight fever, and aching muscles.
Hazardous reactions	Reaction with hydrogen peroxide is strongly exothermic, even in solution.
First aid	Standard treatment for exposure by all routes (see pages 108-122).
Fire hazard	Extinguish fires with dry chemical powder.
Spillage disposal	See general section.

RSC *Chemical Safety Data Sheets* Vol. 2, No. 119, 1989 gives extended coverage.

1214. Tin(IV) chloride

Colourless, fuming liquid; m.p. -33 °C; b.p. 114 °C; reacts with water forming hydrochloric acid.

Limit values	OES short-term 4 mg m^{-3} (as Sn); long-term 2 mg m^{-3} (as Sn).
Toxic effects	The vapour irritates the respiratory system and may cause dyspnoea, coughing, slight fever, and aching muscles. The vapour irritates the eyes. The liquid irritates the eyes and skin and may cause burns. Will result in internal irritation and damage if taken by mouth.
Hazardous reactions	Traces may catalyse delayed decomposition of alkyl nitrates or violent polymerization of ethylene oxide.

First aid	Standard treatment for exposure by all routes (see pages 108-122).
Fire hazard	Extinguish fires with dry chemical powder.
Spillage disposal	See general section.

RSC *Chemical Safety Data Sheets* Vol. 2, No. 120, 1989 gives extended coverage.

1215. Tin(IV) oxide

White crystals; m.p. 1127 ˚C; b.p. sublimes at 1800-1900 ˚C; insoluble in water.

Limit values	OES short-term 4 mg m^{-3} (as Sn); long-term 2 mg m^{-3} (as Sn).
Toxic effects	May be irritant to skin and eyes. Inhalation of dust or fumes causes a benign pneumoconiosis.
Hazardous reactions	Reacts violently or explosively with a variety of inorganic compounds.
First aid	Standard treatment for exposure by all routes (see pages 108-122).
Fire hazard	Extinguish fires with dry chemical powder.
Spillage disposal	See general section.

RSC *Chemical Safety Data Sheets* Vol. 2, No. 121, 1989 gives extended coverage.

1216. Titanium

Dark grey solid; m.p. 1677 ˚C; b.p. approx. 3277 ˚C; insoluble in water.

Toxic effects	May cause skin and eye irritation. May be harmful if inhaled or ingested.
Hazardous reactions	The finely divided metal is pyrophoric and once burning is difficult to extinguish as it burns in both carbon dioxide and nitrogen. Violent reactions with halogens, halocarbons, metal salts, and oxidants.
First aid	Standard treatment for exposure by all routes (see pages 108-122).
Fire hazard	Autoignition temperature 1200 ˚C for solid metal; 250 ˚C for dust. Use class D fire extinguishing agents only. Do NOT use water or dry chemical powder.
Spillage disposal	See general section.

1217. Titanium(II) chloride

Black, crystalline solid; m.p. 475 ˚C in vacuo with decomposition; reacts with water evolving hydrogen.

| **Hazardous reactions** | Ignites readily in air, particularly if moist. |

1218. Titanium(III) chloride

Violet, crystalline solid; m.p. 440 °C with decomposition; reacts violently with water.

Toxic effects   Extremely destructive to skin, eyes, upper respiratory tract, and all mucous membranes. Harmful if inhaled, ingested, or absorbed through the skin. Inhalation may be fatal as a result of spasm, inflammation, and oedema.

Hazardous reactions   Reacts vigorously with air or water, pyrophoric if finely divided.

First aid   Standard treatment for exposure by all routes (see pages 108-122).

Fire hazard   Use only class D fire extinguishing agents. Do NOT use water.

Spillage disposal   See general section.

1219. Titanium(IV) chloride

Colourless, fuming liquid with a pungent odour; m.p. - 30 °C; b.p. 136 °C; soluble in cold water but decomposed by hot water with liberation of hydrogen chloride.

RISKS
Reacts violently with water – Causes burns – Irritating to eyes and respiratory system (R14, R34, R36/37)

SAFETY PRECAUTIONS
Keep container tightly closed and dry – In case of contact with eyes, rinse immediately with plenty of water and seek medical advice (S7/8, S26)

Toxic effects   The vapour irritates the eyes and respiratory system. The liquid irritates the skin and eyes and may cause burns. Assumed to cause severe internal irritation and damage if taken by mouth.

First aid   Standard treatment for exposure by all routes (see pages 108-122).

Fire hazard   Extinguish fires with carbon dioxide, dry chemical powder, or alcohol or polymer foam.

Spillage disposal   See general section.

1220. Titanium dioxide

White solids; m.p. 1855 °C; insoluble in water.

Limit values   OES long-term 10 mg m^{-3} (total inhalable dust), 5 mg m^{-3} (respirable dust).

Toxic effects   Skin irritant. Slight lung fibrosis if inhaled.

First aid   Standard treatment for exposure by all routes (see pages 108-122).

Fire hazard   Use extinguishant appropriate to the surroundings.

Spillage disposal   See general section.

1221. *o*-Tolidine and *o*-tolidine dihydrochloride

The base is a colourless to grey or brown powder; m.p. (base) 129-131 ˚C;(salt) decomposes above 340 ˚C; slightly soluble in water.

RISKS
May cause cancer – Harmful if swallowed (R45, R22)

SAFETY PRECAUTIONS
Avoid exposure – Obtain special instructions before use – If you feel unwell, seek medical advice (show label where possible) (S53, S44)

Toxic effects   The dihydrochloride and its solutions irritate the skin and eyes. There is evidence that *o*-tolidine, through continued absorption, can cause cancer of the bladder. The use of *o*-tolidine and its salts is controlled in the United Kingdom by the Carcinogenic Substances Regulations 1967.

First aid   Standard treatment for exposure by all routes (see pages 108-122).

Fire hazard   Extinguish fires with water spray, carbon dioxide, dry chemical powder, or alcohol or polymer foam.

Spillage disposal   See general section.

RSC *Chemical Safety Data Sheets* Vol. 4b, No. 142, 1991 gives extended coverage of the base.

1222. Toluene

Colourless liquid with characteristic odour; m.p. -95 ˚C; b.p. 111 ˚C; immiscible with water.

RISKS
Highly flammable – Harmful by inhalation (R11, R20)

SAFETY PRECAUTIONS
Keep away from sources of ignition – No Smoking – Avoid contact with eyes – Do not empty into drains – Take precautionary measures against static discharges (S16, S25, S29, S33)

Limit values   OES short-term 150 p.p.m. (560 mg m^{-3}); long-term 50 p.p.m. (188 mg m^{-3}).

Toxic effects   Inhalation of the vapour may cause dizziness, headache, nausea, and mental confusion. The vapour and liquid irritate the eyes and mucous membranes. Absorption through the skin and ingestion would cause poisoning. If the toluene contains more than traces of benzene as an impurity, breathing of vapour over long periods may cause blood disease. Prolonged skin contact may cause dermatitis.

Hazardous reactions   Violent or explosive reactions with a range of oxidants and other substances.

First aid   Standard treatment for exposure by all routes (see pages 108-122).

Fire hazard   Flash point 4 ˚C; Explosive limits 1.3-7.0%; Autoignition temperature 530-600 ˚C. Extinguish small fires with dry powder, carbon dioxide, or alcohol-resistant foam. Large fires are best controlled by the foam.

Spillage disposal   See general section.

RSC *Chemical Safety Data Sheets* Vol. 1, No. 95, 1988 gives extended coverage.

1223. Toluene-2-, -3-, and -4-diazonium salts

Hazardous reactions   Hazardous reactions with ammonium sulfide, hydrogen sulfide, or potassium iodide.

1224. *p*-Toluenesulfonyl chloride

Off-white, crystalline solid; m.p. 69 °C; b.p. 134.5 °C at 10 mm Hg; insoluble in water but moisture sensitive.

Limit values   OES short-term 5 mg m^{-3}.

Toxic effects   Extremely destructive to skin, eyes, upper respiratory tract, and mucous membranes. Harmful if inhaled, ingested, or absorbed through the skin. Inhalation may be fatal as a result of spasm, inflammation, chemical pneumonitis, and pulmonary oedema.

First aid   Standard treatment for exposure by all routes (see pages 108-122).

Fire hazard   Extinguish fires with carbon dioxide, dry chemical powder, or alcohol or polymer foam. Do NOT use water.

Spillage disposal   See general section.

1225. Toluidines

o-Toluidine and *m*-toluidine are red to dark brown liquids; *p*-toluidine consists of pale brown crystals; m.p. (*o*-) -16.4 °C; (*m*-) -31.3 °C; (*p*-) 43.8 °C; b.p. (*o*-) 200.4 °C; (*m*-) 203.4 °C; (*p*-) 200.6 °C; slightly soluble in water.

RISKS
Toxic by inhalation, in contact with skin, and if swallowed – Danger of cumulative effects (R23/24/25, R33)

SAFETY PRECAUTIONS
After contact with skin, wash immediately with plenty of water – Wear protective clothing and gloves – If you feel unwell, seek medical advice (show label where possible) (S28, S36/37, S44)

Limit values   OES short-term 5 p.p.m. (22 mg m^{-3}) (*o*-); long-term 2 p.p.m. (9 mg m^{-3}) (*o*-).

Toxic effects   Excessive breathing of the vapour, ingestion, or absorption through the skin may cause headache, drowsiness, cyanosis, mental confusion, and, in severe cases, convulsions. They are dangerous to the eyes.

Hazardous reactions   Ignite in contact with fuming nitric acid.

First aid   Standard treatment for exposure by all routes (see pages 108-122).

Fire hazard   Flash point (*o-*) + (*m-*) 85 ˚C; (*p-*) 88 ˚C; Explosive limits (*o-*) lower 1.5%; (*m-*) +
                     (*p-*) 1.1-6.6%; Autoignition temperature (*o-*) + (*p-*) 482 ˚C. Extinguish fires with
                     water spray, carbon dioxide, dry chemical powder, or alcohol or polymer foam.

Spillage   See general section.
disposal

RSC *Chemical Safety Data Sheets* Vol. 4b, Nos. 144 and 145, 1991 give extended coverage of
the *o-* and *p-*isomers.

1226. Tolylcoppers

Hazardous   All isomers usually explode strongly on exposure to oxygen at 0 ˚C, or weakly
reactions   above 100 ˚C in vacuo.

1227. Triallyl cyanurate

Colourless liquid or solid; m.p. 27 ˚C; b.p. 162 ˚C at 2 mm Hg; hydrolysed by water forming allyl
alcohol.

Toxic effects   These arise because of the allyl alcohol formed on hydrolysis.

First aid   Standard treatment for exposure by all routes (see pages 108-122).

Fire hazard   Flash point 80 ˚C. Extinguish fires with water spray, dry chemical powder, or
                     foam extinguishants.

Spillage   See general section.
disposal

1228. Triallyl phosphate

Hazardous   Distillation residue exploded.
reactions

1229. Tribenzylarsine

Hazardous   Oxidizes slowly at first in air but becomes violent.
reactions

1230. Tribromosilane

Mobile liquid; m.p. -73.5 ˚C; b.p. 112 ˚C; readily hydrolysed by water.

Toxic effects   Irritant because of the hydrogen bromide liberated on contact with moist
                     tissues.

Hazardous   Usually ignites when poured in air.
reactions

1231. Tributylbismuth

Hazardous   Explodes in oxygen and ignites in air.
reactions

1232. Tributylborane

(n-) Colourless liquid; m.p. (n-) 34 °C; b.p. (n-) 109 °C at 20 mm Hg; (iso-) 188 °C; may decompose with water.

Toxic effects   Irritant to skin, eyes, upper respiratory tract, and mucous membranes. May be harmful by inhalation or ingestion.

Hazardous   Mixture of n- and iso- isomers ignited on exposure to air.
reactions

First aid   Standard treatment for exposure by all routes (see pages 108-122).

Fire hazard   Flash point (n-) -35 °C. Extinguish fires with dry chemical powder.

Spillage   See general section.
disposal

1233. Tributyl phosphate

Colourless liquid; m.p. -79 °C; b.p. 180-183 °C at 22 mm Hg; sparingly soluble in water.

RISKS
Harmful if swallowed (R22)

SAFETY PRECAUTIONS
Avoid contact with eyes (S25)

Limit values   OES short-term 5 mg m^{-3} (all isomers); long-term 5 mg m^{-3} (all isomers).

Toxic effects   Irritant to skin, eyes, upper respiratory tract, and mucous membranes. May be harmful by inhalation, ingestion, or skin absorption. May cause central nervous system stimulation.

First aid   Standard treatment for exposure by all routes (see pages 108-122).

Fire hazard   Flash point 193 °C; Autoignition temperature 410 °C. Extinguish fires with water spray, carbon dioxide, dry chemical powder, or alcohol or polymer foam.

Spillage   See general section.
disposal

1234. Tributylphosphine

Colourless liquid with garlic-like odour. Usually packed under nitrogen to avoid oxidation; b.p. 240 °C; practically insoluble in water.

Toxic effects   Irritates respiratory system and skin; very irritant to eyes; moderately toxic by ingestion.

First aid   Standard treatment for exposure by all routes (see pages 108-122).

Fire hazard Flash point 40 ˚C; Autoignition temperature 200 ˚C. Extinguish fire with foam, dry powder, or carbon dioxide.

Spillage disposal See general section.

1235. Tributyl tin oxide

Clean pale yellow liquid; m.p. below -45 ˚C; b.p. 220-230 ˚C at 10 mm Hg; sparingly soluble in water.

RISKS
Toxic by inhalation, in contact with skin, and if swallowed (R23/24/25)

SAFETY PRECAUTIONS
In case of contact with eyes, rinse immediately with plenty of water and seek medical advice – Take off immediately all contaminated clothing – After contact with skin, wash immediately with plenty of water – If you feel unwell, seek medical advice (show label where possible) (S26, S27, S28, S44)

Limit values OES short-term 0.2 mg m^{-3}; long-term 0.1 mg m^{-3}.

Toxic effects Irritant to skin and eyes. Also irritant to upper respiratory tract producing sore throat, cough, and vomiting.

First aid Standard treatment for exposure by all routes (see pages 108-122).

Fire hazard Flash point above 112 ˚C.

Spillage disposal See general section.

1236. Tricarbonyl(η-cyclopentadienyl)manganese

Pale yellow, crystalline solid with camphoraceous odour; m.p. 76.8-77.1 ˚C but sublimes at 60 ˚C; insoluble in water.

Limit values OES short-term 0.3 mg m^{-3} (as Mn); long-term 0.1 mg m^{-3} (as Mn).

Toxic effects May irritate the skin and can be absorbed through it. Also harmful if ingested or absorbed through the skin. A mild narcotic which can cause kidney damage.

First aid Standard treatment for exposure by all routes (see pages 108-122).

Fire hazard Extinguish fires with water spray, carbon dioxide, dry chemical powder, or alcohol or polymer foam.

Spillage disposal See general section.

1237. Trichloroacetic acid

White, hygroscopic crystals; m.p. 58 ˚C; b.p. 195.5 ˚C at 754 mm Hg; soluble in water.

RISKS
Causes severe burns (R35)

SAFETY PRECAUTIONS
Avoid contact with skin and eyes – In case of contact with eyes, rinse immediately with plenty of water and seek medical advice (S24/25, S26)

Limit values   OES long-term 1 p.p.m. (5 mg m^{-3}) under review.

Toxic effects   On inhalation it is extremely destructive to the upper respiratory tract and could be fatal. Severely irritates the eyes and skin, producing blisters after a latent period, but is not readily absorbed through the skin. Assumed to cause severe irritation and damage if taken by mouth.

Hazardous reactions   Violent reaction with copper and dimethyl sulfoxide.

First aid   Standard treatment for exposure by all routes (see pages 108-122).

Fire hazard   Flash point above 110 ˚C (closed cup). Extinguish fires with carbon dioxide, dry chemical powder, or alcohol or polymer foam.

Spillage disposal   See general section.

RSC *Chemical Safety Data Sheets* Vol. 3, No. 70, 1990 gives extended coverage.

1238. Trichloroacetonitrile

Crystalline solid with odour of chloral and hydrogen cyanide; m.p. 61 ˚C; b.p. 83-84 ˚C.

RISKS
Toxic by inhalation, in contact with skin, and if swallowed (R23/24/25)

SAFETY PRECAUTIONS
If you feel unwell, seek medical advice (show label where possible) (S44)

Toxic effects   Strong irritant to skin, eyes, and upper respiratory tract. May be fatal following inhalation or ingestion. Mutagenic.

First aid   Standard treatment for exposure by all routes (see pages 108-122), with special attention to the treatment for cyanides.

Fire hazard   Extinguish fires with water spray, carbon dioxide, dry chemical powder, or alcohol or polymer foam.

Spillage disposal   See general section. Large spills may need specialist help.

RSC *Chemical Safety Data Sheets* Vol. 4b, No. 146, 1991 gives extended coverage.

1239. Trichloroacetyl chloride

Colourless liquid with acrid pungent odour; b.p. 118 ˚C; reacts with water forming trichloroacetic and hydrochloric acids.

Toxic effects The vapour irritates the eyes and respiratory system. The liquid burns the eyes and skin. Assumed to cause severe internal irritation and damage if taken by mouth.

First aid Standard treatment for exposure by all routes (see pages 108-122).

Fire hazard Extinguish fires with carbon dioxide or dry chemical powder.

Spillage disposal See general section.

1240. Trichloroaniline

Crystalline solid; m.p. 77.5-78.5 ˚C; b.p. 262 ˚C at 46 mm Hg.

RISKS
Toxic by inhalation, in contact with skin, and if swallowed – Danger of cumulative effects (R23/24/25, R33)

SAFETY PRECAUTIONS
After contact with skin, wash immediately with plenty of water – Wear protective clothing and gloves – If you feel unwell, seek medical advice (show label where possible) (S28, S36/37, S44)

Toxic effects May be irritant to skin, eyes, and upper respiratory tract. Mutagenic.

First aid Standard treatment for exposure by all routes (see pages 108-122), with particular attention being paid to the possibility of convulsions and/or shock.

Fire hazard Extinguish fires with water spray, carbon dioxide, dry chemical powder, or alcohol or polymer foam.

Spillage disposal See general section.

RSC *Chemical Safety Data Sheets* Vol. 4b, No. 147, 1991 gives extended coverage.

1241. 1,2,4-Trichlorobenzene

Colourless liquid or solid; m.p. 17 ˚C; b.p. 213 ˚C; insoluble in water.

Limit values OES short-term 5 p.p.m. (40 mg m^{-3}); long-term 5 p.p.m. (40 mg m^{-3}).

Toxic effects Irritant to skin, eyes, upper respiratory tract, and mucous membranes. Harmful if inhaled, ingested, or absorbed through the skin.

First aid Standard treatment for exposure by all routes (see pages 108-122).

Fire hazard Flash point above 110 ˚C. Extinguish fires with water spray, carbon dioxide, dry chemical powder, or alcohol or polymer foam.

Spillage disposal See general section.

1242. 1,1,1-Trichlorobis(chlorophenyl) ethane

White, crystalline solid; m.p. 108.5-109 °C; b.p. 260 °C; insoluble in water.

RISKS
Toxic if swallowed – Possible risk of irreversible effects – Danger of serious damage to health by prolonged exposure (R25, R40, R48)

SAFETY PRECAUTIONS
Do not breathe dust – Wear protective clothing and gloves – If you feel unwell, seek medical advice (show label where possible) (S22, S36/37, S44)

Limit values	OES short-term 3 mg m^{-3}; long-term 1 mg m^{-3}.
Toxic effects	Irritant. Absorption by inhalation, ingestion, or skin absorption can lead to headache, vomiting, sore throat, numbness, malaise, tremors, convulsions, liver damage, and cardiac or respiratory failure.
First aid	Standard treatment for exposure by all routes (see pages 108-122).
Fire hazard	Flash point 72-75 °C. Extinguish fires with water spray, dry chemical powder, or alcohol or polymer foam.
Spillage disposal	See general section.

1243. 1,1,1-Trichloroethane

Colourless liquid; b.p. 74 °C.; insoluble in water.

SAFETY PRECAUTIONS
Do not breathe vapour – Avoid contact with skin and eyes (S23, S24/25)

Toxic effects	The vapour may irritate the eyes and respiratory system; it is narcotic in high concentrations. The liquid may irritate the skin; it irritates the eyes without causing serious damage and must be assumed to be harmful if taken by mouth.
Hazardous reactions	Mixture with potassium may explode on light impact; violent decomposition, with evolution of HCl, may occur when it comes into contact with aluminium, magnesium, or their alloys.
Spillage disposal	See general section.

RSC *Chemical Safety Data Sheets* Vol. 1, No. 96, 1988 gives extended coverage.

1244. 1,1,2-Trichloroethane

Clear, colourless, non-flammable liquid; m.p. -36.7 ˚C; b.p. 113 ˚C; almost insoluble in water.

RISKS
Harmful by inhalation, in contact with skin, and if swallowed (R20/21/22)

SAFETY PRECAUTIONS
Keep container in a well ventilated place (S9)

Limit values OES short-term 20 p.p.m. (90 mg m^{-3}) under review; long-term 10 p.p.m. (45 mg m^{-3}) under review.

Toxic effects High concentrations of vapour are irritant to eyes and upper respiratory tract. Absorption by inhalation, ingestion, or through the skin can cause damage to liver and kidneys.

First aid Standard treatment for exposure by all routes (see pages 108-122).

Fire hazard Explosive limits 6.0-15.5%. Decomposes on heating to give very toxic fumes. Extinguish fires with extinguishant appropriate to the surroundings.

Spillage disposal See general section.

RSC *Chemical Safety Data Sheets* Vol. 1, No. 97, 1988 gives extended coverage.

1245. Trichloroethylene

Colourless, heavy liquid with sweetish chloroform-like odour; m.p. -87 ˚C; b.p. 86.7 ˚C; almost insolube in water.

RISKS
Possible risk of irreversible effects (R40)

SAFETY PRECAUTIONS
Do not breathe vapour – Wear protective clothing and gloves (S23, S36/37)

Limit values MEL short-term 150 p.p.m. (802 mg m^{-3}); long-term 100 p.p.m. (535 mg m^{-3}).

Toxic effects Inhalation of the vapour may cause headache, dizziness, nausea, and, with high concentrations, unconsciousness and death. The vapour and liquid irritate the eyes. Ingestion produces similar effects to inhalation of the vapour.

Hazardous reactions Decomposes with strong alkalies with evolution of spontaneously flammable dichloroacetylene; reacts violently with many metals and other substances. Corrosion products from hydrolysis during large scale distillation led to exothermic decomposition.

First aid Standard treatment for exposure by all routes (see pages 108-122).

Fire hazard Explosive limits 12.5-90%; Autoignition temperature 420 ˚C. Toxic fumes evolved on decomposition. Use water spray to fight fires.

Spillage disposal See general section.

RSC *Chemical Safety Data Sheets* Vol. 1, No. 98, 1988 gives extended coverage.

1246. Trichlorofluoromethane

Colourless liquid below 23 ˚C; m.p. -111 ˚C; b.p. 24.1 ˚C; insoluble in water.

Limit values	OES short-term 1250 p.p.m. (7000 mg m^{-3}); long-term 1000 p.p.m. (5600 mg m^{-3}).
Toxic effects	Mildly irritant to eyes. Skin contact can produce dryness and cryogenic effects. Inhalation produces respiratory distress, cardiotoxicity, asthma, and narcosis.
First aid	Standard treatment for exposure by all routes (see pages 108-122).
Fire hazard	Non-flammable but decomposes on heating to produce highly toxic fumes.
Spillage disposal	See general section.

RSC *Chemical Safety Data Sheets* Vol. 1, No. 99, 1988 gives extended coverage.

1247. Trichloromethanesulfenyl chloride

Oily, yellow liquid; b.p. 149 ˚C with slight decomposition; insoluble in water, slowly hydrolysed.

Toxic effects	The vapour irritates the eyes and respiratory system. The liquid irritates the skin and eyes. May be fatal if inhaled, ingested, or absorbed through the skin.
First aid	Standard treatment for exposure by all routes (see pages 108-122).
Fire hazard	Extinguish fires with carbon dioxide, dry chemical powder, or alcohol or polymer foam.
Spillage disposal	See general section.

1248. Trichloronaphthalene

Colourless to pale yellow solid with aromatic odour; m.p. 92.8 ˚C; b.p. 304-354 ˚C; insoluble in water.

Toxic effects	Skin contact can give rise to chloracne.
First aid	Standard treatment for exposure by all routes (see pages 108-122)
Fire hazard	Flash point 200 ˚C (open cup).

1249. Trichloronitromethane

Slightly oily, colourless liquid with intense penetrating odour; m.p. -64 ˚C; b.p. 112 ˚C; almost insoluble in water.

RISKS
Very toxic by inhalation, in contact with skin, and if swallowed – Irritating to eyes, respiratory system and skin (R26/27/26, R36/37/38)

SAFETY PRECAUTIONS
In case of contact with eyes, rinse immediately with plenty of water and seek medical advice – Wear suitable protective clothing – In case of accident or if you feel unwell, seek medical advice immediately (show label where possible) (S26, S36, S45)

Limit values	OES short-term 0.3 p.p.m. (2 mg m^{-3}); long-term 0.1 p.p.m. (0.7 mg m^{-3}).
Toxic effects	The vapour irritates the respiratory system leading in severe cases to bronchitis and recurrent asthmatic attacks through lung damage; it causes nausea and vomiting. The vapour irritates the eyes severely, causing intense lachrymation. The vapour and liquid irritate the skin. If swallowed, the liquid causes vomiting and diarrhoea.
Hazardous reactions	Above a critical volume, bulk containers can be shocked into detonation; reacts violently with aniline at 145 ˚C, and with alcoholic sodium hydroxide; mixture with 3-bromopropyne is shock- and heat-sensitive explosive.
First aid	Standard treatment for exposure by all routes (see pages 108-122).
Fire hazard	Use extinguishant appropriate to the surroundings.
Spillage disposal	See general section.

RSC *Chemical Safety Data Sheets* Vol. 4b, No. 148, 1991 gives extended coverage.

1250. 2,4,5- and 2,4,6-Trichlorophenols

Colourless crystals or grey flakes with strong, phenolic odours; m.p. 67 and 69 ˚C, respectively; b.p. 252 and 246 ˚C, respectively; both are practically insoluble in water.

RISKS
Harmful if swallowed – Irritating to eyes and skin (R22, R36/38)

SAFETY PRECAUTIONS
Do not breathe dust – Keep container dry (S22, S8)

Toxic effects	Inhalation, ingestion, or skin absorption of the dust or solid may result in lung, liver, or kidney damage; symptoms of poisoning are an increase followed by a decrease in respiratory rate and urinary output, fever, increased bowel action, weakness of movement, collapse, and convulsions. Skin contact may cause dermatitis.
First aid	Standard treatment for exposure by all routes (see pages 108-122).
Fire hazard	Extinguish fires with carbon dioxide, dry chemical powder, or alcohol or polymer foam.
Spillage disposal	See general section.

1251. 1,2,3-Trichloropropane

Colourless liquid; m.p. -14 ˚C; b.p. 156 ˚C; sparingly soluble in water.

RISKS
Harmful by inhalation, in contact with skin, and if swallowed (R20/21/22)

SAFETY PRECAUTIONS
Wear suitable gloves – Wear eye/face protection (S37, S39)

Limit values   OES short-term 75 p.p.m. (450 mg m^{-3}); long-term 50 p.p.m. (300 mg m^{-3}).

Toxic effects   Irritant to skin, eyes, upper respiratory tract, and mucous membranes. Exposure may cause burning sensation, coughing, wheezing, headache, nausea, and vomiting.

First aid   Standard treatment for exposure by all routes (see pages 108-122).

Fire hazard   Flash point 82 ˚C. Extinguish fires with water spray, carbon dioxide, dry chemical powder, or alcohol or polymer foam.

Spillage disposal   See general section.

1252. 2,4,6-Trichloro-s-triazine

Colourless crystals with pungent odour; m.p. 145.8 ˚C; b.p. 190 ˚C; hydrolyses in presence of water, forming hydrochloric acid.

RISKS
Irritating to eyes, respiratory system, and skin (R36/37/38)

SAFETY PRECAUTIONS
After contact with skin, wash immediately with plenty of water (S28)

Toxic effects   The dust irritates the respiratory system, eyes, and skin. Assumed to be irritating and damaging to the alimentary system if taken by mouth.

Hazardous reactions   Violent or explosive reactions with dimethylformamide, dimethyl sulfoxide, 2-ethoxyethanol, methanol, or water.

First aid   Standard treatment for exposure by all routes (see pages 108-122).

Fire hazard   Extinguish fires with carbon dioxide, dry chemical powder, or alcohol or polymer foam.

Spillage disposal   See general section.

1253. 1,1,2-Trichlorotrifluoroethane

Colourless non-combustible liquid; m.p. -35 ˚C; b.p. 47.6 ˚C; insoluble in water.

Limit values   OES short-term 1250 p.p.m. (9500 mg m^{-3}); long-term 1000 p.p.m. (7600 mg m^{-3}).

Toxic effects   Inhalation of vapour produces respiratory distress, cardiotoxicity, bronchopneumonia, and asthma.

First aid   Standard treatment for exposure by all routes (see pages 108-122).

Fire hazard   Decomposes on heating to produce toxic fumes. Use extinguishant appropriate to the surroundings.

Spillage disposal   See general section.

RSC *Chemical Safety Data Sheets* Vol. 1, No. 100, 1988 gives extended coverage.

1254. Triethanolamine

Viscous, colourless liquid or white solid; m.p. 18-21 ˚C; b.p. 190-193 ˚C at 5 mm Hg; soluble in water.

Toxic effects   Irritant to skin, eyes, upper respiratory tract, and mucous membranes.

First aid   Standard treatment for exposure by all routes (see pages 108-122).

Fire hazard   Flash point 185 ˚C. Extinguish fires with carbon dioxide, dry chemical powder, or alcohol or polymer foam.

Spillage disposal   See general section.

1255. Triethylaluminium

Colourless liquid; m.p. -50 ˚C; b.p. 128-130 ˚C at 50 mm Hg.

Toxic effects   Extremely destructive to skin, eyes, upper respiratory tract, and mucous membranes. Harmful if inhaled, ingested, or absorbed through the skin. Inhalation may be fatal as a result of spasm, inflammation, and oedema.

Hazardous reactions   Ignites in air; reacts vigorously or explosively with a variety of substances.

First aid   Standard treatment for exposure by all routes (see pages 108-122).

Fire hazard   Extinguish fires with dry chemical powder. Do NOT use water.

Spillage disposal   See general section.

1256. Triethylamine

Colourless liquid with strong ammoniacal odour; m.p. -115 ˚C; b.p. 89.4 ˚C; slightly soluble in water.

RISKS
Highly flammable – Irritating to eyes and respiratory system (R11, R36/37)

SAFETY PRECAUTIONS
Keep away from sources of ignition – No Smoking – In case of contact with eyes, rinse immediately with plenty of water and seek medical advice – Do not empty into drains (S16, S26, S29)

Limit values	OES short-term 15 p.p.m. (60 mg m^{-3}); long-term 10 p.p.m. (40 mg m^{-3}).
Toxic effects	The vapour irritates the eyes and respiratory system. The liquid irritates the skin and eyes. Assumed to be irritant and poisonous if taken by mouth.
Hazardous reactions	Complex with N_2O_4, containing excess of latter, exploded below 0 °C when free of solvent.
First aid	Standard treatment for exposure by all routes (see pages 108-122).
Fire hazard	Flash point -0.7 °C (open cup); Autoignition temperature 232 °C. Extinguish fire with dry powder, carbon dioxide, or alcohol-resistant foam.
Spillage disposal	See general section.

RSC *Chemical Safety Data Sheets* Vol. 1, No. 101, 1988 gives extended coverage.

1257. Triethylantimony

Hazardous reactions	Inflames in air.

1258. Triethylarsine

Colourless liquid; b.p. 140 °C with slight decomposition; insoluble in water.

Hazardous reactions	Inflames in air.

1259. Triethylbismuth

Liquid with disagreeable odour.

Hazardous reactions	Ignites in air, and explodes at about 150 °C.

1260. Triethylborane

Colourless liquid; m.p. -93 °C; b.p. 950 °C.

Toxic effects	Irritant to skin and eyes. Harmful if inhaled, ingested, or absorbed through the skin.
Hazardous reactions	Ignites in air.
First aid	Standard treatment for exposure by all routes (see pages 108-122).
Fire hazard	Extinguish fires with dry chemical powder.
Spillage disposal	See general section.

1261. Triethylene glycol diacrylate

RISKS
Irritating to eyes and skin – May cause sensitization by skin contact (R36/38, R43)

SAFETY PRECAUTIONS
In case of contact with eyes, rinse immediately with plenty of water and seek medical advice –
After contact with skin, wash immediately with plenty of water (S26, S28)

Toxic effects   Irritant to skin and eyes. Low oral toxicity. Possible animal carcinogen.

First aid   Standard treatment for exposure by all routes (see pages 108-122).

Fire hazard   Extinguish fires with water spray, carbon dioxide, dry chemical powder, or
                      alcohol or polymer foam.

Spillage   See general section.
  disposal

RSC *Chemical Safety Data Sheets* Vol. 3, No. 72, 1990 gives extended coverage.

1262. Triethylenetetramine

Colourless to yellowish, viscous liquid with an ammoniacal odour; m.p. 12 ˚C; b.p. 266-267 ˚C;
miscible with water.

RISKS
Harmful in contact with skin – Causes burns – May cause sensitization by skin contact (R21,
R34, R43)

SAFETY PRECAUTIONS
In case of contact with eyes, rinse immediately with plenty of water and seek medical advice –
Wear suitable protective clothing, gloves, and eye/face protection (S26, S36/37/39)

Toxic effects   It is irritant and corrosive to skin, eyes, upper respiratory tract, and mucous
                      membranes. Inhalation may be fatal following bronchial spasm, inflammation,
                      and oedema. It is readily absorbed through the skin.

First aid   Standard treatment for exposure by all routes (see pages 108-122).

Fire hazard   Flash point 135 ˚C; Autoignition temperature 337.8 ˚C. Extinguish fires with
                      water spray, carbon dioxide, dry chemical powder, or alcohol foam.

Spillage   See general section.
  disposal

RSC *Chemical Safety Data Sheets* Vol. 3, No. 73, 1990 gives extended coverage.

1263. Triethylgallium

Liquid; m.p. -82.3 ˚C; b.p. 142.6 ˚C.

Hazardous   Ignites in air.
  reactions

1264. Triethyl phosphate

Colourless liquid with a mild odour; m.p. -56.4 °C; b.p. 215 °C; soluble in water, slowly hydrolysed at elevated temperatures.

RISKS
Harmful if swallowed (R22)

SAFETY PRECAUTIONS
Avoid contact with eyes (S25)

Toxic effects   Irritant to skin and eyes.

First aid   Standard treatment for exposure by all routes (see pages 108-122).

Fire hazard   Flash point 115.5 °C; Explosive limits 1.7-10.0%; Autoignition temperature 452 °C. Extinguish fires with water spray, carbon dioxide, dry chemcial powder, or alcohol or polymer foam.

Spillage disposal   See general section.

1265. Triethylphosphine

Colourless liquid; b.p. 127.5 °C at 744 mm Hg; insoluble in water.

Toxic effects   May be harmful following inhalation, ingestion, or skin absorption.

Hazardous reactions   Explosive product by reaction of oxygen at low temperature.

First aid   Standard treatment for exposure by all routes (see pages 108-122).

Fire hazard   Flash point -17 °C. Extinguish fires with water spray, carbon dioxide, dry chemical powder, or alcohol or polymer foam.

Spillage disposal   See general section.

1266. Triethynylaluminium

Hazardous reactions   Residue from sublimation of dioxan complex is explosive; trimethylamine complex may also explode on sublimation.

1267. Triethynylantimony

Crystalline solid; m.p. 71-72 °C; decomposed by water.

Hazardous reactions   Explodes on strong friction.

1268. Triethynylarsine

Colourless crystals; m.p. 49-50 ˚C.

Hazardous   Explodes on strong friction.
reactions

1269. 1,3,5-Triethynylbenzene

Colourless, crystalline solid; m.p. 105-107 ˚C.

Hazardous   Exploded on rapid heating and compression.
reactions

1270. Triethynylphosphine

Crystalline solid; b.p. 36-37 ˚C.

Hazardous   Explodes on strong friction and may explode spontaneously on standing.
reactions

1271. Trifluoroacetic acid and anhydride

Colourless liquids with pungent odour; m.p. (acid) -15.36 ˚C; (anhydride) -65 ˚C; b.p. (acid) 71.8 ˚C; (anhydride) 40 ˚C; the acid is miscible with water, and the anhydride reacts with water forming the acid.

RISKS
(≥2%, ≤10%) Irritating to eyes, respiratory system, and skin (R36/37/38)
(>10%) Harmful by inhalation – Causes severe burns (R20, R35)

SAFETY PRECAUTIONS
(≥2%, ≤10%) Do not breathe vapour – In case of contact with eyes, rinse immediately with plenty of water and seek medical advice (S23, S26)
(>10%) – Keep container in a well ventilated place – In case of contact with eyes, rinse immediately with plenty of water and seek medical advice – Take off immediately all contaminated clothing – After contact with skin, wash immediately with plenty of water (S9, S26, S27, S28)

Toxic effects   The vapours irritate the eyes and respiratory system. Inhalation leads to considerable destruction of tissue and can be fatal. The liquids burn the eyes and quickly penetrate the skin to cause deep-seated burns. Assumed to cause severe burning and damage if taken by mouth.

First aid   Standard treatment for exposure by all routes (see pages 108-122).

Fire hazard   Use extinguishant appropriate to the surroundings.

Spillage   See general section.
disposal

RSC *Chemical Safety Data Sheets* Vol. 3, No. 74, 1990 gives extended coverage of trifluoroacetic acid.

1272. Trifluoromethanesulfonic acid

Colourless liquid; b.p. 162 ˚C; miscible with water.

Toxic effects Irritant to skin, eyes, upper respiratory tract, and mucous membranes. Harmful on inhalation, ingestion, or skin absorption. Inhalation may be fatal as a result of spasm, inflammation, and oedema.

Hazardous reactions As the strongest acid known, it can exert powerful catalytic effects, *e.g.* on Friedel-Crafts reactions.

First aid Standard treatment for exposure by all routes (see pages 108-122).

Fire hazard Extinguish fires with carbon dioxide, dry chemical powder, or alcohol or polymer foam.

Spillage disposal See general section.

1273. Trifluoroperoxyacetic acid

Hazardous reactions An extremely powerful oxidizing agent that must be used with great care.

1274. 3,3,3-Trifluoropropyne

Hazardous reactions Liable to explode.

1275. Tri-isobutylaluminium

Clear, colourless liquid; m.p. 4.3 ˚C; decomposes on further heating; reacts violently with water.

Toxic effects Extremely corrosive to all mucous membranes.

Hazardous reactions Powerful reductant supplied in hydrocarbon solvent; undiluted material ignites in air.

Fire hazard Autoignition temperature below 4 ˚C. Extinguish fires with carbon dioxide or dry chemical powder. Do NOT use water or foam.

Spillage disposal See general section.

1276. Tri-isopropylphosphine

Liquid; b.p. 176-178 ˚C.

Hazardous reactions Reacts rather vigorously with most peroxides, ozonides, *N*-oxides, and chloroform.

1277. Trimellitic anhydride

Off-white flakes; m.p. 161-163.5 ˚C; b.p. 240-245 ˚C at 14 mm Hg.

RISKS
Irritating to eyes, respiratory system, and skin – May cause sensitization by inhalation
(R36/37/38, R42)

SAFETY PRECAUTIONS
Do not breathe dust – After contact with skin, wash immediately with plenty of water (S22, S28)

Limit values   OES long-term 0.04 mg m^{-3}.

Toxic effects   Irritant to skin, eyes, upper respiratory tract, and mucous membranes. Inhalation produces an initial irritant response which can be followed by sensitization, longer term irritant effects, and possibly death.

First aid   Standard treatment for exposure by all routes (see pages 108-122).

Fire hazard   Flash point 227 ˚C. Extinguish fires with water spray, carbon dioxide, dry chemical powder, or alcohol or polymer foam.

Spillage   See general section.
  disposal

RSC *Chemical Safety Data Sheets* Vol. 3, No. 75, 1990 gives extended coverage.

1278. Trimercury tetraphosphide

Hazardous   Ignites in chlorine or when warmed in air; mixture with potassium chlorate
  reactions   explodes on impact.

1279. Trimethylaluminium

Colourless liquid; m.p. 15 ˚C; b.p. 125-126 ˚C; reacts with water to liberate flammable and/or explosive gas.

Toxic effects   Irritant and corrosive to skin, eyes, upper respiratory tract, and mucous membranes. Harmful if inhaled, ingested, or absorbed through the skin. Inhalation may be fatal as a result of spasm, inflammation, and oedema.

Hazardous   Extremely pyrophoric.
  reactions

First aid   Standard treatment for exposure by all routes (see pages 108-122).

Fire hazard   Extinguish fires with dry chemical powder. Do NOT use water.

Spillage   See general section.
  disposal

1280. Trimethylamine and solutions

Colourless gas with fishy odour that clings to clothes. Available in liquefied form in cylinders and also in aqueous and ethanolic solutions; m.p. (gas) -117 °C, b.p. (gas) 3 °O; soluble in water.

RISKS
Extremely flammable liquefied gas – Irritating to eyes and respiratory system (R13, R36/37)

SAFETY PRECAUTIONS
Keep away from sources of ignition – No Smoking – In case of contact with eyes, rinse immediately with plenty of water and seek medical advice – Do not empty into drains (S16, S26, S29)

Limit values OES short-term 15 p.p.m. (36 mg m^{-3}); long-term 10 p.p.m. (24 mg m^{-3}).

Toxic effects The vapour irritates the mucous membranes and respiratory system. Inhalation of high concentrations may affect the nervous system. The vapour and solutions irritate the eyes. The solutions may irritate the skin. Assumed to be poisonous if taken by mouth.

First aid Standard treatment for exposure by all routes (see pages 108-122).

Fire hazard Explosive limits (gas) 2.0-11.6%; Autoignition temperature (gas) 190 °C. (a) Gas: as the gas is supplied in a cylinder, turning off the valve will reduce any fire involving it; if possible, cylinders should be removed quickly from an area in which fire has developed. (b) Solutions in water and ethanol: extinguish fire with water spray, foam, dry powder, or carbon dioxide.

Spillage disposal Surplus gas or leaking cylinder can be vented slowly into a water-fed scrubbing tower or column, or into a fume cupboard served with such a tower. Also see general section.

1281. Trimethylamine oxide

Hygroscopic crystalline solid; m.p. 255-257 °C; soluble in water.

Hazardous reactions The oxide exploded during concentration.

1282. Trimethylarsine

Colourless liquid; b.p. 52.8 °C; slightly soluble in water.

Hazardous reactions Inflames in air; reaction with halogens is violent.

1283. Trimethylbismuth

Liquid; m.p. -86 ˚C; b.p. 110 ˚C.

Toxic effects   Toxic by inhalation, ingestion, or skin absorption. Can cause central nervous system depression, encephalopathy, and narcosis.

Hazardous reactions   Ignites in air.

1284. Trimethylborane

Gas; m.p. -161.5 ˚C; b.p. -20.2 ˚C.

Hazardous reactions   Ignites in air.

1285. Trimethylchlorosilane

Colourless, volatile, fuming liquid with pungent odour; b.p. 57 ˚C; reacts violently with water.

Toxic effects   The vapour and fumes are strongly irritant to the eyes, skin, and respiratory system. The liquid burns the skin and eyes and will cause severe damage if taken internally.

First aid   Standard treatment for exposure by all routes (see pages 108-122).

Fire hazard   Flash point -18 ˚C. Extinguish fire with dry sand, dry powder, or carbon dioxide.

Spillage disposal   See general section.

1286. Trimethylgallium

Liquid; m.p. -15.9 ˚C; b.p. 55.7 ˚C; reacts violently with water.

Hazardous reactions   Ignites in air and reacts violently with water.

1287. Trimethylolpropane triacrylate

Viscous, colourless liquid.

RISKS
Irritating to eyes and skin – May cause sensitization by skin contact (R36/38, R43)

SAFETY PRECAUTIONS
Wear eye/face protection (S39)

Toxic effects   Irritant to skin and eyes. Mildly toxic on ingestion.

First aid   Standard treatment for exposure by all routes (see pages 108-122).

Fire hazard   Flash point above 110 °C (closed cup). Extinguish fires with water spray, carbon dioxide, dry chemical powder, or alcohol or polymer foam.

Spillage   See general section.
disposal

RSC *Chemical Safety Data Sheets* Vol. 3, No. 76, 1990 gives extended coverage.

1288. 2,4,4-Trimethylpentene

Colourless liquid; m.p. -106.5 °C; b.p. 102 °C; insoluble in water.

RISKS
Highly flammable (R11)

SAFETY PRECAUTIONS
Keep container in a well ventilated place – Keep away from sources of ignition – No Smoking – Do not empty into drains – Take precautionary measures against static discharges (S9, S16, S29, S33)

Toxic effects   The vapour is slightly irritant at low concentrations, more so and narcotic at high concentrations. The liquid irritates the eyes and may irritate the skin.

First aid   Standard treatment for exposure by all routes (see pages 108-122).

Fire hazard   Flash point 2 °C. Extinguish fire with foam, dry powder, or carbon dioxide.

Spillage   See general section.
disposal

1289. Trimethyl phosphate

Colourless liquid; m.p. -47.1 °C; b.p. 197.2 °C; soluble in water.

Toxic effects   Harmful in inhaled, ingested, or absorbed through the skin. May cause irritation. Animal carcinogen.

Hazardous   Distillation residue exploded.
reactions

First aid   Standard treatment for exposure by all routes (see pages 108-122).

Fire hazard   Extinguish fires with water spray, carbon dioxide, dry chemical powder, or alcohol or polymer foam.

Spillage   See general section.
disposal

1290. Trimethylphosphine

Colourless liquid; b.p. 40-42 °C; insoluble in water.

Hazardous   May ignite in air.
reactions

1291. Trimethyl phosphite

Colourless liquid; m.p. -78 ˚C; b.p. 111-112 ˚C; hydrolysed by water.

Limit values   OES long-term 2 p.p.m. (10 mg m^{-3}).

Toxic effects   High concentrations are extremely destructive to tissues of the upper respiratory tract and mucous membranes. Symptoms of exposure may include burning sensation, coughing, wheezing, laryngitis, headache, nausea, and vomiting. The hydrolysis product (dimethyl hydrogen phosphite) is an animal carcinogen.

Hazardous reactions   Forms explosive mixtures with air.

First aid   Standard treatment for exposure by all routes (see pages 108-122).

Fire hazard   Flash point 27 ˚C. Extinguish fires with carbon dioxide, dry chemical powder, or alcohol or polymer foam.

Spillage disposal   See general section.

1292. Trimethylthallium

Crystalline solid; m.p. 38.5 ˚C.

Hazardous reactions   Liable to explode above 90 ˚C; ignites in air.

1293. Trinitroacetonitrile

Waxy solid; m.p. 41.5 ˚C; b.p. explodes at 220 ˚C; decomposed by water.

1294. 2,2,2-Trinitroethanol

Crystalline solid; m.p. 72 ˚C.

Hazardous reactions   Shock-sensitive explosive which has exploded during distillation.

1295. Trinitromethane

Colourless crystalline solid; m.p. 23 ˚C with explosion; b.p. 45-47 ˚C at 22 mg Hg; soluble in water.

Toxic effects   Irritant to skin, eyes, and all mucous membranes. Inhalation causes headache, nausea, and mild narcosis.

Hazardous reactions   Exploded during distillation; exploded in mixture with an impure ketone; frozen mixtures of trinitromethane and propan-2-ol exploded during thawing.

1296. 2,4,6-Trinitrotoluene

Crystalline solid but material now supplied wetted with not less than 30 wt% water.; m.p. 80.1 ˚C; b.p. explodes at 280 ˚C; sparingly soluble in water.

Limit values	OES long-term 0.5 mg m^{-3}.
Toxic effects	Harmful on inhalation, ingestion, or skin absorption. Experimental teratogen.
Hazardous reactions	The explosion temperature of TNT was reduced by addition of 1% red lead, sodium carbonate, or potassium hydroxide.
First aid	Standard treatment for exposure by all routes (see pages 108-122).
Spillage disposal	See general section.

1297. Triphenylaluminium

Crystalline solid; m.p. 229-232 ˚C; reacts violently with water.

Hazardous reactions	Evolves heat and sparks on contact with water.

1298. Triphenyl phosphate

White, crystalline solid; m.p. 49-50 ˚C; b.p. 245 ˚C at 11 mm Hg; insoluble in water.

Limit values	OES short-term 6 mg m^{-3}; long-term 3 mg m^{-3}.
Toxic effects	Irritant to skin and eyes.
First aid	Standard treatment for exposure by all routes (see pages 108-122).
Fire hazard	Use extinguishant appropriate to the surroundings.
Spillage disposal	See general section.

1299. Triphenyl phosphite

Colourless to pale yellow solid or oily liquid with clean, pleasant odour; m.p. 22-25 ˚C; b.p. 155-160 ˚C at 0.1 mm Hg; insoluble in water.

RISKS
Irritating to eyes and skin (R36/38)

SAFETY PRECAUTIONS
After contact with skin, wash immediately with plenty of water (S28)

Toxic effects	Irritant to skin, eyes, and upper respiratory tract. Inhalation can cause dizziness, interference with vision, dyspnoea, collapse, and death.
First aid	Standard treatment for exposure by all routes (see pages 108-122).

Fire hazard   Flash point 218 ˚C (open cup). Extinguish fires with carbon dioxide or dry chemical powder.

**Spillage   See general section.
disposal**

RSC *Chemical Safety Data Sheets* Vol. 3, No. 77, 1990 gives extended coverage.

1300. Triphenyltin acetate

Crystalline solid; m.p. 120 ˚C; almost insoluble in water.

RISKS
Toxic by inhalation, in contact with skin, and if swallowed (R23/24/25)

SAFETY PRECAUTIONS
Keep out of reach of children – Keep away from food, drink, and animal feeding stuffs – If you feel unwell, seek medical advice (show label where possible) (S2, S13, S44)

Toxic effects   Irritant to skin and eyes. Common symptoms following absorption are headache, nausea, vomiting, and abdominal pain. Experimental teratogen.

First aid   Standard treatment for exposure by all routes (see pages 108-122), but attention should be paid to the possibility of convulsions occurring.

Fire hazard   Extinguish fires with water spray, carbon dioxide, dry chemical powder, or alcohol or polymer foam.

**Spillage   See general section.
disposal**

RSC *Chemical Safety Data Sheets* Vol. 4b, No. 149, 1991 gives extended coverage.

1301. Triphenyltin hydroperoxide

RISKS
Toxic by inhalation, in contact with skin, and if swallowed (R23/24/25)

SAFETY PRECAUTIONS
In case of contact with eyes, rinse immediately with plenty of water and seek medical advice – Take off immediately all contaminated clothing – After contact with skin, wash immediately with plenty of water – If you feel unwell, seek medical advice (show label where possible) (S26, S27, S28, S44)

**Hazardous   Explodes reproducibly at 75 ˚C.
reactions**

1302. Triphenyltin hydroxide

White powder; m.p. 122 ˚C; almost insoluble in water.

RISKS
Toxic by inhalation, in contact with skin, and if swallowed (R23/24/25)

SAFETY PRECAUTIONS
Keep out of reach of children – Keep away from food, drink, and animal feeding stuffs – If you feel unwell, seek medical advice (show label where possible) (S2, S13, S44)

Toxic effects	Severe eye irritant.
First aid	Standard treatment for exposure by all routes (see pages 108-122), with particular attention to the possibility of convulsions occurring.
Fire hazard	Extinguish fires with water spray, carbon dioxide, dry chemical powder, or alcohol or polymer foam.
Spillage disposal	See general section.

RSC *Chemical Safety Data Sheets* Vol. 4b, No. 150, 1991 gives extended coverage.

1303. Trisilane

Colourless liquid; m.p. -117.4 ˚C; b.p. 52.9 ˚C; decomposed by water.

Hazardous reactions	Ignites or explodes in air or oxygen.

1304. Trisilylamine

Liquid; m.p. -105.6 ˚C; b.p. 52 ˚C; reacts vigorously with water.

Hazardous reactions	Ignites in air.

1305. Tritolyl phosphate

Pale brown, almost colourless, liquid; b.p. 410 ˚C with slight decomposition; immiscible with water.

RISKS
Toxic by inhalation, in contact with skin, and if swallowed – Danger of very serious irreversible effects (R23/24/25, R39)

SAFETY PRECAUTIONS
When using do not eat, drink, or smoke – After contact with skin, wash immediately with plenty of water – If you feel unwell, seek medical advice (show label where possible) (S20/21, S28, S44)

Limit values	OES short-term 0.3 mg m^{-3}; long-term 0.1 mg m^{-3}.

Toxic effects	When absorbed through the skin or ingested, tritolyl phosphate may cause serious damage to the nervous and digestive systems. Poisoning may show itself in degrees of muscular pain and paralysis.
First aid	Standard treatment for exposure by all routes (see pages 108-122).
Fire hazard	Extinguish fires with water spray, carbon dioxide, dry chemical powder, or alcohol or polymer foam.
Spillage disposal	See general section.

1306. Trivinylbismuth

Liquid; m.p. -124.5 ˚C; b.p. 158.1 ˚C.

Hazardous reactions	Ignites in air.

1307. Tungsten (and compounds)

(Metal) Grey-black solid; m.p. 3380 ˚C; b.p. approx. 5530 ˚C; insoluble in water.

Limit values	OES short-term 3 mg m^{-3} (soluble compounds), 10 mg m^{-3} (insoluble compounds) (as W); long-term 1 mg m^{-3} (soluble compounds), 5 mg m^{-3} (insoluble compounds) (as W).
Toxic effects	The metal powder is irritant to skin and eyes. Most, if not all, of the compounds are harmful by inhalation, ingestion, and skin absorption; in some cases this can lead to death.
Hazardous reactions	The finely divided metal may ignite on heating with air or on contact with a range of oxidants, usually on heating.
First aid	Standard treatment for exposure by all routes (see pages 108-122).
Fire hazard	Fires involving metal powder should be extinguished using dry chemical powder. In all other cases carbon dioxide, dry chemical powder, or alcohol or polymer foam should be used, although in a few cases any extinguishant suitable to the surroundings could be used.
Spillage disposal	See general section. The metal may be collected for re-use or recycling.

1308. Turpentine

Colourless liquid; b.p. 154-170 ˚C; insoluble in water.

RISKS
Flammable – Harmful by inhalation, in contact with skin, and if swallowed (R10, R20/21/22)

SAFETY PRECAUTIONS
Keep out of reach of children (S2)

Limit values	OES short-term 150 p.p.m. (840 mg m^{-3}); long-term 100 p.p.m. (560 mg m^{-3}).

Toxic effects   Irritant to skin, eyes, and all tissues. Occupational skin diseases are common among workers. Inhalation, ingestion, and skin absorption can also cause systemic damage to kidneys and central nervous system.

First aid   Standard treatment for exposure by all routes (see pages 108-122).

Fire hazard   Flash point 30-44 °C; Explosive limits (lower) 0.8%; Autoignition temperature 240 °C. Extinguish fires with carbon dioxide, dry chemical powder, or foam extinguishants.

Spillage disposal   See general section.

RSC *Chemical Safety Data Sheets* Vol. 1, No. 102, 1988 gives extended coverage.

1309. Uranium

White, crystalline solid; m.p. 1133 °C; b.p. 3925 °C; insoluble in water.

RISKS
Very toxic by inhalation and if swallowed – Danger of cumulative effects (R26/28, R33)

SAFETY PRECAUTIONS
When using do not eat, drink, or smoke – In case of accident or if you feel unwell, seek medical advice immediately (show label where possible) (S20/21, S45)

Hazardous reactions   Storage of foil in closed containers in presence of air and moisture may produce a pyrophoric surface; the metal incandesces or ignites in ammonia, halogen vapours, or carbon dioxide at various temperatures; explosion when carbon tetrachloride extinguisher used on uranium fire.

Spillage disposal   See general section.

1310. Uranium compounds

The commonest uranium compounds encountered in the laboratory are the acetate, nitrate, and double zinc and magnesium acetates, all of which are yellow, crystalline salts soluble in water. Uranium hexafluoride is a colourless or pale yellow, crystalline solid which sublimes readily at about 56 °C. Users in the United Kingdom who stock appreciable quantities of uranium compounds are advised to ascertain their responsibilities under the 1985 legislation for unsealed sources.

RISKS
Very toxic by inhalation and if swallowed – Danger of cumulative effects (R26/28, R33)

SAFETY PRECAUTIONS
When using do not eat, drink, or smoke – In case of accident or if you feel unwell, seek medical advice immediately (show label where possible) (S20/21, S45)

Limit values   OES short-term 0.6 mg m^{-3} (as U) for natural soluble compounds; long-term 0.2 mg m^{-3} (as U) for natural soluble compounds.

Toxic effects   The dust may irritate the lungs and cause retention of uranium in the body with subsequent damage to the kidneys. The vapour of the hexachloride irritates the respiratory system and may injure the kidneys. Assumed to cause internal damage if taken by mouth.

First aid   Standard treatment for exposure by all routes (see pages 108-122).

Spillage   See general section.
  disposal

1311. Urea

White solid; m.p. 133-135 ˚C; decomposes on further heating; soluble in water, slowly hydrolysed.

Toxic effects   Irritant to skin, eyes, upper respiratory tract, and mucous membranes.

Hazardous   Reacts vigorously with some substances.
  reactions

First aid   Standard treatment for exposure by all routes (see pages 108-122).

Fire hazard   Extinguish fires with water spray, carbon dioxide, dry chemical powder, or
                      alcohol or polymer foam.

Spillage   See general section.
  disposal

1312. Valeraldehyde

Liquid; m.p. -92 ˚C; b.p. 102-103 ˚C; very slightly soluble in water.

Toxic effects   Irritating to skin, eyes, upper respiratory tract, and all mucous membranes.
                      Harmful by inhalation, ingestion, or skin absorption.

Fire hazard   Flash point 12 ˚C. Extinguish fires with carbon dioxide, dry chemical powder,
                      or alcohol or polymer foam.

Spillage   See general section.
  disposal

1313. Vanadium compounds

Vanadium pentoxide is a red-brown to dark brown powder. The other compounds most
commonly encountered are the sodium and ammonium vanadates, which are colourless
crystalline solids; m.p. (pentoxide) 670 ˚C; b.p. (pentoxide) 1750 ˚C with decomposition.

Limit values   OES long-term (pentoxide) 0.5 mg m^{-3} (as V) (total inhalable dust), 0.05
                      mg m^{-3} (as V) (fume and respirable dust).

Toxic effects   The dust or fume of vanadium pentoxide causes irritation of the respiratory
                      system, chest constriction, coughing, and the tongue assumes a
                      blackish-green colour. The dust or fume irritates the eyes and may cause
                      conjunctivitis. If taken by mouth vanadium compounds cause vomiting,
                      excessive salivation, and diarrhoea; large doses may damage the nervous
                      system, causing drowsiness, convulsions, unconsciousness, and death.

First aid   Standard treatment for exposure by all routes (see pages 108-122).

Fire hazard   Use extinguishant appropriate to the surroundings.

Spillage   See general section.
  disposal

1314. Vanadium trichloride

Pink, deliquescent, crystalline solid; m.p. decomposes on heating; soluble in water.

Toxic effects	Irritant to skin, eyes, and mucous membranes. Toxic on ingestion.
Hazardous reactions	Reaction with Grignard reagents is almost explosively violent under some conditions.
First aid	Standard treatment for exposure by all routes (see pages 108-122).
Spillage disposal	See general section.

1315. Vanadyl triperchlorate

Hazardous reactions	Explodes above 80 °C, and ignites many organic solvents on contact.

1316. Vanillin

White, crystalline powder with pleasant odour; m.p. 81-83 °C; b.p. 170 °C at 15 mm Hg; soluble in water.

Toxic effects	Moderately toxic by inhalation and ingestion. It is pharmacologically active causing low blood pressure, increased respiration rate, and eventually death due to cardiovascular collapse.
Hazardous reactions	Addition of a little vanillin to a strong solution of trihydrated thallium(III) nitrate in 90% formic acid led to a violent reaction.
First aid	Standard treatment for exposure by all routes (see pages 108-122).
Fire hazard	Extinguish fires with water spray, carbon dioxide, dry chemical powder, or alcohol or polymer foam.
Spillage disposal	See general section.

1317. Vinyl acetate

Colourless liquid; m.p. -92.8 °C; b.p. 73 °C; slightly soluble in water.

RISKS
Highly flammable (R11)

SAFETY PRECAUTIONS
Keep away from sources of ignition – No Smoking – Do not breathe vapour – Do not empty into drains – Take precautionary measures against static discharges (S16, S23, S29, S33)

Limit values	OES short-term 20 p.p.m. (60 mg m^{-3}); long-term 10 p.p.m. (30 mg m^{-3}).

Toxic effects	The vapour may be narcotic when inhaled in high concentrations. The liquid irritates the eyes and may irritate the skin by its defatting action; it is assumed to be harmful if taken by mouth.
Hazardous reactions	Polymerization may accelerate to dangerous extent; the vapour reacts vigorously in contact with silica gel or alumina; unstabilized polymer exposed to oxygen at 50 °C gave interpolymeric peroxide which was explosive.
First aid	Standard treatment for exposure by all routes (see pages 108-122).
Fire hazard	Flash point -8 °C; Explosive limits 2.6-13.4%; Autoignition temperature 427 °C. Extinguish fire with foam, dry powder, or carbon dioxide.
Spillage disposal	See general section.

1318. Vinyl acetate ozonide

Hazardous reactions	Explosive when dry.

1319. Vinyl bromide

Colourless liquid or gas; m.p. -138 °C; b.p. 16 °C; insoluble in water.

RISKS
Extremely flammable liquefied gas (R13)

SAFETY PRECAUTIONS
Keep container in a well ventilated place – Keep away from sources of ignition – No Smoking – Take precautionary measures against static discharges (S9, S16, S33)

Limit values	OES long-term 5 p.p.m. (20 mg m^{-3}).
Toxic effects	Inhalation of vapour in high concentrations may produce dizziness and narcosis. The liquid irritates the eyes and may irritate the skin by its defatting action; it is assumed to be harmful if taken by mouth. In view of the recent observation that vinyl chloride can cause cancer of the liver, it must be assumed that vinyl bromide is likely to behave in a similar manner.
First aid	Standard treatment for exposure by all routes (see pages 108-122).
Fire hazard	Flash point <-8 °C. Extinguish fire with water spray, foam, dry powder, or carbon dioxide.
Spillage disposal	See general section.

1320. Vinyl chloride

Colourless gas with pleasant, sweet odour; m.p. -160 ˚C; b.p. -14 ˚C; slightly soluble in water.

RISKS
May cause cancer – Extremely flammable liquefied gas (R45, R13)

SAFETY PRECAUTIONS
Avoid exposure – Obtain special instructions before use – Keep container in a well ventilated place – Keep away from sources of ignition – No Smoking – If you feel unwell, seek medical advice (show label where possible) (S53, S9, S16, S44)

Limit values   MEL long-term 7 p.p.m..

Toxic effects   The liquid may irritate and burn the skin, the latter due to its freezing action. Inhalation of vapour in high concentrations produces dizziness and narcosis. Exposure to lower concentrations of vapour may lead to loss of feeling in hands and feet. More seriously, it is now proven that exposure to working atmospheres of vinyl chloride may lead to a rare liver cancer, up to 20 years after initial exposure.

Hazardous reactions   Formation of unstable polyperoxide may occur.

First aid   Standard treatment for exposure by all routes (see pages 108-122).

Fire hazard   Flash point -78 ˚C; Explosive limits 4-22%; Autoignition temperature 472 ˚C. As the gas is supplied in a cylinder, turning off the valve will reduce any fire involving it; if possible, cylinders should be removed quickly from an area in which a fire has developed.

Spillage disposal   Surplus gas or leaking cylinder can be vented slowly to air in a safe, open area or gas burnt-off through a suitable burner in a fume cupboard.

1321. Vinyl cyclohexene diepoxide

Colourless liquid; m.p. <55 ˚C; b.p. 227 ˚C; soluble in water.

RISKS
Toxic by inhalation, in contact with skin, and if swallowed – Possible risk of irreversible effects (R23/24/25, R40)

SAFETY PRECAUTIONS
Do not breathe vapour – Avoid contact with skin – If you feel unwell, seek medical advice (show label where possible) (S23, S24, S44)

Limit values   OES long-term 10 p.p.m. (60 mg m^{-3}) under review.

Toxic effects   Severe irritant to skin and eyes. Central nervous system depressant. Mutagen and animal carcinogen.

First aid   Standard treatment for exposure by all routes (see pages 108-122).

Fire hazard   Flash point 110 ˚C; Autoignition temperature 393.3 ˚C. Extinguish fires with water spray, dry chemical powder, or alcohol or polymer foam.

Spillage disposal   See general section.

RSC *Chemical Safety Data Sheets* Vol. 4b, No. 153, 1991 gives extended coverage.

1322. Vinyl fluoride

Colourless gas; m.p. -160.5 °C; b.p. -72.2 °C; insoluble in water.

Toxic effects   Danger of frostbite injury following contact with the liquid. Coughing, dyspnoea, possible systemic, and central nervous system effects following inhalation.

Fire hazard   Explosive limits 2.6-22%. Use water spray to keep cylinders cool or move them out of the immediate area.

1323. Vinyllithium

Pale white solid.

Hazardous   Violently pyrophoric when freshly prepared.
reactions

1324. 2-Vinylpyridine

Colourless liquid rapidly darkening to red-brown due to polymerization. t-Butylcatechol is commonly added to minimize polymer formation; b.p. 158 °C; slightly soluble in water.

Toxic effects   These are not well documented, but animal experiments suggest high toxicity. The vapour irritates the skin, eyes, and respiratory system. The liquid irritates the skin and eyes and must be assumed to be irritant and injurious if taken by mouth.

Hazardous   Polymerization is sometimes spontaneous and may become violent.
reactions

First aid   Standard treatment for exposure by all routes (see pages 108-122).

Fire hazard   Flash point 42 °C. Extinguish fire with foam, dry powder, or carbon dioxide.

Spillage   See general section.
disposal

1325. Xylenes

Colourless liquids with aromatic odour; m.p. -25.2 °C (*o*-); -47.9 °C (*m*-); 13.3 °C (*p*-); b.p. 144 °C (*o*-); 139 °C (*m*-); and 138 °C (*p*-); immiscible with water.

RISKS
(mixtures, m- or p-) Flammable – Harmful by inhalation and in contact with skin – Irritating to skin (R10, R20/21, R38)
(o-) Highly flammable – Harmful by inhalation and in contact with skin – Irritating to skin (R11, R20/21, R38)

SAFETY PRECAUTIONS
(mixtures, m- or p-) Avoid contact with eyes (S25)
(o-) – Keep away from sources of ignition – No Smoking – Avoid contact with eyes – Do not empty into drains (S16, S25, S29)

Limit values   OES short-term 150 p.p.m. (650 mg m^{-3}); long-term 100 p.p.m. (435 mg m^{-3}).

Toxic effects	Inhalation of the vapour may cause dizziness, headache, nausea, and mental confusion. The vapour and liquid irritate the eyes and mucous membranes. Absorption through the skin and ingestion would cause poisoning. Prolonged skin contact may cause dermatitis. If the xylene contains benzene as an impurity, repeated breathing of vapour over long periods may cause blood disease.
Hazardous reactions	Aerobic and nitric acid oxidations of *p*-xylene to terephthalic acid both carry special hazards.
First aid	Standard treatment for exposure by all routes (see pages 108-122).
Fire hazard	Flash point 17 ˚C (*o*-) and 25 ˚C (*m*- and *p*-); Explosive limits Approximately 1-7%; Autoignition temperature 464 ˚C (*o*-); 528 ˚C (*m*-); and 529 ˚C (*p*-). Extinguish fire with foam, dry powder, or carbon dioxide. Large fires are best controlled with alcohol-resistant foam.
Spillage disposal	See general section.

RSC *Chemical Safety Data Sheets* Vol. 1, No. 103, 1988 gives extended coverage.

1326. Xylenols

With the exception of 2,4-xylenol, which is often encountered as a yellow-brown liquid, the commoner xylenols are colourless, crystalline solids; m.p. 48-75 ˚C; b.p. 203-226 ˚C; slightly soluble in water.

RISKS
Toxic in contact with skin and if swallowed – Causes burns (R24/25, R34)

SAFETY PRECAUTIONS
Keep out of reach of children – After contact with skin, wash immediately with plenty of water – If you feel unwell, seek medical advice (show label where possible) (S2, S28, S44)

Toxic effects	The vapour of heated xylenols is irritant to the respiratory system. They irritate or burn the eyes and skin severely. Considerable absorption through the skin or ingestion may cause headache, dizzines, nausea, vomiting, stomach pain, exhaustion, and coma. Repeated inhalation or absorption of small amounts may result in damage to the liver or kidneys.
First aid	Standard treatment for exposure by all routes (see pages 108-122).
Fire hazard	Extinguish fires with water spray, carbon dioxide, dry chemical powder, or alcohol or polymer foam.
Spillage disposal	See general section.

RSC *Chemical Safety Data Sheets* Vol. 4b, No. 154, 1991 gives extended coverage.

1327. Xylidines

Most of the common xylidines are red to dark-brown liquids, although 3,4-xylidine is a pale brown, crystalline solid. Commercial xylidine is a mixture of isomers; b.p. 213- 226 ˚C; slightly soluble in water.

RISKS
Toxic by inhalation, in contact with skin, and if swallowed – Danger of cumulative effects (R23/24/25, R33)

SAFETY PRECAUTIONS
After contact with skin, wash immediately with plenty of water – Wear protective clothing and gloves – If you feel unwell, seek medical advice (show label where possible) (S28, S36/37, S44)

Limit values	OES short-term 10 p.p.m. (50 mg m^{-3}); long-term 2 p.p.m. (10 mg m^{-3}).
Toxic effects	Excessive breathing of the vapour, ingestion, or absorption through the skin may cause headache, drowsiness, cyanosis, mental confusion, and, in severe cases, convulsions and death. The xylidines are dangerous to the eyes. Prolonged exposure to the vapour or slight skin exposures over a period may affect the nervous system and the blood, causing fatigue, loss of appetite, headache, and dizziness.
Hazardous reactions	Ignition on contact with fuming nitric acid.
First aid	Standard treatment for exposure by all routes (see pages 108-122).
Fire hazard	Flash point 96 ˚C (closed cup). Extinguish fires with water spray, carbon dioxide, dry chemical powder, or alcohol or polymer foam.
Spillage disposal	See general section.

RSC *Chemical Safety Data Sheets* Vol. 4b, No. 155, 1991 gives extended coverage.

1328. Yttrium

Dark grey solid; m.p. approx. 1500 ˚C; b.p. approx. 3225 ˚C; insoluble in water.

Limit values	OES short-term 3 mg m^{-3}; long-term 1 mg m^{-3}.
Toxic effects	Irritant to skin and eyes. May be harmful if absorbed into the body.
First aid	Standard treatment for exposure by all routes (see pages 108-122).
Fire hazard	Fires involving yttrium powder should be treated with dry chemical powder. In other cases use extinguishant appropriate to the surroundings.
Spillage disposal	See general section.

1329. Zinc

Silvery metal; m.p. 419 °C; b.p. 907 °C; insoluble in water.

RISKS
Flammable – Contact with water liberates highly flammable gases (H10, H15)

SAFETY PRECAUTIONS
Keep container tightly closed and dry – In case of fire, use class D fire extinguishers. Do NOT use water (S7/8, S43)

Toxic effects	May cause skin and eye irritation. May be harmful if inhaled or ingested.
Hazardous reactions	A zinc dust explosion occurred during sieving of hot dry material. Mixtures of zinc powder or dust with a wide variety of substances can react violently or explosively.
First aid	Standard treatment for exposure by all routes (see pages 108-122).
Fire hazard	Use only class D fire extinguishing agents. Do NOT use water on fires involving zinc dust.
Spillage disposal	See general section. Zinc, if not in dust form, can be swept up, bagged, and saved for re-use or recycling.

1330. Zinc chloride

White, deliquescent powder or lumps; m.p. approx. 290 °C; b.p. 732 °C; soluble in water.

RISKS
Causes burns (R34)

SAFETY PRECAUTIONS
Keep container tightly closed and dry – After contact with skin, wash immediately with plenty of water (S7/8, S28)

Limit values	OES short-term 2 mg m^{-3} (fume); long-term 1 mg m^{-3} (fume).
Toxic effects	A moderate irritant of skin or mucous membranes; major exposure may lead to dermatitis, asthma, and inflammation of the cornea. Repeated or prolonged exposure may give rise to respiratory and gastro-intestinal effects which could be fatal.
First aid	Standard treatment for exposure by all routes (see pages 108-122).
Fire hazard	Use a fire extinguishant appropriate to the surroundings.
Spillage disposal	See general section.

1331. Zinc chromate

Yellow solid; very slightly soluble in water.

RISKS
May cause cancer – Harmful if swallowed – May cause sensitization by skin contact (R45, R22, R43)

SAFETY PRECAUTIONS
Avoid exposure – Obtain special instructions before use – If you feel unwell, seek medical advice (show label where possible) (S53, S44)

Limit values	OES long-term 0.5 mg m^{-3} (as Cr).
Toxic effects	Can cause dermatitis and nasal ulceration. Lung carcinogen.
First aid	Standard treatment for exposure by all routes (see pages 108-122).
Fire hazard	Extinguish fires with water spray, carbon dioxide, dry chemical powder, or alcohol or polymer foam.
Spillage disposal	See general section.

RSC *Chemical Safety Data Sheets* Vol. 4b, No. 156, 1991 gives extended coverage.

1332. Zinc oxide

White powder; m.p. 1975 ˚C; almost insoluble in water.

Limit values	OES short-term 10 mg m^{-3} (fume); long-term 5 mg m^{-3} (fume).
Toxic effects	The powder may be irritant to skin, eyes, and upper respiratory tract. The freshly formed fume can cause metal fume fever resulting in muscular pains, nausea, fever, and chills.
First aid	Standard treatment for exposure by all routes (see pages 108-122).
Fire hazard	Use extinguishant appropriate to the surroundings.
Spillage disposal	See general section.

1333. Zinc permanganate

Dark brown crystalline solid; m.p. 100 ˚C; decomposes on heating; very soluble in water.

Toxic effects	Irritant to skin, eyes, and mucous membranes.
Hazardous reactions	Reactions of solid zinc permanganate with organic compounds are violent.
Spillage disposal	See general section.

1334. Zinc peroxide

Yellow solid; m.p. 150 ˚C with decomposition; reacts exothermically with water.

Toxic effects   Irritant to skin and eyes. Harmful on inhalation or ingestion.

Hazardous reactions   Hydrated peroxide explodes at 212 ˚C; mixtures with aluminium or zinc powders burn brilliantly.

First aid   Standard treatment for exposure by all routes (see pages 108-122).

Fire hazard   Extinguish fires with carbon dioxide or dry chemical powder.

Spillage disposal   See general section.

1335. Zinc phosphide

Cubic, dark grey crystals or powder with a faint phosphorus odour; m.p. 420 ˚C; b.p. 1100 ˚C; insoluble in and reacts with water.

RISKS
Very toxic if swallowed – Contact with acids liberates very toxic gas (R28, R32)

SAFETY PRECAUTIONS
Keep locked up and out of reach of children – When using do not eat, drink, or smoke – Do not breathe dust – After contact with skin, wash immediately with plenty of water – In case of accident or if you feel unwell, seek medical advice immediately (show label where possible) (S1/2, S20/21, S22, S28, S45)

Toxic effects   Severely irritant and corrosive to skin and eyes. Inhalation may produce lung irritation and pulmonary oedema. Ingestion can produce gastro-intestinal upsets, followed by renal and hepatic damage, cardiovascular and respiratory failure. Symptoms may be delayed and death occur upto one week later.

Hazardous reactions   Decomposed violently with acids and other oxidizing agents.

First aid   Standard treatment for exposure by all routes (see pages 108-122), with particular attention to the possibility of lung congestion.

Fire hazard   Extinguish fires with dry chemical powder or alcohol or polymer foam. Do NOT use water or carbon dioxide.

Spillage disposal   See general section.

RSC *Chemical Safety Data Sheets* Vol. 4b, No. 157, 1991 gives extended coverage.

1336. Zirconium

Crystalline metal; m.p. 1852 ˚C; b.p. 4375 ˚C; insoluble in water.

RISKS
Contact with water liberates highly flammable gases (R15)

SAFETY PRECAUTIONS
Keep container tightly closed and dry – In case of fire, use class D extinguisher. Do NOT use water or carbon dioxide (S7/8, S43)

Toxic effects	May cause skin and eye irritation. May be harmful if inhaled or ingested.
Hazardous reactions	Pyrophoric; mixtures of zirconium powder with a wide variety of substances will react violently or explosively. Zirconium powder, damp with 5-10% water may ignite; although water is used to prevent ignition, the powder, once ignited, will burn under water more violently than in air.
First aid	Standard treatment for exposure by all routes (see pages 108-122).
Fire hazard	Extinguish fires involving zirconium dust with class D fire extinguishing agents. Do NOT use water or carbon dioxide. For zirconium in other forms, use extinguishant appropriate to the surroundings.
Spillage disposal	See general section. Zirconium, if not in dust form, may be collected and held for re-use or recycling.

1337. Zirconium dicarbide

Limit values	OES short-term 10 mg m^{-3} (as Zr); long-term 5 mg m^{-3} (as Zr).
Hazardous reactions	Ignites in cold fluorine, in chlorine at 250 ˚C, bromine at 300 ˚C, and iodine at 400 ˚C.

1338. Zirconium dichloride

Black solid; m.p. 350 ˚C with decomposition; decomposed by water giving off hydrogen.

Limit values	OES short-term 10 mg m^{-3} (as Zr); long-term 5 mg m^{-3} (as Zr).
Hazardous reactions	Ignites in air when warm.
Spillage disposal	See general section.

1339. Zirconium tetrachloride

White, crystalline solid; b.p. 300 ˚C with sublimation; reacts violently with water.

Limit values	OES short-term 10 mg m^{-3} (as Zr); long-term 5 mg m^{-3} (as Zr).
Toxic effects	Irritant to skin, eyes, upper respiratory tract, and mucous membranes. Harmful if inhaled, ingested, or absorbed through the skin. Inhalation may be fatal as a result of spasm, inflammation, and oedema.
Hazardous reactions	Ignited lithium metal strip.
First aid	Standard treatment for exposure by all routes (see pages 108-122).
Fire hazard	Extinguish fires with carbon dioxide, dry chemical powder, or alcohol or polymer foam. Do NOT use water.
Spillage disposal	See general section.

INDEX OF CHEMICALS

The numbers in this index refer to the **ITEM NUMBERS** of the chemicals and **not** to page numbers. Entries in lower case are synonyms.

INDEX OF CAS REGISTRY NUMBERS

The numbers in this index refer to the **ITEM NUMBERS** of the chemicals and **not** to page numbers.